Integration of 2D Materials for Electronics Applications

Integration of 2D Materials for Electronics Applications

Special Issue Editors

Filippo Giannazzo
Samuel Lara Avila
Jens Eriksson
Sushant Sonde

MDPI • Basel • Beijing • Wuhan • Barcelona • Belgrade

MDPI

Special Issue Editors

Filippo Giannazzo
Institute for Microelectronics
and Microsystems (CNR-IMM)
Italy

Samuel Lara Avila
Chalmers University of Technology
Sweden

Jens Eriksson
Linköping University
Sweden

Sushant Sonde
The University of Chicago
USA

Editorial Office
MDPI
St. Alban-Anlage 66
4052 Basel, Switzerland

This is a reprint of articles from the Special Issue published online in the open access journal *Crystals* (ISSN 2073-4352) from 2017 to 2018 (available at: https://www.mdpi.com/journal/crystals/special_issues/2d)

For citation purposes, cite each article independently as indicated on the article page online and as indicated below:

LastName, A.A.; LastName, B.B.; LastName, C.C. Article Title. *Journal Name* **Year**, *Article Number*, Page Range.

ISBN 978-3-03897-606-6 (Pbk)
ISBN 978-3-03897-607-3 (PDF)

Contents

About the Special Issue Editors . . . . **vii**

Kyung Ho Kim, Samuel Lara-Avila, Hans He, Hojin Kang, Yung Woo Park, Rositsa Yakimova and Sergey Kubatkin
Thermal Stability of Epitaxial Graphene Electrodes for Conductive Polymer Nanofiber Devices
Reprinted from: *Crystals* **2017**, *7*, 378, doi:10.3390/cryst7120378 . . . . **1**

Amritesh Rai, Hema C. P. Movva, Anupam Roy, Deepyanti Taneja, Sayema Chowdhury and Sanjay K. Banerjee
Progress in Contact, Doping and Mobility Engineering of MoS$_2$: An Atomically Thin 2D Semiconductor
Reprinted from: *Crystals* **2018**, *8*, 316, doi:10.3390/cryst8080316 . . . . **12**

Filippo Giannazzo, Giuseppe Greco, Fabrizio Roccaforte and Sushant S. Sonde
Vertical Transistors Based on 2D Materials: Status and Prospects
Reprinted from: *Crystals* **2018**, *8*, 70, doi:10.3390/cryst8020070 . . . . **96**

Jingyu Li, Xiaozhang Chen, David Wei Zhang and Peng Zhou
Van der Waals Heterostructure Based Field Effect Transistor Application
Reprinted from: *Crystals* **2018**, *8*, 8, doi:10.3390/cryst8010008 . . . . **121**

Fei Hui, Shaochuan Chen, Xianhu Liang, Bin Yuan, Xu Jing, Yuanyuan Shi and Mario Lanza
Graphene Coated Nanoprobes: A Review
Reprinted from: *Crystals* **2017**, *7*, 269, doi:10.3390/cryst7090269 . . . . **144**

Francesco Ruffino and Filippo Giannazzo
A Review on Metal Nanoparticles Nucleation and Growth on/in Graphene
Reprinted from: *Crystals* **2017**, *7*, 219, doi:10.3390/cryst7070219 . . . . **165**

Chiara Musumeci
Advanced Scanning Probe Microscopy of Graphene and Other 2D Materials
Reprinted from: *Crystals* **2017**, *7*, 216, doi:10.3390/cryst7070216 . . . . **205**

Jie Sun, Xuejian Li, Weiling Guo, Miao Zhao, Xing Fan, Yibo Dong, Chen Xu, Jun Deng and Yifeng Fu
Synthesis Methods of Two-Dimensional MoS$_2$: A Brief Review
Reprinted from: *Crystals* **2017**, *7*, 198, doi:10.3390/cryst7070198 . . . . **224**

Ivan Shtepliuk, Tihomir Iakimov, Volodymyr Khranovskyy, Jens Eriksson, Filippo Giannazzo and Rositsa Yakimova
Role of the Potential Barrier in the Electrical Performance of the Graphene/SiC Interface
Reprinted from: *Crystals* **2017**, *7*, 162, doi:10.3390/cryst7060162 . . . . **235**

About the Special Issue Editors

Filippo Giannazzo (Ph.D.) got his Ph.D. in Materials Science from the University of Catania, Italy, in 2002. He joined the Institute for Microelectronics and Microsystems of CNR (IMM-CNR) as a researcher in 2006 and is senior researcher from 2010. He is expert in scanning probe microscopy methods for the characterization of carrier transport properties in advanced materials for micro and nanoelectronics (wide-bandgap semiconductors, heterostructures, dielectrics, organics, 2D materials). He is author of more than 270 papers, 7 book chapters (H-index = 35, Source Scopus) and an international patent. He is frequently invited speaker in national and international conferences. He holds several national and international collaborations with academic institutions and industries. He has been involved in several National and EU projects, and is currently coordinating the FlagERA project "GraNitE". He has been member of the organizing committee of several international conferences, co-chair of two EMRS Fall symposia (2010, 2014) on advanced characterizations, co-organizer of the "International School of Physics and Technology of Matter" (Otranto, 2014). In 2004 he received the SISM award from the Italian Society of Microscopy and in 2014 the Accademia Gioenia "G. P. Grimaldi" award.

Samuel Lara-Avila obtained his Ph.D. at Chalmers University of Technology (Sweden) in 2012, where he is currently appointed as Associate Research Professor at the Quantum Device Physics Laboratory. His research interests include electron transport and light matter interactions in low dimensional-systems such as single-molecules and two-dimensional materials, as well as directed assembly of nanoparticles at surfaces. For his work on graphene, he was awarded the International Union of Pure and Applied Physics (IUPAP) Young Scientist (Early Career) Prize in Fundamental Metrology, in recognition of outstanding contribution to the understanding of quantum electrical transport in epitaxial grapheme, leading to the development of a novel quantum resistance standard. He is author of over 55 papers, 3 book chapters (H-index = 20, source WoS) and two international patents.

Jens Eriksson got his Ph.D. in December 2010 from the Superior School of the University of Catania, Italy. During his PhD studies (2007–2010) he held a Marie-Curie Scholarship as Early Stage Researcher at CNR-IMM, Catania. He joined Linköping University in 2011 as a post-doc, received his habilitation (Docent title) in 2015, and is working as Associate Professor and head of the Applied Sensor Science research group since 2017. His research focus is on novel materials for chemical sensors in the scope of ultra-high sensitivity applications in environmental monitoring. He has over 40 publications (H-index = 9, web of science) within the areas of silicon carbide, 2D-materials, and chemical sensors, and has presented five invited talks at international conferences and twice been session chair at EMRS spring meeting in Lille (2014 and 2016). He is/ has been PI in several projects, with both research- and industrial focus, and is currently coordinating the innovation project "Sensor for faster, cheaper, and easier determination of dioxins in the environment", funded by Sweden's Innovation Agency. He has been member of the organizing committee of two international conferences/workshops.

Sushant S. Sonde, Ph.D., is a Research Scientist at the Institute for Molecular Engineering, University of Chicago and Argonne National Laboratory. With a general theme of 'Materials and Device

Innovation', Dr. Sonde has pursued interest in various material systems and development of viable electronic devices out of them at various high-profile research laboratories in Europe (IMEC, Belgium; CNR-IMM Catania, Italy) and USA (Microelectronics Research Center, UT Austin, Texas; IME-UChicago/Argonne National Laboratory). Most prominent amongst those are high mobility semiconductor materials, 2D materials and Oxide materials. Dr. Sonde's interest and involvement ranges from materials development, materials engineering, advanced material characterization and nanofabrication into proof-of concept devices. He has authored/co-authored various high impact factor research articles in the said fields. Dr. Sonde is recipient of various international awards for his research efforts, that include Oberbuergermeister-Dieter-Goerlitz-Preis (2007) from The City of Deggendorf, Germany; Young Scientist Award (2009) at The European Material Research Society Spring Meeting, Strasbourg, France; Dept. of Energy Research Highlight (2018) at Argonne National Laboratory, Best Paper Awards (2015 and 2017) and 4 filed patents/invention disclosures so far.

Article

Thermal Stability of Epitaxial Graphene Electrodes for Conductive Polymer Nanofiber Devices

Kyung Ho Kim [1,*], **Samuel Lara-Avila** [1,2], **Hans He** [1], **Hojin Kang** [3], **Yung Woo Park** [4,*], **Rositsa Yakimova** [5] **and Sergey Kubatkin** [1]

1   Department of Microtechnology and Nanoscience, Chalmers University of Technology, Gothenburg SE412-96, Sweden; samuel.lara@chalmers.se (S.L.-A.); hanshe@chalmers.se (H.H.); sergey.kubatkin@chalmers.se (S.K.)
2   National Physical Laboratory, Hampton Road, Teddington TW11 0LW, UK
3   Department of Physics and Astronomy, Seoul National University, Seoul 08826, Korea; hkang@phya.snu.ac.kr
4   Institute of Applied Physics, Seoul National University, Seoul 08826, Korea
5   Department of Physics, Chemistry and Biology, Linkoping University, Linkoping SE581-83, Sweden; roy@ifm.liu.se
*   Correspondence: kyungh@chalmers.se (K.H.K.); ywpark@snu.ac.kr (Y.W.P.); Tel.: +46-31-772-5475 (K.H.K.); +82-2-880-6607 (Y.W.P.)

Academic Editor: Helmut Cölfen
Received: 21 November 2017; Accepted: 11 December 2017; Published: 14 December 2017

Abstract: We used large area, monolayer graphene epitaxially grown on SiC (0001) as contact electrodes for polymer nanofiber devices. Our fabrication process, which avoids polymer resist residues on the graphene surface, results in graphene-polyaniline nanofiber devices with Ohmic contacts and electrical conductivity comparable to that of Au-nanofiber devices. We further checked the thermal stability of the graphene contacts to polyaniline devices by annealing up to T = 800 °C, the temperature at which polyaniline nanofibers are carbonized but the graphene electrode remains intact. The thermal stability and Ohmic contact of polymer nanofibers are demonstrated here, which together with the chemical stability and atomic flatness of graphene, make epitaxial graphene on SiC an attractive contact material for future all-carbon electronic devices.

Keywords: graphene; graphene electrodes; epitaxial graphene on SiC; polymer nanofibers; polyaniline nanofibers; carbonization; organic electronics; carbon electronics

1. Introduction

Conductive polymers are promising platforms for the next generation of carbon-based electronics. With these organic materials, the variety of devices that have already been developed span a wide range of applications that include flexible field–effect transistors [1], actuators [2], sensors [3], and nano-optoelectronic devices [4]. For conductive polymers, efficient injection and extraction of charges between the contact electrode and the active channel is often complicated due to the incompatibility between organic channels and inorganic contacts [5,6]. In this sense, carbon-based contacts [5], and particularly graphene, are appealing solutions to interface organic polymers to the outer world and materialize the vision of all-carbon electronics [5,7]. As an electrical contact, graphene offers numerous properties that complement the versatility of electronic polymers, including high electron mobility [8–11], thermal conductivity [12], optical transparency [13,14], tunability of work function [15], and chemical/thermal stability. Furthermore, in combination with metals, graphene could be also used as an interfacial layer to engineer the charge transfer between metal contacts and other carbon-based systems [16]. More generally, graphene as an electrical contact has been proven to be a superior solution in various electronics applications from organic field effect transistors [17–23], organic solar cells [24], organic light emitting diodes [25] to nanoelectromechanical infrared detectors [26], and electrophysiology and neuroimaging [27,28]. In addition to electronics, biosensors [29] and biomedical applications such as point-of-care testing devices [30] use graphene to improve analytical performances.

In practice, additional requirements that have to be met by graphene contact technologies include scalability, reproducibility (e.g., clean surface), and robustness against chemical and thermal treatments during device fabrication. Graphene grown by chemical vapor deposition (CVD) [16–19,25–28,31] and from reduced graphene oxide [24] are somewhat suitable for scalability. CVD graphene has to be transferred to an insulating substrate and the transfer process is prone to leave resist residues and to result in discontinuous graphene layers (i.e., voids) over large scales. An alternative technology is epitaxial graphene grown on the Si face of silicon carbide substrates (G/SiC), which has drawn less attention for contact technology due to the relative higher cost of materials. Nonetheless, as-grown G/SiC is also scalable [32], being a continuous single crystal with its size limited only by the SiC substrate size [33]. Additionally, G/SiC is atomically flat and clean implying that atomically clean interfaces can be readily achieved on this material. Since the SiC substrate is electrically insulating, there is no need to transfer (i.e., contaminate) the graphene layer. The main source of contamination for G/SiC is the microfabrication process that involves organic polymer resists. However, polymer residues can be avoided by using shadow masks or metal masks directly deposited on graphene during fabrication [34–37]. Alternatively, resist residues and other common contaminants of the surface can be removed using scalable methods such as high temperature annealing [38].

In this paper, we demonstrate the suitability of G/SiC as an electrical contact for polymer nanofibers, a low dimensional carbon system. We patterned a large area of G/SiC using a metal protection mask to ensure that the G/SiC surface is free of resist residues that degrade the nanofiber/graphene interface. For the organic channel, polyaniline (PANI) nanofibers were contacted on G/SiC and we found that the quality of contact is comparable to that of Au electrodes. We further checked the thermal stability of the device by annealing it at 800 °C under argon flow and upon annealing, we found that the graphene electrodes remained operational and the PANI nanofibers were carbonized as confirmed by current-voltage (I-V) characterization and Raman spectroscopy.

2. Results and Discussion

2.1. Characterization of Graphene Electrodes

The as-grown G/SiC, characterized by the express optical microscopy method [39], is homogeneous monolayer graphene with about 10–15% bilayer domain inclusions [32]. Figure 1 is the schematic illustration of the fabrication process of the G/SiC electrode (see Methods), where the key step is the deposition of an aluminum protection layer on the as-grown material. This Al layer is removed in the last fabrication process, and its role is to prevent graphene from directly contacting organic resist that degrades the graphene-nanofiber interface. Together with G/SiC electrodes, we have fabricated Hall bars to enable the electrical characterization of the graphene layer. Hall measurement of the G/SiC shows that the electron mobility is of the order of ~1000 cm^2/Vs and the electron carrier density is ~4×10^{12} cm^{-2} at 300 K. The high electron concentration is consistent with the charge transfer from the surface donor state of SiC to G/SiC reported previously [40,41].

Figure 2a is the optical microscope image of a graphene electrode pattern with a length (width) of 10 μm (1 μm). The G/SiC pattern is discernable from SiC and we found a few inclusions of bilayer (BL) domains (seen as darker stripes) in the monolayer (ML) G/SiC. Figure 2b is the I-V characteristics of the graphene lead before and after annealing. Both of the I-V of each lead are linear and the adjacent leads are electrically insulating before and after annealing. The decrease of resistance in G/SiC leads after annealing can be attributed to either desorption of species from the graphene surface or by a modified contact resistance between Au and G/SiC after the thermal annealing step [42]. Statistics on the resistivity of G/SiC leads before annealing show that the average resistivity of 11 leads is ~11 kΩ/square. In more detail, the average two probe resistivity of 7 G/SiC leads of width 1 μm (length 10 μm or 20 μm) was 13 kΩ/square and that of 4 G/SiC leads with width 2 μm (length 100 μm) was 8 kΩ/square. The higher resistivity of 1 μm width G/SiC can be attributed to the roughness of edges and charge inhomogeneity arising from bilayer domains [43], which presumably has a greater impact on the narrower G/SiC leads.

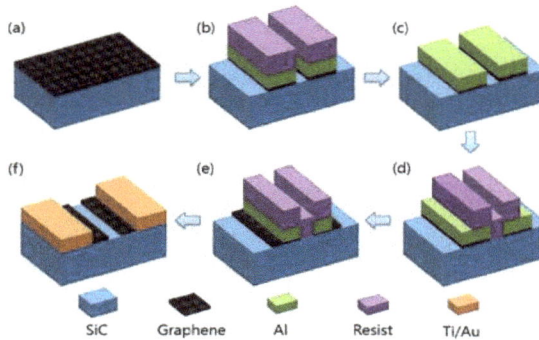

Figure 1. The schematic illustration of the fabrication process of the G/SiC electrode: (**a**) As-grown epitaxial graphene on SiC (G/SiC); (**b**) An aluminum protection layer was first deposited on G/SiC, and this was followed by electron beam lithography (EBL) and successive graphene etching in oxygen plasma; (**c**) Resist is removed with organic solvents; (**d**) A second EBL step for defining global Ti/Au contacts (**e**) Al removal by wet etching; (**f**) Deposition of Ti/Au global contacts on G/SiC electrodes and lift-off in organic solvents.

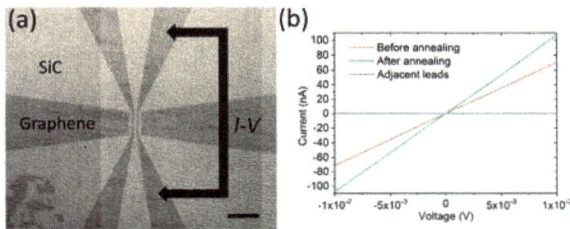

Figure 2. Thermal stability of graphene electrodes. (**a**) The optical microscope image of the G/SiC electrode with width (length) 1 μm (10 μm). Scale bar: 10 μm; (**b**) The linear current-voltage (I-V) characteristics of the G/SiC lead marked by arrows in (**a**) before and after annealing at T = 800 °C. The adjacent leads are insulated before and after annealing and the resistance of the G/SiC lead decreased after the T = 800 °C annealing.

2.2. Characterization of Graphene-Nanofiber Devices before and after Thermal Annealing Step

In order to assess the quality of graphene as a contact for polymer nanofibers, we chose polyaniline (PANI) as the conductive channel medium. PANI nanofibers have a unique acid/base doping/dedoping chemistry that is reversibly switchable from the doped state to the dedoped state by exposure to hydrochloric acid and ammonia [44–46]. Together with the enhanced surface to volume ratio in nanofiber morphology, PANI nanofibers are also promising for gas sensing applications [1,47,48]. Besides, the carbonization of polymers by pyrolysis [49–58] shows potential for applications such as a fuel cell [53] and catalyst [56,57], and PANI produces nitrogen containing conducting carbons after pyrolysis [52–58]. On the as-fabricated G/SiC electrode, a suspension of solution containing PANI nanofibers were dispersed (see Method) and we observed that fibers readily form an Ohmic contact to graphene electrodes. Furthermore, the thermal stability of epitaxial graphene electrodes allows thermal processes at elevated temperatures to be carried out. Indeed, we annealed the device up to T = 800 °C and found that the contact between graphene and fibers remain Ohmic. We performed the thermal annealing cycle under continuous argon flow to prevent oxidation of organic species. This method allowed us to investigate not only the thermal stability of the PANI nanofiber-G/SiC devices but also to explore the electron transport properties of carbonized polymer nanofibers in general [59]. Figure 3a,b show the AFM topography of PANI nanofibers contacted on G/SiC electrodes before and after T = 800 °C annealing, respectively. Upon high temperature annealing, the G/SiC electrode remains intact and most of the PANI nanofibers were preserved as shown in Figure 3b. Comparison of Figure 3a,b at the same area before and after annealing, shows that the overall shape of the nanofibers is retained; however, both the width and the height of PANI nanofibers are significantly reduced to about 50% after annealing (Figure 3c). This is consistent with previous reports that PANI undergoes dehydrogenation and cross-linking of adjacent chains upon high temperature pyrolysis, and that the weight of polyacetylene (PA) films/fibers [49–51] and PANI films/tubes [52–58] is reduced after pyrolysis while retaining the fibril morphology. I-V characteristics of the PANI nanofibers on G/SiC electrodes before annealing show that the adjacent G/SiC leads are electrically connected due to the PANI nanofibers contacting the two adjacent G/SiC electrodes. The device shows linear and symmetric I-V characteristics of PANI nanofibers on G/SiC before and after annealing, with the resistance increased about 10 times upon annealing. The symmetric and linear I-V is consistent with previous reports regarding annealed PANI nanofibers at 800 °C [59].

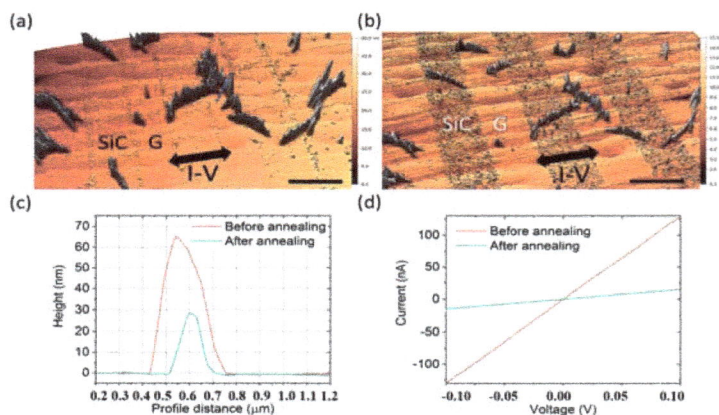

Figure 3. Characterization of graphene-nanofiber devices before and after the thermal annealing step. (**a**) Atomic force microscopy (AFM) topography image of G/SiC electrodes contacting polyaniline (PANI) nanofibers, where graphene leads are indicated by G.; (**b**) AFM topography image of (**a**) after thermal annealing at T = 800 °C. The graphene leads remain intact and morphology of PANI nanofibers are preserved. Scale bar: 2 µm; (**c**); The reduction in size of PANI nanofibers after annealing is compared in the AFM height profile of the region indicated by blue lines in (**a,b**). Both the width (320 nm to 190 nm) and height (65 nm to 28 nm) are reduced after annealing; (**d**) I-V characteristics of the adjacent graphene electrodes before and after annealing. Between the two electrodes in which I-V was measured, three PANI nanofibers are contacted in total (Device G4, see Figure S7). After annealing, the resistance typically increases to 10 times.

We verified the integrity of the devices, including the graphene contacts, after the thermal annealing step by Raman spectroscopy and found that PANI fibers undergo carbonization but graphene remains essentially intact. Figure 4 shows the Raman spectroscopy (λ = 638 nm) measured on bundles of

PANI nanofibers (Figure 4a) and of G/SiC (Figure 4b) before and after annealing. We found substantial changes in the PANI nanofiber after annealing. In the pristine form, the Raman spectra of PANI nanofibers show complex peaks that indicates PANI nanofibers. Raman spectroscopy on the annealed PANI nanofiber bundles shows that the PANI nanofibers become amorphous carbon nanofibers as confirmed by the broad D (1353 cm^{-1}) and G bands (1590 cm^{-1}) of graphite (Figure 4b) [49–59]. In contrast, the G/SiC remained intact after annealing as shown in Figure 4b. Figure 4b displays the Raman spectra of the pristine, annealed G/SiC, and the etched SiC region as a reference. The Raman spectra on G/SiC includes contributions both from the bulk SiC substrate and the so-called buffer layer. Therefore, correcting the Raman spectra of G/SiC by subtracting the spectrum of SiC substrate may introduce artifacts due to the contribution of the substrate [60]. The presence of G and 2D peaks before and after annealing means that the G/SiC remains intact after annealing [60,61]. The thermal stability of graphene is comparable to that of oxides such as Sr_2RuO_4 (stable at 900 °C) [62] and olivine (stable at 500 °C) electrodes [63].

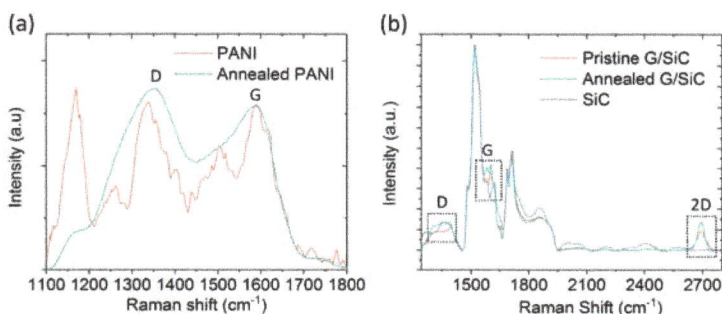

Figure 4. Raman spectroscopy before and after annealing (**a**) Raman spectroscopy on a bundle of PANI nanofibers before and after 800 °C annealing. After annealing, the complex peaks in PANI nanofibers turned to two broad peaks marked by D and G bands. The intensity of PANI is normalized with respect to the maximum value of D band in annealed PANI nanofibers; (**b**) Raman spectroscopy of the pristine graphene, annealed graphene and SiC. Dotted boxes indicate the vicinity of D, G, and 2D peaks. The intensity is normalized by the highest peak of Raman spectra measured on the SiC substrate.

2.3. Comparison of Graphene with Gold as a Contact for PANI Nanofibers

We benchmarked graphene as a contact for polymer nanofibers against gold, which is the standard contact metal for these materials. Figure 5a shows the AFM topography of a Ti/Au electrode deposited on a Si/SiO$_2$ (300 nm) substrate and a PANI nanofiber contacted on Au electrodes. The conductivity and height of PANI nanofibers measured on both G/SiC and Au electrodes of this study range from 0.5–5 S/cm and 50–110 nm, respectively. Figure 5b compares the conductivity of PANI nanofibers on graphene electrodes (G1–G4) (see Methods and Figures S5–S7) to that on Au electrodes (Au1–Au6) (see Methods and Figures S1–S4). The conductivity of PANI nanofibers on G/SiC electrodes (0.5–2.3 S/cm) was slightly lower than that on Au (1.2–5 S/cm); however, this is comparable with the conductivity of PANI nanofibers measured on Au electrodes reported in the literature [64].

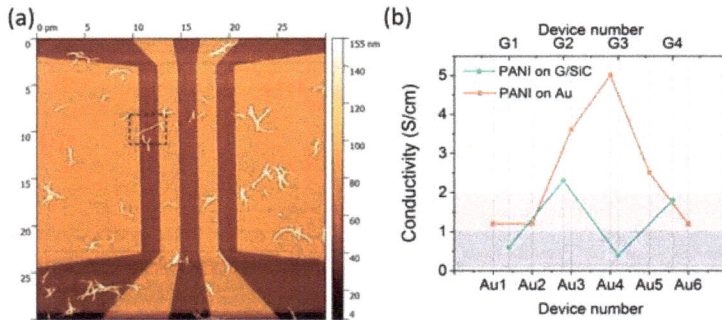

Figure 5. Comparison of graphene and gold as contact for PANI nanofibers. (**a**) AFM topography of PANI nanofibers contacted on Au electrodes (Au4, Figure S2). The contact of a PANI nanofiber contacted by Au electrodes is indicated by a dotted box; (**b**) Conductivity of PANI nanofibers measured on both Au (Au1–Au6, Figures S1–S4) and graphene electrodes (G1–G4, Figures S5–S7). The blue (red) shaded region is the conductivity of PANI nanofibers (annealed PANI nanofibers at 800 °C) measured on Au electrodes in Ref [64] (Ti/Au bottom contact electrode in Ref. [59]).

3. Materials and Methods

3.1. Growth of Epitaxial Graphene on SiC

The graphene was purchased from Graphensic AB. The crystallographic orientation of the 4H-SiC substrate is (0001) which provides large terraces and minimizes bilayer inclusions. The graphene fabrication process includes standard two-step cleaning procedure including HF solution dipping prior to loading into the growth reactor. The latter consists of a vertical radio frequency (RF) heated graphite crucible placed in a quartz tube with a thermal insulation between their walls. Upon reaching base vacuum in the range of 10^{-6} mbar, heating is performed until 2000 °C and this temperature is held for 5 min. After that the RF generator is switched off and the graphene wafer is cooled down to room temperature. The wafer is subjected to microscopy examination to check the graphene morphology and after that, to further processing steps.

3.2. Fabrication of Graphene and Au Electrodes

3.2.1. Fabrication of Graphene Electrodes

Graphene electrodes in Figure 2 and of devices G1–G4 were fabricated on the as-grown graphene on the Si face of the 4H-SiC surface. For the first step, Al (20 nm) was deposited to avoid resist residue and the standard electron beam lithography (EBL) using e-beam resist ARP-6200 (Allresist, Strausberg, Germany) was performed on top of Al. After developing the e-beam resist, a MF-319 photodeveloper (Dow Europe, Horgen, Switzerland) was used for the wet etch of Al underneath and the exposed graphene was dry-etched using oxygen plasma (Figure 1b). After dissolving the remaining resist in organic solvent mr-REM-400 (Micro resist Tech., Berlin, Germany) (Figure 1c), the second EBL was employed for global Ti/Au (5/100 nm) contacts to the G/SiC leads for wire bonding. Before depositing Ti/Au for the global contact, Al was wet-etched using MF-319 photodeveloper (Figure 1e) to ensure contact between graphene and Ti/Au.

3.2.2. Fabrication of Gold Electrodes

Au electrodes in devices Au1–Au6 were fabricated by standard EBL using a poly (methylmethacrylate) (PMMA) (MicroChem, Westborough, MA, USA) double layer mask on Si/SiO$_2$ (300 nm) substrates. We used the same electrode design that was used for graphene electrodes and Ti/Au (5/50 nm) was evaporated on the patterned PMMA and lifted off in organic solvent acetone. The thickness of Ti/Au (5/50 nm) was chosen to be comparable with the height of typical PANI nanofibers.

3.3. Synthesis of Polyaniline Nanofibers and Contacting to Graphene and Au Electrodes

PANI nanofibers were synthesized using a known synthesis protocol [44–46]. 0.08 mmol of aniline (Sigma-Aldrich, St. Louis, MO, USA) was dissolved in 10 mL of 1 M HCl and a catalytic amount of *p*-phenylenediamine (5 mg) (Sigma-Aldrich, St. Louis, MO, USA) in a minimal amount of methanol was added into the aniline solution. 0.2 mmol of ammonium peroxidisulfate (Sigma-Aldrich, St. Louis, MO, USA) was dissolved in 10 mL of 1 M HCl and the two prepared solutions were rapidly mixed for 10 s and left for one day. A droplet of the suspension of the PANI nanofibers doped by hydrochloric acid was deposited on both the G/SiC and Au electrodes and blow-dried. Then we inspected these under optical and atomic force microscope and selected those devices in which single fibers are contacted. The probability of finding such devices is low, and we presented 6 devices in total (3 graphene contacts and 3 gold contacts) and also presented 4 devices corresponding to three or four polymer nanofibers (1 graphene contact and 3 gold contacts). The AFM and I-V curves of the nanofibers on graphene (G1–G4) and on Au (Au1–Au6) are described in detail in the Supplementary Materials.

3.4. Electrical Characterization, Raman Spectroscopy and Carbonization

Electrical characterization of G/SiC electrodes, PANI nanofibers on G/SiC and Au electrodes, and the annealed devices was carried out using the Semiconductor Characterization System (SCS) parameter analyzer (Keithley Instruments, Solon, OH, USA) at room temperature under ambient conditions in both two-terminal and four-terminal configurations. Raman spectroscopy measurement was performed under ambient conditions using a Raman spectrometer equipped with a spot size ~1 µm (λ = 638 nm) (Horiba Scientific, Longjumeau, France). The signal acquisition time was one minute and averaged 5 times due to the relatively small signal of the graphene compared with the signal from the SiC substrate. The annealing took place in a tube furnace at 800 °C for one hour under argon flow with automated ramping rate of 1 °C/min in both heating and cooling steps.

4. Conclusions

In conclusion, we used epitaxial graphene on SiC as Ohmic contacts to polymer nanofibers. We showed that G/SiC-PANI devices exhibit a conductivity comparable to that of PANI nanofibers on Au electrodes. Thermal annealing of the G/SiC-PANI nanofiber device showed that the device is intact after 800 °C annealing and that the PANI nanofibers become amorphous carbons with reduced height and width, making epitaxial graphene contacts promising for applications that require operation at high temperature. While the thermal stability of G/SiC is comparable to that of other materials, graphene offers additional properties such as chemical stability and atomic flatness that make it an attractive platform as a substrate and contact material for future all-carbon devices.

Supplementary Materials: The following are available online at www.mdpi.com/2073-4352/7/12/378/s1, Figure S1: Device Au1–Au3 (a) Atomic force microscope topography of PANI contacted between Au contacts 1-2, 2-3, and 3-4 (Au1, Au2, Au3, respectively); (b) Current-Voltage characteristics of PANI nanofibers contacted between contact 1-2 (Au1), 2-3 (Au2), 3-4 (Au3), and four-probe measurement; Figure S2: Device Au4 (a) Atomic force microscope topography of PANI contacted between Au contacts 1-2 (Au4); (b) Current-Voltage characteristics of the PANI nanofiber contacted between contacts 1-2 (Au4); Figure S3: Device Au5 (a) Atomic force microscope topography of PANI contacted between Au contacts 1-2 (Au5); (b) Current-Voltage characteristics of the PANI nanofiber contacted between contacts 1-2 (Au5); Figure S4: Device Au6 (a) Atomic force microscope topography of PANI contacted between Au contacts 1-2 (Au6); (b) Current-Voltage characteristics of the PANI nanofibers contacted between contacts 1-2 (Au1); Figure S5: Device G1 (a) AFM phase of PANI contacted between G/SiC contacts 1-2 (G1). We checked that the electrodes (1) and (2) were electrically insulating before nanofiber deposition. ((2) and (3) were electrically shorted due to incomplete graphene etching as shown in the AFM phase image); (b) Current-Voltage characteristics of the PANI nanofiber contacted between contacts 1-2 (G1) before and after T = 800 °C annealing. In this device, the electrical resistance decreased after annealing; Figure S6: G2 (a) AFM phase of PANI contacted G/SiC contact 1-2 (G2). We checked that the electrodes (1) and (2) were electrically insulating before nanofiber deposition. (b) Current-Voltage characteristics of the PANI nanofiber contacted between contact 1-2 (G2) before T = 800 °C annealing. After annealing the nanofiber was cut and not conductive; Figure S7: G3 and G4 AFM topography (a) and phase (b) of PANI contacted G/SiC on contact 1-2 (G3), 2-3 (G4), and 3-4. The device shown in Figure 3 is G4 and among the three PANI nanofibers in G4, the nanofiber in Figure 3 is in the middle of the electrode. We checked that the electrodes (1), (2), (3), and (4) were electrically insulating each other before nanofiber deposition. (c) and (d) are the AFM topography and phase after T = 800 °C annealing, respectively; (e) Current-Voltage characteristics of the PANI nanofiber contacted between contacts 1-2 (G3), 2-3 (G4), and 3-4 before T = 800 °C annealing. (f) Current-Voltage characteristics of the PANI nanofiber contacted between contacts 1-2 (G3), 2-3 (G4), and 3-4 after T = 800 °C annealing. Scale bars in (a)–(d) are 10 um; Table S1: Summary of PANI-Au devices (Au1–Au6) in height, source-drain distance, and conductivity; Table S2: Summary of PANI-G/SiC devices (G1–G6) in height, source-drain distance and conductivity.

Acknowledgments: This work was jointly supported by the Swedish-Korean Basic Research Cooperative Program of the National Research Foundation (NRF) NRF-2017R1A2A1A18070721, the Swedish Foundation for Strategic Research (SSF) IS14-0053, GMT14-0077, RMA15-0024, Swedish Research Council, Knut and Alice Wallenberg Foundation, and Chalmers Area of Advance NANO. Partial support was provided by the GRDC (2015K1A4A3047345), the FPRD of BK21 from the NRF through the Ministry of Science, ICT Future Planning (MSIP), Korea.

Author Contributions: Sergey Kubatkin, Yung Woo Park, Samuel Lara-Avila and Kyung Ho Kim conceived and designed the experiments; Kyung Ho Kim and Hans He performed the experiments and Kyung Ho Kim analyzed the data; Hojin Kang contributed to polyaniline synthesis; Rositsa Kang developed the process for G/SiC growth; Kyung Ho Kim and Samuel Lara-Avila wrote the paper. All authors reviewed the manuscript.

Conflicts of Interest: The authors declare no conflict of interest. The funding sponsors had no role in the design of the study; in the collection, analyses, or interpretation of data; in the writing of the manuscript, and in the decision to publish the results.

References

1. Chen, D.; Lei, S.; Chen, Y. A single polyaniline nanofiber field effect transistor and its gas sensing mechanisms. *Sensors* **2011**, *11*, 6509–6516. [CrossRef] [PubMed]

2. Baker, C.O.; Shedd, B.; Innis, P.C.; Whitten, P.G.; Spinks, G.M.; Wallace, G.G.; Kaner, R.B. Monolithic actuators from flash-welded polyaniline nanofibers. *Adv. Mater.* **2008**, *20*, 155–158. [CrossRef]

3. Chen, X.; Wong, C.K.Y.; Yuan, C.A.; Zhang, G. Nanowire-based gas sensors. *Sens. Actuators B Chem.* **2013**, *177*, 178–195. [CrossRef]

4. Yu, H.; Li, B. Wavelength-converted wave-guiding in dye-doped polymer nanofibers. *Sci. Rep.* **2013**, *3*, 1674. [CrossRef] [PubMed]

5. Park, J.U.; Nam, S.; Lee, M.S.; Lieber, C.M. Synthesis of monolithic graphene–graphite integrated electronics. *Nat. Mater.* **2011**, *11*, 120–125. [CrossRef] [PubMed]

6. Liu, W.; Wei, J.; Sun, X.; Yu, H. A Study on graphene—metal contact. *Crystals* **2013**, *3*, 257–274. [CrossRef]

7. Lin, Y.M.; Valdes-Garcia, A.; Han, S.J.; Farmer, D.B.; Meric, I.; Sun, Y.N.; Wu, Y.Q.; Dimitrakopoulos, C.; Grill, A.; Avouris, P.; et al. Wafer-scale graphene integrated circuit. *Science* **2011**, *332*, 1294–1298. [CrossRef] [PubMed]

8. Neto, A.H.C.; Guinea, F.; Peres, N.M.R.; Novoselov, K.S.; Geim, A.K. The electronic properties of graphene. *Rev. Mod. Phys.* **2009**, *81*, 109–162. [CrossRef]

9. Banszerus, L.; Schmitz, M.; Engels, S.; Dauber, J.; Oellers, M.; Haupt, F.; Watanabe, K.; Taniguchi, T.; Beschoten, B.; Stampfer, C. Ultrahigh-mobility graphene devices from chemical vapor deposition on reusable copper. *Sci. Adv.* **2015**, *1*, e1500222. [CrossRef] [PubMed]

10. Yager, T.; Webb, M.J.; Grennberg, H.; Yakimova, R.; Lara-Avila, S.; Kubatkin, S. High mobility epitaxial graphene devices via aqueous-ozone processing. *Appl. Phys. Lett.* **2015**, *106*, 063503. [CrossRef]

11. Boyd, D.A.; Lin, W.H.; Hsu, C.C.; Teague, M.L.; Chen, C.C.; Lo, Y.Y.; Chan, W.Y.; Su, W.B.; Cheng, T.C.; Chang, C.S.; et al. Single-step deposition of high-mobility graphene at reduced temperatures. *Nat. Commun.* **2015**, *6*, 6620. [CrossRef] [PubMed]

12. Balandin, A.A.; Ghosh, S.; Bao, W.; Calizo, I.; Teweldebrhan, D.; Miao, F.; Lau, C.N. Superior thermal conductivity of single-layer graphene. *Nano Lett.* **2008**, *8*, 902–907. [CrossRef] [PubMed]

13. Nair, R.R.; Blake, P.; Grigorenko, A.N.; Novoselov, K.S.; Booth, T.J.; Stauber, T.; Peres, N.M.R.; Geim, A.K. Fine structure constant defines visual transparency of graphene. *Science* **2008**, *320*, 1308. [CrossRef] [PubMed]

14. Bonaccorso, F.; Sun, Z.; Hasan, T.; Ferrari, A.C. Graphene photonics and optoelectronics. *Nat. Photonics* **2010**, *4*, 611–622. [CrossRef]

15. Mansour, A.E.; Said, M.M.; Dey, S.; Hu, H.; Zhang, S.; Munir, R.; Zhang, Y.; Moudgil, K.; Barlow, S.; Marder, S.R.; et al. Facile doping and work-function modification of few-layer graphene using molecular oxidants and reductants. *Adv. Funct. Mater.* **2017**, *27*, 1602004. [CrossRef]

16. Hong, S.K.; Song, S.M.; Sul, O.; Cho, B.J. Reduction of metal-graphene contact resistance by direct growth of graphene over metal. *Carbon Lett.* **2013**, *14*, 171–174. [CrossRef]

17. Lee, S.; Jo, G.; Kang, S.J.; Park, W.; Kahng, Y.H.; Kim, D.Y.; Lee, B.H.; Lee, T. Characterization on improved effective mobility of pentacene organic field-effect transistors using graphene electrodes. *Jpn. J. Appl. Phys.* **2012**, *51*, 02BK09. [CrossRef]

18. Liu, W.; Jackson, B.L.; Zhu, J.; Miao, C.; Park, Y.; Sun, K.; Woo, J.; Xie, Y. Large scale pattern graphene electrode for high performance in transparent organic single crystal field-effect transistors. *ACS Nano* **2010**, *4*, 3927–3932. [CrossRef] [PubMed]

19. Cao, Y.; Liu, S.; Shen, Q.; Yan, K.; Li, P.; Xu, J.; Yu, D.; Steigerwald, M.L.; Nuckolls, C.; Liu, Z.; et al. High-performance photoresponsive organic nanotransistors with single-layer graphenes as two-dimensional electrodes. *Adv. Funct. Mater.* **2009**, *19*, 2743–2748. [CrossRef]

20. Park, J.K.; Song, S.M.; Mun, J.H.; Cho, B.J. Graphene gate electrode for MOS structure-based electronic devices. *Nano Lett.* **2011**, *11*, 5383–5386. [CrossRef] [PubMed]

21. Di, C.A.; Wei, D.; Yu, G.; Liu, Y.; Guo, Y.; Zhu, D. Patterned graphene as source/drain electrodes for bottom-contact organic field-effect transistors. *Adv. Mater.* **2008**, *20*, 3289–3293. [CrossRef]

22. Pang, S.; Tsao, H.N.; Feng, X.; Mullen, K. Patterned graphene electrodes from solution-processed graphite oxide films for organic field-effect transistors. *Adv. Mater.* **2009**, *21*, 3488–3491. [CrossRef]

23. Henrichsen, H.H.; Bøggild, P. Graphene electrodes for n-type organic field-effect transistors. *Microelectron. Eng.* **2010**, *87*, 1120–1122. [CrossRef]

24. Wang, X.; Zhi, L.; Müllen, K. Transparent, conductive graphene electrodes for dye-sensitized solar cells. *Nano Lett.* **2008**, *8*, 323–327. [CrossRef] [PubMed]

25. Jo, G.; Choe, M.; Cho, C.Y.; Kim, J.H.; Park, W.; Lee, S.; Hong, W.K.; Kim, T.W.; Park, S.J.; Hong, B.H.; et al. Large-scale patterned multi-layer graphene films as transparent conducting electrodes for GaN light-emitting diodes. *Nanotechnol.* **2010**, *21*, 175201. [CrossRef] [PubMed]

26. Qian, Z.; Hui, Y.; Liu, F.; Kang, S.; Kar, S.; Rinaldi, M. Graphene–aluminum nitride NEMS resonant infrared detector. *Microsyst. Nanoeng.* **2016**, *2*, 16026. [CrossRef]

27. Kuzum, D.; Takano, H.; Shim, E.; Reed, J.C.; Juul, H.; Richardson, A.G.; de Vries, J.; Bink, H.; Dichter, M.A.; Lucas, T.H.; et al. Transparent and flexible low noise graphene electrodes for simultaneous electrophysiology and neuroimaging. *Nat. Commun.* **2014**, *5*, 5259. [CrossRef] [PubMed]

28. Park, D.W.; Brodnick, S.K.; Ness, J.P.; Atry, F.; Krugner-Higby, L.; Sandberg, A.; Mikael, S.; Richner, T.J.; Novello, J.; Kim, H.; et al. Fabrication and utility of a transparent graphene neural electrode array for electrophysiology, in vivo imaging, and optogenetics. *Nat. Protoc.* **2016**, *11*, 2201–2222. [CrossRef] [PubMed]

29. Vashist, S.K.; Luong, J.H.T. Recent advances in electrochemical biosensing schemes using graphene and graphene-based nanocomposites. *Carbon* **2015**, *84*, 519–550. [CrossRef]

30. Vashist, S.K.; Luppa, P.B.; Yeo, L.Y.; Ozcan, A.; Luong, J.H.T. Emerging Technologies for Next-Generation Point-of-Care Testing. *Trends Biotechnol.* **2015**, *33*, 692–705. [CrossRef] [PubMed]

31. Zhu, Y.; Sun, Z.; Yan, Z.; Jin, Z.; Tour, J.M. Rational design of hybrid graphene films for high-performance transparent electrodes. *ACS Nano* **2011**, *5*, 6472–6479. [CrossRef] [PubMed]

32. Yager, T.; Lartsev, A.; Yakimova, R.; Lara-Avila, S.; Kubatkin, S. Wafer-scale homogeneity of transport properties in epitaxial graphene on SiC. *Carbon* **2015**, *87*, 409–414. [CrossRef]

33. Yazdi, G.; Iakimov, T.; Yakimova, R. Epitaxial graphene on SiC: A review of growth and characterization. *Crystals* **2016**, *6*, 53. [CrossRef]

34. Shih, F.Y.; Chen, S.Y.; Liu, C.H.; Ho, P.H.; Wu, T.S.; Chen, C.W.; Chen, Y.F.; Wang, W.H. Residue-free fabrication of high-performance graphene devices by patterned PMMA stencil mask. *AIP Adv.* **2014**, *4*, 67129. [CrossRef]

35. Kybert, N.J.; Han, G.H.; Lerner, M.B.; Dattoli, E.N.; Esfandiar, A.; Charlie Johnson, A.T. Scalable arrays of chemical vapor sensors based on DNA-decorated graphene. *Nano Res.* **2014**, *7*, 95–103. [CrossRef]

36. Yong, K.; Ashraf, A.; Kang, P.; Nam, S. Rapid stencil mask fabrication enabled one-step polymer-free graphene patterning and direct transfer for flexible graphene devices. *Sci. Rep.* **2016**, *6*, 24890. [CrossRef] [PubMed]

37. Hsu, A.; Wang, H.; Kim, K.K.; Kong, J.; Palacios, T. Impact of graphene interface quality on contact resistance and RF device performance. *IEEE. Electron. Device Lett.* **2011**, *32*, 1008–1010. [CrossRef]

38. Xie, W.; Weng, L.T.; Ng, K.M.; Chan, C.K.; Chan, C.M. Clean graphene surface through high temperature annealing. *Carbon* **2015**, *94*, 740–748. [CrossRef]

39. Yager, T.; Lartsev, A.; Mahashabde, S.; Charpentier, S.; Davidovikj, D.; Danilov, A.; Yakimova, R.; Panchal, V.; Kazakova, O.; Tzalenchuk, A.; et al. Express optical analysis of epitaxial graphene on SiC: Impact of morphology on quantum transport. *Nano Lett.* **2013**, *13*, 4217–4223. [CrossRef] [PubMed]

40. Kopylov, S.; Tzalenchuk, A.; Kubatkin, S.; Fal'Ko, V.I. Charge transfer between epitaxial graphene and silicon carbide. *Appl. Phys. Lett.* **2010**, *97*, 112109. [CrossRef]

41. Janssen, T.J.B.M.; Tzalenchuk, A.; Yakimova, R.; Kubatkin, S.; Lara-Avila, S.; Kopylov, S.V.; Fal'Ko, V.I. Anomalously strong pinning of the filling factor $\nu = 2$ in epitaxial graphene. *Phys. Rev. B.* **2011**, *83*, 233402. [CrossRef]

42. Leong, W.S.; Nai, C.T.; Thong, J.T.L. What does annealing do to metal-graphene contacts? *Nano Lett.* **2014**, *14*, 3840–3847. [CrossRef] [PubMed]

43. Yager, T.; Lartsev, A.; Cedergren, K.; Yakimova, R.; Panchal, V.; Kazakova, O.; Tzalenchuk, A.; Kim, K.H.; Park, Y.W.; Lara-Avila, S.; et al. Low contact resistance in epitaxial graphene devices for quantum metrology. *AIP Adv.* **2015**, *5*, 087134. [CrossRef]

44. Tran, H.D.; Wang, Y.; D'Arcy, J.M.; Kaner, R.B. Toward an understanding of the formation of conducting polymer nanofibers. *ACS Nano* **2008**, *2*, 1841–1848. [CrossRef] [PubMed]

45. Huang, J.; Kaner, R.B. The intrinsic nanofibrillar morphology of polyaniline. *Chem. Commun.* **2006**, *0*, 367–376. [CrossRef] [PubMed]

46. Tran, H.D.; Norris, I.; D'Arcy, J.M.; Tsang, H.; Wang, Y.; Mattes, B.R.; Kaner, R.B. Substituted polyaniline nanofibers produced via rapid initiated polymerization. *Macromolecular* **2008**, *41*, 7405–7410. [CrossRef]

47. Wu, Z.; Chen, X.; Zhu, S.; Zhou, Z.; Yao, Y.; Quan, W.; Liu, B. Enhanced sensitivity of ammonia sensor using graphene/polyaniline nanocomposite. *Sens. Actuators B Chem.* **2013**, *178*, 485–493. [CrossRef]

48. Crowley, K.; Smyth, M.; Killard, A.; Morrin, A. Printing polyaniline for sensor applications. *Chem. Pap.* **2013**, *67*, 771–780. [CrossRef]

49. Goto, A.; Kyotani, M.; Tsugawa, K.; Piao, G.; Akagi, K.; Yamaguchi, C.; Matsui, H.; Koga, Y. Nanostructures of pyrolytic carbon from a polyacetylene thin film. *Carbon* **2003**, *41*, 131–138. [CrossRef]

50. Kyotani, M.; Matsushita, S.; Nagai, T.; Matsui, Y.; Shimomura, M.; Kaito, A.; Akagi, K. Helical carbon and graphitic films prepared from iodine-doped helical polyacetylene film using morphology-retaining carbonization. *J. Am. Chem. Soc.* **2008**, *130*, 10880–10881. [CrossRef] [PubMed]

51. Matsushita, S.; Akagi, K. Macroscopically aligned graphite films prepared from iodine-doped stretchable polyacetylene films using morphology-retaining carbonization. *J. Am. Chem. Soc.* **2015**, *137*, 9077–9087. [CrossRef] [PubMed]

52. Bober, P.; Trchová, M.; Morávková, Z.; Kovářová, J.; Vulić, I.; Gavrilov, N.; Pašti, I.A.; Stejskal, J. Phosphorus and nitrogen-containing carbons obtained by the carbonization of conducting polyaniline complex with phosphites. *Electrochem. Acta* **2017**, *246*, 443–450. [CrossRef]

53. Maiyalagan, T.; Viswanathan, B.; Varadaraju, U.V. Nitrogen containing carbon nanotubes as supports for Pt-Alternate anodes for fuel cell applications. *Electrochem. Commun.* **2005**, *7*, 905–912. [CrossRef]

54. Rozlívková, Z.; Trchová, M.; Exnerová, M.; Stejskal, J. The carbonization of granular polyaniline to produce nitrogen-containing carbon. *Synth. Met.* **2011**, *161*, 1122–1129. [CrossRef]

55. Trchová, M.; Konyushenko, E.N.; Stejskal, J.; Kovářová, J.; Ćirić-Marjanović, G. The conversion of polyaniline nanotubes to nitrogen-containing carbon nanotubes and their comparison with multi-walled carbon nanotubes. *Polym. Degrad. Stab.* **2009**, *94*, 929–938. [CrossRef]

56. Quílez-Bermejo, J.; González-Gaitan, C.; Morallón, E.; Cazorla-Amorós, D. Effect of carbonization conditions of polyaniline on its catalytic activity towards ORR. Some insights about the nature of the active sites. *Carbon* **2017**, *119*, 62–71. [CrossRef]

57. Shen, W.; Fan, W. Nitrogen-containing porous carbons: synthesis and application. *J. Mater. Chem. A* **2013**, *1*, 999–1013. [CrossRef]

58. Mentus, S.; Ćirić-Marjanović, G.; Trchová, M.; Stejskal, J. Conducting carbonized polyaniline nanotubes. *Nanotechnology* **2009**, *20*, 245601. [CrossRef] [PubMed]

59. Kim, K.H.; Lara-Avila, S.; Kang, H.; He, H.; Eklöf, J.; Hong, S.J.; Park, M.; Moth-Poulsen, K.; Matsushita, S.; Akagi, K.; et al. Apparent power law scaling of variable range hopping conduction in carbonized polymer nanofibers. *Sci. Rep.* **2016**, *6*, 37783. [CrossRef] [PubMed]
60. Fromm, F.; Oliveira, M.H.; Molina-Sánchez, A.; Hundhausen, M.; Lopes, J.M.J.; Riechert, H.; Wirtz, L.; Seyller, T. Contribution of the buffer layer to the Raman spectrum of epitaxial graphene on SiC(0001). *New J. Phys.* **2013**, *15*, 043031. [CrossRef]
61. Lee, D.S.; Riedl, C.; Krau, B.; Klitzing, K.V.; Starke, U.; Smet, J.H.; Festkörperforschung, M.; Stuttgart, D. Raman spectra of epitaxial graphene on SiC and of epitaxial graphene transferred to SiO_2. *Nano Lett.* **2008**, *8*, 4320–4325. [CrossRef] [PubMed]
62. Takahashi, R.; Lippmaa, M. Thermally Stable Sr_2RuO_4 Electrode for Oxide Heterostructures. *ACS Appl. Mater. Interfaces* **2017**, *9*, 21314–21321. [CrossRef] [PubMed]
63. Park, K.Y.; Kim, H.; Lee, S.; Kim, J.; Hong, J.; Lim, H.D.; Park, I.; Kang, K. Thermal structural stability of multi-component olivine electrode for lithium ion batteries. *CrystEngComm* **2016**, *18*, 7463–7470. [CrossRef]
64. Choi, A.J. Magneto Resistance of One-Dimensional Polymer Nanofibers. Ph.D. Thesis, Seoul National University, Seoul, Korea, February 2012.

Review

Progress in Contact, Doping and Mobility Engineering of MoS₂: An Atomically Thin 2D Semiconductor

Amritesh Rai * [ID], Hema C. P. Movva [ID], Anupam Roy, Deepyanti Taneja, Sayema Chowdhury and Sanjay K. Banerjee

Microelectronics Research Center, Department of Electrical and Computer Engineering, The University of Texas at Austin, Austin, TX 78758, USA; hemacp@utexas.edu (H.C.P.M.); anupam@austin.utexas.edu (A.R.); dtaneja@utexas.edu (D.T.); sayemac88@utexas.edu (S.C.); banerjee@ece.utexas.edu (S.K.B.)

* Correspondence: amritesh557@utexas.edu; Tel.: +1-614-530-9557

Received: 30 January 2018; Accepted: 19 May 2018; Published: 6 August 2018

Abstract: Atomically thin molybdenum disulfide (MoS₂), a member of the transition metal dichalcogenide (TMDC) family, has emerged as the prototypical two-dimensional (2D) semiconductor with a multitude of interesting properties and promising device applications spanning all realms of electronics and optoelectronics. While possessing inherent advantages over conventional bulk semiconducting materials (such as Si, Ge and III-Vs) in terms of enabling ultra-short channel and, thus, energy efficient field-effect transistors (FETs), the mechanically flexible and transparent nature of MoS₂ makes it even more attractive for use in ubiquitous flexible and transparent electronic systems. However, before the fascinating properties of MoS₂ can be effectively harnessed and put to good use in practical and commercial applications, several important technological roadblocks pertaining to its contact, doping and mobility (μ) engineering must be overcome. This paper reviews the important technologically relevant properties of semiconducting 2D TMDCs followed by a discussion of the performance projections of, and the major engineering challenges that confront, 2D MoS₂-based devices. Finally, this review provides a comprehensive overview of the various engineering solutions employed, thus far, to address the all-important issues of contact resistance (R_C), controllable and area-selective doping, and charge carrier mobility enhancement in these devices. Several key experimental and theoretical results are cited to supplement the discussions and provide further insight.

Keywords: two-dimensional (2D) materials; transition metal dichalcogenides (TMDCs); molybdenum disulfide (MoS₂); field-effect transistors (FETs); Schottky barrier (SB); tunneling; contact resistance (R_C); doping; mobility (μ); scattering; dielectrics

Contents

1. Introduction .... [3]

2. Projected Performance of 2D MoS₂ .... [5]

3. Major Challenges in Contact, Doping and Mobility Engineering of 2D MoS₂ .... [7]

    3.1. The Schottky Barrier and the van der Waals (vdW) Gap .... [7]

    3.2. Contact Length Scaling, Doping and Extrinsic Carrier Scattering .... [10]

    3.3. Tackling the Major Challenges .... [11]

4. Contact Work Function Engineering .... [11]

    4.1. N-Type Work Function Engineering .... [12]

    4.2. P-Type Work Function Engineering .... [14]

5. Effect of Stoichiometry, Contact Morphology and Deposition Conditions .... [14]

6. Electric Double Layer (EDL) Gating .... [16]

7. Surface Charge Transfer Doping .... [18]

    7.1. Charge Transfer Electron Doping .... [19]

    7.2. Charge Transfer Hole Doping .... [21]

8. Use of Interfacial Contact 'Tunnel' Barriers .... [23]

9. Graphene 2D Contacts to MoS₂ .... [26]

10. Effects of MoS₂ Layer Thickness .... [30]

11. Effects of Contact Architecture (Top versus Edge) .... [34]

12. Hybridization and Phase Engineering .... [37]

13. Engineering Structural Defects, Interface Traps and Surface States .... [40]

14. Role of Dielectrics in Doping and Mobility Engineering .... [44]

    14.1. Dielectrics as Dopants .... [45]

    14.2. Mobility Engineering with Dielectrics: Role of High-κ .... [47]

    14.3. Limitations of High-κ Dielectrics and Advantages of Nitride Dielectric Environments ... [50]

15. Substitutional Doping of 2D MoS₂ .... [55]

    15.1. Hole Doping by Cation Substitution .... [57]

    15.2. Electron Doping by Cation Substitution .... [58]

    15.3. Electron and Hole Doping by Anion Substitution .... [59]

    15.4. Towards Controlled and Area-Selective Substitutional Doping .... [62]

16. Conclusions and Future Outlook .... [64]

1. Introduction

The isolation and characterization of graphene, an atomically thin layer of carbon atoms arranged in a hexagonal lattice, in 2004 by Geim and Novoselov ushered in the era of two-dimensional (2D) atomically thin layered materials [1]. This all-important discovery came at the backdrop of a continuous ongoing quest by the semiconductor industry to search for new semiconducting materials, engineering techniques and efficient transistor topologies to extend "Moore's Law"—an observation made in the 1960s by Gordon Moore which stated that the number of transistors on a complementary metal-oxide-semiconductor (CMOS) microprocessor chip and, hence, the chip's performance, would double every two years or so [2–4]. In effect, this law led to the shrinking down of conventional CMOS transistors (down into the nm regime) to enhance their density and performance on the chip [5–10]. However, in the past decade or so, the performance gains derived due to dimensional scaling have been severely offset by the detrimental short-channel effects (SCE) that cause high OFF-state leakage currents (due to loss of effective gate control over the charge carriers in the semiconducting channel and inability of the gate to turn the channel fully OFF) leading to higher static power consumption and heat dissipation (i.e., wasted power), which have dire implications for Moore's Law [11–16]. With continued scaling (sub-10 nm regime), the SCE effect will get far worse and even state-of-the-art CMOS transistor architectures designed to enhance gate controllability (such as MuGFETs, UTB-FETs, FinFETs, etc.) will face serious challenges in minimizing the overall power consumption. Hence, the need of the hour is an appropriate transistor channel material that allows for a high degree of gate controllability at these ultra-short dimensions [17–20]. In this light, graphene has been thoroughly researched for its remarkable properties, such as 2D atomically thin nature, extremely high carrier mobilities, superior mechanical strength, flexibility, optical transparency, and high thermal conductivity, that can be useful for a wide range of device applications [21–23]. While graphene can allow for excellent gate controllability due to its innate atomic thickness, a major drawback of graphene is its "semi-metallic" nature and, hence, the absence of an electronic "band-gap" (E_g)—a necessary attribute any material must possess to be considered for electronic/optoelectronic device applications. Hence, a graphene transistor cannot be turned "OFF" [24,25].

Graphene's shortcomings led to the search for alternative materials with similar yet complementary properties. This led to the emergence of a laundry list of 2D layered materials ranging from insulators to semiconductors and metals [26,27]. Among these 2D materials, the family of transition metal dichalcogenides (TMDCs) has garnered the most attention [28]. These TMDCs are characterized by the general formula MX_2 where M represents a transition metal (M = Mo, W, Re, etc.) and X is a chalcogen (X = S, Se, Te) [29,30]. Analogous to graphene, these layered 2D TMDCs can be isolated down to a single atomic layer from their bulk form. A TMDC monolayer can be visualized as a layer of transition metal atoms sandwiched in-between two layers of chalcogen atoms (of the form X-M-X) with strong intra-layer covalent bonding, whereas the inter-layer bonding between two adjacent TMDC layers is of the van der Waals (vdW) type (Figure 1a schematically illustrates the 3D crystal structure of molybdenum disulfide or MoS_2, the prototypical TMDC). Moreover, depending on the specific crystal structure and atomic layer stacking sequence (1T, 2H or 3R), these TMDCs can have metallic, semiconducting or superconducting phases [29,30]. Of particular interest is the subset of semiconducting 2D TMDCs as they offer several promising advantages over conventional 3D semiconductors (Si, Ge and III-Vs) such as: (i) inherent ultra-thin bodies enabling enhanced electrostatic gate control and carrier confinement versus 3D bulk semiconductors (this can help mitigate SCE in ultra-scaled FETs based on 2D TMDCs as their ultra-thin bodies can allow significant reduction of the so-called characteristic "channel length (L_{CH}) scaling" factor "λ", given by $\lambda = \sqrt{(t_{OX} t_{BODY} \varepsilon_{BODY})/\varepsilon_{OX}}$, where t_{OX} and t_{BODY} are the thicknesses of the gate oxide and channel, respectively, and ε_{OX} and ε_{BODY} are their respective dielectric constants; a simple relationship for the scaling limit of FETs, i.e., minimum length required to prevent SCE, is given by $L_{CH} > 3\lambda$) (Figure 1c shows the schematic cross sections of the gate-channel regions of FETs employing bulk 3D and 2D semiconducting channels and compares their electrostatic carrier confinements) [31]; (ii) availability of a wide range of sizeable band-gaps and diverse band-alignments [32]; and (iii) lack of surface "dangling bonds" unlike conventional 3D semiconductors (Figure 1b schematically compares the surface of bulk 3D and 2D materials) allowing for the formation of pristine defect-free interfaces (especially 2D/2D vdW interfaces) [33]. These attributes make the semiconducting 2D TMDCs extremely promising for future "ultra-scaled" and "ultra-low-power" devices [30,31,33–39]. Among the semiconducting 2D TMDCs, MoS$_2$ has been the most popular and widely pursued material by the research community owing to its natural availability and environmental/ambient stability. Like most semiconducting TMDCs, MoS$_2$ is characterized by a thickness-dependent band-gap as has been verified both theoretically and experimentally: in its bulk form, it has an indirect band-gap of ~1.2 eV, whereas in its monolayer form, the band-gap increases to ~1.8 eV due to quantum confinement effects and is direct (Figure 1d illustrates the band-structure evolution of MoS$_2$ with decreasing layer thickness) [40–44]. This band-gap variability, together with high carrier mobilities, mechanical flexibility, and optical transparency, makes 2D MoS$_2$ extremely attractive for practical nano- and optoelectronic device applications on both rigid and flexible platforms [45–51].

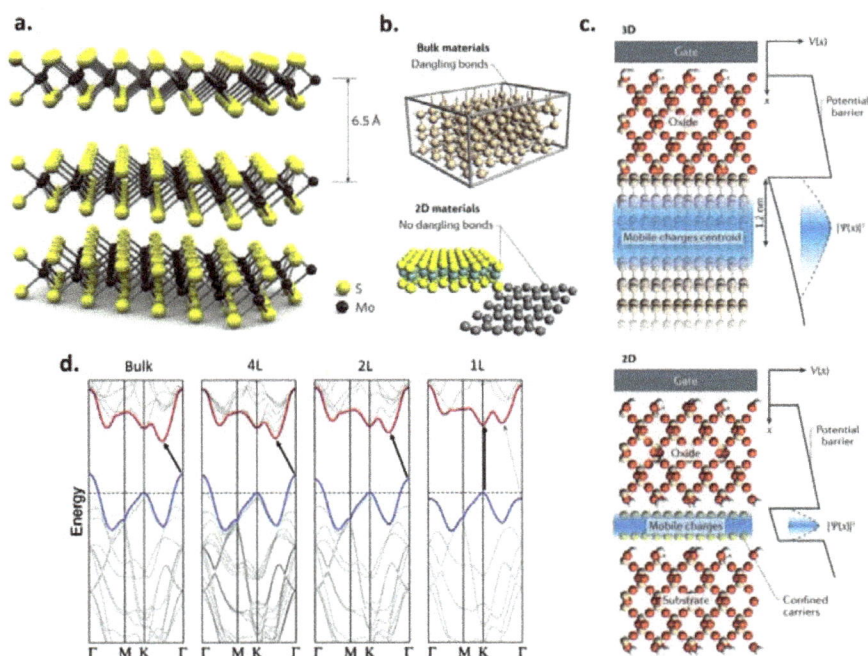

Figure 1. (**a**) 3D schematic of the crystal structure of semiconducting 2H MoS$_2$, the prototypical TMDC, showing stacked atomic layers. Atoms in each layer are covalently bonded, whereas a vdW gap exists between adjacent layers with an interlayer separation of ~0.65 nm. Adapted with permission from [40]. Copyright Springer Nature 2011. (**b**) Schematic illustration of bulk 3D (**top**) versus 2D materials (**bottom**) showing the absence of surface dangling bonds in the latter. (**c**) Schematic illustration of the carrier confinement and electrostatic gate coupling in bulk 3D (**top** schematic) versus 2D semiconducting materials (**bottom** schematic) when used as the channel material in a conventional FET architecture. 2D semiconductors offer much better gate control and enhanced carrier confinement, as opposed to 3D semiconductors, owing to their innate atomic thickness. (**b**,**c**) Adapted with permission from [35]. Copyright Springer Nature 2016. (**d**) Band-structure evolution of MoS$_2$ from bulk to monolayer (1L) showing the transition from an indirect to a direct band-gap (as indicated by the solid black arrow). Adapted with permission from [41]. Copyright 2010 American Chemical Society.

MoS$_2$ can also be combined with conventional 3D semiconductors (such as Si and III-Vs), other 2D materials (e.g., TMDCs or graphene), and 1D and 0D materials to form various 2D/3D, 2D/2D, 2D/1D and 2D/0D vdW heterostructure devices, respectively, enabling a wide gamut of functionalities [52–59]. Indeed, several device applications such as ultra-scaled FETs [60–63], digital logic [64–67], memory [68–71], analog/RF [72–75], conventional diodes [76–79], photodetectors [80–83], light emitting diodes (LEDs) [84–87], lasers [88,89], photovoltaics [90–93], sensors [94–97], ultra-low-power tunneling-devices such as tunnel-FETs (TFETs) [98–101], and piezotronics [102,103], among several others, have been demonstrated using 2D MoS$_2$ (either on exfoliated MoS$_2$ flakes or synthesized MoS$_2$ films), highlighting its promise and versatility. Concurrently, massive research effort has been devoted to solving various key technical challenges, such as large-area wafer-scale synthesis using techniques like chemical vapor deposition (CVD) and its variants (such as metal–organic CVD or MOCVD), van der Waals (vdW) epitaxy, [104–107], reduction of parasitic contact resistance (R$_C$), and enhancement of charge carrier mobility (μ), that can improve the operational efficiency of these devices and allow MoS$_2$-based circuits and systems to become technologically and commercially relevant. The focus of this review paper is to give a comprehensive overview of the progress made in the contact, doping and mobility engineering techniques for MoS$_2$, which collectively represent one of the most significant technological bottlenecks for 2D MoS$_2$ technology.

2. Projected Performance of 2D MoS$_2$

To realize low-power and high-performance electronic/optoelectronic devices based on 2D semiconducting TMDC materials, several key parameters, such as contact resistance (R$_C$), channel/contact doping (n- or p-type) and charge carrier mobility (for both electrons and holes), need to be effectively engineered to harness the maximum intrinsic efficiency from the device [31,35,36,38,39]. In the case of MoS$_2$, excluding the effect of any external factors, its calculated/predicted intrinsic performance is indeed extremely promising. Firstly, the quantum limit to contact resistance (R$_{Cmin}$) for crystalline semiconducting materials in the 2D limit is determined by the number of conducting modes in the semiconducting channel which, in turn, is connected to the 2D sheet carrier density (n$_{2D}$, in units of 10^{13} cm^{-2}) as R$_{Cmin}$ = $26/\sqrt{n_{2D}}$ Ω·μm (Figure 2a depicts this quantum limit in a plot of R$_C$ versus n$_{2D}$) [108–111]. For n$_{2D}$ = 10^{13} cm^{-2}, this yields an R$_{Cmin}$ of 26 Ω·μm, which is well below the projected maximum allowable parasitic source/drain (S/D) resistances for high-performance Si CMOS technology (for example, 80 Ω·μm for multiple-gate FET technology) as per the ITRS requirements for the year 2026 [112]. Thus, 2D MoS$_2$ has the potential of meeting the R$_C$ requirements if a sheet carrier density of ~10^{13} cm^{-2} or higher is realized in the contact regions by doping or other means. Secondly, the predicted room temperature (RT, i.e., 300 K) phonon-limited, or "intrinsic", electron mobility for monolayer MoS$_2$ falls in the range of 130–480 cm^2/V-s [113–116]. On the other hand, the predicted phonon-limited hole mobility for monolayer MoS$_2$ is supposed to be as high as 200–270 cm^2/V-s [115,117]. Moreover, the calculated saturation velocities (v_{sat}) of electrons and holes in monolayer MoS$_2$ are 3.4–4.8 × 10^6 and 3.8 × 10^6 cm/s, respectively [115]. This makes MoS$_2$ extremely promising for various semiconductor device applications and gives it a distinct advantage for use in thin-film transistor (TFT) technologies as its predicted carrier mobilities are higher than conventional TFT materials such as organic and amorphous semiconductors as well as metal oxides (Figure 2b compares the mobility of TMDCs against various other semiconducting materials) [118–120]. In fact, MoS$_2$ offers channel mobilities that are comparable to single-crystalline Si [121]. Moreover, MoS$_2$ can potentially outperform conventional 3D semiconductor devices at aggressively scaled channel lengths (L$_{CH}$ < 5 nm) thanks to its excellent electrostatic integrity [122,123], finite band-gap, and preserved carrier mobilities even at sub-nm thickness (monolayer MoS$_2$ thickness ~0.65 nm), unlike 3D semiconductors that can experience severe mobility degradation (due to scattering from dangling bonds, interface states, atomic level fluctuations, surface roughness, etc.) and a large band-gap increase (due to quantum confinement effects) with dimensional/body thickness scaling below ~5–10 nm [35,36,124–126]. Thus, the high predicted mobilities and saturation velocities, coupled with its atomically thin nature, high optical transparency and mechanical flexibility, makes 2D MoS$_2$ very attractive for applications in ultra-scaled CMOS technologies as well as in flexible nanoelectronics and flexible "smart" systems [74,118,127–129].

The projected performance potential of MoS$_2$ transistors has also been investigated by several research groups and compared to conventional CMOS devices for applicability in future technology nodes. For example, the performance of double-gated monolayer MoS$_2$ FETs was theoretically examined (in the presence of intrinsic phonon scattering) and compared to ultra-thin body (UTB) Si FETs by Liu et al., with results showing that MoS$_2$ FETs can have a 52% smaller drain-induced barrier lowering (DIBL) and a 13% smaller subthreshold swing (SS) than 3-nm-thick-body Si FETs at an L_{CH} of 10 nm with the same gating [123]. This favorable performance and better scaling potential of monolayer MoS$_2$ FETs compared to UTB Si counterparts was attributed to its atomically thin body (~0.65 nm thick) and larger effective mass that can suppress direct source-to-drain tunneling at ultra-scaled dimensions. Moreover, the performance of MoS$_2$ FETs was found to fulfill the requirements for high-performance logic devices at the ultimate scaling limit as per the ITRS targets for the year 2023 [123]. Through rigorous dissipative quantum transport simulations, Cao et al. found that bilayer MoS$_2$ FETs can indeed meet the high-performance (HP) requirement (i.e., the ON-state current drive capability) up to the 6.6 nm node as per the ITRS. Moreover, they showed that with proper choice of materials and device structure engineering, MoS$_2$ FETs can meet both the HP and low-standby-power (LP, i.e., good subthreshold electrostatics in the OFF-state) requirements for the sub-5 nm node as per the ITRS projections for the year 2026 [130]. Another recent simulation study by Smithe et al. revealed that, if the predicted saturation velocity of monolayer MoS$_2$ can be experimentally realized (i.e., $v_{sat} > 3 \times 10^6$ cm/s), then MoS$_2$ FETs can potentially meet the required ON-currents (while meeting the OFF-current requirements) for both HP and LP applications at scaled ITRS technology nodes below 20 nm (Figure 2c compares the projected ON-currents of monolayer MoS$_2$ FETs against ITRS requirements for different MoS$_2$ v_{sat} and field-effect mobility (μ_{FE} or μ_{eff}) values, as a function of gate length "L") [131]. While these performance projections are extremely encouraging, it must be kept in mind that these calculations of contact resistance, mobilities, and FET performances assume an ideal or a near-ideal scenario wherein the 2D MoS$_2$ under consideration is pristine with a defect-free crystal structure, and its material/device properties are evaluated in the absence of extrinsic carrier scattering sources and while considering ideal contact electrodes (i.e., Ohmic contacts). In practice, several non-idealities and inherent challenges exist that can have a detrimental effect on the key performance metrics, adversely affecting the overall MoS$_2$ device performance.

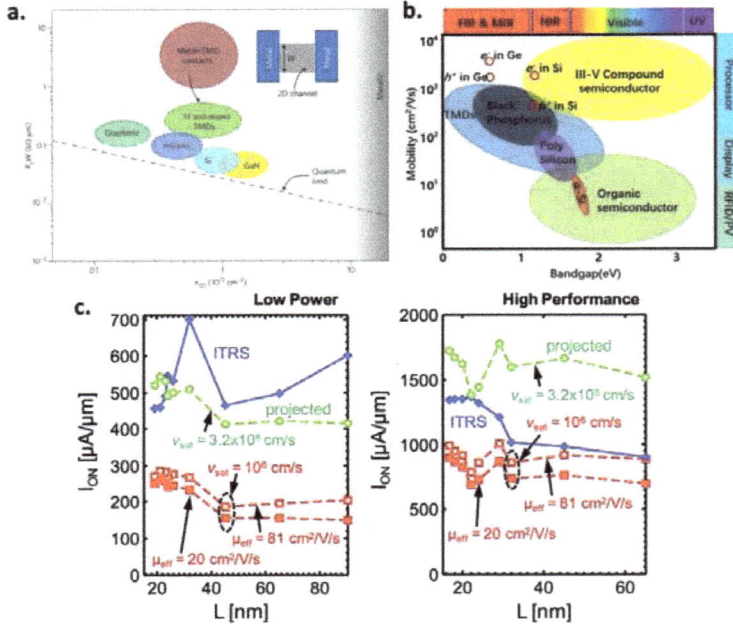

Figure 2. (a) Contact resistance (R_C) plotted as a function of the 2D sheet carrier density (n_{2D}) showing the respective contact resistances of various semiconducting materials (Si, III-Vs, graphene, and TMDCs). The red dashed line represents the quantum limit to R_C. Top right inset shows the schematic top view of a basic transistor configuration. Adapted with permission from [111]. Copyright Springer Nature 2014. (b) Plot of carrier mobility versus band-gap for various semiconducting materials used in technological applications such as processors, displays, RFIDs and photovoltaics. TMDCs have a distinct advantage over poly/amorphous Si and organic semiconductors, and their mobilities are comparable to that of single-crystalline Si. Adapted with permission from [171]. Copyright 2017 John Wiley and Sons. (c) Projected ON-current performance versus gate length L of monolayer MoS_2 FETs compared against low-power (LP) (**left plot**) and high-performance (HP) (**right plot**) ITRS requirements. ITRS requirements are shown in blue with fixed I_{OFF} = 10 pA μm^{-1} for LP and 100 nA μm^{-1} for HP. Simulations in red use v_{sat} = 10^6 cm s^{-1}, with solid symbols for CVD-grown MoS_2 (μ_{FE} = 20 cm^2 V^{-1} s^{-1}) and open symbols for exfoliated MoS_2 (μ_{FE} = 81 cm^2 V^{-1} s^{-1}). The green curve shows projections for MoS_2 FETs using both the higher mobility value (i.e., 81 cm^2 V^{-1} s^{-1}) and higher v_{sat} = 3.2 × 10^6 cm s^{-1}, that meet ITRS requirements for both LP and HP applications for gate lengths L < 20 nm. Adapted with permission from [131]. Copyright 2016 IOP Publishing.

3. Major Challenges in Contact, Doping and Mobility Engineering of 2D MoS$_2$

3.1. The Schottky Barrier and the van der Waals (vdW) Gap

One of the biggest issues confronting MoS_2-based devices is the presence of a Schottky barrier (SB) at the interface between MoS_2 and the contact metal electrode. This results in a "non-Ohmic" or a Schottky electrical contact characterized by an energy barrier, called the Schottky barrier height (SBH or Φ_{SB}), that hinders the injection of charge carriers into the device channel [132]. Consequently, this notable SBH leads to a large R_C and a performance degradation (e.g., low field-effect mobilities) in two-terminal MoS_2 devices since a large portion of the applied drain bias gets dropped across this R_C [133,134]. The presence of the SBH in MoS_2 devices has been experimentally verified by several research groups [134–139], and these barriers are thought to be formed due to strong Fermi level pinning (FLP) effects at the contact metal/MoS$_2$ interface [110,132,140]. Detailed microscopic and spectroscopic studies on natural MoS$_2$ flakes revealed high concentrations of defects and impurities, such as sulfur vacancies (SVs) and subsurface metal-like impurities, which are thought to be responsible for the strong FLP [141–144]. These SV defects/impurities lead to a large background n-doping in the MoS$_2$ and introduce unwanted energy levels or "mid-gap states" closer to the conduction band edge (CBE) within its band-gap that ultimately governs the location of the charge neutrality level where the metal Fermi level gets pinned resulting in fixed barrier heights at the contact/MoS$_2$ interface [145–147]. Further insight on the possible origin of this FLP effect was shed by theoretical calculations based on density functional theory (DFT). Kang et al. reported that interactions between certain metals and MoS$_2$ can lead to the formation of a "metal/MoS$_2$ alloy" at the contact interface with a much lower work function than unalloyed MoS$_2$. This leads to an abnormal FLP as if the MoS$_2$ is contacted to a low work function metal [148]. Gong et al., on the other hand, claimed that the FLP mechanism at metal/MoS$_2$ interfaces is unique and distinctively different from traditional metal-semiconductor junctions. According to their calculations, the FLP at the metal/MoS$_2$ interface is a result of two simultaneous effects: first, a modification of the metal work function by interface dipole formation due to the charge redistribution at the interface and, second, by the formation of mid-gap states originating from Mo d-orbitals, that result from the weakening of the intralayer S-Mo bonds due to the interfacial interaction, and the degree thereof, between the metal and the S atom orbitals [149]. A qualitatively similar result was obtained by Farmanbar et al. where they studied the interaction between a wide range of metals and MoS$_2$ using DFT and found that this MoS$_2$/metal interaction leads to the formation of interface states due to perturbation of the MoS$_2$ electronic band-structure, with energies in the MoS$_2$ band-gap that pin the metal Fermi level below its CBE. The extent of this interfacial interaction depends on whether the metal is physisorbed (i.e., weakly adsorbed) or chemisorbed (i.e., strongly adsorbed) on the MoS$_2$ surface, resulting in a small or large density of interface states, respectively. Moreover, the authors showed that by artificially enlarging the physical distance between MoS$_2$ and the metal, these interface states vanished [150]. Experimentally, this physical separation can be achieved by inserting suitable interfacial tunnel barriers or buffer layers in-between the MoS$_2$ and the contact metal (more on interfacial contact tunnel barriers is discussed in Section 8). Additionally, Guo et al. suggested that the strongly pinned SBHs at the metal/2D MoS$_2$ interface arises due to strong bonding between the contact metal atoms and the TMDC chalcogen atoms [151], in accordance with the age-old theory of metal-induced gap states (MIGS) established for metal contacts to conventional bulk 3D semiconductors [152–154].

Regardless of the exact underlying physical mechanism involved, FLP is an undesired effect as it leads to fixed SBHs at metal/MoS$_2$ interfaces. It is for this very pinning effect that most metal-contacted MoS$_2$ FETs typically show unipolar n-type behavior as the metal Fermi level gets pinned near the CBE of MoS$_2$ irrespective of the metal work function [135,136,155,156]. In addition to degrading the device performance due to large R$_C$, the reduced tunability of the SBH due to FLP is detrimental towards realizing both n-type and p-type Ohmic contacts to MoS$_2$ desirable for CMOS applications [110]. Besides SBH, another relevant parameter associated with these Schottky barriers is the width of its depletion region in the semiconductor channel or, simply, the Schottky barrier width (SBW). The SBW is largely dependent on the extent of semiconductor "band-bending" in the 2D TMDC/MoS$_2$ channel under the electrode contacted region [157]. Both the SBH and the SBW together determine the charge injection in the 2D MoS$_2$ channel. While SBH governs the extent of thermionic emission of carriers "over" the barrier, SBW determines the extent of thermionic field emission (i.e., thermally-assisted tunneling) and/or field emission (i.e., direct tunneling) "through" the width of this barrier due to the quantum mechanical tunneling of charge carriers (Figure 3a shows the band-alignment at the metal/2D TMDC interface under different gating conditions and illustrates the different charge carrier injection mechanisms) [110,132,158,159]. Hence, both the SBH and SBW must be minimized to achieve efficient injection of charge carriers (electrons or holes) from the contact into the semiconducting MoS$_2$ channel. Additionally, the FLP-induced SBH has been found to depend strongly on the MoS$_2$ layer thickness (especially in the limit of 1–5 layers) since the electronic band-structure of MoS_2 undergoes a drastic change as its thickness is decreased (recall that band-gap increases with decreasing MoS_2 thickness), leading to a modification of its electron affinity and relative shifts in its band edge positions (i.e., CBE and valence band edge or VBE) in the energy-momentum (or E-k) space [44,160]. Owing to these factors, thinner MoS_2 with a larger band-gap typically yields a larger SBH with metal contacts as will be discussed later. This effect is particularly important for devices based on direct band-gap monolayer MoS_2 for optoelectronic applications. Finally, in addition to the SB, there are several other important issues that require careful consideration. In an ideal scenario, the surface of TMDCs has an absence or at least a dearth of dangling bonds and, thus, MoS_2 does not tend to form interfacial covalent bonds with the as-deposited contact electrodes. Hence, the metal/MoS_2 interface is characterized by the presence of a van der Waals (vdW) gap, especially in the top contact geometry (which is most common). This vdW gap acts like an additional "tunnel barrier" for the charge carriers in series with the inherent metal/MoS_2 SB (as shown in Figure 3a) and can increase the overall R_C [110,134,148]. Moreover, this vdW gap-induced tunnel barrier also manifests itself in multilayer MoS_2 devices as additional "interlayer" resistors (since adjacent MoS_2 atomic layers are also separated by a vdW gap) and can have implications on the overall device performance. Therefore, for purely electronic applications, the thickness of MoS_2 must be carefully chosen for optimum device performance as will be discussed in more detail later. Some elegant ways to overcome this vdW gap issue are to realize "hybridized" top contacts and/or "edge contacts" (that have a greater degree of orbital interaction with the MoS_2 atoms/bonds resulting in a more intimate contact having lower R_C) instead of the regular top contacts [110,161], and these solutions are discussed in more detail later on along with their promises and inherent challenges.

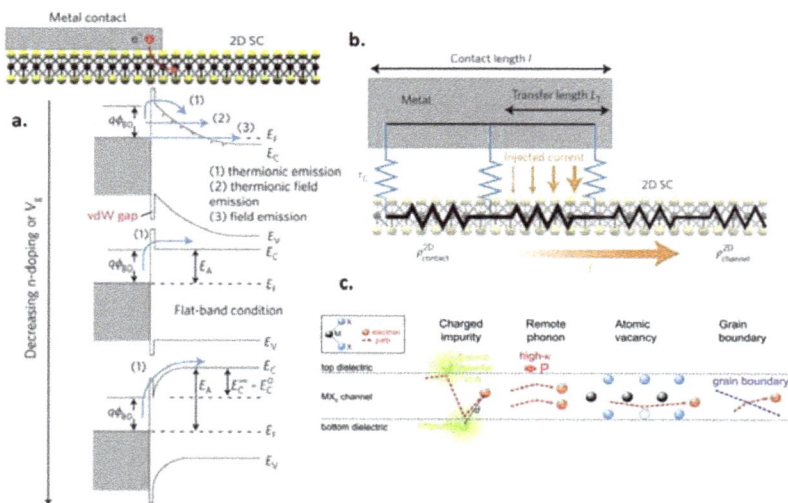

Figure 3. (a) Energy band diagram of the n-type contact/MoS_2 interface under different gating (electrostatic n-doping) conditions depicting the different charge injection mechanisms/paths from the metal into the MoS_2 channel across the SB. $q\Phi_{B0}$ represents the SBH. Thermionic emission is represented by Path (1), thermionic field emission by Path (2) and field emission by Path (3) as shown in the top band diagram for the case of maximum n-doping or maximum gate voltage V_g (that causes maximum downward band-bending). The additional tunnel barrier due to the vdW gap is also shown (marked by the red text). The lateral distance through which the carriers "tunnel" through in Paths (2) and (3) represents the SBW. As V_g decreases (i.e., n-doping decreases), the band-bending decreases and charge injection is governed by thermionic emission only, as shown by Path (1) in the middle and

bottom energy band diagrams. (**b**) Schematic illustration of the contact length (L_C), transfer length (L_T) and current injection (or the "current crowding" effect) near the metal contact/2D TMDC interface edge. The different resistive components at play are marked in the resistor network model (note: in the figure, ρ_C is depicted at r_C, L_C is depicted as l, and TMDC is depicted as SC). (**a**,**b**) Adapted with permission from [110]. Copyright Springer Nature 2015. (**c**) Schematic illustration of the various extrinsic charge carrier scattering mechanisms in a 2D TMDC/MX_2 channel. The black and blue balls denote the M and X atoms, respectively. The orange balls and corresponding orange dashed arrows denote the electrons and their paths in the channel, respectively. Change in the direction of the carrier path denotes a scattering event. The green balls and the smeared green areas denote the charged impurities and their scattering potentials, respectively. The red arrow denotes the polar phonon in the top dielectric. Hollow blue circle represents atomic vacancies which tend to form in both natural and synthetic chalcogenides. Blue dashed line represents grain boundaries (GBs) which are typically present in synthetic chalcogenides. Adapted from [160] with permission of The Royal Society of Chemistry.

3.2. Contact Length Scaling, Doping and Extrinsic Carrier Scattering

A major problem arises when we consider "contact length scaling" for MoS_2. Contact length (L_C) scaling is required when we consider designing aggressively scaled ultra-short-channel devices based on any semiconductor, because L_C must be shrunk by a similar factor as the channel length (L_{CH}) as it will determine the final device footprint/density and can lead to chips with smaller area and faster speeds [162,163]. However, while scaling L_{CH} decreases the channel resistance (R_{CH}), scaling L_C increases R_C in 2D TMDCs. These two effects are contradictory to each other and device performance will ultimately be limited by R_C for aggressively scaled devices [164]. L_C scaling issue mainly arises from the fact that in 2D TMDCs like MoS_2, the transfer length (L_T)—i.e., the average length over which the charge carriers move in the semiconductor before being transferred to the contact electrode (also referred to as the "current crowding" effect at metal/semiconductor contacts) [165–167]—is often large (Figure 3b shows the schematic illustration of this current crowding effect at the metal/2D TMDC junction using a resistor network model). For example, L_T = 600 nm for monolayer MoS_2 [157] and 200 nm for six-layer MoS_2 with Ti contacts [167]. If the L_C is scaled below L_T (i.e., $L_C << L_T$), then R_C increases as per the relation $R_C = \rho_C/L_C$ where ρ_C is the specific contact resistivity [note that R_C is independent of L_C when $L_C >> L_T$ and is then given by the relation $R_C = \sqrt{(\rho_C \, \rho_{SH})}$ where ρ_{SH} is the sheet resistance of the semiconducting channel underneath the contact] [110,168]. Therefore, for ultra-short-channel FETs (targeting the sub-10 nm node) based on 2D MoS_2, it is extremely important to minimize ρ_C or, in other words, minimize L_T [since $L_T = \sqrt{(\rho_C/\rho_{SH})}$] to achieve low R_C. This is important because the R_C of any FET must only be a small fraction (~20%) of the total FET resistance (i.e., $R_{CH} + 2R_C$) for the transistor to operate properly while ensuring that its current-voltage (I-V) behavior is primarily determined by the intrinsic channel resistance R_{CH} [110,112]. Hence, it is imperative that R_C must scale (i.e., reduce) together with both L_{CH} and L_C before MoS_2-based FETs can come anywhere close to rivaling the performance of state-of-the-art Si and III-V device analogs (for reference, the R_C values reported for most TMDC/MoS_2 FETs to date are about an order of magnitude higher than in today's Si Fin-FET technologies where R_C is well below 100 $\Omega \cdot \mu m$) [110,111,132]. Now, the ρ_C is strongly dependent on the SBH among other factors, hence minimizing or eliminating the SBH is a guaranteed way to alleviate the R_C issue in MoS_2 FETs. Next, the ultra-thin nature of the 2D MoS_2 makes it incredibly challenging to employ conventional CMOS-compatible doping techniques (ion implantation or high-temperature diffusion) to perform controlled and area-selective doping to control the carrier type (n or p) and carrier concentration (ranging from degenerate in the source/drain contact regions to non-degenerate in the channel region) in MoS_2 FETs, especially at the monolayer limit [169]. This is primarily because the atomically thin MoS_2 lattice is highly susceptible to structural damage and etching which, for example, is typically unavoidable in the ion implantation process [170]. Lastly, MoS_2 devices typically show much lower intrinsic carrier mobilities in experiments than the predicted phonon-limited values, implying the existence of extrinsic carrier scattering sources. Thus, it is important to eliminate or minimize the effect of these extrinsic charge carrier scattering mechanisms, such as substrate remote phonons, surface roughness, charged impurities, intrinsic structural defects (e.g., SVs), interface charge traps (D_{it}) and grain boundary (GB) defects (Figure 3c schematically illustrates some prominent extrinsic charge carrier scattering mechanisms), that can severely degrade the mobility in MoS_2-based devices [160,171–179].

3.3. Tackling the Major Challenges

To achieve low-power, high-performance and ultra-scaled devices based on 2D MoS_2, it is highly necessary to come up with effective solutions to alleviate the various problems, as mentioned above, that have an adverse effect on key device performance metrics. It is worth noting that solutions to several of these problems are intertwined and solving one can alleviate the other. As an obvious case, reduction of the SB (either by minimization of the SBH or thinning of the SBW) lowers the R_C and effectively improves the charge injection efficiency and the field-effect mobility (μ_{FE}) of the MoS_2 FETs. Reduction of the SBH can lead to a reduced specific contact resistivity ρ_C. With area-selective and controlled doping, one can potentially realize degenerately doped S/D contact regions in MoS_2, just like in the conventional Si-CMOS case, to achieve Ohmic contacts. Realization of edge contact to few- or multilayer MoS_2, such that each individual layer of the stack is independently contacted, can not only help in eliminating the vdW gap-induced tunnel barriers, it can also be useful in terms of contact scaling and overall device area/footprint reduction. Unsurprisingly, therefore, there has been an extensive research effort in the past few years to explore effective solutions for mitigating the challenges associated with the contact, doping and mobility engineering of 2D MoS_2 devices. These solutions are categorically discussed in the various sections below, highlighting several insightful experimental and theoretical results reported thus far. The reader should note that, although the discussion is focused on MoS_2, majority of these issues, along with their underlying concepts and engineering solutions, are readily applicable to other members of the semiconducting 2D TMDC family (e.g., $MoSe_2$, WS_2, and WSe_2) as well.

4. Contact Work Function Engineering

A very straightforward approach to minimize the SBH for either electrons or holes has been through "work function" (Φ_M) engineering of the contact electrodes. In an ideal scenario, without any FLP, Fermi level of low Φ_M contacts (typically $\Phi_M < 4.5$ eV) can align closer to the CBE of MoS_2 (since the electron affinity of MoS_2 is about 4.2 eV) resulting in smaller SBH for electrons and, likewise, the Fermi level of large Φ_M contacts (typically $\Phi_M > 5$ eV) can align closer to the VBE of MoS_2 resulting in smaller SBH for hole injection. This is known as the Schottky–Mott rule or the Schottky limit, wherein the SBH at any metal/semiconductor junction can be determined by the difference between the metal's work function and the semiconductor's electron affinity [110,158,180]. One would assume then, that by choice of a proper metal work function, it would be possible to eliminate the SBH and realize purely Ohmic contacts. In reality, however, hardly any metal/semiconductor (MS) junctions (including those for traditional bulk or 3D semiconductors) follow this rule due to the FLP effect, and the Fermi level at the MS interface is typically pinned at the interface state energy (referred to as the Bardeen limit of pinning) arising due to MIGS [152–154,181,182]. Hence, the contact Fermi level lies somewhere in-between the Schottky limit (i.e., no pinning) and the Bardeen limit (i.e., perfect pinning) depending on the severity of the FLP, which ultimately determines the SBH [180]. Strategies to achieve Fermi level "depinning" can, therefore, be important to realize true Ohmic contacts by virtue of contact work function engineering alone (as will be discussed later). However, even in the presence of strong FLP effect, as observed in 2D MoS_2 (due to reasons described before), and despite the fact that metals typically get pinned near the CBE of MoS_2 resulting in the largely observed n-type device behavior, it has been shown that the magnitude of the SBH at the contact/MoS_2 interface can be directly correlated to the work function of the contact metal. Efforts to achieve p-type injection in MoS_2 via work function engineering are also discussed.

4.1. N-Type Work Function Engineering

For n-type few-layer MoS$_2$ devices, Das et al. showed that low work function metals such as scandium (Sc, Φ_M = 3.5 eV) and titanium (Ti, Φ_M = 4.3 eV) yield a lower SBH for electron injection into the MoS$_2$ conduction band, resulting in a lower R_C, than higher work function metals such as nickel (Ni, Φ_M = 5.0 eV) and platinum (Pt, Φ_M = 5.9 eV) [135]. From a detailed temperature-dependent study that accounted for both thermionic emission over the SBH and thermally-assisted tunneling through the SBW, the authors extracted the true SBH (i.e., Φ_{SB} extracted at the flatband voltage) to be ~30 meV, ~50 meV, ~150 meV, and ~230 meV for Sc, Ti, Ni, and Pt, respectively, clearly suggestive of the strong FLP near the CBE of MoS$_2$ (the Fermi level pinning factor S = dΦ_{SB}/dΦ_M was around 0.1 indicative of strong pinning). Moreover, the extracted field effect mobilities were found to be 21, 90, 125, and 184 cm^2/V-s for Pt, Ni, Ti, and Sc contacts, respectively, clearly highlighting the detrimental effect of large SBHs on both the R_C and the ON-state device performance (Figure 4a,b show the expected and true metal Fermi level line-up with the MoS$_2$ electronic bands, respectively, with Sc providing the best electron injection) [135]. Similar to the case of Sc contacts, Liu et al. showed that low work function Ti could also be used as an efficient n-type contact for few-layer (5–15 layers) MoS$_2$. Using Ti, they achieved a low R_C of 0.8 kΩ·µm and, based on theoretical calculations, surmised that Ti can heavily dope the MoS$_2$ surface leading to a good contact. Moreover, the authors emphasized upon the importance of MoS$_2$ layer thickness, post-contact "annealing" and realizing "edge contacts" to enhance the performance of few-layer MoS$_2$ devices with Ti contacts [183]. In particular, edge contacts to few- or multilayer devices are more promising because each individual layer in the few-layer device can be independently contacted from the side (more on the effects of MoS$_2$ layer thickness on the SBH and carrier mobility is discussed in Section 10, while Section 11 discusses the advantages of making side or "edge contacts" to few- or multilayer MoS$_2$).

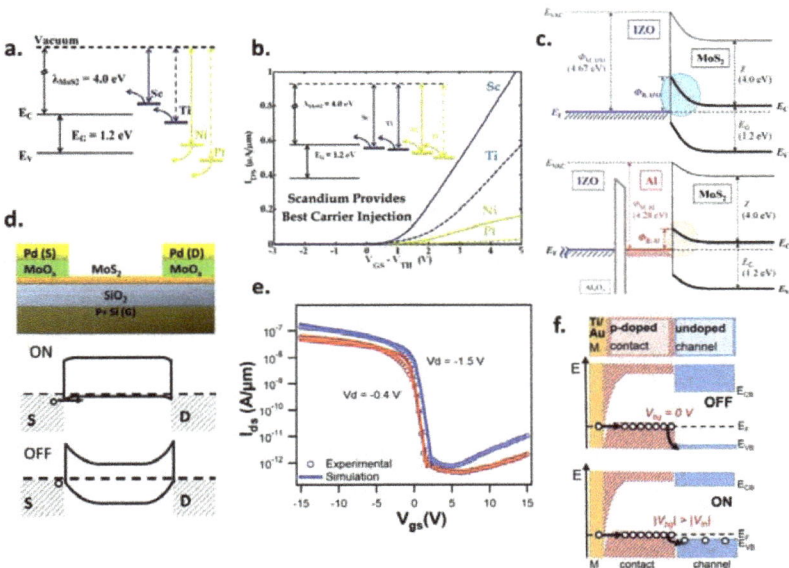

Figure 4. (a) Expected line-up of metal Fermi levels (Sc, Ti, Ni, Pt) with the electronic bands of few-layer MoS$_2$ considering the difference between the electron affinity of MoS$_2$ and the work function of the corresponding metal. (b) Transfer characteristics (I_{DS} versus V_{GS}; linear scale) of a 10 nm thick MoS$_2$ back-gated FET with Sc, Ti, Ni, and Pt metal contacts at 300 K for V_{DS} = 0.2 V. The FET shows n-type behavior with all metals, including high Φ_M Ni and Pt, with low Φ_M Sc clearly providing the best

carrier injection. Inset shows the true metal Fermi level line-up with the MoS_2 bands taking Fermi level pinning (FLP) into account. (**a,b**) Adapted with permission from [135]. Copyright 2013 American Chemical Society. (**c**) Energy band diagrams between IZO (**top**) and IZO/Al (**bottom**) contact electrodes and the MoS_2 channel. The MoS_2-Al-IZO contact includes a thin tunnel layer of amorphous Al_2O_3 due to surface oxidation of Al. The SB in each case is depicted by the colored dashed circular area. Sandwiching a thin layer of low Φ_M Al helps minimize the SBH as depicted in the band diagrams. Adapted from [184]. (**d**) Schematic of an MoS_2 back-gated PFET with high work function MoO_x contacts (**top**) together with the qualitative band diagrams for the ON and OFF states of the MoS_2 PFET (**bottom**). (**e**) Transfer curves of the MoS_2 PFET clearly showing good p-type behavior with high Φ_M MoO_x contacts. (**d,e**) Adapted with permission from [187]. Copyright 2014 American Chemical Society. (**f**) Band profiles explaining the working principle in the OFF- (**top**) and the ON-states (**bottom**) of the MoS_2 PFET with 2D/2D contacts. Holes are injected from a metal (M) into a degenerately p-doped MoS_2 contact layer through a highly transparent interface (i.e., negligible SB). Hole injection from the degenerately p-doped contact layer across the 2D/2D interface into the undoped MoS_2 channel is modulated by the back gate voltage. The bands in the undoped MoS_2 channel near the degenerately p-doped contact can be freely modulated thanks to the weak vdW interaction at the 2D/2D interface. Adapted with permission from [190]. Copyright 2016 American Chemical Society.

In another work, Hong et al. combined thin layers of low work function aluminum (Al, Φ_M = 4.06–4.26 eV) sandwiched in-between MoS_2 and indium zinc oxide (IZO), a transparent conducting oxide having a large work function (Φ_M ~5.14 eV), to realize high-performance and transparent multilayer MoS_2 FETs. The low work function Al contact led to a much reduced SBH resulting in a 24-fold increase of the field-effect mobility (from 1.4 cm^2/V-s in MoS_2/IZO to 33.6 cm^2/V-s in MoS_2/Al/IZO), three orders of magnitude enhancement in the ON/OFF current ratio, robust current saturation and linear output characteristics in these MoS_2 FETs (Figure 4c explains the SBH lowering due to the insertion of low Φ_M Al in-between IZO and MoS_2 via band diagrams). Moreover, the transparent IZO S/D electrodes allowed a transmittance of 87.4% in the visible spectrum [184]. Recently, in a major push towards large-area fully transparent MoS_2 electronics, Dai et al. demonstrated aluminum-doped zinc oxide (AZO) transparent contacts deposited via atomic layer deposition (ALD), with tunable conductivity and work function, to make Ohmic contacts to CVD-grown MoS_2. The work function and resistivity of the AZO film could be tuned by changing the Zn:Al subcycle ratio during the ALD growth process and optimized AZO films with a combination of low resistivity and low work function (Φ_M ~4.54 eV, similar to Ti) were chosen as contacts. Overall, the AZO-contacted CVD MoS_2 FETs showed promising performance with linear output characteristics at RT (suggesting Ohmic-like contacts), a μ_{FE} of 4.2 cm^2/V-s, low threshold voltage (V_{th}) of 0.69 V, low SS of 114 mV/decade, large ON/OFF ratio >10^8, and an average visible-range transmittance of 85% for fully transparent MoS_2 FETs on glass substrates (with AZO S/D and gate contacts, and HfO_2 gate dielectric) [185]. To achieve more effective n-type work function engineered contacts, mitigating the deleterious effects of strong FLP at the 3D metal/2D semiconductor interface, Liu et al. suggested the use of surface engineered 2D "MXenes" as a potential SB-free n-type metal contact to MoS_2. 2D MXenes are a class of metal carbides/nitrides with the general formula $M_{n+1}X_nT_x$ (where M is an early transition metal, X is C and/or N, T represents a surface terminating group, and n = 1–3) that can make a vdW contact to MoS_2 having an inherent vdW gap. This weak vdW interaction can suppress the formation of gap states at the interface leading to a weaker FLP than conventional 3D metal contacts. Moreover, based on first principles calculations, the authors showed that MXenes having "OH" as the surface terminating group can have very low work functions (<3 eV) due to surface dipole effects, even lower than that of Sc metal, leading to Ohmic contacts [186].

4.2. P-Type Work Function Engineering

For realizing p-type MoS_2 devices, Chuang et al. used substoichiometric molybdenum trioxide (MoO_x, x < 3), an extreme high work function transition metal oxide with Φ_M = 6.6 eV, in the S/D

Crystals **2018**, 8, 316

contacts and demonstrated efficient hole injection in the MoS_2 valence band, as opposed to high work function metals that typically showed n-type behavior due to strong FLP. The efficacy of MoO_X as a hole injector was attributed not only to its high Φ_M, but also to its better interface properties (such as lower tendency to form MIGS than elemental metals) that caused a lower degree of FLP. Using MoO_x, the authors could demonstrate MoS_2 PFETs (essential for realizing CMOS-type devices together with MoS_2 NFETs) with ON/OFF ratios ~10^4, and MoS_2 Schottky diodes with asymmetric MoO_x and Ni contacts. The SBH for holes was extracted to be ~310 meV for MoO_x/MoS_2 contacts (Figure 4d,e show the FET schematic as well as the qualitative band diagrams, and the p-type transfer curves for the MoO_x-contacted MoS_2 FETs, respectively) [187]. A detailed theoretical investigation by McDonnell et al. further revealed that the work function of MoO_x should be sufficient to provide an Ohmic hole contact to MoS_2 (provided carbon impurities and Mo^{5+} concentration at the interface can be carefully controlled) [188]. Like high work function MoO_x, high work function graphene oxide (GO, Φ_M ~5–6 eV) has also been proposed as an efficient hole injector in monolayer MoS_2. Theoretically, the p-type SBH at the MoS_2/GO interface can be made smaller by increasing the oxygen concentration and the fraction of epoxy functional groups in GO (which increases its Φ_M). Compared to MoO_x, GO can be promising as it is easier to fabricate, and its production methods are simpler and inexpensive [189].

More recently, an extremely promising experimental approach to realize low-resistance p-type Ohmic contacts to MoS_2 FETs was demonstrated by Chuang et al. where they utilized a "2D/2D" vertical heterostructure contact strategy [190]. In their approach, the undoped semiconducting MoS_2 channel is contacted in the S/D regions by degenerately p-doped $Mo_{0.995}Nb_{0.005}S_2$ [the degenerately p-doped MoS_2 was obtained by substitutional doping of MoS_2 using niobium (Nb) during the crystal growth process; more on substitutional doping of MoS_2 is discussed in Section 15]. The work function difference between the undoped and the degenerately p-doped MoS_2 creates a band offset across the 2D/2D vdW interface. This band offset can be electrostatically tuned by a back gate voltage owing to the weak interlayer vdW interaction at the 2D/2D junction, essentially resulting in a negligible SBH in the ON-state of the FET (Figure 4f illustrates the working principle of MoS_2 PFETs with 2D/2D contacts via band diagrams). Note that the vdW interface also promotes weaker FLP by suppressing the formation of interface gap states. The authors reported field-effect hole mobilities as high as 180 cm^2/V-s at RT, observation of a metal-insulator transition (MIT) in the temperature-dependent conductivity, and linear output characteristics down to 5 K in their p-type MoS_2 FETs with these low-resistance 2D/2D contacts [190]. Finally, an alloyed 2D metal/MoS_2 contact scheme, similar to recent reports on 2D tungsten diselenide (WSe_2) where a $NbSe_2/W_xNb_{1-x}Se_2/WSe_2$ contact interface was realized (here $NbSe_2$ is a metallic 2D TMDC), could also be used to facilitate p-type MoS_2 FETs. Such alloyed 2D junctions have been shown to have atomically sharp vdW interfaces with both reduced interface traps and SBH, and can help maximize the electrical reliability of 2D devices [191]. In the same context of 2D/2D vdW contacts, and as described in the previous section for n-type contacts, Liu et al. also predicted that 2D MXenes with "O" surface terminations can yield a p-type SB-free contact to MoS_2, as some of the O-terminated MXenes can have a rather high work function that is even higher than that of elemental Pt [186]. Additionally, in a separate theoretical study, Liu et al. predicted that 2D niobium disulfide (NbS_2), a 2D TMDC metal with a high work function (>6 eV), can be a promising 2D electrode for achieving low SBH for p-type contacts to MoS_2 while combining all the advantages associated with a vdW interface and weak FLP [140].

5. Effect of Stoichiometry, Contact Morphology and Deposition Conditions

While work function engineering of the contacts seems a simple and straightforward approach to realize either n- or p-type contacts with low SBHs to MoS_2 (taking FLP into account of course), there are other reports that reveal that contact work function engineering alone is not always a good predictor for forming high-quality electrical contacts to MoS_2. For instance, McDonnell et al., in their study of structural defects on MoS_2, found that both n- and p-type regions can exist at different sites on the same MoS_2 sample. The n-type regions were found to be S-deficient (S/Mo ratio ~1.8:1), whereas the S/Mo stoichiometry in the p-type regions was 2.3:1, indicating that these regions were either S-rich or Mo-deficient. These variations in the structural defect density can strongly impact the observed n- or p-type I-V characteristics in MoS_2 devices irrespective of the contact metal (Figure 5a,b show both n- and p-type behavior, respectively, with the same Au contacts at different MoS_2 locations) [141]. This nanoscale spatial inhomogeneity on the MoS_2 surface was further elucidated by Giannazzo et al. where they used high resolution conductive atomic force microscopy (CAFM) to study the spatial variations in the SBH (Φ_{SB}) and local resistivity (ρ_{loc}). They found an excellent correlation between the Φ_{SB} and ρ_{loc} values, with low (high) ρ_{loc} regions corresponding to low (high) Φ_{SB} regions (see Figure 5c), and concluded that the low resistivity/low SBH regions were a result of n-type SV clusters on the MoS_2 surface [192]. Yuan et al. highlighted the importance of metal/MoS_2 interface morphology, and the thermal conductivity of the metal, on the performance of MoS_2 FETs. They compared monolayer and few-layer MoS_2 devices with Ag and Ti contacts, both having a similar low work function (Φ_M = 4.3 eV) and showed that devices with Ag contacts had 60× larger ON-state currents than those with Ti contacts. This was attributed to the significantly smoother and denser topography of Ag films on MoS_2 owing to the excellent wettability of Ag on MoS_2 as well as to the higher thermal conductivity of Ag (~20× larger than Ti) that can enhance the heat dissipation efficiency and, hence, prevent heat-induced mobility degradation in MoS_2 FETs (Figure 5d shows a comparison of the transfer characteristics between identical MoS_2 FETs with Ag and Ti contacts) [193].

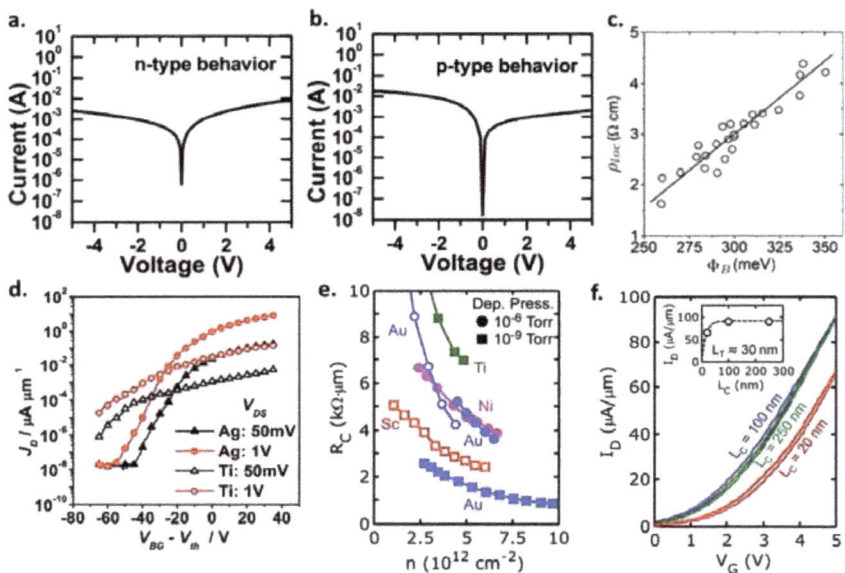

Figure 5. (**a,b**) Electrical characteristics of Au contacts deposited on a single piece of MoS_2 showing both n- (**a**) and p-type (**b**) behavior at different locations, confirming the stoichiometry-dependent doping variation in MoS_2. Adapted with permission from [141]. Copyright 2014 American Chemical Society. (**c**) Plot of local resistivity (ρ_{loc}) as a function of SBH as determined from conductive AFM measurements performed on different MoS_2 sample locations. A direct correlation is observed, confirming the existence of nanoscale inhomogeneities on the surface of MoS_2. Adapted with permission from [192]. Copyright 2015 by the American Physical Society. (**d**) Transfer curves of few-layer MoS_2 FETs with similar work function Ag and Ti contacts showing a clear performance enhancement (higher ON-currents and higher ON/OFF ratios) in the case of Ag contacts. Adapted with permission from [193]. Copyright 2015 American Chemical Society. (**e**) Measured R_C versus n_{2D} for MoS_2 FETs with multiple contact metals deposited under different deposition pressures clearly showing

that lower deposition pressures lead to lower R_C due to cleaner interfaces. The cleanest UHV-Au contacts reach R_C ~740 $\Omega \cdot \mu m$ at n_{2D} ~10^{13} cm^{-2} after the metal electrode resistance is subtracted. This value is even lower than the R_C achieved with low work function Sc contacts. (**f**) Measured transfer characteristics of identical MoS$_2$ FETs with UHV-deposited Au contacts having L_C = 20, 100 and 250 nm (L_{CH} = 40 nm and V_{DS} = 1 V). Inset: Measured current (hollow circles) and simulated current (dashed line) versus L_C at V_{GS} = 5 V, from which a transfer length L_T of ~30 nm is extracted. From the transfer curves, it is evident that when $L_C \ll L_T$ for metal/MoS$_2$ contacts, the device performance degrades due to increase in the R_C originating from the current crowding effect. (**e,f**) Adapted with permission from [164]. Copyright 2016 American Chemical Society.

The excellent wettability and morphology of Ag on MoS$_2$ was further exploited by Kim et al. to demonstrate, for the first time, low-cost inkjet-printed Ag S/D electrodes on large-area CVD-grown monolayer MoS$_2$ FETs, using a commercial nanoparticle-type Ag ink and a drop-on-demand printer. The favorable surface interaction between Ag and MoS$_2$ makes Ag-based printable inks highly compatible for enabling inkjet-printed electrodes on MoS$_2$, a process that is promising for large-area and low-cost MoS$_2$-based thin-film electronics [194]. English et al. revealed the importance of metal deposition conditions and showed that gold (Au), a high work function metal (Φ_M = 5.1 eV), deposited under ultra-high vacuum (UHV) conditions (base pressure ~10^{-9} Torr) yielded a cleaner, higher quality and air-stable (over 4 months) metal/MoS$_2$ contact with a low R_C of ~740 $\Omega \cdot \mu m$, that was even lower than the R_C achieved using low Φ_M metals, such as Sc, Ti and Ni, on MoS$_2$ (Figure 5e compares R_C versus n_{2D} for various metals deposited on MoS$_2$ under two different deposition base pressures) [164,195]. The authors also studied the effects of MoS$_2$ FET scaling and found that the R_C starts dominating the overall device performance below L_{CH} = 90 nm. Moreover, the effects of L_C scaling were also analyzed and, as expected, a current degradation of 30% was observed when L_C became less than L_T due to increase in the R_C when $L_C \ll L_T$, as explained earlier in Section 3.2 (Figure 5f shows the transfer characteristics of MoS$_2$ FETs evaluated at varying contact lengths L_C) [164]. The importance of base vacuum pressure while depositing contacts (especially low Φ_M reactive metals) on MoS$_2$ was also highlighted by McDonnell et al. where they studied Ti contacts deposited under high vacuum (HV, ~10^{-6} mbar) and UHV (~10^{-9} mbar) using X-ray photoelectron spectroscopy (XPS). Under HV, an interfacial TiO$_2$ layer is formed due to the oxidation of Ti, whereas metallic Ti is deposited under UHV that can react with the MoS$_2$ to form less conductive Ti$_x$S$_y$ and metallic Mo at the interface [196]. Similarly, Smyth et al. performed an intensive XPS study to reveal the interfacial chemistry between high work function (Au and Ir) and low work function (Cr and Sc) metals deposited on MoS$_2$ under HV and UHV deposition ambient. They found that while Au does not react with MoS$_2$ regardless of the reactor ambient, Ir leads to interfacial reactions with MoS$_2$ under both HV and UHV. In contrast, both Cr and Sc lead to interfacial reactions under UHV. Additionally, Sc is rapidly oxidized, whereas Cr is only partially oxidized when deposited under HV conditions [197]. Thus, it is evident that the deposition chamber ambient or base pressure can strongly influence the contact/MoS$_2$ interface chemistry and, ultimately, the SB and R_C in MoS$_2$ devices.

6. Electric Double Layer (EDL) Gating

Several groups have also demonstrated the concept of electric double layer (EDL) gating on 2D MoS$_2$ devices using a variety of liquid, solid and gel-based "electrolytes" that serve as the gating medium. In a typical EDL gating approach, an ionic liquid (IL) or a solid polymer electrolyte (PE) is drop-casted on top of an MoS$_2$ FET (typically back-gated) covering the entire FET area along with its S/D contacts. The electrolyte is electrostatically gate-controlled through a top electrode or a side electrode pre-fabricated near the device channel (Figure 6a shows the schematic illustration of the EDL gating approach on MoS$_2$ FETs). When a positive (negative) voltage is applied on the gate electrode, mobile negative (positive) ions in the electrolytic medium accumulate near the gate electrode, whereas positive (negative) ions accumulate near the MoS$_2$ channel, leading to the formation of an EDL at the interfaces between the IL/PE electrolyte and solid surfaces (i.e., the gate electrode and the MoS$_2$

surface). At the MoS$_2$ interface, this results in the induction of either electrons or holes in the channel (depending on the gate bias polarity) essentially doping the channel either n- or p-type (Figure 6b schematically illustrates the formation of the EDL at the electrolyte/solid interfaces) [120,198–203]. A major advantage of the EDL gating/doping technique is that extremely high sheet carrier densities (on the order of n$_{2D}$ ~10^{14} cm^{-2}; much higher than the carrier densities achievable in MoS$_2$ FETs gated using solid dielectrics, e.g., SiO$_2$ or high-κ dielectrics) along with broad carrier density tunability can be realized in the channel due to the large geometrical capacitances and highly efficient gating afforded by the thin EDL layer. Moreover, the doping-induced high carrier densities cause a large band-bending in the MoS$_2$ channel which is beneficial for minimizing the R$_C$ at the MoS$_2$/contact interface due to substantial reduction of the Schottky-depletion width (or the SBW) allowing for easy injection of carriers in the channel via tunneling. This results in increased FET carrier mobilities (μ_{FE}). Although the EDL technique is promising for investigating the electronic transport properties of MoS$_2$, it has some major drawbacks which make it unsuitable for practical device applications. For example, the ionic liquids are unstable, sensitive to moisture, and chemically reactive. Hence, the device measurements must be carried out under high vacuum and at low-temperatures. Moreover, both the liquid and solid electrolytes are physically bulky (several microns thick) and cannot be scaled to nanoscale dimensions [120,198–203].

Despite these limitations, the use of EDL gating on MoS$_2$ has shown some interesting device behavior. The first report of EDL gating using an ionic liquid on MoS$_2$ was by Zhang et al. where they demonstrated ambipolar operation in thin MoS$_2$ flakes characterized by large ON-state conductivities and ON/OFF ratios >10^2 for both the electron and hole branches. The n$_{2D}$ reached 1.0 and 0.75 × 10^{14} cm^{-2} for electrons and holes at | V$_G$ | = 3 V, respectively, while their maximum Hall mobilities were 44 and 86 cm^2/V-s, respectively [198]. Perera et al. reported ambipolarity, significantly higher electron mobilities (~60 cm^2/V-s at 250 K) and near ideal SS (~50 mV/decade at 250 K) in ionic liquid gated MoS$_2$ FETs as compared to comparable back-gated MoS$_2$ FETs (Figure 6c shows the ambipolar behavior in the transfer characteristics of the IL-gated MoS$_2$ FET). They observed an increase in the electron mobility from ~100 to 220 cm^2/V-s as the temperature was lowered from 180 K to 77 K. This performance enhancement was primarily attributed to the reduction of the SB at the S/D contact interface by the enhanced MoS$_2$ band-bending due to EDL doping [200]. The use of a solid PE as an EDL gate on monolayer MoS$_2$ was shown by Lin et al. where they used a PE consisting of poly(ethylene oxide) (PEO) and lithium perchlorate (LiClO$_4$) as a SB reducer and a channel mobility booster. In this case, the PEO serves at the polymer base, whereas the Li$^+$ and ClO$_4^-$ ions serve as the mobile ionic dopants. A three order of magnitude enhancement in the electron mobility (from 0.1 to 150 cm^2/V-s) was achieved that was attributed to the reduction of the contact SB as well as to an "ionic screening" effect. Moreover, PE-gated devices showed a near ideal SS (~60 mV/decade at RT, implying high gating efficiency) and high ON/OFF ratios (~10^6) [199]. A similar PE gating approach, using PEO polymer medium and cesium perchlorate (CsClO$_4$) as the ion source (Cs$^+$ and ClO$_4^-$), was used by Fathipour et al. which yielded an R$_C$ of 200 Ω·μm (comparable to the best R$_C$ reports on MoS$_2$) and high current densities (~300 μA/μm) in MoS$_2$ NFETs [204]. Recently, a "2D electrolyte" capable of electrostatically doping the surface of MoS$_2$ was introduced by Liang et al. The electrolyte is only 0.5–0.7 nm thick and consists of an atomically thin cobalt crown ether phthalocyanine (CoCrPc) and LiClO$_4$ molecules, such that one CoCrPc molecule can solvate one Li$^+$ ion. In this technique, the CoCrPc is deposited on the 2D MoS$_2$ surface by drop-casting and annealing to form an ordered array. The Li$^+$ ion location with respect to the CoCrPc/MoS$_2$ interface can be modulated by a gate bias, similar to the conventional EDL approach, to dope the MoS$_2$ by inducing image charges on the MoS$_2$ surface, with n$_{2D}$ as high as ~10^{12} cm^{-2} (Figure 6d schematically illustrates the concept of 2D CoCrPc-based electrolytes and the relative movement of the Li$^+$ ion with respect to the CoCrPc/MoS$_2$ interface). Moreover, the 2D electrolyte shows "bistability", with the extent of n-doping (either more or less) dependent on the magnitude and polarity of the external gate bias [205]. This work is indeed promising as it shows that electrolytes can be scaled to atomically thin dimensions and can be used for adjustable doping/gating of 2D MoS_2, but the ambient stability of the 2D electrolyte is still under scrutiny.

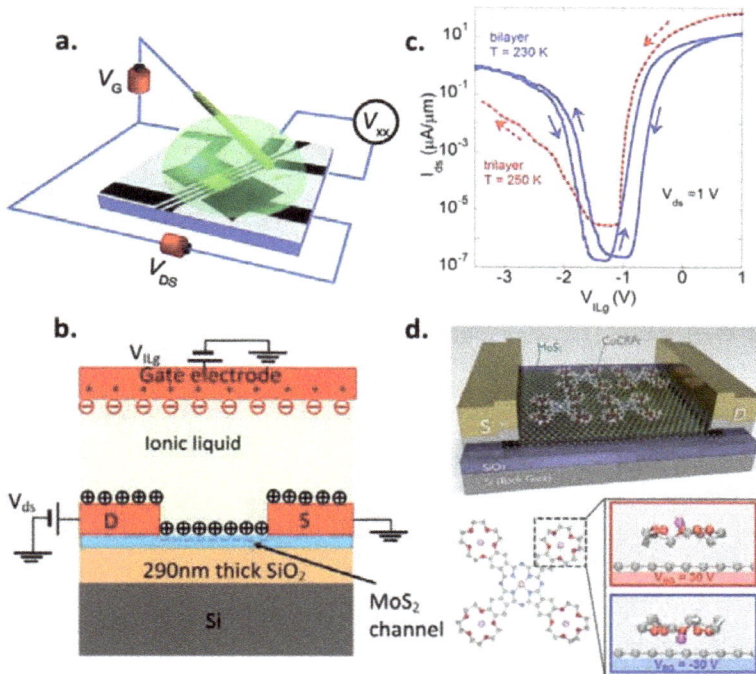

Figure 6. (a) 3D schematic of an ionic liquid (IL) gated device. The transparent light green blob represents the ionic liquid that is contacted by a narrow gate electrode on the top. Side gates fabricated close to the device channel represent another popular EDL gating scheme. Adapted with permission from [198]. Copyright 2012 American Chemical Society. (b) Schematic illustration of the working principle of an IL-gated MoS_2 FET showing the formation of an electric double layer (EDL) in close proximity of the MoS_2 channel when a voltage is applied on the gate electrode. (c) Transfer characteristics of representative bilayer and trilayer MoS_2 IL-gated FETs measured at the drain-source bias V_{DS} of 1 V. Ambipolarity is observed owing to the extremely high geometrical capacitance of the EDL layer which enables large MoS_2 band-bending, SB/R_C reduction and accumulation of both electrons and holes in the MoS_2 channel. (b,c) Adapted with permission from [200]. Copyright 2013 American Chemical Society. (d) Schematic of an MoS_2 FET with a 2D electrolyte (**top**) and the atomic structure of a CoCrPc molecule showing the four crown ethers (CE) as well as two states of the CE/Li^+ molecular complex under applied gate biases with opposite polarity (**bottom**). The Li^+ ions (represented by solid pink spheres) embedded in these CoCrPc molecules can move either towards or farther away from the MoS_2 channel surface depending on the applied back gate bias and cause doping via image charge formation. Adapted from [205] with permission of the Electrochemical Society.

7. Surface Charge Transfer Doping

Surface charge transfer doping, utilizing various chemical/molecular reagents and sub-stoichiometric high-κ oxides, has been investigated as an alternative method to achieve controllable channel doping as well as access region doping to alleviate the SB/R_C issue in MoS_2 FETs. In this approach, depending on the electron affinity/work function of the adsorbed or deposited interfacial specie, electrons either get donated to or accepted from the MoS_2 surface resulting in n-type or p-type doping, respectively. This technique typically involves heavily doping the contact/access regions of the MoS$_2$ FET with electrons/holes which renders the SB transparent due to substantial "thinning" of the SBW (due to large band-bending in the highly doped MoS$_2$ near or underneath the contact). Thus, the charge carriers can easily "tunnel" through the narrow SBW into the channel resulting in Ohmic contacts. Conceptually, this approach is similar to that used in conventional Si CMOS technology where the S/D regions are degenerately doped by donor (e.g., P and As) or acceptor (e.g., B) species to facilitate carrier tunneling and low R$_C$ for n- and p-type contacts, respectively, at the metal-semiconductor contact interface [206–208].

7.1. Charge Transfer Electron Doping

Initial studies on MoS$_2$ devices utilized strong electron-donating reactive chemical species such as polyethyleneimine (PEI) [209] and reactive group-I metals such as potassium (K) [210]. Although successful electron doping, and an improvement in the MoS$_2$ FET performance (by reduction of the sheet/contact/access resistances), was achieved using these techniques, the doping reagents used were unstable under ambient conditions and, hence, practically unfeasible [209,210]. The first air- and vacuum-stable n-type charge transfer doping of MoS$_2$ was subsequently demonstrated by Kiriya et al. using benzyl viologen (BV), an electron donor organic compound having one of the highest reduction potentials. Using BV doping, the authors obtained an electron sheet density of ~1.2 × 10^{13} cm^{-2}, which corresponds to the degenerate limit for MoS$_2$ as well as a 3× reduction in the R$_C$ of MoS$_2$ FETs (Figure 7a,b show the schematic illustration of the BV doping process, and performance enhancement in the transfer curves of the MoS$_2$ FET after BV doping, respectively). Moreover, the BV dopant molecules could be reversibly removed by immersion in toluene, thereby promoting controlled and selective-area doping [211]. In an interesting experimental and theoretical study, Rai et al. demonstrated the use of sub-stoichiometric high-κ oxides, such as TiO$_x$ (x < 2), HfO$_x$ (x < 2) and Al$_2$O$_x$ (x < 3), as air-stable n-type charge transfer dopants on monolayer MoS$_2$. This high-κ oxide doping effect, arising due to interfacial-oxygen-vacancies in the high-κ oxide, could be used as an effective way to fabricate high-κ-encapsulated top-gated MoS$_2$ FETs with selective doping of the S/D access regions to alleviate the R$_C$ issue, merely by adjusting the interfacial high-κ oxide stoichiometry (Figure 7c shows the R$_C$ of a back-gated monolayer MoS$_2$ FET, extracted using the transfer length measurement or "TLM" method, as a function of back gate bias before and after sub-stoichiometric TiO$_x$ doping) [212,213]. The underlying doping mechanism is similar for all high-κ oxides and involves the creation of donor states/bands near the CBE of MoS$_2$ by the uncompensated interfacial metal atoms of the sub-stoichiometric high-κ oxides. Moreover, this doping effect is absent in the case of purely stoichiometric high-κ oxides as has been verified both experimentally as well as theoretically using DFT calculations [212–214] (Figure 7d compares the DFT band-structures and atom-projected-density-of-states, AP-DOS, for both an oxygen-rich, i.e., stoichiometric, and an oxygen-deficient TiO$_x$/MoS$_2$ interface confirming the n-doping effect only in the latter case). Using this doping technique, the authors reported an R$_C$ as low as 180 Ω·μm in TiO$_x$-encapsulated monolayer MoS$_2$, that is among the lowest reported R$_C$ values for monolayer MoS$_2$ FETs. The extracted transfer length L$_T$ reduced from 145 nm before doping to 15 nm after TiO$_x$ doping highlighting the effectiveness of heavily doping the MoS$_2$ near the contact regions to drive down the L$_T$ which is important for ultra-scaled devices. Moreover, an enhancement in both the μ_{FE} and intrinsic mobility was observed, strongly indicating that this high-κ doping effect plays an important role in boosting the electron mobility in high-κ-encapsulated MoS$_2$ FETs.

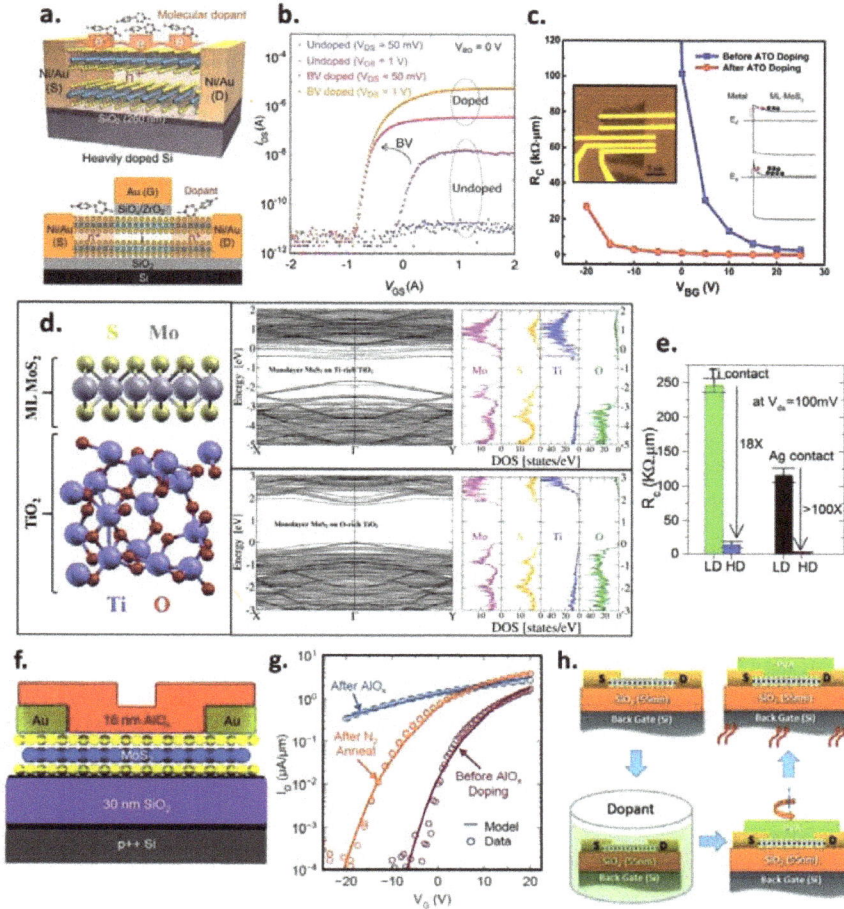

Figure 7. (**a**) Schematic illustration of MoS$_2$ FETs used for benzyl viologen (BV) surface charge transfer doping. Top schematic illustrates BV doping of a bare back-gated MoS$_2$ FET, whereas the bottom schematic illustrates the self-aligned BV doping of S/D access regions in a top-gated MoS$_2$ FET. (**b**) Transfer characteristics of the top-gated device before (blue and purple) and after (pink and orange) BV treatment at V$_{DS}$ = 50 mV and 1 V. The substrate was grounded (V$_{BG}$ = 0 V) during the measurements. A clear enhancement of the FET performance is seen after BV doping. (**a,b**) Adapted with permission from [211]. Copyright 2014 American Chemical Society. (**c**) Extracted R$_C$ versus V$_{BG}$ before (blue) and after (red) amorphous TiO$_x$ (ATO) doping measured using a TLM structure. The R$_C$ shows a strong gate dependence before doping (Schottky behavior) and a weak gate dependence after doping (Ohmic behavior). Left inset: Optical micrograph of the as-fabricated transfer line method (TLM) structure. Right inset: Qualitative band diagrams of the metal/MoS$_2$ interface before (**top**) and after (**bottom**) ATO doping showing increased band-bending in the MoS$_2$ after doping leading to SBW reduction and enhanced electron injection into the channel via tunneling. (**d**) **Left schematic**: Supercell showing the composite crystal structure of the monolayer MoS$_2$/TiO$_2$ interface used in the DFT simulations. **Top right**: Band-structure and atom-projected-density-of-states (AP-DOS) plots for MoS$_2$-on-sub-stoichiometric TiO$_x$ case. In the presence of interfacial oxygen "O" vacancies, electronic states/bands from the uncompensated Ti atoms are introduced near the CBE of monolayer MoS$_2$ causing

the Fermi level (represented by the 0 eV energy level on the y-axis) to get pinned above the conduction band indicating strong n-doping. **Bottom right**: Band-structure and AP-DOS plots for the MoS_2-on-stoichiometric TiO_2 case. No doping effect is seen in this case and the Fermi level remains pinned at the VBE of MoS_2. (**c,d**) Adapted with permission from [212]. Copyright 2015 American Chemical Society. (**e**) Extracted R_C for lightly doped (LD) and heavily doped (HD) top-gated MoS_2 FETs with Ti and Ag S/D contacts using sub-stoichiometric HfO_x as the top dielectric. A drastic decrease in R_C is obtained (18× for Ti-contacted FETs; >100× for Ag-contacted FETs) after heavy HfO_x doping. (**e,c**) (**right inset**) Adapted with permission from [215]. Copyright 2017 IEEE. (**f**) Schematic of a back-gated CVD-grown monolayer MoS_2 FET with a top AlO_x doping layer and Au contacts. (**g**) Semilog transfer curves of the MoS_2 FET before and after AlO_x deposition, and after N_2 anneal. A significant n-doping effect is seen after AlO_x deposition accompanied with an SS degradation. The N_2 anneal helps restore the SS while maintaining the n-doping effect of AlO_x due to conversion of "deep-level traps" into "shallow-level donors". (**f,g**) Adapted with permission from [216]. Copyright 2017 IEEE. (**h**) Schematic illustration of the PVA coating process for n-doping on back-gated MoS_2 FETs. Adapted from [217] with permission of The Royal Society of Chemistry.

In similar reports, both McClellan et al. and Alharbi et al. demonstrated the efficacy of n-doping by sub-stoichiometric high-κ oxides in improving the performance of MoS_2 FETs. Alharbi et al. used sub-stoichiometric HfO_x as the top gate dielectric in FETs fabricated on CVD-grown monolayer MoS_2 and achieved an R_C as low as ~480 Ω·μm under heavy HfO_x doping (>100× improvement than the light HfO_x doping case) and a mobility of ~64 cm^2/V-s (Figure 7e shows the improvement in R_C for both Ti- and Ag-contacted top-gated MoS_2 FETs under light and heavy HfO_x doping). Moreover, the top-gated geometry allowed effective control over the channel resulting in an SS of ~125 mV/decade and an ON/OFF ratio > 10^6 [215]. McClellan et al., on the other hand, utilized AlO_x encapsulation to n-dope back-gated monolayer MoS_2 FETs and achieved an R_C of ~480 Ω·μm, a μ_{FE} of ~34 cm^2/V-s and a record ON-current of 700 μA/μm. A key step in their approach was annealing of the MoS_2 devices in an N_2 ambient after AlO_x encapsulation, which helped restore the SS and μ_{FE} by converting the "deep-level traps" at or near the AlO_x/MoS_2 interface into "shallow-level donors" (Figure 7f,g show the back-gated MoS_2 FET schematic, and the effect of AlO_x doping, as well as N_2 post-annealing on the FET transfer curves, respectively) [216]. Besides n-doping using sub-stoichiometric high-κ oxide encapsulation, poly(vinyl-alcohol) (PVA) polymeric coatings can also be used as strong n-type dopants for MoS_2 as shown by Rosa et al. They showed a 30% reduction in the R_C and the sheet resistance (R_{SH}) was reduced from 161 kΩ sq^{-1} to 20 kΩ sq^{-1} after PVA doping (Figure 7h schematically illustrates the PVA coating process). The non-covalent and non-destructive PVA doping increased the carrier concentration without any μ_{FE} degradation, with the μ_{FE} actually increasing with dopant concentration (from 20 to 28 cm^2/V-s for 0 to 1% PVA). Moreover, the PVA doping efficiency was enhanced after a dehydration anneal (as H_2O molecules were found to hinder the electron transfer from the PVA to the MoS_2 surface) which led to the best MoS_2 device performance in this study. Finally, the authors showed that encapsulating the PVA coating with an ALD-grown Al_2O_3 film can make it robust against the environment with long-lasting doping effects [217]. Other reports on surface charge transfer n-doping of MoS_2 include air-stable doping using hydrazine [218], p-toluene sulfonic acid [219], black phosphorous quantum dots [220], and self-assembled oleylamine (OA) networks [221]. In the case of OA doping, n_{2D} as high as 1.9×10^{13} cm^{-2} at zero gate bias was achieved without any μ_{FE} degradation, along with a 5× reduction in R_C.

7.2. Charge Transfer Hole Doping

For p-type charge transfer doping of MoS_2, Choi et al. reported the use of $AuCl_3$ solution (spin-coated on the MoS_2 FETs) which acts as an effective electron acceptor due to its large positive reduction potential. The mechanism involves formation of Au nano-aggregates through the reduction of $AuCl_4^-$ ions in the solution by receiving electrons from the MoS_2 layer, thereby leading to a significant p-doping of the MoS_2 [76]. The same $AuCl_3$ doping method was used by Liu et al. to realize high-performance MoS$_2$ PFETs with high hole mobilities (68 cm^2/V-s at RT, 132 cm^2/V-s at 133 K), low contact resistance (2.3 kΩ·μm) and ON/OFF ratios >10^7 (Figure 8a shows the transfer curves of the AuCl$_3$-doped back-gated MoS$_2$ PFETs) [222]. The authors also employed "graphene buffer layers" in the contact regions of their AuCl$_3$-doped MoS$_2$ PFETs to demonstrate further reduction of R$_C$ for hole injection into the MoS$_2$ valence band. This is because AuCl$_3$ not only p-dopes the MoS$_2$ causing an upward band-bending in the channel, thereby, reducing the SBW for hole injection from the contact, but it can also p-dope the graphene contact layer causing its Fermi level to move downwards and align closer to the MoS$_2$ VBE, thereby, reducing the SBH for holes. Moreover, the Fermi level in graphene can also be electrostatically tuned giving it an inherent advantage over regular metal contacts (more on graphene contacts to MoS$_2$ is discussed later in Section 9). Tarasov et al. reported controlled n- and p-doping of large-area (>10 cm^2) highly uniform trilayer MoS$_2$ films using stable molecular reductants (such as dihydrobenzimidazole derivatives and benzimidazoline radicals) and oxidants (such as "Magic Blue"), respectively. They achieved high doping densities up to 8 × 10^{12} cm^{-2} and work function modulation up to ±1 eV [223]. Similarly, Sim et al. demonstrated a highly effective and stable doping mechanism based on thiol-based molecular functionalization (note: thiol molecules are organosulfur compounds containing an -SH group) that makes use of the sulfur vacancies in MoS$_2$. In this approach, the -SH terminated end of the thiol molecules get tightly chemisorbed on these MoS$_2$ SV sites, and these thiol molecules act as either donors or acceptors depending upon the nature of the functional groups attached to them (e.g., NH$_2$ for n-doping and F-containing groups for p-doping) (Figure 8b shows the schematic representation of this thiol-based molecular doping approach on MoS$_2$ FETs). A significant enhancement and reduction in the carrier concentration was observed for n- (Δn = +3.7 × 10^{12} cm^{-2}) and p-doping (Δn = −1.8 × 10^{11} cm^{-2}), respectively, using this technique [224]. A very recent report by Min et al. introduced a novel way to realize p-type MoS$_2$ FETs via charge transfer between MoS$_2$ and wide band-gap n-type InGaZnO (IGZO) films (E$_g$ = 3.1 eV) deposited on top of these thick MoS$_2$ flakes (E$_g$ = 1.2 eV) [225]. High work function Pt metal contacts and prolonged ambient thermal annealing at 300 °C were crucial for the realization of these PFETs. In this approach, the prolonged 300 °C anneal causes the IGZO to become a more intrinsic semiconductor (i.e., reduction in its electron carrier density) as the O-vacancies in IZGO (responsible for its n-type doping) get filled by the O atoms in air. This increases the work function (i.e., lowering of the Fermi level) of the IGZO film, thereby, causing an interfacial transfer of electrons from the MoS$_2$ flakes to the IGZO since the equilibrium Fermi level of the MoS$_2$/IGZO system must remain constant. In other words, lowering of the Fermi level in IGZO also drags down the Fermi level in the MoS$_2$ due to charge transfer, causing electron depletion in the MoS$_2$ layer. This process continues until the MoS$_2$ gets heavily depleted of electrons or, in other words, accumulated with holes, eventually resulting in a superior p-type FET performance with Pt-contacts having high hole mobilities of 24.1 cm^2/V-s (Figure 8c,d show the evolution of this p-doping process with increasing ambient annealing time, and the MoS$_2$/IGZO band diagram explaining the p-doping mechanism, respectively). Moreover, the IGZO serves as an encapsulation layer and imparts long term air stability to the device (MoS$_2$ PFETs maintained most of their performance even after 142 days in ambient). With proper choice of contact metals and annealing duration, the authors were also able to demonstrate CMOS-inverter operation on the same MoS$_2$ flake. Thus, the MoS$_2$/IGZO heterojunctions represent a promising and practical approach towards realizing stable MoS$_2$ PFETs necessary for enabling CMOS-applications based on 2D MoS$_2$ [225].

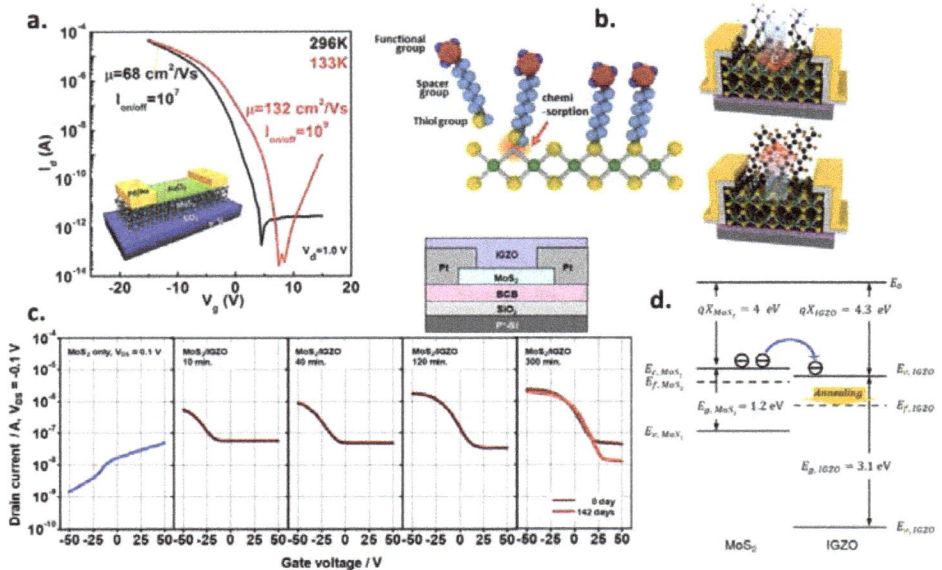

Figure 8. (**a**) Semilog transfer characteristics of an AuCl$_3$-doped MoS$_2$ PFET at RT and 133 K, clearly showing enhanced p-type performance with high hole mobilities and high ON/OFF ratios. Inset shows the schematic of a back-gated MoS$_2$ FET spin-coated with AuCl$_3$ solution. Adapted with permission from [222]. Copyright 2016 John Wiley and Sons. (**b**) Schematic illustrations of the chemisorption of functionalized thiol molecules (containing -SH groups) onto MoS$_2$ sulfur vacancies (**left**), and of the charge transfer doping of back-gated MoS$_2$ FETs with MEA-terminated (**top right**) and FDT-terminated (**bottom right**) thiol molecules, respectively, used for the molecular functionalization on MoS$_2$. The MEA molecule has an NH$_2$ functional group and causes n-doping of the MoS$_2$, whereas the FDT molecule has a fluorocarbon functional group and causes p-doping. Adapted with permission from [224]. Copyright 2015 American Chemical Society. (**c**) Transfer characteristics of an MoS$_2$ only (leftmost blue curve) and the MoS$_2$/IGZO heterojunction FET (maroon curves) with Pt S/D contacts annealed in the ambient for 10, 40, 120, and 300 min at 300 °C, clearly showing gradual enhancement of the p-doping with increasing annealing time due to gradually increased electron transfer from the MoS$_2$ flake to the IGZO capping layer. The final p-doped state of the MoS$_2$ FET was maintained even after 142 days (red curve) showing great ambient stability. Schematic illustration of the IGZO-capped back-gated MoS$_2$ FET shown on the top right corner above the figure. (**d**) Energy band diagram of the MoS$_2$ channel and the IGZO film explaining the charge transfer p-doping process. The ambient annealing lowers the Fermi level of the IGZO film due to decreased n-doping in the IGZO layer, as indicated by the broad yellow arrow. In the MoS$_2$/IGZO heterostructure, this causes transfer of electrons from the conduction band of MoS$_2$ to the conduction band of the IGZO film to maintain the equilibrium Fermi level. Thus, due to this electron transfer, Fermi level of the MoS$_2$ gradually decreases/lowers (i.e., moves towards the VBE of MoS$_2$) with ambient annealing, eventually leading to the MoS$_2$ film becoming strongly p-type. (**c,d**) Adapted with permission from [225]. Copyright 2018 American Chemical Society.

8. Use of Interfacial Contact "Tunnel" Barriers

Contact engineering utilizing ultra-thin interfacial "tunnel barriers" has been employed as another promising way to reduce the SBH and R$_C$ in MoS$_2$ devices. This method, widely explored for engineering the contact resistivity in conventional 3D semiconductor FETs based on Si and Ge [226–232], is based on the incorporation of an ultra-thin insulating material (such as 2D hexagonal boron nitride or hBN, and oxides such as TiO_2) in-between the MoS_2 and the contact electrode, to effectively realize a metal-insulator-semiconductor (MIS) configuration at the contact. This thin interfacial insulating "buffer" layer in the MIS structure serves as a "Fermi level de-pinning (FLDP) layer" by increasing the physical separation between the MoS_2 and the contact electrode owing to its finite thickness, thereby, breaking or minimizing the metal/MoS_2 interfacial interaction responsible for the creation of mid-gap interface states that cause FLP [149–151]. Once this depinning is achieved, the contact work function can effectively be chosen to line up with or closer to the CBE or VBE of MoS_2 in accordance with the Schottky–Mott rule. This approach can help to significantly lower or eliminate the SBH and allow for easy tunneling of the charge carriers through the ultra-thin interfacial barrier into the MoS_2 bands/channel (Figure 9a,b show the qualitative band diagrams of a metal/MoS_2 contact interface with/without an interfacial TiO_2 tunnel barrier, and 3D schematic illustrations of FETs incorporating these interfacial tunnel barriers in their contact regions, respectively). However, in this approach, one has to be mindful of the thickness of the inserted tunnel barrier. It must be thick enough to suppress the metal/MoS_2 interfacial interaction and FLP, yet thin enough to ensure a high tunneling probability at the MoS_2 band edges [226,232]. If the barrier becomes too thick, then the tunneling resistance of the carriers through the barrier will increase significantly, offsetting the advantages gained due to decrease of the R_C (or ρ_C) via SBH reduction (as illustrated in Figure 9d).

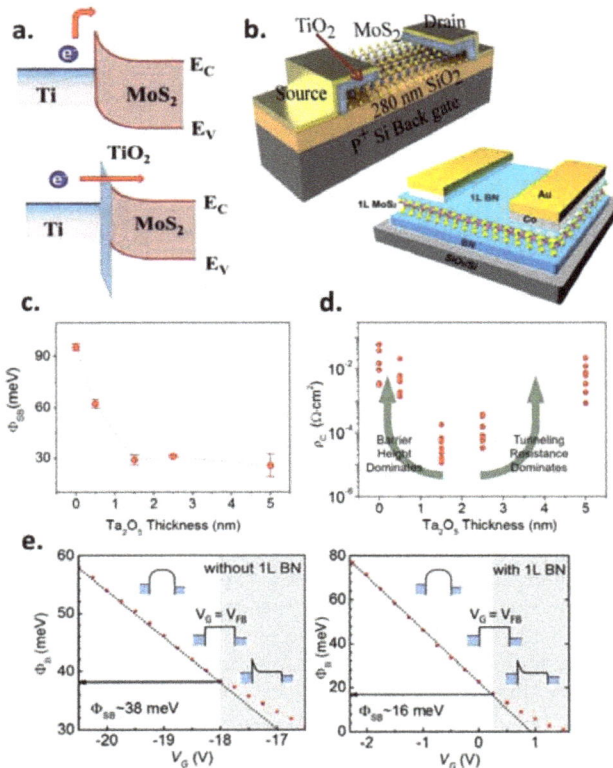

Figure 9. (**a**) Qualitative band diagrams illustrating the effect of an interfacial contact tunnel barrier (e.g., TiO_2) on the charge carrier injection in the MoS_2 channel. A significant SBH exists in the case of conventional metal (e.g., Ti)/MoS_2 contacts and charge is injected primarily due to thermionic emission (**top**). However, the presence of a TiO_2 contact tunnel barrier (**bottom**) minimizes the SBH by

alleviating the FLP effect at the metal/MoS$_2$ interface by effectively creating a physical separation between the two and, hence, minimizing the interfacial reactions between the metal and MoS$_2$ that can create mid-gap states. Easier charge injection can, therefore, be achieved via tunneling (**bottom**). Adapted with permission from [234]. Copyright 2014 IEEE. (**b**) Schematic illustrations of back-gated MoS$_2$ FETs with interfacial contact tunnel barriers. Top schematic: TiO$_2$ barrier only in the S/D contact regions. Adapted with permission from [236]. Copyright 2016 American Chemical Society. Bottom schematic: monolayer hBN barrier covering the entire MoS$_2$ surface (i.e., contact + channel regions). (**c**) Plot of extracted SBH as a function of interfacial Ta$_2$O$_5$ thickness. Lowest SBH is realized for an optimized thickness of ~1.5 nm and remains mostly constant with further increase in the Ta$_2$O$_5$ thickness. (**d**) Plot of specific contact resistivity (ρ_C) as a function of Ta$_2$O$_5$ thickness. Lowest ρ_C is achieved for an optimum thickness around 1.5 nm. For thicknesses <1.5 nm, ρ_C increases due to increase in the SBH (since Ta$_2$O$_5$ is not thick enough to prevent the metal/MoS$_2$ interaction and FLP). For thicknesses >1.5 nm, ρ_C increases due to increase in the tunneling resistance (since Ta$_2$O$_5$ becomes way too thick). Thus, choosing the optimum thickness of the interfacial tunnel barrier is important. (**c**,**d**) Adapted with permission from [237]. Copyright 2016 American Chemical Society. (**e**) Extracted SBH (Φ_{SB}) as a function of gate voltage (V_G) for Co/MoS$_2$ (**left plot**) and Co/1L hBN/MoS$_2$ (**right plot**) contact interfaces. The Φ_{SB} value at the flatband condition (i.e., $V_G = V_{FB}$) represents the true SBH. The extracted flatband SBH in the case of Co/1L hBN/MoS$_2$ is much lower, signifying the importance of having ultra-thin hBN as an interfacial contact tunnel barrier. (**e**) and bottom schematic of (**b**) Adapted with permission from [241]. Copyright 2017 American Chemical Society.

One of the first reports utilizing this approach was by Chen et al. who used a thin magnesium oxide (MgO) barrier (2 nm thick) between ferromagnetic cobalt (Co) electrodes and monolayer MoS$_2$ which resulted in the reduction of the SBH for electrons by as much as 84% [233]. Park et al. reported the use of TiO$_2$ and Al$_2$O$_3$ interfacial FLDP layers and showed that TiO$_2$ resulted in a 5× decrease of R_C at the metal/MoS$_2$ channel interface, with a corresponding increase in the drain current and mobility of the FET. The authors attributed the enhanced R_C decrease in the case of TiO$_2$ to reduction in the SBH (Φ_{SB} reduced from 180 meV to about 90 meV) due to a combined effect of Fermi level de-pinning and stronger dipole effects of the interfacial TiO$_2$ layer than the Al$_2$O$_3$ layer [234]. A similar approach was used by Kaushik et al. where they used ultra-thin TiO$_2$ ALD interfacial layers and demonstrated a 24× reduction in the R_C and a low constant Φ_{SB} of 40 meV in MoS$_2$ FETs irrespective of the contact metal [235,236]. However, they attributed this improvement mainly to the interfacial n-doping effect of TiO$_2$ arising due to a charge transfer mechanism which renders the TiO$_2$/MoS$_2$ interface metallic. This is similar to the n-doping effect observed by Rai et al. at the TiO$_x$/MoS$_2$ interface [212]. These results suggest that TiO$_2$ can be promising as an interfacial contact tunnel barrier due to a combined effect of FLDP and n-doping. Lee et al. demonstrated the use of Ta$_2$O$_5$ as thin interfacial tunneling layers (1.5 nm thick) between CVD-synthesized few-layer MoS$_2$ films and the metal contacts. Using this approach, the extracted Φ_{SB} was reduced from 95 meV in devices without any Ta$_2$O$_5$ to about 29 meV in devices containing Ta$_2$O$_5$ (Figure 9c shows the extracted SBH as a function of Ta$_2$O$_5$ barrier thickness). Moreover, the authors presented a statistical study on over 200 devices made on large area MoS$_2$ films (>4 cm^2) and reported a three orders of magnitude reduction in the specific contact resistivity (ρ_C) and about two orders of magnitude increase in the ON-current of the devices by insertion of the thickness-optimized Ta$_2$O$_5$ layer (Figure 9d shows the dependence of ρ_C on the interfacial Ta$_2$O$_5$ thickness, clearly highlighting the importance of selecting the optimum tunnel barrier thickness to achieve low ρ_C) [237].

In addition to ultra-thin insulating oxides, other 2D materials such as graphene and a monolayer of insulating hBN were also proposed as effective 2D insertions or buffer layers to alleviate the n-type SBH at the metal/MoS$_2$ interface [238,239]. Experimentally, Wang et al. reported the use of ultra-thin CVD-synthesized hBN (thickness = 0.6 nm) as an interfacial tunneling layer to reduce the SBH and realize high mobility MoS$_2$ NFETs. In comparison to oxides, the atomically thin nature of hBN can have advantages in terms of offering relatively small tunneling resistance. The authors achieved a small SBH of 31 meV in MoS_2 FETs with hBN/Ni/Au contacts as well as a high μ_{FE} of 73 cm^2/V-s (321.4 cm^2/V-s) and an output current of 330 μA/μm (572 μA/μm) at RT (77 K) [240]. Similarly, Cui et al. utilized cobalt (Co) with a monolayer (1L) of hBN as the tunnel barrier to realize low-temperature Ohmic contacts to monolayer MoS_2. The authors extracted a flatband SBH of 16 meV for the Co/1L hBN/MoS_2 case, a reduction from 38 meV for the Co/MoS_2 case (as shown in Figure 9e), and reported the best low-temperature MoS_2 contacts to date, with an R_C value of 3.0 kΩ·μm at 1.7 K extracted at a carrier density of only 5.3 × 10^{12} cm^{-2} [241]. This drastic R_C improvement led to the observation of interesting quantum oscillations in monolayer MoS_2 devices at much lower carrier densities compared to previous works. In addition to the role of monolayer hBN as a tunnel barrier, a critical factor that led to the enhanced behavior and greatly reduced SBH in these Co/hBN-contacted MoS_2 devices was the strong interaction between the hBN and Co that led to a lowering of the latter's work function from 5.0 eV (for pure Co) to 3.3 eV [241], in excellent agreement with theoretical predictions made by Farmanbar et al. [238]. The use of an additional MoS_2 layer itself as an interfacial buffer layer has also been suggested by Chai et al. that can not only help prevent the interfacial reactions between the contact metal and the MoS_2 channel layer (by preventing any unwanted band-structure modification of the channel MoS_2 layer, thereby, preserving its semiconducting property), but can also lead to a reduced n-type SBH with proper choice of a low work function metal [242]. Finally, while most of the experimental/theoretical studies utilizing an interfacial tunnel barrier with MoS_2 have focused on decreasing the n-type SBH, theory predicts the same approach can be useful for mitigating the p-type SBH as well by carefully choosing or modifying the buffer layers. For example, Farmanbar et al. revealed that using a monolayer of high work function metallic NbS_2 (Φ_M ~6 eV) as the buffer layer could give a barrierless or Ohmic p-type contact to MoS_2 irrespective of the contact metal [243]. Similarly, Musso et al. predicted a fluorographene (C_2F) buffer layer to yield an Ohmic p-type contact to MoS_2 with high work function Pt as the contact metal [244]. Another study by Su et al. suggested that the SBH for both electrons and holes in the metal/hBN/MoS_2 contact geometry can be decreased or even completely eliminated by doping the hBN buffer layer with high concentrations of Li (electron-poor) and O (electron-rich) dopants, respectively, and that this effect can be more pronounced when the doped-hBN buffer layer spreads all over the MoS_2 device surface. Moreover, the authors predicted that both the intrinsic nature of the MoS_2 and the weak FLP effects at the metal/hBN/MoS_2 interface are preserved irrespective of the dopant type and concentration [245].

9. Graphene 2D Contacts to MoS_2

The wonder material graphene, a 2D semimetal composed of a single sheet of carbon atoms arranged in a honeycomb lattice, has also been explored as an alternative 2D contact material for 2D MoS_2. The remarkable properties of graphene have already been well studied and reported [1,246–248]. Owing to its unique band-structure with a linear Dirac-like spectrum [249], the charge carriers in graphene mimic relativistic particles and can effectively move at the speed of light [250]. This leads to extremely high charge carrier mobilities for both electrons and holes in graphene [251,252]. Moreover, unlike regular bulk metals, graphene's unique band-structure allows its Fermi level position (or, in other words, its work function) to be easily tuned around its "Dirac" point or the charge neutrality point (i.e., the point at which the conduction and valance bands of graphene meet each other in the momentum space) by an external doping source (electrostatic doping, chemical doping, etc.) leading to the accumulation of both electrons and holes in graphene depending on the doping polarity [232,253]. The superior electrical properties of graphene, therefore, make it an attractive choice for use as an atomically thin 2D vdW electrical contact to MoS_2. It can be used as an independent contact to MoS_2 or as an insertion between MoS_2 and conventional metal contacts (Figure 10a schematically illustrates MoS_2 FETs with graphene contacts in these two possible configurations). This latter case closely resembles the FLDP approach using insulating interfacial tunnel barriers (such as oxides or hBN) as described previously in Section 8. In this scenario, however, graphene is an electrically active semimetal that helps promote a strong electronic coupling between the metal and the MoS_2 despite the increased physical separation, and while maintaining a vdW-type interaction (since vdW gaps exist on either side of the inserted graphene layer), between the two. Contrary to the case of regular metal/MoS_2 contacts where the metal Fermi level typically gets pinned near the CBE of MoS_2 irrespective of the metal work function, the metal/graphene/MoS_2 contact (or simply an independent graphene/MoS_2 contact) can enable more efficient carrier injection into the MoS_2 channel. For metal/graphene/MoS_2 contacts, this is due in part to the physical separation created between the metal and the MoS_2 layer by the inserted graphene sheet, thereby, promoting FLDP by minimizing the metal/MoS_2 interfacial interaction that otherwise can lead to unwanted interface or mid-gap states (i.e., MIGS). However, the primarily mechanism responsible for the enhanced carrier injection in MoS_2 FETs with metal/graphene or independent graphene contacts is the dynamic tunability of the graphene Fermi level (due to gate bias-induced electrostatic doping, a combination of electrostatic and chemical doping, etc.) that can enable it to easily move up or down and align closer to the MoS_2 CBE or VBE, thereby, reducing the SBH for either electrons or holes, respectively [222,253,254].

The first such report of a metal/graphene hetero-contact on MoS_2 was by Du et al. where they demonstrated few-layer MoS_2 back-gated NFETs with Ti/graphene top contacts showing drain currents >160 µA/µm at 1 µm gate length with an ON/OFF ratio of 10^7. Compared to MoS_2 FETs without the graphene interlayer, the authors observed a 2.1× improvement in the ON-resistance and a 3.3× improvement in the R_C with Ti/graphene hetero-contacts. This performance enhancement was attributed to the fact that the positive back gate bias not only electrostatically n-doped the MoS_2, but n-doped the sandwiched graphene layer in the contact regions as well. Due to this electrostatic n-doping, the graphene Fermi level shifted upwards, moving further beyond the regular Ti-MoS_2 pinning level, leading to SBH (or R_C) reduction and more efficient electron injection into the MoS_2 conduction band (Figure 10b,c compare the extracted R_C as a function of back gate bias V_{BG}, and the output characteristics of back-gated MoS_2 FETs, both with and without graphene contacts, respectively) [255]. Leong et al. demonstrated MoS_2 NFETs with "nickel-etched-graphene" electrodes that yielded an R_C as low as 200 Ω·µm, a two orders of magnitude improvement over pure Ni electrodes as well as a 3× improvement in the μ_{FE} (from 27 to 80 cm^2/V-s). The authors found a bilayer graphene (BLG) insertion to be more effective than a single graphene layer. In addition to the electrostatic tunability of the graphene Fermi level, this large R_C reduction was attributed to two main factors: first, to the reduction in SBH thanks to a significantly smaller work function of the Ni-BLG electrodes (4.08 eV as compared to 5.5 eV for pure Ni) resulting due to the strong interaction between Ni and BLG and, second, to a Ni-catalyzed etching treatment of the BLG prior to stacking the electrode stack on MoS_2. This etching treatment of the BLG created a vast density of nano-sized pits with reactive zigzag edges that covalently bonded to the Ni, thereby, enhancing carrier tunneling and minimizing the resistance at the Ni/BLG interface (Figure 10d–f show the R_C versus gate voltage for MoS_2 FETs with different Ni-graphene contact configurations, the 3D schematic illustration of the Ni-etched-graphene contact to MoS_2, and the output characteristics of Ni-contacted MoS_2 FETs both with and without etched-graphene insertions, respectively; these results clearly reveal the significantly enhanced performance achieved in the case of MoS_2 NFETs with Ni-etched-graphene electrodes) [256].

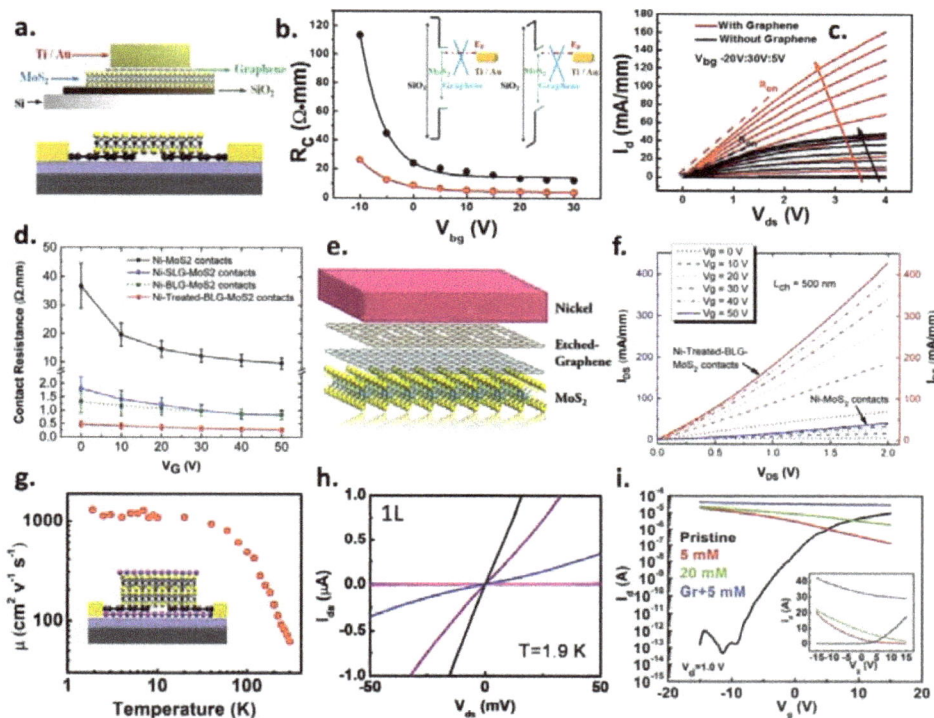

Figure 10. (a) Schematic representations of MoS$_2$ FETs with graphene contacts inserted in-between the MoS$_2$ and the metal (top) or used as an independent back contact (bottom). (b) Plot of R$_C$ versus back gate voltage V$_{BG}$ for MoS$_2$ NFETs with Ti/graphene (red curve) and Ti (black curve) contacts. Ti/graphene contact clearly yields much lower R$_C$ for electron injection thanks to the Fermi level tunability of graphene by the positive gate bias. Inset shows qualitative band diagrams illustrating the gate tunability of graphene's Fermi level and explaining the working principle of MoS$_2$/graphene contacts in the OFF (left inset) and ON states (right inset) of the device. Positive gate bias not only causes downward band-bending in the MoS$_2$, but helps move the graphene Fermi level upwards by doping the graphene with electrons (right inset). (b) and top schematic of (a) Adapted with permission from [254]. Copyright 2014 IEEE. (c) Output characteristics of a Ti-contacted MoS$_2$ FET with (red curves) and without (black curves) graphene insertion in the contact region. The FET with Ti/graphene contact shows enhanced currents and clearly outperforms the FET with regular Ti contacts. Adapted with permission from [255]. Copyright 2014 IEEE. (d) R$_C$ plotted as a function of back gate V$_{BG}$ for different Ni-graphene contact configurations to MoS$_2$ FETs. The best performance (i.e., lowest R$_C$ of ~200 Ω·μm) was obtained from the FET with Ni-Treated BLG-MoS$_2$ contacts (red curve). (e) 3D schematic illustration of the Ni-etched-graphene contact electrodes to MoS$_2$. The etched graphene layer facilitates a lower resistance interface between Ni and the graphene due to the formation of reactive zigzag edges that enhances carrier tunneling. (f) Output characteristics of Ni-contacted MoS$_2$ NFETs with (red curves) and without (blue curves) treated bilayer graphene (BLG) insertions. The FET with Ni-Treated-BLG contacts clearly outperforms the FET with direct Ni contacts (~10× improvement in the ON-current due to reduced R$_C$). (d–f) Adapted with permission from [256]. Copyright 2015 American Chemical Society. (g) Mobility versus temperature for an hBN-encapsulated MoS$_2$ NFET with graphene back contacts showing an extrinsic (i.e., two-point) μ_{FE} up to 1300 cm^2/V-s at low-T (1.9 K)

thanks to the zero SBH afforded by the gate-controlled graphene contacts. (**h**) Output characteristics of a back-gated monolayer MoS_2 NFET at 1.9 K showing linear I_D-V_D behavior at higher positive gate biases (gate voltage ranges from −60 to 80 V in 20 V steps) confirming the Ohmic nature of graphene contacts even at such low temperatures. (**g,h**) and bottom schematic of (**a**) Adapted with permission from [257]. Copyright 2015 American Chemical Society. (**i**) Comparison of the semilog transfer characteristics of Pd-contacted back-gated MoS_2 FETs, with (blue curve) and without (black, red and green curves) interfacial graphene layers in the contact regions, under high $AuCl_3$ doping concentrations. Before $AuCl_3$ doping, the pristine Pd-contacted FET shows typical n-type behavior with strong gate modulation (black curve). Under high $AuCl_3$ doping concentrations of 5 mM (red curve) and 20 mM (green curve), the same FET shows p-type behavior with slight gate modulation due to heavy p-doping of the MoS_2 by $AuCl_3$ (with the current levels being higher for the 20 mM case than 5 mM, as expected). In contrast, for the Pd/graphene-contacted FET, an even higher hole current level with practically no gate modulation is achieved at the lower doping concentration of 5 mM (blue curve) as compared to the Pd-contacted FET with the higher doping concentration of 20 mM (green curve). The comparative results clearly reveal the added advantage of using tunable graphene contacts in providing enhanced carrier injection (in this case, holes) in the MoS_2 channel as compared to direct metal contacts. Inset shows the linear transfer characteristics. Adapted with permission from [222]. Copyright 2016 John Wiley and Sons.

Liu et al. further utilized the Fermi level tunability of graphene to demonstrate Schottky barrier-free Ohmic contacts to MoS_2 FETs in the ON-state under a proper gate voltage. The authors extracted an SBH of zero at positive gate biases, thereby, confirming the efficacy of graphene in realizing perfectly "matched" contacts to MoS_2. Moreover, having realized barrier-free contacts, the authors demonstrated linear output I-V characteristics and a record high extrinsic (i.e., two-terminal) μ_{FE} up to 1300 cm^2/V-s at cryogenic temperatures down to 1.9 K in their MoS_2 NFETs (Figure 10g,h show the temperature-dependence of μ_{FE}, and linear output characteristics even at 1.9 K, respectively, for MoS_2 FETs with graphene contacts) [257]. More recently, Singh et al. employed graphene contacts in their dual-gated monolayer MoS_2 NFETs with Al_2O_3 as the top gate dielectric. The output characteristics with different top gate voltages indicated an Ohmic-like contact between graphene and MoS_2. They extracted an extrinsic μ_{FE} of 71 cm^2/V-s in the back-gated configuration (without any Al_2O_3 deposition), whereas a boosted extrinsic μ_{FE} of 131 cm^2/V-s in the top-gated configuration [258]. Likewise, low resistance graphene/MoS_2 contacts were utilized by both Cui et al. [259] and Lee et al. [260] in their studies of hBN-encapsulated MoS_2 devices that showed high Hall- and two-terminal electron mobilities, respectively. However, in all these reports utilizing a combination of both graphene contacts and dielectric encapsulation on MoS_2 FETs, the high-κ Al_2O_3 and the hBN dielectric environments could also play a significant role in enhancing the MoS_2 carrier mobility in addition to the lower R_C (or SBH) afforded by graphene contacts (more on the effects of dielectric environment, such as hBN and high-κ dielectrics, on MoS_2 carrier mobility enhancement is discussed later in Section 14). It is interesting to note that despite the Fermi level tunability of graphene contacts, most experimental reports to date demonstrate the efficacy of graphene in making better n-type contacts to MoS_2, but reports of graphene as a p-type contact to MoS_2 are, in general, lacking. This, again, is possibly because the Fermi level of graphene aligns closer to the CBE of MoS_2 under equilibrium to maintain charge neutrality, as dictated by the interface/mid-gap states, similar to the scenario of Fermi level pinning at 3D metal/MoS_2 interfaces. Additionally, the work function of pristine graphene (i.e., when its Fermi level is exactly at the Dirac point) is ~4.3–4.4 eV [140] which is close to the electron affinity of MoS_2 (~4.0–4.2 eV) [135,140]. Hence, owing to these factors, one can expect the n-type SBH at the graphene/MoS_2 interface to not be substantially large to begin with. Moreover, this n-type SBH can be minimized further thanks to the electrostatic n-doping of the graphene/MoS_2 interface by positive gate biases which not only helps move the graphene Fermi level upwards to align even closer to the MoS_2 CBE, but causes downward band-bending in the MoS_2 as well (recall that downward band-bending helps reduce the SBW at the contact/MoS$_2$ interface leading to enhanced carrier injection).

In principle, however, the Fermi level of graphene can potentially also be tuned deep into its valence band to align closer to the VBE of MoS$_2$ resulting in a small p-type SBH for hole injection. To achieve this, electrostatic p-doping of the graphene/MoS$_2$ interface alone may not be sufficient enough (since moving the graphene Fermi level from near the MoS$_2$ CBE all the way down towards the MoS$_2$ VBE, or causing a large upward band-bending in the MoS$_2$ to minimize the SBW for hole injection, would require extremely large negative gate biases that may be practically unfeasible) and additional doping/work function tuning mechanisms in the contact regions may be required. One such report of back-gated MoS$_2$ FETs with chemically p-doped graphene/MoS$_2$ contact regions was by Liu et al. [222] as already highlighted in Section 7.2. The authors used AuCl$_3$ solution as a surface charge transfer p-dopant to demonstrate high-performance unipolar MoS$_2$ PFETs as well as ambipolar MoS$_2$ FETs with hole dominated transport (note that AuCl$_3$ causes p-doping in both MoS$_2$ and graphene). In their study, bare back-gated MoS$_2$ FETs (i.e., FETs employing only gate electrostatic doping via the back gate) with metal/graphene contacts showed n-type behavior as expected. However, upon AuCl$_3$ chemical doping, the FET polarity switched to p-type. Moreover, the AuCl$_3$-doped MoS$_2$ PFETs with metal/graphene contacts showed enhanced p-type performance (i.e., lower R$_C$ and higher μ_{FE} for holes) than AuCl$_3$-doped MoS$_2$ PFETs with direct metal contacts. The comparative analysis done in this study unambiguously proved that a combination of both chemical and electrostatic p-doping was required to cause a large downward shift of the graphene Fermi level which resulted in reduced R$_C$ (and SBH) and enhanced p-type injection in metal/graphene-contacted MoS$_2$ FETs as compared to regular metal-contacted MoS$_2$ FETs (Figure 10i compares the transfer characteristics of Pd-contacted back-gated MoS$_2$ FETs, both with and without an interfacial graphene layer in the contact regions, under high AuCl$_3$ dopant concentrations, clearly revealing that for a given AuCl$_3$ concentration, the presence of an interfacial graphene layer facilitates enhanced hole injection in the MoS$_2$ FET than direct metal contacts due to the combined effects of chemical as well as electrostatic p-doping in aligning the graphene Fermi level closer to the MoS$_2$ VBE) [222]. In addition to the commonly used top or bottom contact configuration for graphene contacts to MoS$_2$ wherein the basal planes of graphene and MoS$_2$ overlap, lateral graphene "edge contacts" to MOCVD-grown MoS$_2$ were also demonstrated by Guimaraes et al. [261] and its implications are discussed in Section 11 that talks about the effects of contact architecture. Finally, the use of substitutionally-doped graphene has also been theoretically predicted to yield better contacts to MoS$_2$ with lower R$_C$ by Liu et al. In principle, graphene can be doped with elements such as B and N that have fewer and more electrons than C, respectively. Hence, B doping can increase the graphene work function (i.e., by making it hole-doped), while N doping can reduce its work function (i.e., by making it electron-doped). In particular, the authors showed that a high N doping concentration (C/N ratio ~20:1) can result in the work function of graphene to be as low as that of low work function metal Sc, thereby, making it a promising contact for electron injection in MoS$_2$ NFETs [140].

10. Effects of MoS$_2$ Layer Thickness

The layer thickness of MoS$_2$ can play an important role in determining the magnitude of the contact SBH, contact resistivity (ρ_C) as well as the overall carrier mobility (μ_{FE}) of the device. One of the first studies on the effect of MoS$_2$ layer thickness on the SBH and specific contact resistivity (ρ_C) was done by Li et al. on back-gated MoS$_2$ NFETs with Au top contacts [262]. They found two interesting and contrasting effects: first, for MoS$_2$ thicker than five layers (i.e., in the bulk or 3D limit), the ρ_C increased slightly with increasing MoS$_2$ thickness; and, second, the ρ_C increased sharply with reducing MoS$_2$ thickness below five layers (i.e., in the 2D limit). The first effect was attributed to the added thickness or, in other words, the added resistance of the "inactive" upper MoS$_2$ layers (i.e., layers further away from the back gate dielectric interface) which do not actively participate in the lateral current flow from the source to the drain of the FET, but through which the injected carriers must pass through orthogonally to reach the lower "active" MoS_2 layers primarily responsible for the lateral current flow (since majority of the gate-induced carriers are located in the lower layers). The added resistance of the upper "inactive" MoS_2 layers originates from the added "interlayer" resistances in the current flow path, as described later in this section. The second effect, on the other hand, was attributed to the quantum confinement-induced electronic structure modification and band-gap (E_g) increase of MoS_2 (Figure 11a shows the plot of ρ_C as a function of MoS_2 layer thickness for Au-contacted MoS_2 FETs, depicting the two contrasting effects in the 2D and 3D limits). As the MoS_2 layers reduce from five to one layer, its E_g increases from 1.2 eV for 5L to 1.8 eV for 1L [44] leading to a corresponding gradual increase in the SBH at the Au/MoS_2 contact interface due to relative changes in the MoS_2 band edge positions with respect to the Au work function (i.e., Fermi level position of Au). A quantitative relationship between the n-type SBH and MoS_2 layer thickness was established, with the extracted SBH increasing from 0.3 to 0.6 eV by merely reducing the MoS_2 thickness from five to one layer (Figure 11b,c show the plot of extracted SBH versus MoS_2 band-gap, and a qualitative illustration of the increase in SBH with reducing MoS_2 thickness for Au-contacted MoS_2 devices, respectively) [262]. Kwon et al. further confirmed the thickness-dependent SBH in Al-contacted MoS_2 devices with thinner MoS_2 devices yielding a larger SBH for electrons (Φ_{SB} ~50 meV for 1L) than the thicker ones (Φ_{SB} ~7 meV for 3L). Moreover, the layer-dependent SBH had a clear manifestation in the extracted R_C that showed an increase with decreasing MoS_2 thickness [263]. Although these studies of layer-dependent SBH have focused on n-type MoS_2 devices, the same trend and reasoning should also hold true for p-type MoS_2 devices as well. The reader should note that the layer-dependent SBH for a given metal/MoS_2 contact is more dominant in the 2D limit (i.e., 1–5 layers) since the MoS_2 band-gap changes drastically only in the 2D regime. In contrast, the band-gap remains constant at ~1.2 eV for thicker MoS_2 (i.e., >5 layers) and, hence, the SBH would remain largely constant at a relatively low value for devices made on thicker MoS_2 films for a given metal contact. The contribution of the SBH to the overall device R_C for a given metal/MoS_2 system can, therefore, be minimized beyond a critical MoS_2 layer thickness.

In a multilayer MoS_2 device with conventional top contacts, however, the effective R_C is governed not only by the resistance due to the SBH (R_{SB}), but by an unusual out-of-plane "interlayer resistance" (R_{int}) in-between the individual MoS_2 layers as well. This R_{int} is nothing but a direct manifestation of the presence of interlayer "vdW gaps". The implications of this unique R_{int} that manifests in multilayer MoS_2 devices have been analyzed in detail by Das et al. employing a resistor network model based on Thomas–Fermi charge screening (which relates to the charge screening length λ) and interlayer coupling (which captures the effect of R_{int}) [264]. Due to the finite charge screening length in MoS_2 (λ_{MoS2} ~7–8 nm or 12 layers), the centroid of the current flow distribution (or the current "HOT-SPOT") in a multilayer MoS_2 device can migrate dynamically between its individual layers (i.e., moving either closer to or farther away from the S/D electrodes along the vertical axis) depending on the applied electrostatic gate bias, effectively determining the number of these "in-series" interlayer resistors (i.e., the total R_{int}) involved along the current path from the source to the drain of the MoS_2 FET [265]. For example, if the "HOT-SPOT" is located closer to the S/D electrodes, then fewer interlayer resistors would be involved along the current flow path (Figure 11e schematically depicts the resistor network model as applicable to multilayer MoS_2 as well as the dynamic migration of the current "HOT-SPOT" as a function of both MoS_2 layer thickness and gate overdrive voltage, in the back-gated FET configuration). This R_{int} contribution adds to the Schottky barrier resistance R_{SB}, leading to an effective total R_C (or ρ_C) for the multilayer MoS_2 system that limits the extracted μ_{FE}. Note that this "HOT-SPOT" migration can only happen within the thickness range set by the charge screening length of MoS_2. Thus, it would be disadvantageous to have MoS_2 devices with too many layers or thicknesses greater than its charge screening length of ~7–8 nm, as, ultimately, the device performance will be severely limited by the additional interlayer resistors involved. Experimentally, the extracted value of R_{int} between two adjacent MoS_2 layers has been estimated to be ~2.0 kΩ·μm by Na et al. [266], and a recent experimental study by Bhattacharjee et al. reported a net R_{int} of 1.53 kΩ·μm for 5–7 nm thick MoS_2 [267]. Thus, the number of these individual interlayer resistors involved in the charge transport can have a significant impact on parameters such as μ_{FE} and ON-currents of few-layer or multilayer MoS_2 FETs.

The carrier mobility of monolayer MoS_2, on the other hand, is, in general, lower than few-layer MoS_2 films due to the deleterious effects of various extrinsic charge scattering mechanisms (such as charged impurities, surface roughness, remote interfacial phonons from the dielectric) that are at play at the MoS_2/dielectric interface [113,176,177,268,269]. This is because the strength of the "screening" against these extrinsic scattering mechanisms is naturally weaker in monolayers (since both surfaces of the monolayer are directly exposed to its surroundings) as compared to few-layers where the outer MoS_2 layers effectively "shield" the inner-lying layers making them less susceptible to the external environment. A simple calculation of scattering rates as a function of MoS_2 layer thickness by Li et al., taking into account the typical carrier scattering sources (e.g., intrinsic phonons and charged impurities), sheds important insight into the increased susceptibility of carrier scattering in thinner MoS_2 films as compared to thicker ones (Figure 11f shows the calculated carrier scattering rates for various extrinsic sources as a function of MoS_2 layer thickness, clearly showing that the overall scattering increases with decreasing thickness) [269]. Moreover, monolayer MoS_2 FETs suffer from a larger SBH issue as described earlier. Hence, keeping all these factors in mind, one can surmise that there exists an "ideal" MoS_2 layer thickness to guarantee the best device performance in terms of reduced net R_C (considering effects of both R_{SB} and R_{int}), minimal external carrier scattering and increased μ_{FE}. Indeed, the dependence of carrier mobility on the MoS_2 layer thickness has been experimentally studied and a non-monotonic trend was revealed by Das et al. wherein the maximum mobility was achieved for a flake thickness of around 8 nm or 12 atomic layers for a given metal contact. Below 8 nm, μ_{FE} reduces primarily due to carrier scattering, whereas above 8 nm, μ_{FE} reduces due to increased overall R_{int}. (Figure 11d shows the plot of extracted μ_{FE} versus flake thickness for MoS_2 FETs with different metal contacts with peak μ_{FE} values occurring around 8 nm) [135,264,265]. A similar trend was reported by Li et al. where the maximum mobility was achieved for a 14-layer MoS_2 device [269]. Lin et al. further corroborated this MoS_2 thickness effect by showing that the optimal MoS_2 thickness range for maximum mobility (or, in other words, maximum device performance) was somewhere within 5–10 atomic layers [270], in good agreement with other reports.

Recently, a relationship between MoS_2 film thickness and its charge transfer surface doping was also elucidated by Rosa et al. using both experiments and semi-classical modeling. They studied the n-doping of back-gated MoS_2 FETs via PVA coating (that can enhance MoS_2 FET performance, as highlighted in Section 7.1) while varying the MoS_2 layer thickness, and found that the penetration depth of the carriers induced by PVA doping was approximately 5.2 nm from the PVA/MoS_2 interface (Figure 11g shows the calculated cross-sectional profile of the electron density for two different MoS_2 layer thicknesses). Thus, for MoS_2 films thicker than 5.2 nm, the dopant-induced charge would be farther away from the MoS_2/gate dielectric interface resulting in a poor gate electrostatic control and degradation of the FET performance (mainly in terms of compromised ON/OFF ratios). Conversely, for thinner MoS_2 films (<5.2 nm), the dopant-induced charge will be much closer to the gate interface resulting in better electrostatic control of the device channel, enhanced charge depletion capability of the gate and improved ON/OFF ratios even in the presence of doping [271]. Therefore, for designing high-performance electronic devices based on MoS_2, its layer thickness must be considered as it can have pronounced effects on important device parameters such as the overall R_C (considering effects of both SBH and R_{int}) and μ_{FE} as well as on the charge carrier distribution in the device. However, for optoelectronic device applications where monolayer MoS_2 is necessitated due to the direct band-gap requirement, alternative approaches to mitigate R_C (due to larger SBH) and μ_{FE} degradation (due to increased external carrier scattering) must be used. Finally, it is interesting to note the effect of MoS_2 layer thickness scaling on its charge carrier effective masses (since the effective masses depend on the MoS_2 band-structure which, in turn, varies with the layer thickness) as highlighted by Yun et al. In particular, the effective mass of electrons in the lowest lying conduction band valley reduces from 0.551 to 0.483 m_o (where m_o is the free electron mass) as MoS_2 is thinned from bulk to monolayers [272].

Since carrier mobility is inversely proportional to its effective mass, this implies that monolayer MoS_2 will have a higher intrinsic electron mobility than multilayers. However, in practice, the monolayer MoS_2 device mobility is far worse due to various external effects as already pointed out.

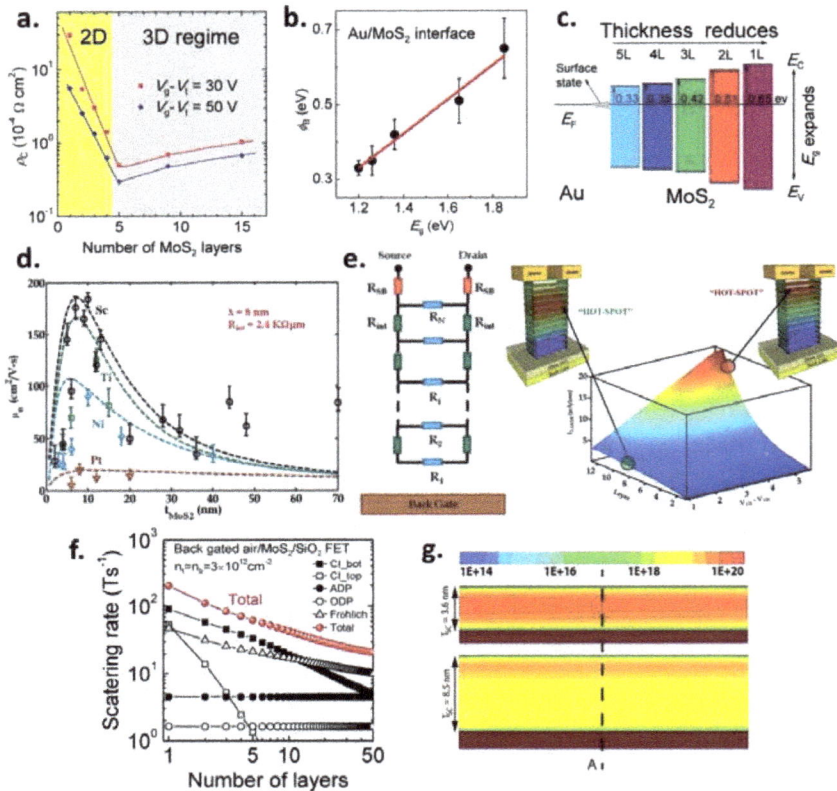

Figure 11. (**a**) Plot of specific contact resistivity (ρ_C) versus MoS_2 layer thickness for Au-contacted back-gated MoS_2 NFETs at two different gate overdrive voltages. Below five layers, ρ_C increases sharply due to increase in the SBH which is directly correlated to the band-gap increase of MoS_2 with decreasing thickness. Above five layers, the ρ_C increases only slightly due to the added resistance of the upper "inactive" MoS_2 layers through which the injected carriers have to pass through orthogonally to reach the lower "active" layers of the FET (note that the SBH height remains constant above five layers). Based on this plot, one can effectively delineate the 2D and 3D regimes for MoS_2-based devices. (**b**) Plot of extracted SBH at the Au/MoS_2 interface as a function of MoS_2 band-gap. The SBH height increases linearly with increasing band-gap (i.e., reducing MoS_2 thickness from bulk down to a monolayer). (**c**) Evolution of the MoS_2 band edge positions with respect to the Fermi level position at the Au/MoS_2 contact interface as the MoS_2 thickness gradually reduces from five layers to one layer. The n-type SBH increases as the band-gap increases. (**a**–**c**) Adapted with permission from [262]. Copyright 2014 American Chemical Society. (**d**) Plot of extracted mobility (μ_{FE}) as a function of MoS_2 layer thickness for MoS_2 FETs with different metal contacts. A non-monotonic trend in the μ_{FE} is evident with the maximum μ_{FE} achieved for a thickness of ~8 nm for each metal/MoS_2 contact. A good match between experimental μ_{FE} values (marked spots) and calculated μ_{FE} values (dashed lines) can be seen. μ_{FE} calculations were done based on a model considering effects of both Thomas–Fermi charge screening

and interlayer resistance (R_{int}). (**e**) **Left schematic**: Depiction of the resistor network model that can be used to describe the various resistance components, i.e., due to the SBH (R_{SB}, shown in red), interlayer resistances (R_{int}, shown in green) and intra-layer resistances (R_N, shown in blue), that dominate the charge transport in a back-gated multilayer MoS_2 FET. (**d**) and **left schematic** of (**e**) Adapted with permission from [264]. Copyright 2013 John Wiley and Sons. **Right schematic**: Illustration of the dynamic vertical migration of the current "HOT-SPOT", or the centroid of the current flow distribution, in a multilayer MoS_2 FET as a function of both MoS_2 layer thickness and the gate overdrive voltage. The "HOT-SPOT" location effectively determines the number of interlayer resistors (or the net R_{int}) involved in the charge transport. Note that this "HOT-SPOT" migration can happen only within the thickness range set by the MoS_2 charge screening length (λ_{MoS2}). For thicknesses above λ_{MoS2} of ~7–8 nm, R_{int} becomes the dominant R_C contributor. Adapted with permission from [265]. Copyright 2013 American Chemical Society. (**f**) Calculated scattering rates as a function of MoS_2 layer thickness clearly showing that the net scattering increases as the layer thickness decreases. Moreover, the external scattering sources such as charged impurities (filled and hollow squares in the plot) dominate for thinner layers. Adapted with permission from [269]. Copyright 2013 American Chemical Society. (**g**) Cross-sectional representation of the calculated electron density profile for two different MoS_2 thicknesses (orange regions represent the highest densities) showing that the PVA doping-induced charge resides primarily in a ~5 nm thick region near the MoS_2/PVA interface. Adapted from [271].

11. Effects of Contact Architecture (Top versus Edge)

The contact architecture can also be important in designing high-quality contacts to MoS_2 FETs with low R_C (Figure 12a schematically illustrates the common contact architectures used in MoS_2 FETs). Typically, the most commonly utilized configuration is the top contact geometry where the metal contact is deposited on top of the basal plane of MoS_2. As mentioned earlier, this sort of topography leads to a vdW gap-induced tunnel barrier in-between the largely inert MoS_2 basal plane (due to lack of surface dangling bonds) and the contact electrode [110]. This vdW gap contributes an additional tunneling resistance to the overall R_C. One approach to mitigate this vdW gap prevalent in the top contact geometry is using metals that can bond or "hybridize" better with the underlying MoS_2 surface (hybridization is discussed in more detail in Section 12). Another approach to resolve this issue could be to use "side" or "edge contacts" to MoS_2. It has been shown theoretically that edge contacts can lead to much better charge carrier injection in the MoS_2 layer as opposed to top contacts [148,161]. This is because MoS_2 edges are chemically more active due to the presence of unsaturated chemical bonds and, thus, can form stronger bonds with a shorter bonding distance (i.e., smaller physical separation) with the contact metals, thereby, minimizing or even eliminating the vdW gap-induced tunnel barrier (Figure 12a also shows the calculated electronic interaction at the Au/MoS_2 contact interface for both the top and edge contact configurations as well as a 3D schematic illustration of the MoS_2/metal atoms in an edge-contacted geometry). Edge contacts to graphene FETs have indeed shown extremely low R_C values as compared to conventional top contacts, thereby, proving the efficacy of this contacting scheme [273–275]. Moreover, it can be inferred that this edge contact geometry is particularly attractive for few-layer or multilayer devices based on MoS_2 or other 2D semiconducting materials wherein each individual layer can be independently contacted from the sides. This will help eliminate the adverse effects of the interlayer resistance R_{int} (as described in the previous section) since the injected charge carriers will no longer have to "hop" from one layer to the other across the interlayer vdW gaps (as would be the case in top-contacted multilayer MoS_2 devices), instead the carriers can be laterally injected into all the constituent MoS_2 layers simultaneously [183,276,277]. Furthermore, edge contacts have the inherent advantage of minimizing the overall contact volume and area (since an overlap between the bulk metals and MoS_2 is no longer required) and, thus, are promising from the scaling point of view [261]. It is also worth noting that the edge contact geometry should not be limited by the "current crowding" effect that comes into play in the top or bottom contact geometry (and which results in an exponential R_C increase when $L_C \ll L_T$, as described in Section 3.2). Experimentally, however, realizing pure edge contacts to MoS$_2$ has been challenging due to the close proximity required and stringent fabrication requirements (for instance, the contact metal may fail to contact all the MoS$_2$ edges due to processing-induced voids etc.), and there have only been a few reports.

Yoo et al. reported multilayer MoS$_2$ NFETs with DC-sputtered molybdenum (Mo) S/D contacts and confirmed the presence of both top and edge contacts via cross-sectional TEM imaging. They found that the electrical performance of their 96-layer or 67 nm thick MoS$_2$ FET (μ_{FE} ~24 cm^2/V-s) was similar to few-layer (3–5L) MoS$_2$ FETs with Mo top contacts (μ_{FE} ~26–27 cm^2/V-s). This clearly suggests that the detrimental effect of R$_{int}$ was greatly minimized even though the MoS$_2$ was 67 layers thick, thanks to the formation of edge contacts (due to conformal Mo deposition by DC-sputtering) that injected carriers deep into the constituent layers of the multilayer MoS$_2$ device [278]. Chai et al. developed a "passivation first, metallization second" technique for fabricating pure edge contacts to two types of back-gated MoS$_2$-based heterostructures wherein the MoS$_2$ was first encapsulated in-between dielectrics: Al$_2$O$_3$/MoS$_2$/SiO$_2$ and hBN/MoS$_2$/hBN. The authors then performed a plasma etching step to create vertical trenches in these heterostructures to expose the MoS$_2$ edges for contact formation and revealed that the SF$_6$ plasma is better suited than CF$_4$ plasma as the former created a smooth side wall profile with less residues. Although the MoS$_2$ device performance achieved in this study was not great (possibly due to discontinuities at the metal/MoS$_2$ contact edges that impeded carrier injection) with the process needing further optimization, the fabrication procedure presented represents a promising way to make edge contacts to fully-encapsulated 2D TMDC-based devices [279]. Guimaraes et al. reported a scalable bottom-up technique where they grew wafer-scale monolayer MoS$_2$ using MOCVD from the patterned edges of CVD-grown graphene on SiO$_2$ substrates that resulted in seamless MoS$_2$/graphene 1D Ohmic edge contacts (Figure 12b schematically compares 2D metal/MoS$_2$ top contacts with 1D graphene/MoS$_2$ edge contacts) [261]. This approach combines the advantages of having gate-tunable graphene contacts along with the edge contact geometry. Although the extracted SBH was low (Φ_{SB} ~24 meV at V$_{BG}$ = 10 V; 4 meV at V$_{BG}$ = 60 V), the average R$_C$ was found to be 30 k$\Omega \cdot \mu$m which is much higher than the lowest R$_C$ values reported in literature for monolayer MoS$_2$ devices. This is because, although the graphene edge contacts provide better carrier injection into the MoS$_2$ layer, these carriers are injected through a significantly reduced contact area. Regardless, MoS$_2$ FETs with 1D graphene edge contacts outperformed FETs with 2D metal contacts, and showed Ohmic behavior down to liquid helium temperatures (i.e., ~2 K) with the two-probe μ_{FE} ranging between 10–30 cm^2/V-s. Moreover, these promising device results with graphene edge contacts were obtained while maintaining minimal electrode volume and contact area, a key advantage afforded by 1D graphene edge contacts when it comes to device scaling (Figure 12c,d show the R$_C$ comparison for different contact geometries used in this study, i.e., edge graphene, top graphene or top metal, and the sheet conductance versus gate voltage for top-gated MoS$_2$ FETs comparing 1D graphene edge contacts to 2D metal top contacts, respectively) [261].

The benefits of forming side or edge contacts to multilayer MoS$_2$ FETs were further highlighted by Zheng et al. who compared FETs made on CVD-grown multilayer MoS$_2$ with "gradually shrinking" basal planes to FETs made on similarly thick exfoliated MoS$_2$ having abrupt edges. The gradually shrinking basal planes created "terraced" edges on the multilayer CVD MoS$_2$ such that each layer was exposed to facilitate a good edge contact formation with the electrode. In contrast, the exfoliated multilayer MoS$_2$ with abrupt edges could only be contacted from the top (Figure 12e schematically illustrates the top-contacted exfoliated multilayer MoS$_2$ with abrupt edges as well as the edge-contacted CVD-grown multilayer MoS$_2$ with terraced edges). As expected, the edge-contacted CVD MoS$_2$ having terraced edges (with contacts to each layer) outperformed the top-contacted MoS$_2$ in terms of both μ_{FE} and ON-currents for any given layer thickness (Figure 12f shows a statistical comparison of μ_{FE} and ON-currents as a function of layer thickness between CVD-grown MoS$_2$ with terraced edges and exfoliated MoS$_2$ with abrupt edges) [280]. The results clearly reveal the efficacy of forming edge contacts to multilayer MoS$_2$ devices. Moreover, it is instructive to note that, for both mono- and multilayer MoS$_2$ devices, the effective edge contact length can be increased by cutting zigzag or jagged edges in the MoS_2 film that can help lower the R_C even further via edge injection. Furthermore, a combination of both top and edge contacts to multilayer MoS_2 (keeping in mind the area constraints, of course, since the contact length must be scaled together with device dimensional scaling) could also be a promising approach to maximize the area for charge injection while eliminating the deleterious effects of R_{int}. Finally, besides top and edge contacts, use of "bottom" contacts, with independent electrostatic gating of the contact regions via a strongly coupled top gate (that can provide strong electrostatic doping and band-bending in the MoS_2 contact regions, leading to SBW narrowing and, thus, reduction of the SBW tunneling resistance), could also be beneficial in minimizing R_C as shown by Movva et al. in the case of WSe_2 PFETs with Pt back contacts [281].

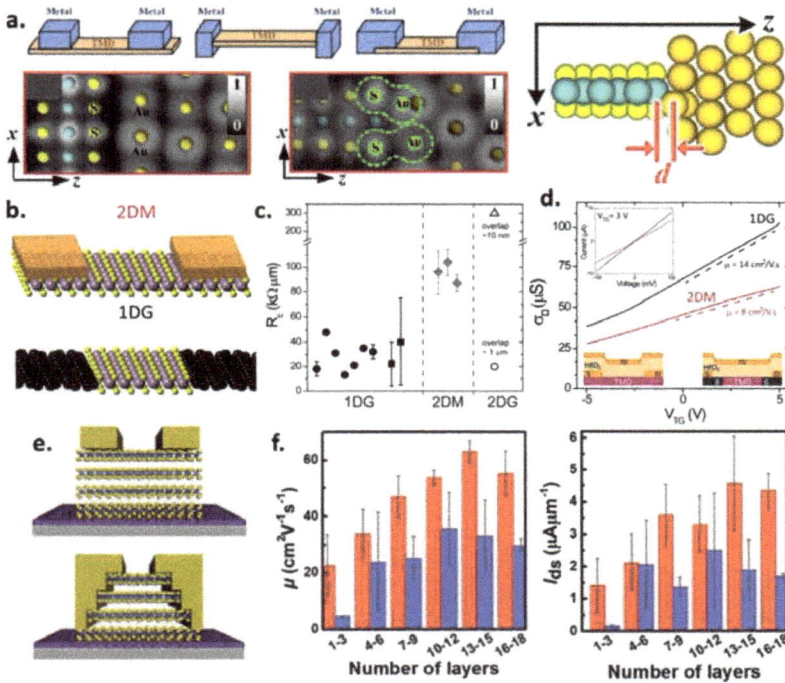

Figure 12. (a) **Top left panel**: Schematics of some common TMDC/metal contact configurations: top contacts, edge contacts and a combination of both top and edge contacts. **Bottom left panel**: Plots of calculated electron localization function (ELF) for both top- (**left plot**) and edge-contacted (**right plot**) MoS_2 with Au metal electrodes. ELF closer to 1 (i.e., light-colored areas in the plot) indicates higher probability of finding an electron. It can clearly be seen that edge-contacted MoS_2 has overlapping electron orbitals between the MoS_2 and the Au metal atoms (as indicated by the dashed green contours in the right plot). In contrast, the top-contacted MoS_2 clearly shows a vdW gap between the MoS_2 and the metal atoms (as shown by the dark region in the left plot). **Right schematic**: Schematic illustration of the metal/MoS_2 atomic interaction in the edge-contacted MoS_2 configuration. Adapted from [148]. (b) Schematic comparison of top-contacted MoS_2 with a metal (2DM implies a 2D metal top contact) and edge-contacted MoS_2 with graphene (1DG implies a 1D graphene edge contact). (c) Comparison of R_C as a function of different contact configurations to MoS_2 [i.e., 1DG, 2DM and 2D graphene top contacts (2DG)]. Lowest R_C is achieved with 1DG contacts to MoS_2. Note that R_C for the 2DG contact with a ~1 μm overlap is comparable to 1DG contacts, but at the expense of a large overlap area. (d) Top gate voltage (V_{TG}) dependence of the two-terminal sheet conductance ($\sigma\square$) measured from MoS_2 devices

with 1DG contacts (black curve) and with 2DM electrodes (red curve; contact dimensions 23 μm × 22 μm × 55 nm). **Top inset**: Output I-V characteristics for the two devices at V_{TG} = 3 V. **Bottom insets**: Cross-sectional MoS$_2$ device schematics with 2DM (**bottom left schematic**) and 1DG (**bottom right schematic**) contacts. The TMDC channel, graphene contacts (g), source (S), drain (D), and top gate electrodes (TG) with the HfO$_2$ gate dielectric layer are shown. (**b–d**) Adapted with permission from [261]. Copyright 2016 American Chemical Society. (**e**) Schematic illustrations of the contact geometries used in multilayer exfoliated MoS$_2$ flakes (**top schematic**) and CVD-grown MoS$_2$ flakes with gradually shrinking basal planes or "terraced" edges (**bottom schematic**). (**f**) Statistics of electron mobility (**left plot**) and ON-currents (**right plot**) measured at V_G = 40 V and V_D = 1 V as a function of number of MoS$_2$ layers. The red columns represent data from CVD-grown MoS$_2$ flakes having terraced edges, whereas the blue columns from exfoliated MoS$_2$ flakes. Devices made on side-contacted CVD flakes with terraced edges clearly outperform the devices made on exfoliated flakes for all layer thicknesses, highlighting the advantage of making electrical contacts to each individual layer in a multilayer MoS$_2$ device. (**e,f**) Adapted with permission from [280]. Copyright 2017 John Wiley and Sons.

12. Hybridization and Phase Engineering

The concept of hybridization essentially implies enhancing the chemical interaction (through covalent bonding or, in other words, stronger atomic orbital overlapping) between the contact metal atoms and the MoS$_2$ to form a more intimate contact interface [51,161]. Edge contacts to MoS$_2$ (as described in the previous section) are an example of such hybridized contacts due to the enhanced chemical interaction between the unsaturated bonds at the MoS$_2$ edges and the contact metal atoms. However, pure metal edge contacts to MoS$_2$ with good electrical quality have been experimentally difficult to realize (as of yet). Alternatively, hybridization can also be invoked in a top contact geometry by carefully selecting the right contact metal for a given 2D TMDC material, and/or by carefully optimizing the contact fabrication steps, leading to elimination of the vdW gap-induced tunnel barriers [148,161,282,283]. For example, metals having a lower lattice mismatch with MoS$_2$ can potentially induce stronger hybridization by enabling a "closer physical proximity" between the metal and MoS$_2$ sulfur atoms that can maximize their orbital overlapping. Stronger hybridization can also be invoked by metals having "*d* orbital electrons" as they can allow for a higher probability of orbital interaction or "mixing" with the *d* orbitals of the MoS$_2$ system since the MoS$_2$ band edges are primarily made up of Mo *d* orbitals (Figure 13a shows the basic concept of hybridization by schematically illustrating the interfacial atomic arrangements in top-contacted MoS$_2$ with Au and Mo contacts, with the latter representing the hybridized case) [148,161]. With these views in mind, Kang et al. reported the use of Mo as a high-performance n-type contact to mono- and few-layer MoS$_2$ FETs and provided a physical understanding of the underlying mechanism through intensive DFT calculations. Comparing Mo contacts to Ti, the authors demonstrated Mo-contacted few-layer MoS$_2$ FETs with smaller R_C (~2 kΩ·μm), higher ON-currents (271 μA/μm), and higher mobilities (~27 cm^2/V-s), which were much better than those obtained using Ti contacts. It was revealed that although Mo has a higher work function (Φ_M = 4.5 eV) than Ti (Φ_M = 4.3 eV), it has a high degree of "atomic *d* orbital overlapping" with the underlying MoS$_2$ which results in an intimate top contact [284]. Their DFT calculations revealed that S atoms in MoS$_2$ are dragged by the Mo metal atoms to form Mo-S interface bonds resulting in modification of the MoS$_2$ band-structure, rendering it metallic underneath the Mo contacts (Figure 13b shows the DFT-calculated band-structure and the partial-density-of-states "PDOS" plot for the Mo-MoS$_2$ system showing the presence of metallized states). Moreover, it was found that electrons associated with the overlap states in the Mo-MoS$_2$ system are not localized, implying that the Mo contact does not degrade the conductivity of the underlying MoS$_2$. Furthermore, a small SBH of 0.1 eV was calculated for the Mo-MoS$_2$ system which was much lower than that for the Ti-MoS$_2$ system (Φ_{SB} ~0.33 eV) despite Mo being a higher work function metal [284].

Recently, a new method of forming hybridized Mo contacts to MoS$_2$, together with large-area CVD integration, was demonstrated by Song et al. Compared to conventional CVD growth methods that provide MoS_2 films (by sulfurization of deposited Mo layers, etc.) without any device processing, the authors pre-deposited Au top contact metal (having defined areas) on Mo films deposited on sapphire substrates to serve as the S/D electrodes following the CVD sulfurization process, thus, readily creating top-contacted MoS_2 structures for FET fabrication. In this novel strategy, the pre-deposited Au effectively serves as a "mask" during the CVD sulfurization process, resulting in significantly impeded sulfurization of the Mo directly underneath the Au metal. In contrast, the Mo in the exposed channel region gets fully converted (or sulfurized) to MoS_2. Therefore, this method results in seamless and hybridized edge contacts between the channel MoS_2 and the metallic Mo (i.e., unsulfurized Mo) underneath the Au electrodes (Figure 13c schematically compares the MoS_2 device fabrication process flows for Au contact metal deposition both after and before sulfurization of the as-deposited Mo films as represented by Method A and Method B, respectively; the qualitative band diagrams explaining the two resultant metal/MoS_2 contact interfaces are also shown highlighting the superiority of Method B in making hybridized edge contacts to MoS_2). Top-gated FET measurements revealed that this fabrication strategy provides better contact performance than the traditional metal-on-CVD MoS_2 approach [285]. Although the overall device performance was under-par (possibly due to poor quality of the CVD MoS_2), this strategy is promising as it combines the advantage of using Mo as a hybridized contact to MoS_2 together with the advantages of an edge contact geometry (see previous section for a detailed discussion on edge contacts to MoS_2). Another recent report by Abraham et al. showed the benefits of using annealed Ag contacts on few-layer (5–14 layers) MoS_2 FETs [286]. As described in Section 5, Ag has been shown to be a good electrical contact to MoS_2 owing to its high wettability and thermal conductivity [193]. In this study, however, the authors further revealed that annealing (250–300 °C) the as-deposited Ag contacts on few-layer MoS_2 promotes diffusion of the Ag atoms into the MoS_2 lattice causing them to "locally dope" the MoS_2 layers underneath the S/D contact electrodes. The authors extracted an R_C of 200–700 $\Omega \cdot \mu m$ for few-layer MoS_2 devices with annealed Ag contacts that ranks among the best reported R_C values for MoS_2 devices. This method is particularly attractive for few-layer MoS_2 devices as the Ag dopant atoms can diffuse through and mitigate the deleterious effects of the interlayer resistance "R_{int}" (as described in Section 10) by enhancing the interlayer coupling [286]. Moreover, this work also highlights the importance of "post-contact annealing" which is a common technique to improve the adhesion and electrical quality of metal-semiconductor contacts.

An extremely promising approach to mitigate the SB issue was demonstrated by Kappera et al. where they utilized a novel "phase-engineering" technique to convert the MoS_2 in the contact regions from its semiconducting "2H" phase to an environmentally stable metallic "1T" phase [287]. This method involves selective treatment of the semiconducting 2H phase of MoS_2 by an organolithium chemical, namely, n-butyl lithium, which converts the 2H phase into the metallic 1T phase resulting in a seamless in-plane edge contact to MoS_2 (Figure 13d shows the 3D schematic as well as the EFM and TEM images of the 2H/1T MoS_2 contact interface). The mechanism involves the intercalation of Li ions in the MoS_2 lattice that causes the phase transition, and stabilization of the resulting 1T phase by electron donation from the organolithium compound during intercalation [287,288]. The authors achieved a record low R_C of 200 $\Omega \cdot \mu m$ at zero gate bias (see Figure 13d for the resistance versus L_{CH} plot used to extract the R_C via TLM method) using this technique which resulted in high-performance MoS_2 NFETs with excellent current saturation, high mobilities (~50 cm^2/V-s), high drive currents (~100 $\mu A/\mu m$), low SS (<100 mV/decade), and high ON/OFF ratios (>10^7). Moreover, it was shown that the FET performance was highly reproducible and independent of the S/D contact metal, implying that the carrier injection in the MoS_2 channel was controlled by the 1T/2H interface. The low R_C was attributed to the atomically sharp interface between the two MoS_2 phases, and to the fact that the work function of the metallic 1T phase matches well with the CBE energy of the semiconducting 2H phase, thereby, resulting in a negligible SBH at the 1T/2H interface. A detailed investigation of the atomic mechanism of this semiconducting-to-metallic phase transition in MoS_2 (due to intralayer atomic plane gliding which involves a transversal displacement of one of the S atom planes) has been performed by Lin et al. [288]. Moreover, this phase engineering strategy has also been successfully demonstrated in FETs made on CVD-grown MoS$_2$ [289]. Although this 1T/2H contact technique is extremely promising, the stability and performance of the metallic 1T phase under high-performance device operation is yet to be determined.

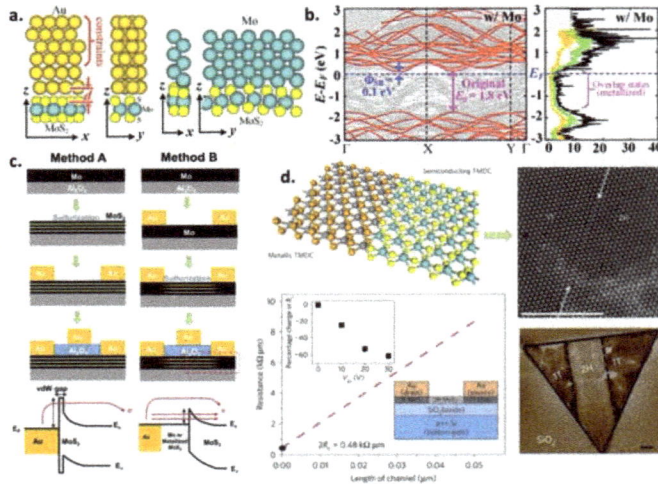

Figure 13. (a) Schematic representation of top-contacted MoS$_2$ with Au (**left schematic**) and Mo (**right schematic**) electrodes showing different atomic views. A vdW gap "d" exists at the Au/MoS$_2$ interface, but is absent at the Mo/MoS$_2$ interface due to strong hybridization between Mo atoms and MoS$_2$. Adapted from [148]. (b) **Left**: DFT-calculated band-structure of the Mo/MoS$_2$ system. For reference, the original band-structure of MoS$_2$ without the Mo contact (shown in red) is superimposed on the Mo/MoS$_2$ band-structure (shown in grey) such that the old and new sub-bands align. The Schottky barrier (Φ_{SB}) is marked in blue and is calculated to be 0.1 eV for the Mo/MoS$_2$ system. **Right**: The corresponding partial-density-of-states (PDOS) plot showing the metallized overlap states in the Mo/MoS$_2$ system. Adapted from [284], with the permission of AIP Publishing. (c) Schematic illustration of the two different fabrication process flows for making electrical contacts to CVD MoS$_2$. Method A represents the conventional process where the Au electrode is deposited after the complete sulfurization of the Mo film into MoS$_2$. Method B represents the process where the Au electrodes are pre-deposited on Mo films before the sulfurization step which yields readymade hybridized edge contacts after sulfurization. The resultant top-gated MoS$_2$ FETs from Method B show better electrical performance than those fabricated using Method A. Corresponding qualitative band diagrams explaining the carrier injection mechanisms are presented below and clearly show the advantages of Method B over Method A in realizing CVD MoS$_2$ FETs with superior electrical contacts (i.e., hybridized Mo edge contacts with no vdW gap). Adapted with permission from [285]. Copyright 2017 IOP Publishing. (d) Phase-engineered metallic contacts to MoS$_2$. **Top left**: Schematic illustration of the seamless "in-plane" contact between metallic and semiconducting phases of a TMDC material. Adapted with permission from [110]. Copyright Springer Nature 2015. **Top right**: High-resolution TEM image of the phase boundary (indicated by the white arrows) between the 1T and 2H phases in a monolayer MoS$_2$ nanosheet (scale bar = 5 nm). **Bottom right**: Electrostatic force microscopy (EFM) phase image of a monolayer MoS$_2$ nanosheet showing the contrast between the 2H and locally patterned 1T MoS$_2$ phases. **Bottom left**: Plot of total resistance versus L$_{CH}$ for R$_C$ determination of 1T/MoS$_2$ contacts using the TLM method at zero applied gate bias. Extracted R$_C$ value of ~200 $\Omega \cdot \mu$m is among the lowest R$_C$ values reported on MoS$_2$. Top inset shows the percentage decrease in R$_C$ with positive gate bias. Bottom inset shows the schematic of a back-gated MoS$_2$ FET with phase-engineered 1T contacts. (**Top right, bottom left** and **bottom right**) Adapted with permission from [287]. Copyright Springer Nature 2014.

13. Engineering Structural Defects, Interface Traps and Surface States

In addition to all the intrinsic phonon scattering mechanisms that inevitably set an upper bound on the MoS$_2$ charge carrier mobility [113–117], the performance/mobility of MoS$_2$ FETs is often dominated by several extrinsic carrier scattering factors, such as structural defects, interface traps and surface states, leading to carrier localization and lower experimental mobilities than the predicted phonon-limited values [171–179,290,291]. Moreover, the scattering problem becomes worse for monolayer or ultra-thin MoS$_2$ devices (due to lack of efficient screening) since charge carriers in it are more susceptible to getting scattered by impurities present at the MoS$_2$/dielectric interface(s) as well as those residing within the MoS$_2$ lattice (see discussion in Section 10 that talks about the effects of MoS$_2$ layer thickness on device performance). One of the main structural defects in MoS$_2$ are the sulfur vacancies (SVs) that can act as charged impurities (CIs), charge trapping as well as short-range scattering centers by introducing localized gap states [292–294]. Moreover, as described in Section 3.1, SVs are also responsible for the strong contact FLP effect that leads to uncontrolled and large SBH for carrier injection [142,143,147,295]. Therefore, passivating these SVs is important to enhance the MoS$_2$ device performance. Yu et al. reported a facile, low-temperature thiol (-SH) chemistry to repair these SVs and improve the monolayer MoS$_2$/dielectric interface, resulting in significant reduction of charged impurities and interfacial traps [296]. They treated both sides of the monolayer MoS$_2$ using the chemical (3-mercaptopropyl) trimethoxysilane, abbreviated as "MPS", under mild annealing. In this approach, not only the S atoms from the MPS molecules passivate the chemically reactive SVs, but the trimethoxysilane groups in MPS react with the SiO$_2$ substrate to form a self-assembled-monolayer (SAM) that can effectively passivate the MoS$_2$/SiO$_2$ interface as well (Figure 14a,b illustrate the chemical structure of the MPS molecule, and the SV passivation process in MoS$_2$ by the MPS sulfur atom, respectively). After MPS treatment, back-gated monolayer MoS$_2$ NFETs showed a much higher RT mobility (~81 cm^2/V-s, which is among the highest reported μ_{FE} values for monolayer MoS$_2$) than the untreated samples. Using TEM analysis, the density of SVs was found to be reduced from ~6.5 × 10^{13} cm^{-2} in as-exfoliated samples to ~1.6 × 10^{13} cm^{-2} in MPS-treated samples. Moreover, using theoretical modeling, the authors were able to extract the densities of CIs and interface trap states (D$_{it}$) that showed lower values in MPS-treated samples than the as-exfoliated ones (CIs reduced from 0.7 × 10^{12} to 0.24 × 10^{12} cm^{-2}, and D$_{it}$ reduced from 8.1 × 10^{12} to 5.22 × 10^{12} cm^{-2}, after MPS treatment), highlighting the importance of passivating the SV defects in MoS$_2$ to minimize carrier scattering and to improve the MoS$_2$ device performance (Figure 14c compares the monolayer MoS$_2$ FET conductivity before and after MPS chemical treatment at RT) [296].

Another promising approach to passivate the SVs in monolayer MoS$_2$ was reported by Amani et al. using an air-stable, solution-based chemical treatment by an organic superacid, namely, bis(trifluoromethane) sulfonamide or TFSI, resulting in a near-unity photoluminescence (PL) yield and minority carrier lifetime enhancement, showing great promise for MoS$_2$-based optoelectronic devices [297]. The potential of SAMs, with the right "end-terminations", in passivating the substrate interface and reducing D$_{it}$ was highlighted by Najmaei et al. where they studied back-gated MoS$_2$ FETs on SAM-modified SiO$_2$ substrates having different functional groups or end-terminations, such as amine (-NH$_2$), methyl (-CH$_3$), fluoro (-CF$_3$), and thiol (-SH), and compared their performances with FETs fabricated on pristine and hydroxylated (i.e., -OH functionalized) SiO$_2$ substrates. From the back-gated transfer curves, it was revealed that the hysteresis (caused due to charge trapping and detrapping by the D$_{it}$) was significantly reduced in FETs made on -SH, -NH$_2$, and -CF$_3$-terminated substrates, implying that their D$_{it}$ was much lower as compared to FETs made on -CH$_3$, -SiO$_2$, and -OH terminated substrates (Figure 14d shows the ball-and-stick models of different SAMs used in the experiment as well as the conductance versus back gate voltage for the MoS$_2$ FET made on -SH-terminated SiO$_2$ substrate) [298]. It was also shown by Giannazzo et al. that D$_{it}$ at the MoS$_2$/oxide interface can be reduced by carrying out a "temperature-bias" annealing process on MoS$_2$ FETs. The authors demonstrated an improvement in the subthreshold behavior and a significant decrease in the electrical hysteresis upon subjecting their as-fabricated back-gated MoS$_2$ FETs to a 200 °C anneal under a positive gate bias ramp (0 to +20 V), directly correlated to a decrease in the D_{it} (note that D_{it} worsens the SS of FETs in addition to causing hysteresis) from ~9 × 10^{11} to ~2 × 10^{11} eV^{-1} cm^{-2} after annealing (Figure 14e shows the semilog transfer characteristics of a back-gated MoS$_2$ FET before and after temperature-bias annealing) [299]. Forming gas annealing could be another effective way to passivate D_{it} and enhance the MoS$_2$ device performance (due to lowering of the SS and increase in the μ_{FE}) as shown by both Bolshakov et al. [300,301] and Young et al. [302].

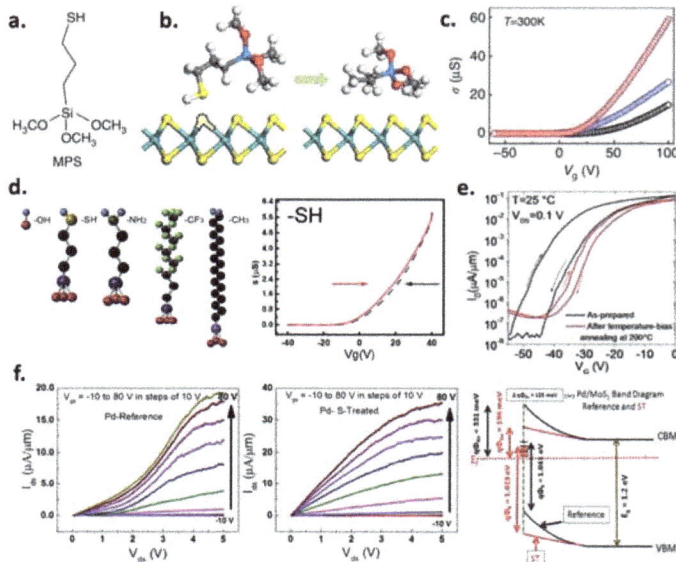

Figure 14. (**a**) Chemical structure of an MPS molecule showing the thiol (-SH) end-termination. (**b**) Schematic illustration of the MoS$_2$ SV passivation mechanism by the sulfur atom (yellow balls) derived from the -SH group of the MPS molecule. The SV in the MoS$_2$ layer is marked by the black-dashed contour. (**c**) Comparison of conductivity versus back gate voltage V_G for FETs made on as-exfoliated (black curve), top side-treated (blue curve) and double side-treated (red curve) monolayer MoS$_2$ by MPS at RT. It is evident that the double-side treated FET shows the highest conductivity due to SV passivation on both the top and bottom surfaces of the monolayer MoS$_2$. (**a–c**) Adapted with permission from [296]. Copyright Springer Nature 2014. (**d**) **Left schematic**: The ball-and-stick models for SAMs with different functional groups used in the substrate surface functionalization along with their orientations relative to overlying MoS$_2$ monolayers. **Right**: Conductance versus back gate voltage for the monolayer MoS$_2$ FET on SiO$_2$ treated with the -SH terminated SAM showing highly reduced hysteresis. Adapted with permission from [298]. Copyright 2014 American Chemical Society. (**e**) RT transfer characteristics of an MoS$_2$ FET showing a significant reduction in the hysteresis after bias-temperature annealing (red curves) at 200 °C due to passivation of interface traps (D_{it}). Adapted with permission from [299]. Copyright 2016 John Wiley and Sons. (**f**) Output characteristics of Pd-contacted back-gated MoS$_2$ NFETs with (**left**) and without (**center**) (NH$_4$)$_2$S chemical sulfur treatment (ST) clearly showing Schottky behavior with poor current saturation and Ohmic behavior with excellent current saturation, respectively. The ST enables reliable Ohmic contacts with reduced variability in the n-type SBH even with high work function metal contacts such as Pd. **Right schematic**: Qualitative band diagrams of the Pd/MoS$_2$ contact interface before (black lines) and after (red lines) ST, illustrating the SBH reduction in the sulfur-treated MoS$_2$ case. Adapted with permission from [303]. Copyright 2016 IEEE.

Bhattacharjee et al. on the other hand, developed a novel sulfur treatment (ST) method for engineering the surface states on MoS_2 via ammonium sulfide [(NH$_4$)$_2$S] chemical treatment to systematically improve the contact performance and reliability of few-layer MoS_2 NFETs with stable high work function metals such as Ni and Pd [303]. The sulfur-treated devices showed consistent Ohmic behavior with good current saturation, reduced SBHs and R_C as opposed to untreated reference samples that showed variable contact behavior (ranging from Schottky to Ohmic) with poor current saturation. It is to be noted that controlling the contact variability of FETs is important for improving their overall yield. However, in contrast to SV passivation and reduction of surface/interface states, the main underlying mechanism responsible for the reliable Ohmic behavior in these sulfur-treated devices was attributed to the removal (either by etching or sulfurization) of spatially non-uniform molybdenum suboxide (MoO$_x$) species (that can form due to surface oxidation, particularly at the SV sites) from the MoS_2 surface due to their reaction with the (NH$_4$)$_2$S chemical [303]. It was suggested that the removal of these suboxide species led to strongly and reliably pinned n-type contacts, with the pinning entirely governed by the surface states, even for high work function metals (i.e., Ni and Pd) with their n-type SBH comparable to that obtained using low work function metals such as Ti (Figure 14f compares the output characteristics of Pd-contacted MoS_2 FETs with and without ST, and shows the band diagram of the Pd/MoS_2 contact interface before and after ST explaining the SBH reduction). Using this sulfur treatment technique, and in conjunction with an optimized e-beam-evaporated HfO$_2$ dielectric, the authors also demonstrated top-gated MoS_2 FETs with a high μ_{FE} of 62 cm^2/V-s and a low average SS of 72 mV/decade. The relatively low SS was attributed to a pristine HfO$_2$/MoS_2 interface having a low D_{it} [304].

Besides sulfur treatments, oxygen/ozone treatment has also been explored to repair the structural defects/SVs in MoS_2. Nan et al. used a mild oxygen (O$_2$) plasma treatment to improve the mobility of MoS_2 devices by an order of magnitude. This was attributed to the passivation of localized states originating from the SVs (that serve as carrier scattering centers) by the incorporation of oxygen ions that chemically bond with the MoS_2 at these SV sites. However, the plasma power and exposure time must be carefully controlled as excessive treatment may damage the MoS_2 lattice (either by physical damage or by excessive MoO$_3$ formation due to oxidation) and deteriorate the material quality [305]. Another novel report of mobility enhancement in multilayer MoS_2 NFETs was by Guo et al. where they used the synergistic effects of ultraviolet (UV) exposure and ozone (O$_3$) plasma treatment. The authors showed an abnormal enhancement, up to an order of magnitude, in the FET mobility (from 2.76 cm^2/V-s to 27.63 cm^2/V-s) and attributed this to the passivation of interface traps/scattering centers as well as to an n-doping effect arising due to the photo-generated excess carriers during the UV/ozone plasma treatment. In this approach, negatively charged oxygen ions get incorporated in the MoS_2 lattice at the SV sites (similar to the mechanism reported by Nan et al. [305]) during the O$_3$ plasma treatment which are simultaneously neutralized by the excess photo-generated holes due to the UV exposure. This results in an aggregate of electrons in the MoS_2 lattice effectively causing n-doping and downward band-bending in the MoS_2 near the contact regions, thereby, narrowing the SBW and reducing the n-type R_C. Moreover, the D_{it} reduced from 1.53×10^{13} cm^{-2} to 5.59×10^{12} cm^{-2} after the UV-O$_3$ treatment signifying the passivation of interface traps/scattering centers (due to O incorporation and increased free carriers) that can further enhance the carrier mobility (Figure 15a illustrates this UV-O$_3$ passivation/doping process of MoS_2 FETs and explains the mechanism via band diagrams) [306]. Again, the reader should note that, while controlled O$_2$/O$_3$ treatment can passivate the structural defects and interface traps in MoS_2 devices, excessive O$_2$/O$_3$ exposure can be harmful for the device performance. This was further revealed by Leonhardtl et al. who studied the effect of oxidants (by air and water exposure) both in the channel and contact regions of MOCVD-grown MoS_2 FETs, and concluded that ambient exposure and MoS_2 layer oxidation must be minimized in order to improve the R_C and μ_{FE} [307]. Finally, while chemical passivation of structural defects and traps are beneficial for improving the MoS_2 device performance, introduction of controlled physical damage in the contact regions via argon (Ar) plasma treatment has also been shown to be beneficial for improving the R_C at the metal/MoS_2 interface (possibly due to the creation of unsaturated/reactive MoS_2 edge sites that can bond better with the contact metal atoms and/or due to the etching/removal of any superficial oxide layers) [308,309].

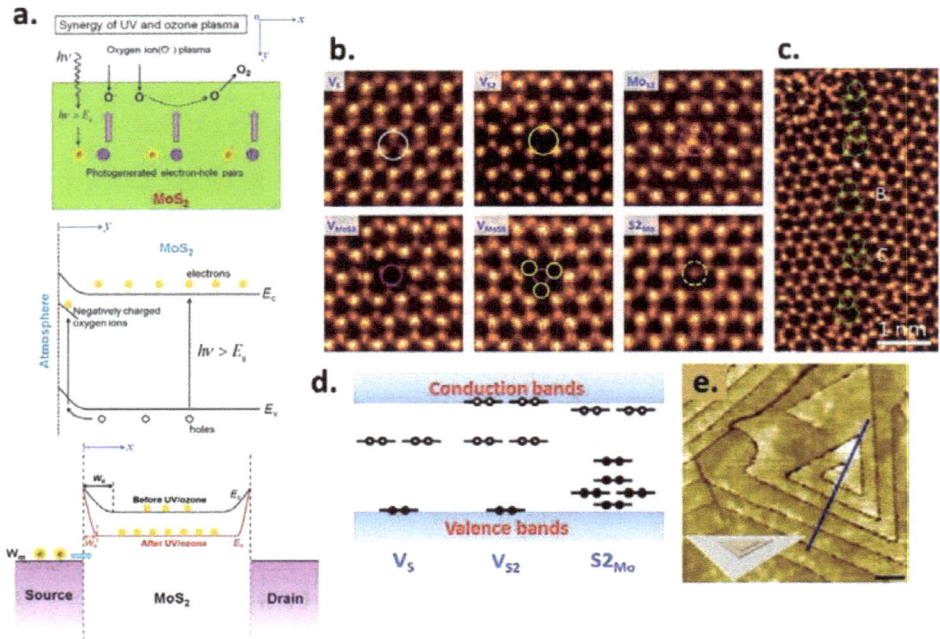

Figure 15. (**a**) **Top schematic**: Illustration of the photo-generated excess electron-hole pairs and neutralization of the incorporated oxygen ions in the MoS_2 lattice by the excess holes under the synergistic effects of UV illumination and ozone (O_3) plasma treatment. **Middle and bottom schematics**: Energy band diagrams illustrating the effects of UV-O_3 treatment on MoS_2. UV photons with energy greater than the MoS_2 band-gap generate excess electrons and holes, with the latter neutralizing the incorporated negatively charged oxygen ions in the MoS_2 lattice (**middle schematic**). **Bottom schematic** shows the band diagram along the MoS_2 FET channel showing the SBH formed at the S/D contact electrodes. After the neutralization of negatively charged O ions by the excess photo-generated holes, the leftover excess electrons cause an effective n-doping of the MoS_2 channel causing stronger downward band-bending in the MoS_2 near the contact interface, thereby narrowing the SBW (see CBE profile after UV-O_3 treatment, shown in red) and enabling more efficient electron injection. Adapted from [306], Copyright 2017, with permission from Elsevier. (**b**) Atomic resolution annular dark field (ADF) images of various intrinsic point defects present in CVD-grown monolayer MoS_2. The nomenclature of these defects (V_S, V_{S2}, $S2_{Mo}$, etc.) stems from the exact nature of their crystalline structure. (**c**) High resolution STEM-ADF image showing a grain boundary (GB) defect in synthetic monolayer MoS_2 comprising various dislocation centers. (**d**) Schematic representation of the DFT-calculated band diagram of MoS_2 showing the defect levels introduced in its band-gap due to various intrinsic point defects shown in (**b**). These defect levels or mid-gap states can act as charge trapping as well as charge scattering centers leading to carrier mobility degradation in synthetic MoS_2 FETs. (**b**–**d**) Adapted with permission from [312]. Copyright 2013 American Chemical Society. (**e**) Scanning tunneling microscopy (STM) image showing the presence of "intra-domain" periodic defects, arranged as concentric triangles, within an individual CVD MoS_2 domain. Adapted from [314], with the permission of AIP Publishing.

It is also instructive to note that in addition to the commonly observed structural defects in 2D MoS$_2$ films derived from either naturally occurring or synthetically grown (via chemical vapor transport, CVT, etc.) bulk MoS$_2$ crystals, there also exists a set of intrinsic structural defects uniquely associated with synthetically grown large-area 2D MoS$_2$ nanosheets. From a commercial viewpoint, it is imperative to realize wafer-scale growth of uniform 2D MoS$_2$ films (either on rigid or flexible substrates) with tunable, application-specific thicknesses for any MoS$_2$-based technology to become scalable and practically viable [107]. Hence, various synthetic routes, such as CVD [104,106], ALD [310], and vdW epitaxy [105] among others, utilizing a diverse set of precursor materials and growth conditions, have been explored to grow wafer-scale MoS$_2$ films [311]. However, these synthesized MoS$_2$ films typically have a polycrystalline nature (i.e., they are formed by the coalescence of several MoS$_2$ domains) and typically contain a rich variety of unique point defects (e.g., antisite defects, vacancy complexes of Mo with three nearby sulfurs or disulfur pairs, etc.) and a diverse set of inter-domain dislocation cores and grain boundaries (GBs) that can introduce localized mid-gap states which, in turn, can scatter and/or trap the charge carriers (Figure 15b–d show the atomic resolution images of some common intrinsic point defects, a high resolution STEM image of a GB defect showing dislocation centers, and the DFT-calculated electronic band-structure showing the mid-gap defect levels introduced by various intrinsic point defects in synthetically grown MoS$_2$ films, respectively). These defects can have dire consequences on the electrical performance of devices derived from synthesized MoS$_2$ films [312,313]. Moreover, synthetic growth techniques can also result in MoS$_2$ surface contamination and substrate property modification (giving rise to charged impurities and/or interface traps) and can also cause growth-induced strain in the MoS$_2$ films [313]. Furthermore, in addition to the various "inter-domain" dislocations and GBs, another distinctive type of narrowly spaced (~50 nm apart) "intra-domain" GBs or periodic defects, arranged in the form of concentric triangles, have also been reported in CVD-grown MoS$_2$ films by Roy et al. (Figure 15e shows an STM image revealing these intra-domain periodic defects in CVD MoS$_2$ films) [314]. While several of the defect-passivation techniques described earlier in this section can also be applied to these synthetic MoS$_2$ films, it is highly necessary to optimize the synthetic growth process itself to achieve defect-free, single-crystalline and pure MoS$_2$ films over commercial wafer-scale substrates that will enable the integration of large-scale 2D MoS$_2$-based devices and circuits.

14. Role of Dielectrics in Doping and Mobility Engineering

The role of dielectrics is of paramount importance in the development and integration of high-performance MoS$_2$ devices. Dielectrics can play a critical role in doping the MoS$_2$, thereby, enhancing its carrier mobility, and in passivating/protecting the device channel against ambient exposure. Traditionally, MoS$_2$ devices/FETs have been largely demonstrated on SiO$_2$ substrates in a back-gated configuration. However, a wide range of dielectrics (such as technologically relevant high-κ dielectrics and 2D hBN) have been explored as substrates and superstrates to enable both top- and dual-gated MoS$_2$ devices, and to engineer various critical device parameters [169,315,316]. Quite obviously, tremendous progress has been made in understanding the underlying growth mechanisms, deposition methods and fabrication techniques (including various pre- and post-deposition processes/treatments such as MoS$_2$ surface pre-functionalization) to integrate several dielectrics on MoS$_2$ (typically via the ALD method). Special attention has been given to ensure that these dielectrics maintain a high degree of uniformity/conformity on the MoS$_2$ surface, and that they display good electrical quality (i.e., minimal "pinhole" defects, low current leakage, high stability, low density of traps at the dielectric/MoS$_2$ interface, etc.) as well as nanometer scalability (to reduce their effective oxide thickness "EOT" for enhanced gate control over the MoS$_2$ channel). The interested reader is directed to various literature reports for further reading on these topics [317–333]. The focus of the following discussion is on the influence of dielectric engineering on important device parameters, such as doping, carrier mobility, ON-currents and contact resistance, and how dielectric engineering can be used to help enhance the performance of MoS$_2$-based devices.

14.1. Dielectrics as Dopants

The role of dielectrics as n-type charge transfer dopants on MoS_2 is discussed above in Section 7.1. Sub-stoichiometric high-κ oxides (such as HfO_x, AlO_x or TiO_x) dope the MoS_2 owing to their interfacial-oxygen-vacancies and this interesting property can be utilized to selectively and controllably dope the MoS_2 regions by merely varying the high-κ oxide stoichiometry (Figure 16a shows the back-gated transfer characteristics of a monolayer MoS_2 FET before and after sub-stoichiometric ALD HfO_x deposition, demonstrating significant n-doping effect in the latter case). This "high-κ doping effect" has been utilized to achieve very low contact resistances in monolayer MoS_2 devices, and has also been suggested as the primary mechanism responsible for the enhancement of both field-effect (μ_{FE}) and intrinsic mobilities (i.e., two-point and four-point mobilities, respectively) in high-κ-encapsulated MoS_2 devices [212–216]. There are also other reports where dielectric engineering has been utilized to dope the MoS_2. For example, Li et al. demonstrated a technique to dope MoS_2 by functionalization of the underlying SiO_2 dielectric surface using self-assembled monolayers (SAMs) having functional groups with different dipole moments. In this technique, the MoS_2 can either be hole- or electron-doped depending on the polarity and strength of the electrostatic interaction between the MoS_2 and the SAM-modified SiO_2 substrate. The authors reported a Fermi level modulation in monolayer MoS_2 of more than 0.45–0.47 eV using this approach [334]. A similar dipole-induced doping effect was also observed by Najmaei et al. in their study of MoS_2 FETs on SAM-modified SiO_2 substrates. They concluded that with the right choice of the end-termination/functional group of the SAM (e.g., -SH terminated), one can get complementary benefits in the MoS_2 device performance by simultaneously passivating the interface traps D_{it} (leading to reduced hysteresis in the device I-V) and enhancing the channel carrier density. For example, in the case of SAM having -SH terminations, the negative dipoles of the -SH groups help push the electrons from the MoS_2/SiO_2 interface into the MoS_2 channel, thereby, causing an n-doping effect (Figure 16b shows the comparison of both sheet conductance versus back gate voltage and the extracted mobility μ_{FE} for back-gated MoS_2 FETs fabricated on SAM-modified SiO_2 substrates having different end-terminations or functional groups) [298]. For reference, a detailed review of SAM-induced electrical property tuning of MoS_2 has been done by Lee et al. [335].

Recently, a novel and extremely promising n-doping technique for MoS_2 based on dipole interaction was reported by Park et al. using phosphorous silicate glass (PSG) as the back gate dielectric. They achieved wide-range controllable n-doping on trilayer and bulk MoS_2 using the PSG insulating layer, with the sheet doping density ranging between 3.6×10^{10} and 8.3×10^{12} cm^{-2}. This was achieved through careful design of the PSG substrate with special emphasis on the weight percentage of P atoms in the PSG layer which determined the starting concentration of the polar P_2O_5 molecules responsible for the doping. Moreover, a "three-step" thermal and optical activation process was employed to improve the PSG/MoS_2 interface properties (reduction of the PSG surface roughness enabled a more intimate contact with the MoS_2) as well as to control the final concentration of the polar P_2O_5 molecules at the PSG/MoS_2 interface which ultimately determined the doping levels in MoS_2 via electrostatic dipole interactions (Figure 16c,d show a schematic illustration of this three-step controllable doping process of MoS_2 by the PSG substrate, and show the transfer characteristics of back-gated MoS_2 FETs on PSG substrates highlighting the wide-range doping tunability achieved via a combination of thermal and optical activation as well as via tuning of the weight percentage of P atoms in the PSG substrate, respectively) [336]. More specifically, in this method, the negative poles of the polar P_2O_5 molecules (made up of electronegative O atoms) are aligned towards the PSG surface and they attract and "hold" the positively-charged holes from the overlying MoS_2 layer at the interface region, thereby, n-doping the MoS_2 body. This doping effect was found to be independent of the MoS_2 thickness and was limited only to the extent of dipole interaction at the PSG/MoS_2 interface. The PSG substrate doping method is very promising from a technological viewpoint, since achieving a wide-range doping capability, spanning the non-degenerate and degenerate regimes, is critical for designing MoS_2-based electronic and optoelectronic devices with useful and tailored properties [336,337]. It was also shown by Joo et al. that hBN, a popular 2D layered insulator (more on the usage and advantages of hBN is discussed later in Section 14.3), can also have an electron doping and SB minimization effect when used as a substrate for monolayer MoS$_2$ devices. The authors found that, unlike the conventional SiO$_2$ substrates, hBN can induce an "excess" electron doping concentration of ~6.5 × 10^{11} cm^{-2} at RT (~5 × 10^{13} cm^{-2} at high temperature), thereby, n-doping the MoS$_2$ resulting in lowering of the effective SB and R$_C$ (due to reduction of the SBW thanks to the doping-induced stronger band-bending at the contact/MoS$_2$ interface). Moreover, a 4× enhancement in the μ_{FE} as well as an early emergence of metal-insulator-transition (MIT) was observed in MoS$_2$ FETs fabricated on hBN substrates (Figure 16e shows the calculated excess electron doping concentration as a function of temperature for SiO$_2$ and hBN substrates, and compares the schematic band diagrams of the Schottky barrier at the metal/MoS$_2$ interface as well as the extracted SBH as a function of back gate voltage for MoS$_2$ FETs fabricated on both SiO$_2$ and hBN substrates). Furthermore, in addition to the substrate doping effect, it was suggested that the inserted hBN in-between the MoS$_2$ and SiO$_2$ can lead to a pronounced "dipole alignment effect" between the positive fixed charges of SiO$_2$ and the negative image charges in the contact metal, resulting in a reduced effective work function of the contact metal and, consequently, a lower effective Φ_{SB} [338].

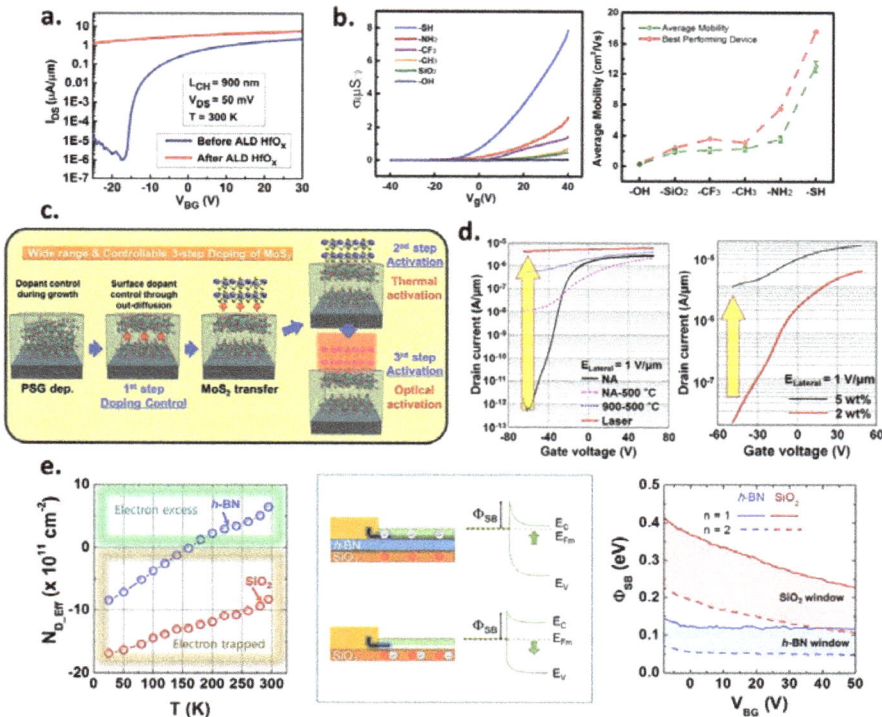

Figure 16. (a) Transfer characteristics (at RT) of a back-gated monolayer MoS$_2$ NFET before (blue curve) and after (red curve) sub-stoichiometric HfO$_x$ deposition (x ~1.56) showing strong n-doping. Adapted with permission from [213]. Copyright 2015 IEEE. (b) Left: Comparison of sheet conductance versus back gate voltage for MoS$_2$ FETs on SiO$_2$ substrates modified with SAMs having different functional groups/end-terminations (as shown in the legend). It can be seen that different functional groups cause different levels of n-doping in the MoS$_2$ depending upon the magnitude and polarity of their dipole moments. Right: Average mobility for MoS$_2$ FETs on different SAM-modified SiO$_2$ substrates.

The mobility increases continuously from the -OH-modified to the -SH-modified substrates and this effect can be attributed to a higher dipole-induced n-doping of the MoS_2 as well as to an enhanced passivation of the interface traps (thus, reduced carrier scattering) as we move from -OH to -SH-modified SiO_2 substrates. Adapted with permission from [298]. Copyright 2014 American Chemical Society. (c) Schematic illustration of the wide-range and controllable "three-step" doping process of MoS_2 by PSG substrates. The doping effect in MoS_2 takes place at the MoS_2/PSG interface via electrostatic interactions with the dipoles of the polar P_2O_5 molecules present at the PSG surface. The doping strength can be controlled via the weight percentage of P in the PSG (which determines starting concentration of the polar P_2O_5 molecules) as well as by performing additional "thermal" and "optical" activation steps (which modifies the interfacial electrostatic interaction between the polar P_2O_5 molecules and the MoS_2). (d) **Left**: Transfer characteristics of back-gated MoS_2 NFETs on PSG substrates showing the wide-range n-doping tunability (i.e., from non-degenerate to degenerate) achieved using a combination of thermal annealing (dashed-pink and dotted-blue curves) and optical exposure (red curve) steps. **Right**: Transfer characteristics showing the n-doping tunability by altering the weight percentage of P atoms in the PSG substrate. (c,d) Adapted with permission from [336]. Copyright 2015 American Chemical Society. (e) Excess electron doping of MoS_2 by hBN substrates. **Left**: Analytically calculated excess electron doping concentration (N_{D_Eff}) as a function of temperature for MoS_2 on hBN (blue circles) and SiO_2 (red circles) substrates showing enhanced n-doping effect at higher temperatures (>165 K) in the former case. **Middle schematic**: Band diagrams of the contact/MoS_2 interface showing the effect of excess electron doping by hBN substrates as compared to SiO_2. As is evident, electron doping by hBN leads to a narrower SBW and, hence, a reduced effective SBH. **Right**: Extracted SBH (Φ_{SB}) as a function of back gate bias for MoS_2 FETs on SiO_2 and hBN substrates. The SBH extracted on hBN substrates is ~3× smaller than that on SiO_2 thanks to the n-doping effect of hBN (n = 1,2 denotes the ideality factor used in the thermionic emission current equation for SBH extraction). Adapted with permission from [338]. Copyright 2016 American Chemical Society.

14.2. Mobility Engineering with Dielectrics: Role of High-κ

Dielectric engineering has been widely utilized to "boost" the mobility of charge carriers in MoS_2-based devices. The reader should note that the dielectric-induced doping of the MoS_2 (as described in the previous section, due to sub-stoichiometric high-κ oxide doping, dipole interaction effects, etc.) can, by itself, help enhance both the peak Hall (μ_{Hall}) and the peak field-effect (μ_{FE}) mobility in ultra-thin MoS_2 devices within a given gate and drain biasing range. This is because it is well known that doping-induced increased carrier densities in the device channel can provide better "screening" against various external carrier scattering sources and can also help reduce the SBW tunneling distance (due to enhanced MoS_2 band-bending) for efficient charge injection [173,212,213,339]. However, even without considering their propensity for doping the MoS_2 via sub-stoichiometric surface charge transfer or their capability to provide enhanced electrostatically-induced carrier densities in the device channel (due to the much higher gate capacitances they offer), high-κ dielectrics can also help mitigate the deleterious Coulombic interaction between charge carriers in low-dimensional (i.e., 2D and 1D) semiconductors and their surrounding charged impurities (CIs) [340]. For 2D MoS_2 devices, these CIs typically reside at the MoS_2/dielectric interface and can originate from various kinds of incorporated residues and adsorbates (gaseous or chemical) during device processing (note that the widely used SiO_2 substrate for MoS_2 devices is highly prone to the adsorption of these CIs due to its highly reactive/hydrophilic surface). Moreover, these CIs can also originate from the intrinsic structural defects such as SVs (as highlighted in Section 13) and/or trapped ionic species in the MoS_2 host lattice. Whatever their source, CIs serve as major scattering centers by giving rise to localized electric fields that can interact strongly with, and perturb the motion of, the MoS_2 charge carriers. Furthermore, the scattering effect of CIs is much more pronounced at low temperatures (<100 K) and can significantly degrade the low temperature mobilities in MoS_2 devices. The high-κ dielectrics, thanks to their large ionic polarizability, can effectively cancel out or "screen" the local electric fields generated by these CIs, thereby, minimizing the scattering effect of CIs on the charge carriers. Note that higher "κ" values increased polarizability of the high-κ dielectric, leading to improved dielectric screening of CIs (Figure 17a illustrates the effect of a "high-κ" environment in minimizing the spread of the Coulombic potential or the localized electric field generated by the charged impurities) [178,290,341,342].

Employing high-κ dielectrics to offset the effect of CIs, Li et al. demonstrated the use of an HfO_2/Al_2O_3 high-κ dielectric stack to fabricate dual-gated MoS_2 NFETs on SiO_2 substrates and showed that the high-κ stack enhanced the RT electron mobility (from 55 to 81 cm^2/V-s) as well as enabled high drain currents at low-T (~660 µA/µm at 4.3 K), while effectively eliminating the self-heating-induced negative differential resistance (NDR) effect owing to its higher thermal conductivity as compared to SiO_2. Moreover, by doing pulsed I-V and low frequency 1/f noise measurements, the authors confirmed a higher interface quality at the Al_2O_3/MoS_2 top interface than the MoS_2/SiO_2 bottom interface with a ~2× reduction in the oxide trap density in the former (Figure 17b compares the temperature-dependent transconductance "g_m" and the oxide trap density "N_{ot}" between MoS_2 FETs with the HfO_2/Al_2O_3 high-κ stack and bare MoS_2 FETs on SiO_2) [343]. Xu et al. introduced a novel dielectric "stack" substrate, comprising $Si/SiO_2/ITO$ (indium tin oxide)/Al_2O_3, that combined the benefits of dielectric screening and high gate capacitance offered by the high-κ Al_2O_3 together with enhanced gate controllability, thanks to the conductive ITO films that served as the gate electrode. They demonstrated back-gated MoS_2 NFETs with high mobilities (~62 cm^2/V-s), record low SS (62 mV/decade), and ON/OFF ratios >10^7. Moreover, using these enhanced device characteristics, the authors demonstrated MoS_2 photodetectors with the best reported photoresponsivity [344]. Similarly, Yu et al. reported monolayer MoS_2 back-gated FETs on HfO_2 and Al_2O_3 substrates showing RT mobilities of 148 and 113 cm^2/V-s, representing an 85% and a 41% improvement over FETs on SiO_2 substrates, respectively. The authors, through experimental and rigorous theoretical modeling, demonstrated the efficacy of high-κ dielectrics over SiO_2, and of higher-κ HfO_2 (κ ~17) over Al_2O_3 (κ ~10), in providing improved screening against CI scattering (Top plot of Figure 17c compares the temperature-dependent mobility of MoS_2 FETs on SiO_2, Al_2O_3 and HfO_2 substrates, at a fixed sheet carrier density, showing a good match between experimental data and the theoretical model employed by the authors). Moreover, the authors theoretically calculated the dependence of the CI-limited mobility (i.e., after subtracting the contribution of phonon scattering) on the MoS_2 sheet carrier density (n_{2D}) for MoS_2 FETs on SiO_2, Al_2O_3 and HfO_2 substrates and found that while the mobility increased with increasing n_{2D} for all dielectric substrates, the mobility values were highest in the case of HfO_2, followed by Al_2O_3, and lowest for SiO_2 (Bottom plot of Figure 17c shows the calculated CI-limited as a function of n_{2D}). A similar dependence was observed by the authors in their experimental data, further validating their theoretical model [345]. These results make sense as increased sheet carrier densities in the MoS_2 channel would provide additional screening against the CI scattering centers, in addition to the high-κ screening effect of the various dielectrics (with HfO_2 being more effective than Al_2O_3 due to its higher κ value). Likewise, Ganapathi et al. reported back-gated multilayer MoS_2 NFETs on HfO_2 substrates with record drain current (180 µA/µm) and transconductance (75 µS/µm) for an L_{CH} of 1 µm and achieved a 2.5× higher μ_{FE} as compared to the FET on SiO_2 substrate (Figure 17d shows the transfer characteristics comparing the drain current and mobility for multilayer MoS_2 FETs on both HfO_2 and SiO_2 substrates) [346].

Besides commonly used ALD-deposited high-κ dielectrics on MoS_2 such as HfO_2 and Al_2O_3, researchers have also resorted to integrating several other high-κ dielectrics in MoS_2 devices using novel approaches showing interesting device results. Chamlagain et al. presented a new strategy to integrate tantalum pentoxide (Ta_2O_5) high-κ dielectric (κ ~15.5) into MoS_2 FETs via chemical transformation and mechanical assembly of reactive 2D tantalum disulfide (TaS_2) layers. At elevated temperatures, mono- and multilayer TaS_2 transforms into atomically flat, spatially uniform and nearly defect-free Ta_2O_5 via thermal oxidation. This approach enabled the integration of high-quality Ta_2O_5 dielectric in both back- and top-gated MoS_2 FETs that displayed low SS values and nearly hysteresis-free transfer characteristics, suggestive of an ultra-clean and high-quality MoS_2/Ta_2O_5 interface with minimal interface traps (extracted D_{it} value was relatively low ~1.2×10^{12} cm^{-2} eV^{-1}) (Figure 17e shows the transfer characteristics of a back-gated MoS_2 FET on Ta_2O_5 dielectric displaying negligible hysteresis and an SS of 64 mV/decade). This authors also reported high-performance top-gated MoS_2 FETs with high RT mobilities (>60 cm^2/V-s), near ideal SS (~61 mV/decade) and pronounced drain current saturation using Ta_2O_5. This approach opens a novel way to integrate high-quality high-κ dielectrics on MoS_2 via chemical transformation of their reactive 2D material precursors, thereby, circumventing the complexities involved in the ALD deposition of high-κ dielectrics. Moreover, this approach is compatible with large-area synthesis techniques such as CVD and can also be readily extended to common high-κ oxides such as HfO_2 (by chemical transformation of reactive 2D hafnium diselenide (HfSe$_2$)) [347]. Zirconium oxide (ZrO$_2$) represents another potential high-κ dielectric candidate as was demonstrated by Kwon et al. where they used sol-gel processed ZrO$_2$ (κ ~22) as the back gate dielectric in MoS_2 FETs (together with conductive ITO as the back gate electrode) and showed enhancement in the MoS_2/ZrO_2 interface quality as well as a 2.5× improvement in the μ_{FE} of MoS_2 FETs on ZrO$_2$ versus those on SiO$_2$ [348].

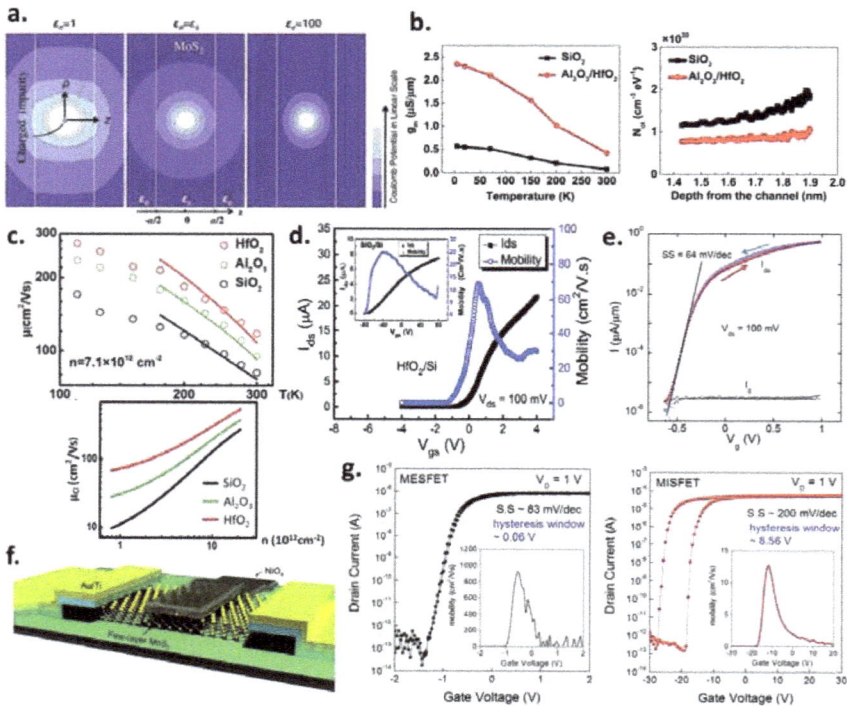

Figure 17. (**a**) The calculated Coulomb potential contours due to a charged impurity (CI) located inside MoS_2 for three different surrounding dielectric environments: $\kappa = 1$ (**left**), $\kappa = 7.6$ (same as MoS_2, **center**) and $\kappa = 100$ (**right**). The spread of the localized electric potential/field of the CI gets strongly damped in higher-κ environments, minimizing its scattering effect on the MoS_2 charge carriers. Adapted from [290]. (**b**) **Left**: Comparison of the temperature-dependent transconductance (g_m) between a back-gated MoS_2 FET on SiO$_2$ (black curve) and a top-gated MoS_2 FET using a dual high-κ dielectric stack (red curve). The g_m is much higher in the latter case employing high-κ dielectrics. **Right**: Oxide trap density versus depth from the channel, derived from the low-frequency 1/f noise measurements,

clearly showing lower trap densities in the dual high-κ dielectric stack as compared to SiO_2. Adapted with permission from [343]. Copyright 2017 American Chemical Society. (**c**) **Top**: Mobility versus temperature (at a fixed carrier density) for MoS_2 FETs on HfO_2, Al_2O_3 and SiO_2 substrates. While FETs with both high-κ dielectrics show mobility enhancement over SiO_2, highest mobilities are achieved in the case of HfO_2 due to the enhanced CI screening effect of higher-κ HfO_2 ($\kappa \sim 17$) than Al_2O_3 ($\kappa \sim 10$). **Bottom**: Log-log plot showing the calculated RT CI-limited mobility versus sheet carrier density (at a fixed CI density) for monolayer MoS_2 FETs on the three different dielectrics. Highest mobilities are achieved in the case of HfO_2 which outperforms Al_2O_3 which, in turn, outperforms SiO_2. Adapted with permission from [345]. Copyright 2015 John Wiley and Sons. (**d**) Mobility (blue curve) and drain current (black curve) as a function of gate bias for back-gated MoS_2 FETs on HfO_2 substrate. Inset shows the data for a similar FET on SiO_2. Much higher peak mobility and drain current values are achieved in the MoS_2 FET on HfO_2 than that on SiO_2. Adapted with permission from [346]. Copyright 2016 IEEE. (**e**) Transfer characteristics of a back-gated MoS_2 FET on high-κ Ta_2O_5 substrate (derived from thermal oxidation of reactive 2D TaS_2) showing negligible hysteresis and a low SS ~ 64 mV/decade, thereby, confirming a high-quality interface between MoS_2 and Ta_2O_5 with minimal traps. Adapted with permission from [347]. Copyright 2017 IOP Publishing. (**f**) 3D schematic of a few-layer MoS_2 MESFET with Schottky-contacted NiO_x top gate. (**g**) Comparison of transfer characteristics between a four-layer MoS_2 "MES" FET with NiO_x top gate (**left plot**, black curves) and a four-layer MoS_2 "MIS" FET with Al_2O_3 top gate (**right plot**, red curves). The MESFET displays much better SS and reduced hysteresis than the MISFET, confirming the superior NiO_x/MoS_2 interface quality that minimizes carrier scattering leading to a much higher peak mobility in the MESFET (inset of each plot shows mobility versus gate voltage). (**f**,**g**) Adapted with permission from [349]. Copyright 2015 American Chemical Society.

While integration of these insulating high-κ dielectrics in top-gated MoS_2 FETs represents the conventional metal-insulator-semiconductor (MIS) FET configuration, Lee et al. demonstrated for the first time a metal-semiconductor (MES) FET on MoS_2 using semi-transparent and conductive NiO_x dielectric as the Schottky-contacted gate electrode which makes a vdW interface with the MoS_2 (Figure 17f illustrates the 3D schematic of the MoS_2 MESFET with Schottky-contacted NiO_x top gate). In this rather unconventional approach towards designing MoS_2 FETs, the authors demonstrated few-layer (~ 10 layers) MoS_2 MESFETs with NiO_x gate to have high intrinsic-like electron mobilities ranging between 500–1200 cm^2/V-s. The NiO_x/MoS_2 MESFETs had low threshold voltages, minimal gate bias-induced hysteresis and displayed a sharp SS, showing a significant improvement over comparable MoS_2 MISFETs made using ALD Al_2O_3 top gate dielectric (Figure 17g compares the transfer characteristics of the MESFET and the MISFET device, showing a much improved hysteresis and SS behavior in the case of the MESFET). Using the high intrinsic electron mobilities, the authors demonstrated their MESFETs to work as a high-speed and highly sensitive phototransistor. The improved MESFET mobilities was mainly attributed to the unique and pristine nature of the vdW MoS_2/NiO_x Schottky interface having a large vdW gap of 3.31 Å as revealed by DFT (even larger than the gaps of 1.6 Å and 2.6 Å at MoS_2 interfaces with common metals such as Ti and Au, respectively) together with negligible interface and bulk traps [349]. Moreover, the channel sheet carrier density in the MESFET was found to be 2–3 orders of magnitude lower than that in the gate-controlled MISFET devices. Hence, the low n_{2D} as well as interface traps in the MESFET structure allowed for scattering-minimized transport without the deleterious effects of the ON-state gate electric field on the carrier mobility as is the case in MISFET devices. However, although the NiO_x/MoS_2 MESFETs showed much improved electron mobilities, their maximum ON-currents were much lower ($\sim 60\times$) than the Al_2O_3/MoS_2 MISFETs owing to the low n_{2D} in the MESFET channel [349].

14.3. Limitations of High-κ Dielectrics and Advantages of Nitride Dielectric Environments

While high-κ dielectrics can be promising for enhancing the carrier mobility in MoS_2-based devices (via dielectric screening of CIs, improved interface quality over SiO_2, reduced oxide trapped charges, etc.), an extremely important point to note is that this performance enhancement at RT is only nominal and still far below the true intrinsic potential of MoS_2. Recall that the RT electron mobility (which is the most relevant mobility number for practical device applications) of monolayer MoS_2 is predicted to be as high as ~480 cm^2/V-s [113] (as described earlier in Section 2). Moreover, the mobility enhancement due to high-κ screening is possible only when the MoS_2 carrier mobility is strongly limited by charged impurities (i.e., the CI density is high, typically >10^{12} cm^{-2}, which limits the MoS_2 carrier mobility to values well below 100 cm^2/V-s due to Coulombic scattering). In other words, in the absence or dearth of CIs, high-κ dielectrics would no longer be useful for improving the device performance of clean MoS_2 samples any further. This is because high-κ dielectrics are a major source of surface optical (SO) phonons or remote optical phonons, which can serve as a major extrinsic carrier scattering source. These SO-phonon modes originate from the oscillations of the polarized metal-oxide bonds (note that these are the same polarized bonds that provide the screening against CIs) in the high-κ dielectric and typically have low activation energies such that these SO-phonon modes can easily be activated at RT. Moreover, in contrast to 3D semiconductors with thicker channels, the ultra-thin 2D channel of TMDCs such as MoS_2 (especially in the monolayer case) is highly susceptible to its surrounding dielectric environment. These SO-phonon modes can, therefore, easily couple to the MoS_2 channel and scatter the charge carriers [171,290,345,350]. Hence, at the MoS_2/high-κ dielectric interface, there is always a competition between the detrimental SO-phonon scattering effect and the advantageous CI screening effect on the carrier mobility, and SO-phonon scattering ultimately becomes the dominant mobility-limiting factor for 2D MoS_2 devices in the limit of decreasing charged impurities. That is to say, while the RT mobility in devices made on ultra-clean MoS_2 will naturally be much higher than devices made on "impure" MoS_2 (i.e., MoS_2 having a large density of CIs), encapsulating the ultra-clean MoS_2 devices in a high-κ dielectric environment would not lead to a further enhancement in their mobility. Instead, the high-κ dielectric would degrade the mobility of ultra-clean MoS_2 devices due to SO-phonon scattering (though the SO-phonon-limited mobilities in ultra-clean MoS_2 devices would still typically be relatively much higher than the highest mobilities achievable in highly impure MoS_2 devices even after high-κ dielectric screening). At this point, it is instructive to note that phonon scattering, in general, is the primary mobility-limiting mechanism in semiconductor devices at higher temperatures (>100 K), wherein the mobility follows a power law dependence on temperature, $\mu \propto T^{-\gamma}$ (the exponent "γ" can be regarded as the "mobility degradation factor" and its value depends on the dominating phonon scattering mechanism, due to either intrinsic or extrinsic SO-phonons or their combination) [113].

This mobility degradation effect due to high-κ SO-phonon scattering also holds true for charge carriers in graphene as shown by Konar et al. [351] as well as for charge carriers in the inversion layer of conventional Si MOSFETs as shown by Fischetti et al. [352]. The magnitude of SO-phonon scattering in MoS_2 is directly (inversely) proportional to the κ-value (SO-phonon energy) of the surrounding dielectric media. Now, for SiO_2 and other commonly used high-κ dielectrics, their SO-phonon energies are as follows (listed in ascending order in units of meV): HfO_2(12.4) < ZrO_2(16.67) < Al_2O_3(48.18) < SiO_2(55.6) [351,353]. In general, the magnitude of the SO-phonon energy of a dielectric is inversely related to its dielectric constant or κ-value. Thus, from this trend, it can clearly be inferred that at RT (i.e., when the thermal energy kT/q ~26 meV), the SO-phonon modes of HfO_2 and ZrO_2 can readily be activated/excited in comparison to the SO-phonon modes of Al_2O_3 and SiO_2. Consequently, HfO_2 would cause the worst SO-phonon scattering of the MoS_2 charge carriers, and SiO_2 the least, among the dielectrics considered (Figure 18a shows the calculated electron mobility in monolayer MoS_2 as a function of the κ-value for different dielectric environments at RT and 100 K, both with and without considering the effects of SO-phonon scattering). Detailed theoretical investigations of temperature-dependent charge transport in MoS_2 in the presence of both CI and SO-phonon scattering by Ma et al. and Yu et al. shed further insight into the dependence of carrier mobility on these extrinsic scattering sources, while providing effective guidelines for selecting the most favorable dielectric environment to extract the maximum mobility from MoS_2 under varying extrinsic conditions (i.e., mobility as close to the truly intrinsic phonon-limited values for MoS_2) [290,345]. For example, calculation of RT field-effect mobility for monolayer MoS_2 devices on different dielectric substrates as a function of CI density by Yu et al. revealed a "critical" CI density of ~0.3 × 10^{12} cm^{-2} above which the mobility was strongly limited by the CIs (this corresponds to the "impure" regime in which high-κ dielectrics can be beneficial for enhancing the mobility by effectively screening the scattering effect of these CIs) and below which the mobility was limited by phonons (this corresponds to the "ultra-clean" regime where high-κ dielectrics are no longer useful due to the detrimental effect of their SO-phonons) (Figure 18b shows the calculated RT mobility of monolayer MoS_2 as a function of CI density "n_{CI}" for MoS_2 on various dielectric substrates, illustrating the phonon-limited and CI-limited transport regimes) [345]. Qualitatively similar results, showing a crossover between the two transport regimes in the plot of mobility versus CI density, were also obtained by Ma et al. [290].

Considering the above discussions, it becomes clear that to achieve the maximum MoS_2 device performance, ultra-clean samples (i.e., those having CI densities well below 10^{12} cm^{-2}) and low-κ dielectrics (as opposed to high-κ HfO_2, ZrO_2, Al_2O_3, etc.) having higher SO-phonon activation energies (to minimize SO-phonon scattering) must be integrated together. To achieve ultra-clean samples, the structural and electronic quality of synthetically grown large-area MoS_2 as well as the device processing/fabrication steps must be carefully optimized to minimize the CI density. Regarding choice of dielectrics, nitride-based wide-band-gap dielectrics such as 2D hexagonal boron nitride (hBN) and aluminum nitride (AlN), both having E_g ~6 eV [354], hold the most promise since they are both medium-κ dielectrics (κ ~5 for hBN and ~9 for AlN) and have much higher SO-phonon energies (hBN: ~93 meV, AlN: ~81 meV) compared to the other high-κ dielectrics discussed above. Therefore, both hBN and AlN can offer an optimized combination of high gate capacitances (required for better electrostatic control over the device channel) and high carrier mobilities (required for achieving high ON-state currents) that are essential for realizing high-performance FETs based on MoS_2 [290]. Moreover, 2D hBN, in particular, has been demonstrated to be an ideal dielectric for 2D materials as opposed to conventional oxides owing to its highly crystalline structure (hBN has a similar hexagonal lattice as graphene and MoS_2), atomically smooth surface, mechanical flexibility, lack of dangling bonds and charge traps, and ability to form pristine 2D/2D interfaces [118,355,356].

Indeed, there have been several reports demonstrating MoS_2 FETs with hBN dielectrics. Cui et al. reported a vdW heterostructure device platform wherein the MoS_2 layers were fully encapsulated within hBN to minimize external scattering due to charged impurities and remote SO-phonons, while gate-tunable graphene was used as the contacts to MoS_2. Using this approach, the authors extracted a low-temperature mobility of 1020 cm^2/V-s for monolayer MoS_2 and 34,000 cm^2/V-s for 6-layer MoS_2 at 4 K, with the values being up to two orders of magnitude higher than what was reported previously (Figure 18c shows the temperature-dependent Hall electron mobility for fully hBN-encapsulated MoS_2 devices having different number of MoS_2 layers). Theoretical fit to the experimental data revealed the interfacial long-range CI density of ~6 × 10^9 cm^{-2} that was about two orders of magnitude lower than the CI density typically obtained for graphene devices on SiO_2 [259]. This confirmed the superior interfacial quality in the sandwiched $hBN/MoS_2/hBN$ structure having minimal CIs. Moreover, owing to the substantially high low-T mobilities, the authors demonstrated the first-ever observation of Shubnikov–de Hass (SdH) oscillations (i.e., quantum oscillations due to Landau-level quantization of the cyclotron motion of charge carriers, observed in high purity 2D systems [357]) in their hBN-encapsulated monolayer MoS_2 device. However, even with such clean samples and hBN dielectrics, the maximum RT MoS_2 electron mobility was only 120 cm^2/V-s and the exponent γ (extracted from μ ~$T^{-\gamma}$) ranged 1.9–2.5, suggesting the existence of phonon scattering sources [259]. A similar fully hBN-encapsulated and graphene-contacted MoS_2 device platform was utilized by Lee et al. where they found via Raman and photoluminescence measurements that the double-sided hBN encapsulation imparts extremely high stability to the MoS_2 device, even at high temperatures (200 °C), with negligible degradation even after four months of ambient exposure [260]. The RT μ_{FE} of an hBN-encapsulated 3-layer MoS_2 FET was extracted to be 69 cm^2/V-s, much higher than the values obtained from un-encapsulated (7 cm^2/V-s) and HfO$_2$-encapsulated (18 cm^2/V-s) MoS$_2$ FETs on SiO$_2$ substrates with regular metal contacts. Moreover, negligible hysteresis and absence of "memory steps" were observed in the I-V transfer characteristics of the hBN-encapsulated MoS$_2$ devices as compared to un-encapsulated and HfO$_2$-encapsulated MoS$_2$ devices, thanks to the ultra-clean MoS$_2$/hBN interface with reduced density of charge traps (Figure 18d compares the hysteresis as well as environmental stability of hBN-encapsulated MoS$_2$ devices against un-encapsulated or HfO$_2$-encapsulated ones via comparison of their transfer characteristics, and also compares the high-temperature stability of various device parameters such as μ and V_{th} in these devices). Furthermore, the hBN-encapsulated devices showed no degradation or breakdown even at a high drain current density of ~6 × 10^7 A/cm^2 affirming their stability over HfO$_2$-encapsulated devices having a lower breakdown current density of 4.9 × 10^7 A/cm^2 [260]. Recently, Xu et al. reported sandwiched hBN/MoS$_2$/hBN devices where the metal contacts were deposited after selective etching of the top hBN layer using O$_2$ plasma to expose the underlying MoS$_2$ in the contact regions. The low-T Ohmic contacts achieved via this process (R_C ~0.5 kΩ·μm), together with the high interface quality afforded by the hBN, resulted in a high low-T μ_{FE} of 14,000 cm^2/V-s and Hall mobility (μ_{Hall}) of 9,900 cm^2/V-s at 2 K. The extracted RT device mobilities in this study, however, were only ~50 cm^2/V-s [358]. It is interesting to observe that in most of the experimental studies on hBN-encapsulated MoS$_2$ devices, while record-high low-T electron mobilities well in excess of 1000 cm^2/V-s have been achieved, the technologically relevant RT mobility still lags behind the best reported values for MoS$_2$ FETs fabricated using other dielectrics (e.g., an RT μ_{FE} of ~150 cm^2/V-s was reported for monolayer MoS$_2$ FET on HfO$_2$ substrate by Yu et al. [345]). This could be due to a multitude of factors such as differences in the material or electronic quality of the starting MoS$_2$ (i.e., structural defects, densities of CIs, traps, etc. in synthesized versus exfoliated MoS$_2$) used in these isolated experiments, the innate material quality of the hBN itself (with perhaps lower than expected SO-phonon energies etc.), and other differences such as quality of the S/D electrical contacts and processing-induced impurities/defects.

Nonetheless, together with the "excess electron doping" and the "dipole alignment effect" induced by hBN [338] as described earlier in Section 14.1, these theoretical and experimental results clearly highlight the advantages of using pristine 2D hBN as an ideal dielectric for MoS$_2$-based electronics over commonly used SiO$_2$ and other high-κ dielectrics, as hBN can afford much lower densities of interface traps and charged impurities, lower surface roughness scattering as well as much better immunity against SO-phonon scattering. Moreover, the innate atomically thin nature of 2D hBN can allow for ultimate gate dielectric scaling that can lead to a much enhanced electrostatic control over 2D MoS$_2$ device channels. Furthermore, combined with semi-metallic and gate-tunable graphene, hBN can help enable high-quality "all-2D" MoS$_2$-based devices and circuits for large-scale flexible nano- and optoelectronics, as shown by Roy et al. [359]. In addition to hBN, experimental evidence of the benefits of using aluminum nitride (AlN) as an alternative nitride-based dielectric for MoS$_2$ devices was also recently reported by Bhattacharjee et al. They compared the performance of identical MoS$_2$ FETs fabricated on SiO$_2$, Al$_2$O$_3$, HfO$_2$ and AlN substrates, with the MoS$_2$-on-AlN FETs outperforming its counterparts. The MoS$_2$-on-AlN FET displayed a μ_{FE} of 46.3 cm^2/V-s and a saturation drain current density of 160 μA/μm (for an L_{CH} of 1 μm), which compare favorably against the highest reported values for MoS$_2$ FETs. Temperature-dependent μ_{FE} calculations revealed the mobility degradation factor (γ) to be lowest for the FET on AlN (γ = 0.88) as compared to all other dielectrics (γ = 1.21, 1.32, and 1.80 for SiO$_2$, Al$_2$O$_3$, and HfO$_2$, respectively). Since phonon scattering is the dominant scattering mechanism at high temperatures and $\mu \propto T^{-\gamma}$ (as described earlier in this section), it is no surprise that AlN affords the lowest phonon scattering owing to its relatively high SO-phonon activation energy of ~81 meV (Figure 18e compares the output characteristics as well as the extracted μ_{FE} and γ values as a function of the SO-phonon energy for MoS$_2$ FETs fabricated on the four different dielectrics used in this study, with FETs on AlN showing the best performance). Moreover, the authors also demonstrated MoS$_2$ FETs encapsulated in an all-nitride dielectric environment comprising hBN/MoS$_2$/AlN structures from which an RT μ_{FE} as high as ~73 cm^2/V-s was extracted, displaying an improvement over the bare MoS$_2$-on-AlN FETs) [360]. Furthermore, the deposition of AlN can be done using CMOS-compatible processes (e.g., MOCVD in this case) making AlN an attractive substrate for large-scale integration of devices and circuits based on 2D MoS$_2$. Finally, besides hBN and AlN, another CMOS-compatible nitride-based dielectric, namely, silicon nitride (Si$_3$N$_4$), having a band-gap of 5.1 eV, dielectric constant of 6.6 and a high SO-phonon energy value (~110 meV) [361], has also been demonstrated as a suitable dielectric for MoS$_2$ FETs [362].

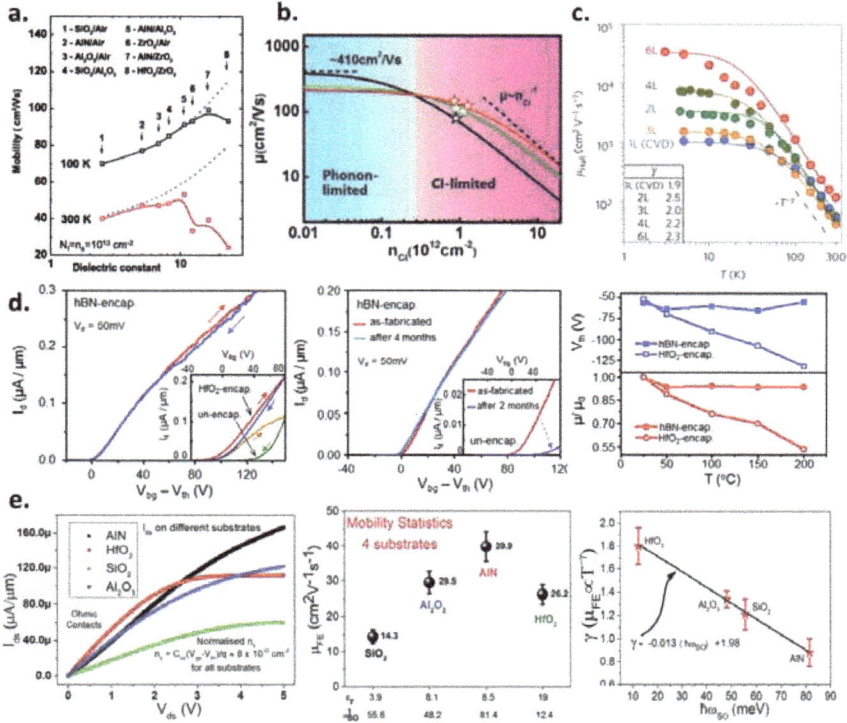

Figure 18. (a) Calculated electron mobility of monolayer MoS$_2$ at RT (300 K) and 100 K as a function of dielectric constant of its surrounding media (calculated at a fixed n$_{2D}$ and CI density). The solid black and red curves represent the net mobility after combining the scattering effects from charged impurities, intrinsic phonons and SO-phonons. Dashed blue lines represent the calculated mobility without considering the effect of SO-phonon scattering. Numbers 1 to 8, as marked on the curves, represent different pairs of dielectrics. As is evident, the MoS$_2$ mobility is drastically reduced in the presence of SO-phonon scattering for higher-κ environments, with the effect being more pronounced at RT. Adapted from [290]. (b) Calculated RT MoS$_2$ mobility as a function of CI density (n$_{CI}$) for MoS$_2$ on SiO$_2$ (black curve), Al$_2$O$_3$ (green curve) and HfO$_2$ (red curve) substrates. Star symbols represent experimental data from the study. From the plot, two transport regimes are clearly evident on either side of the critical n$_{CI}$ ~0.3 × 10^{12} cm^{-2}. When n$_{CI}$ is high, the MoS$_2$ mobility is CI-limited and the high-κ dielectric screening of CIs can be useful in enhancing the mobility in this regime (shaded pink region). However, when n$_{CI}$ is low, high-κ dielectrics can no longer enhance the mobility any further. In this regime, the mobility is phonon-limited (shaded blue region) and lower-κ dielectrics with higher SO-phonon energies are advantageous. Adapted with permission from [345]. Copyright 2015 John Wiley and Sons. (c) Temperature-dependent Hall electron mobilities extracted from fully hBN-encapsulated MoS$_2$ devices with different number of MoS$_2$ layers. The low-T mobility reaches ~1000 cm^2/V-s for

monolayer and ~34,000 cm^2/V-s for 6-layer MoS$_2$, thanks to the high-quality hBN dielectric environment with minimal traps and charged impurities. Inset shows the extracted γ value for the various devices. Adapted with permission from [259]. Copyright Springer Nature 2015. (**d**) **Left**: Back-gated transfer characteristics of an hBN-encapsulated 3-layer MoS$_2$ device showing negligible hysteresis as compared to un-encapsulated and HfO$_2$-encapsulated MoS$_2$ FETs (shown in the inset), indicating the ultra-clean interfaces afforded by hBN. **Middle**: Transfer characteristics of the hBN-encapsulated MoS$_2$ device showing no current degradation even after four months. In contrast, an un-encapsulated MoS$_2$ device shows significant current degradation (as shown in the inset) after two months. **Right**: Comparison of μ/μ_0 (where μ_0 is mobility at RT) and V_{th} between HfO$_2$- and hBN-encapsulated MoS$_2$ devices as a function of increasing temperatures. The hBN-encapsulated device shows a much enhanced stability, whereas the HfO$_2$-encapsulated device shows large variability in its V_{th} and μ/μ_0 values at higher T. Adapted with permission from [260]. Copyright 2015 American Chemical Society. (**e**) **Left**: Normalized output characteristics of MoS$_2$ FETs on SiO$_2$ (green curve), Al$_2$O$_3$ (blue curve), HfO$_2$ (red curve) and AlN (black curve) substrates showing the highest saturation drain current in the case of AlN. **Middle**: Extracted average RT μ_{FE} for MoS$_2$ FETs on four different dielectric substrates as a function of their SO-phonon energies. The highest μ_{FE} is achieved for FETs on AlN substrates which has the lowest scattering effect due to its relatively high SO-phonon energy (~81 meV). **Right**: Extracted mobility degradation factor "γ" as a function of the dielectric SO-phonon energy for MoS$_2$ FETs on different dielectrics. As expected, γ is highest for HfO$_2$ (which has the lowest SO-phonon energy ~12 meV) and lowest for AlN (which has the highest SO-phonon energy). Adapted with permission from [360]. Copyright 2016 John Wiley and Sons.

15. Substitutional Doping of 2D MoS$_2$

There have also been several reports of "substitutional doping" of MoS$_2$ wherein both the Mo cation and the S anion atoms have been substituted by appropriate "donor" or "acceptor" dopant atoms to yield either n-type or p-type MoS$_2$, respectively. In conventional CMOS technology, substitutional doping using ion implantation is the method of choice for controllably doping selected areas of the semiconductor wafer (either Si, Ge or III-Vs) to fabricate complementary FETs and realize complex circuits with desired performances. The ion implantation technique is also used to selectively and degenerately dope the S/D regions of the FET to realize Ohmic n- and p-type contacts for NMOS (i.e., n$^+$-p-n$^+$) and PMOS (i.e., p$^+$-n-p$^+$) device configurations, respectively, as well as to realize various bipolar devices, such as LEDs and photodetectors, for optoelectronic applications [207,363–375]. The ion implantation process is known to induce surface damage and amorphization in the as-implanted semiconductor crystals which requires further annealing to "activate" the implanted dopants and to minimize residual damage [170,376–381]. However, owing to the atomically-thin nature of 2D MoS$_2$ (recall that an MoS$_2$ monolayer is only ~0.65 nm thick), it is extremely challenging to employ the conventional ion implantation technique to dope MoS$_2$ (or 2D materials in general) as the process can induce irreparable surface damage and etching of the MoS$_2$ layers. As described earlier in this review, the traditional approaches for n- and p-doping of MoS$_2$ have employed techniques such as surface charge transfer doping (via adsorption or encapsulation of electron-donating or electron-accepting species), gate electrostatic doping using highly capacitive dielectrics or liquid/solid electrolytes, and doping via electrostatic dipole interactions at the MoS$_2$/dielectric interface. However, practical and stable doping requires "substitution" of a given fraction of the host lattice atoms by the dopant atoms wherein the latter covalently bonds with other atoms in the host lattice. While little progress has been made in the controlled and area-selective "top-down" substitutional doping of MoS$_2$ (as described in Section 15.4), most substitutional doping efforts on MoS$_2$ have relied on the incorporation of dopant atoms during the "bottom-up" or in-situ synthetic growth process (e.g., CVD) which may provide controlled, but inevitably unselective, doping of the entire MoS$_2$ film [169,382,383].

The reader should note that although the focus of this discussion is on the incorporation of "electron-rich" or "hole-rich" dopant atoms that lead to a pronounced n- or p-type doping effect in MoS$_2$, one can also achieve covalent substitution of the host MoS$_2$ atoms by "isoelectronic" or, in other words, MoS$_2$-like atomic species to yield different MoS$_2$-based alloys [384–387]. Indeed, there are several reports where the isoelectronic substitution process, using both transition metal atoms (e.g., W) or chalcogen atoms (e.g., Se), has been carried out on MoS$_2$ resulting in alloyed ternary TMDC species of the form Mo$_{1-x}$W$_x$S$_2$ or MoS$_{2(1-x)}$Se$_{2x}$, that essentially represent a fusion between MoS$_2$ and WS$_2$ or MoS$_2$ and MoSe$_2$, respectively [388–394]. While isoelectronic doping/alloying provides no extra electrons or holes, it represents an important avenue for tuning the band-gap, band-structure, band-edge positions and the carrier effective mass in these MoS$_2$-based alloys via composition tuning [395–398]. For example, Chen et al. demonstrated tunable band-gap emission in monolayer Mo$_{1-x}$W$_x$S$_2$ ranging from 1.82 eV (at x = 0.20) to 1.99 eV (at x = 1) [394]. Similarly, Mann et al. demonstrated band-gap tuning in the range of 1.88 eV (i.e., pure MoS$_2$) to 1.55 eV (i.e., pure MoSe$_2$) in composition-tuned MoS$_{2(1-x)}$Se$_{2x}$ monolayers (Figure 19a shows the evolution of the photoluminescence spectra in composition-tuned MoS$_{2(1-x)}$Se$_{2x}$ monolayers) [390]. Recently, a quaternary alloy of Mo$_x$W$_{1-x}$S$_{2y}$Se$_{2(1-y)}$ was also reported with tunable band-gaps ranging from 1.61 eV to 1.85 eV [399]. Analogous to the case of composition-dependent band-gap tuning in conventional III–V semiconductor alloys [400–403], the band-structure and band-gap engineering of composition-tuned monolayer MoS$_2$-based alloys is extremely promising for enabling 2D optoelectronic applications with tailored properties. Moreover, composition-engineering can be combined with techniques such as "pressure-engineering" to enable a wide variety of band-alignments in these 2D alloys, as shown in the case of Mo$_{1-x}$W$_x$S$_2$ monolayers by Kim et al. [404].

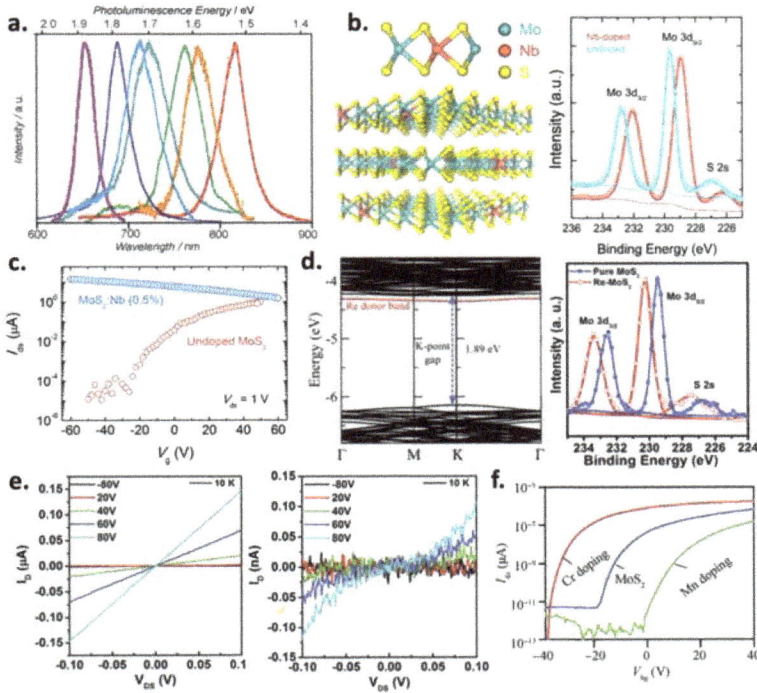

Figure 19. (a) Normalized PL spectra at RT for monolayer MoS$_{2(1-x)}$Se$_{2x}$ alloy films with different compositions showing the band-gap tunability from 1.88 eV (pure MoS$_2$) to 1.55 eV (pure MoSe$_2$). Adapted

with permission from [390]. Copyright 2013 John Wiley and Sons. (**b**) **Left schematic**: 3D cross-sectional illustration of Nb-doped few-layer MoS_2 wherein the Nb dopant atoms replace the Mo host atoms in the MoS_2 lattice. **Right**: XPS spectra of the Mo 3d core level peaks as a function of electron binding energy as measured from the Nb-doped (red) and undoped MoS_2 (light blue). A clear shift in the Mo 3d peaks towards lower binding energies is observed after Nb doping confirming the lowering of the MoS_2 Fermi level due to p-type doping. (**c**) Transfer characteristics of undoped and Nb-doped MoS_2 films. The undoped film shows typical n-type behavior, whereas the Nb-doped film shows degenerate p-type behavior. (**b**,**c**) Adapted with permission from [406]. Copyright 2014 American Chemical Society. (**d**) **Left**: DFT-calculated electronic band-structure of Re-doped MoS_2 showing the presence of Re donor bands/levels close to the CBE of MoS_2 confirming the n-type substitutional doping. **Right**: XPS spectra of the Mo 3d core level peaks as a function of electron binding energy measured from the Re-doped (red) and undoped MoS_2 (blue). In this case, the Mo 3d peaks shift towards higher binding energies after Re doping (opposite to the case of Nb-doped p-type MoS_2) confirming the upshift of the MoS_2 Fermi level due to n-type doping. (**e**) Comparison of output characteristics acquired at 10 K between a Re-doped MoS_2 FET (left plot) and an undoped MoS_2 FET (right plot). The metal-contacted Re-doped MoS_2 device clearly exhibits linear I-V behavior even at 10 K confirming Ohmic contacts, whereas the metal-contacted undoped MoS_2 device exhibits a non-linear and noisy I-V behavior indicative of Schottky contacts. The results clearly confirm the n-doping-induced SBW reduction in metal-contacted Re-doped MoS_2 films leading to lower R_C and enhanced carrier injection even at 10 K. (**d**,**e**) Adapted with permission from [410]. Copyright 2016 John Wiley and Sons. (**f**) Back-gated transfer characteristics for pure MoS_2 (blue curve), Cr-doped MoS_2 (red curve) and Mn-doped (green curve) MoS_2 FETs showing the relative doping effects of Cr and Mn atoms. While Cr shows an n-type doping effect, Mn shows a p-type doping effect. Adapted with permission from [418]. Copyright 2017 IOP Publishing.

15.1. Hole Doping by Cation Substitution

The first report demonstrating the in-situ CVD substitutional doping of MoS_2 was by Laskar et al. where they p-doped MoS_2 using niobium (Nb) atoms. In this approach, the Nb atoms replace the Mo cations in the MoS_2 host lattice and act as efficient electron acceptors because they have one less valence electron than the Mo atoms (note that Nb lies to the left of Mo in the periodic table). The authors showed that the crystalline nature of MoS_2 is preserved after Nb doping and reported reasonable RT mobilities (8.5 cm^2/V-s) at high hole doping densities (3.1×10^{20} cm^{-3}), and a low R_C (0.6 $\Omega \cdot mm$) for p-type conduction (since the high hole doping concentration would help reduce the SBW for hole injection, favoring hole tunneling at the contacts) [405]. A similar in-situ Nb doping approach was reported by Suh et al. where they obtained a degenerate hole density of 3×10^{19} cm^{-3} in their MoS_2 thin films and confirmed the p-doping via XPS, TEM and electrical measurements among others (Figure 19b,c show the 3D schematic of a Nb-doped MoS_2 lattice along with the XPS spectra depicting the shift of the Mo 3d core level peaks associated with p-type doping, and the electrical transfer characteristics of an MoS_2 FET before/after Nb doping, respectively). Moreover, the authors demonstrated gate-tunable current rectification in MoS_2 p-n homojunctions by combining their p-type MoS_2 films with undoped intrinsically n-type MoS_2 films [406]. A detailed study on Nb-doped p-type MoS_2 FETs was reported by Das et al. where they studied the effects of high doping concentration ($\sim 3 \times 10^{19}$ cm^{-3}) and flake thickness on the MoS_2 PFET performance, revealing important insights on the doping constraints of 2D MoS_2. They found that under heavy doping, even ultra-thin 2D semiconductors cannot be fully depleted and may behave as a 3D semiconductor when used in a FET configuration [407].

Mirabelli et al. studied the back-gated FET behavior of Nb-substituted highly p-doped 10 nm thick MoS_2 flakes and highlighted the importance of high Nb doping levels in improving the metal/MoS_2 contact resistance [408]. The hole concentration after doping was extracted to be 4.3×10^{19} cm^{-3} from Hall-effect measurements. Although the FET ON/OFF ratio was compromised due to the uniform high hole doping throughout the MoS_2 contact and channel regions (the extracted MoS_2

depletion region thickness was only 4.7 nm, much less than the 10 nm flake thickness, resulting in high OFF-state currents), the authors extracted the specific contact resistivity (ρ_C) for holes to be 1.05×10^{-7} $\Omega \cdot cm^2$ from TLM measurements which was even lower than the ρ_C for electrons as reported by English et al. (5×10^{-7} $\Omega \cdot cm^2$ using UHV metal deposition [164]) and Kang et al. (2.2×10^{-7} $\Omega \cdot cm^2$ using hybridized Mo contacts [284]), confirming that heavy doping of MoS_2 can be an effective way to drive down the ρ_C (due to SBW thinning in the contact regions). The authors further stated that such high doping levels can also be promising for realizing MoS_2-based "junctionless" FETs, the device architecture of which can help achieve both low leakage and higher immunity against short-channel effects [408]. Besides Nb, zinc (Zn) was also demonstrated to be a p-type dopant in MoS_2 by Xu et al. The authors reported the growth of mm-scale monolayer and bilayer Zn-doped MoS_2 films through a one-step CVD process wherein the Zn concentration was determined to be 1–2% via XPS analysis. Zn was found to suppress the n-type conductivity in MoS_2 FETs, and its stability and p-type accepter nature was confirmed by DFT calculations (a 2% Zn doping level was found to introduce acceptor states right above the MoS_2 VBE). Moreover, the authors showed a p-type transfer behavior in Zn-doped MoS_2 FETs upon annealing in a sulfur environment, highlighting the importance of sulfur vacancy elimination (recall that native SVs result in unintentional background n-doping of the MoS_2, as described in Section 3.1) in addition to transition-metal doping for achieving large-area p-type CVD-MoS_2 films [409].

15.2. Electron Doping by Cation Substitution

In contrast to the in-situ p-doping of MoS_2 by Mo cation substitution, Gao et al. reported the in-situ n-doping of monolayer MoS_2 by substitution of Mo with rhenium (Re) atoms. Note that Re has seven valence electrons as compared to six valence electrons in Mo and, hence, donates an extra electron to the MoS_2 lattice when substituted at the Mo atom site. The authors confirmed the n-doping via XPS and PL measurements as well as DFT calculations (Figure 19d shows the DFT-calculated band-structure of Re-doped MoS_2 depicting the presence of Re donor bands near the MoS_2 CBE and the XPS spectra depicting the shift of the Mo 3d core level peaks associated with n-type doping). Unlike the heavy or degenerate doping reported in studies of Nb-doped MoS_2, the authors could achieve non-degenerate behavior in their Re-doped MoS_2 films as demonstrated by the clear gate modulation observed in the output characteristics of back-gated Re-doped MoS_2 FETs. Moreover, in stark contrast to the undoped MoS_2 NFETs that displayed a strong non-linear Schottky-type I-V behavior at temperatures <100 K (since, at low-T, the charge carriers have insufficient thermal energy to overcome the SBH present at the contact/MoS_2 interface), the Re-doped MoS_2 NFETs displayed linear output characteristics even at temperatures as low as 10 K implying Ohmic nature of the S/D contacts (due to the doping-induced reduction of the SBW that facilitated efficient carrier injection into the MoS_2 channel via tunneling). Hence, these results clearly demonstrate the efficacy of Re doping on the n-type R_C reduction in MoS_2 FETs (Figure 19e compares the low-T output characteristics between an undoped and a Re-doped MoS_2 NFET clearly showing Schottky and Ohmic I-V behavior, respectively) [410]. Hallam et al. also demonstrated the scalable synthesis of Re-doped MoS_2 films with electron concentrations in the range of 5×10^{17} to 9×10^{17} cm^{-3} as determined via Hall effect measurements and supported by DFT calculations. In their approach, the authors used "thermally-assisted conversion" (TAC), similar to the method used by Lasker et al. to synthesize Nb-doped p-type MoS_2 films [405], to convert interleaved Mo-Re films into Re-doped MoS_2 via high-temperature sulfurization [411]. However, the electron mobility of their Re-doped MoS_2 films was low, ranging between 0.1 to 0.7 cm^2/V-s, implying the presence of various carrier scattering sources (including scattering from the incorporated Re dopant atoms).

Zhang et al. shed further light on the criticality of the substrate surface chemistry as well as the substrate-dopant reaction during the substitutional doping of MoS_2. They demonstrated the successful incorporation of manganese (Mn) atoms (up to 2 at.%) into monolayer MoS_2 supported on inert graphene substrates using in-situ vapor phase deposition and confirmed the band-structure modification of MoS_2 via PL and XPS measurements. However, the authors found that the Mn doping of MoS_2 grown on traditional substrates, such as SiO_2 and sapphire, was highly inefficient due to the reactive nature of their surfaces that caused the Mn atoms to bond with the substrate instead of being incorporated into the MoS_2 lattice [412]. Thus, the surface chemistry of the MoS_2 substrate can play an important role and must be considered while carrying out the substitutional doping process. It is worth noting that doping the MoS_2 with atoms of magnetic elements, such as paramagnetic Mn, antiferromagnetic Cr, and ferromagnetic Fe or Co, can pave the way towards realization of 2D "dilute magnetic semiconductors" having high Curie temperatures (above 300 K) as predicted by theory [413,414]. These dilute magnetic semiconductors can exhibit both ferromagnetism as well as useful semiconducting properties that can have promising implications for spintronic applications [415–417]. Hence, in a push towards this goal, several attempts to introduce magnetic dopants in 2D MoS_2 via Mo cation substitution have been carried out, along with detailed studies of the transport properties of the resultant films. Huang et al. reported in-situ Mn- and Cr-doped MoS_2 films via CVD. Detailed FET measurements revealed Mn to have a p-type doping effect (via suppression of n-type conduction), whereas Cr was found to have an enhanced n-type doping effect in MoS_2 in comparison to the undoped control sample (Figure 19f compares the back-gated transfer characteristics of undoped, Mn-doped and Cr-doped MoS_2 NFETs showing the relative doping effects of Cr and Mn). The μ_{FE} of electrons for the undoped, Mn-doped, and Cr-doped devices was extracted to be 15, 7 and 12 cm^2/V-s, respectively [418]. Wang et al. compared the properties of MoS_2 and iron-doped MoS_2 films grown by the CVT method and revealed via Hall effect measurements that although both samples were n-type, the Fe-doped MoS_2 exhibited a higher electron concentration (revealing Fe as an n-dopant in MoS_2) than the undoped MoS_2. Moreover, the RT mobilities of the undoped and Fe-doped samples were extracted to be 79 and 49 cm^2/V-s, respectively, with the lower mobility in Fe-doped MoS_2 attributed to carrier scattering by lattice imperfections and defects introduced by the Fe doping process [419]. Similarly, Li et al. synthesized large-scale cobalt-doped bilayer MoS_2 nanosheets that exhibited n-type transport behavior [420].

15.3. Electron and Hole Doping by Anion Substitution

In addition to substitution of Mo atoms, doping via substitution of the sulfur (S) anion has also been investigated for MoS_2. Yang et al. reported a novel and simple chloride-based molecular doping technique wherein the chlorine (Cl) atoms covalently attach to the Mo atoms at the sulfur vacancy (SV) sites in the MoS_2 lattice upon treatment with 1,2-dichloroethane (DCE) at RT. Since the Cl atom has an extra valence electron than the S atom (note that Cl lies to the right of S in the periodic table), it donates its extra electron to the MoS_2 lattice when substituted at the S atom sites resulting in n-type doping. Using this doping approach, an R_C as low as 500 $\Omega \cdot \mu m$ was extracted for Ni-contacted few-layer MoS_2 NFETs via TLM analysis, and the low R_C was attributed to the high electron doping density in the MoS_2 ($\sim 2.3 \times 10^{19}$ cm^{-3}) that causes increased band-bending at the contact/MoS_2 interface leading to a significant reduction of the SBW, thereby, facilitating electron tunneling. Significant improvements in the extracted transfer length (L_T) and specific contact resistivity (ρ_C) were also observed after Cl doping, with the L_T reducing from 590 nm to 60 nm and the ρ_C reducing from 3×10^{-5} to 3×10^{-7} $\Omega \cdot cm^2$ (showing two orders of magnitude improvement). Recall from the discussion in Section 3.2 that both L_T and ρ_C must be minimized to alleviate the L_C scaling issue in 2D MoS_2 FETs such that ultra-scaled FETs with low R_C can be realized. Along these lines, the authors demonstrated high-performance 100 nm channel Cl-doped MoS_2 NFETs with a high ON-current of 460 $\mu A/\mu m$, μ_{FE} of 50-60 cm^2/V-s, high ON/OFF ratio of 6.3×10^5, and long term environmental stability (Figure 20a compares the TLM-extracted R_C as well as the output characteristics of undoped and Cl-doped 100 nm channel MoS_2 NFETs, showing a much enhanced performance for the latter) [421]. Moreover, the Cl doping technique can also be applied to other 2D semiconducting TMDCs such as WS_2 [422]. For p-type substitutional doping of MoS_2 via S anion substitution, Qin et al. demonstrated the synthesis of nitrogen-doped MoS_2 nanosheets (note that N belongs to group 15 in the periodic table and has one less valence electron than S which is in group 16) using a simple and cost-effective sol-gel process utilizing molybdenum chloride (MoCl$_5$) and thiourea as the starting materials. Although no devices were realized, the authors successfully demonstrated controlled wide-range tunability of the N concentration from 5.8 at.% to 7.6 at.% simply by adjusting the ratio of MoCl$_5$ and thiourea [423]. Azcatl et al. demonstrated a different approach for the covalent nitrogen doping of MoS$_2$ employing a remote N$_2$ plasma surface treatment technique which resulted in chalcogen substitution of the S atoms by the N atoms. The N concentration could be controlled via the N$_2$ plasma exposure time, and electrical characterization of N-doped MoS$_2$ FETs revealed signs of p-doping of the MoS$_2$ (through positive shift in the V$_{th}$), consistent with theoretical predictions and XPS results (Figure 20b shows the 3D schematic of an N-doped MoS$_2$ lattice illustrating the covalent substitution of S atoms by the N atoms as well as the electrical transfer characteristics of a back-gated MoS$_2$ FET alluding to the p-type behavior induced after N doping). Moreover, the authors also reported the first-ever evidence of "compressive strain" induced in the MoS$_2$ lattice upon substitutional doping and established a correlation between the N doping concentration and the resultant compressive strain in MoS$_2$ via DFT calculations [424].

Incorporation of oxygen (O) in the MoS$_2$ lattice was also shown to cause a p-type doping effect by Neal et al. and was attributed to the formation of acceptor states, about ~214 meV above the MoS$_2$ VBE, by the incorporation of high work function MoS$_x$O$_{3-x}$ clusters in the MoS$_2$ lattice. Going against the notion that oxygen exposure only helps in passivating the SVs in MoS$_2$ leading to a decrease in the background electron concentration, thereby, indicating an apparent p-type doping effect, this work provided evidence that oxygen atoms can independently cause p-doping of the MoS$_2$ when substituted at the S atom sites [425]. A very similar work by Giannazzo et al. further confirmed the local substitutional p-doping effect of O atoms via conductive atomic force microscopy (CAFM) measurements. They used "soft" O$_2$ plasma treatments to modify the top surface of multilayer MoS$_2$ resulting in the formation of high work function MoO$_x$S$_{2-x}$ localized alloy clusters. Hence, in these localized regions, the MoS$_2$ band-structure was modified resulting in a gradual downward shift of the Fermi level towards its VBE with increasing O content, as was also verified via DFT band-structure calculations (Left plot of Figure 20d illustrates the variation of the MoS$_2$ Fermi level with respect to its VBE as a function of increasing oxygen concentration as calculated via DFT). In effect, this localized oxygen functionalization of MoS$_2$ leads to the coexistence of "microscopic" n-type doped (i.e., having a low work function and small SBH for electron injection) and p-type doped (i.e., having a high work function and small SBH for hole injection) regions within a larger "macroscopic" MoS$_2$ region (Figure 20c illustrates the extracted n-type SBH map acquired over a small section of the O-functionalized 2D MoS$_2$ surface via CAFM measurements, revealing the coexistence of both small and large n-type SBH regions, with the large n-type SBH regions essentially representing regions having a small p-type SBH for holes) [426]. Utilizing this nanoscale SBH tailoring approach, the authors demonstrated back-gated multilayer MoS$_2$ FETs with selective oxygen functionalization only in the S/D contact regions. This enabled MoS$_2$ FETs showing ambipolar operation thanks to the coexistence of small SBH regions or low resistance paths for both electrons and holes within the metal-contacted S/D regions, facilitating injection of both types of carriers into the MoS$_2$ channel while using the same contact metal. The extracted μ_{FE} values for electrons and holes were 11.5 and 7.2 cm^2/V-s, respectively (Middle schematic and right plot of Figure 20d illustrate the 3D schematic of a Ni-contacted back-gated MoS$_2$ FET with O-functionalized contact regions, and transfer characteristics of the FET showing ambipolar behavior, respectively).

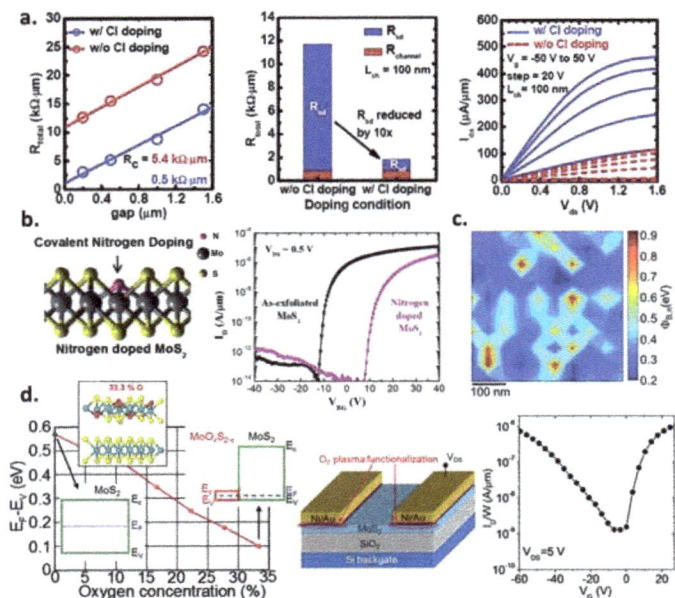

Figure 20. (**a**) **Left**: Total resistance (R_{total}) versus L_{CH} (gaps) used to extract the R_C via TLM analysis in Ni-contacted MoS$_2$ with (blue line) and without (red line) chloride doping. The extracted R_C after Cl doping is as low as 0.5 k$\Omega \cdot \mu$m. **Middle**: R_{total} of a 100 nm channel length MoS$_2$ FET for different doping conditions comparing the relative contributions of the two resistance components, R_{sd} (i.e., R_C) and $R_{channel}$. The R_{total} of the FET is reduced from 11.7 k$\Omega \cdot \mu$m to 1.85 k$\Omega \cdot \mu$m primarily due to significant reduction in the R_C (~10×) after Cl doping (recall that R_C must scale with L_{CH} to extract the intrinsic performance of MoS$_2$ FETs at ultra-short channel lengths). **Right**: Output characteristics of the 100 nm L_{CH} MoS$_2$ FET with (solid blue curves) and without (dashed red curves) Cl doping showing tremendous enhancement in the ON-current level (460 μA/μm from 100 μA/μm) after Cl doping thanks to the R_C reduction. Adapted with permission from [421]. Copyright 2014 IEEE. (**b**) **Left schematic**: Representation of the MoS$_2$ lattice showing covalent N atom (shown in pink) substitution at the S anion site. **Right**: Comparison of transfer characteristics of an as-exfoliated MoS$_2$ FET with an N-doped MoS$_2$ FET, with the latter showing a clear reduction of the n-type behavior, as reflected by the positive V_{th} shift, due to counter p-doping by the incorporated N atoms. Adapted with permission from [424]. Copyright 2016 American Chemical Society. (**c**) 2D maps of the local n-type SBH for electrons as extracted via CAFM measurements on the O$_2$ plasma-treated MoS$_2$ surface. Nanoscale clusters of large n-type SBH regions (orange red spots), representing regions with small p-type SBH for hole injection, are evident which result due to the formation of high work function MoO$_x$S$_{2-x}$ species on the topmost surface of MoS$_2$ via covalent O substitution of S atoms. (**d**) **Left**: Relative position of the Fermi level with respect to the MoS$_2$ VBE (i.e., E_F–E_V) as a function of the incorporated oxygen content in the topmost layer of MoS$_2$, determined via DFT calculations. With increasing O concentration, the Fermi level moves closer to the VBE of MoS$_2$ (giving rise to a p-type nature in these localized O-functionalized regions) along with a corresponding decrease in the band-gap (see top right inset) due to formation of high work function MoO$_x$S$_{2-x}$ clusters. The schematic at the top of the plot illustrates the multilayer MoS$_2$ lattice showing substitutional O atoms (shown in red) in its topmost layer. **Middle schematic**: Illustration of a Ni-contacted back-gated MoS$_2$ FET with selective O$_2$ plasma treatment in its S/D contact regions. **Right**: Transfer characteristics of a representative Ni-contacted MoS$_2$ FET with O-functionalized S/D contact regions showing ambipolar I-V behavior due to the coexistence of localized regions with small n- and p-type SBHs underneath the contact metal that facilitate injection of both electrons and holes in the MoS$_2$ channel, respectively. (**c,d**) Adapted with permission from [426]. Copyright 2017 American Chemical Society.

15.4. Towards Controlled and Area-Selective Substitutional Doping

A breakthrough in the controlled and area-selective p-doping of few-layer MoS_2 was reported by Nipane et al. who used a novel and CMOS-compatible plasma immersion ion implantation (PIII) process using phosphorous (P) atoms [427]. P lies to the left of S in the periodic table and, hence, acts as an acceptor in the MoS_2 lattice due to its electron deficient nature. In this method, the MoS_2 flakes were exposed to an inductively-coupled phosphine (PH_3)/He plasma inside a PIII chamber either before or after S/D contact patterning to achieve area-selective P implants (Figure 21a schematically illustrates the P ion implantation process on MoS_2 inside the PIII chamber). Various characterization techniques (Raman, AFM, XRD, etc.) were employed to identify suitable PIII processing conditions (such as implant energy and dose) to achieve low surface damage and minimal etching of the MoS_2. Back-gated FETs fabricated on P-implanted MoS_2 with varying implant energies and doses showed clear evidence of p-type conduction, ranging from non-degenerate to degenerate behavior, and the p-doping was also verified experimentally via XPS (Figure 21b,c shows the back-gated transfer characteristics for MoS_2 FETs with degenerate and non-degenerate P doping levels, and the XPS spectra depicting the shift of the Mo 3d core level peaks towards lower binding energies after p-doping, respectively). The peak μ_{FE} for holes was extracted to be 8.4 cm^2/V-s and 137.7 cm^2/V-s for the degenerate and non-degenerate doping cases, respectively. Moreover, the authors also demonstrated air-stable lateral homojunction p-n diodes with high rectification ratios (as high as 2×10^4) and good ideality factors (n ~1.2) by selectively p-doping regions of the MoS_2 (Figure 21d schematically illustrates the process used to fabricate lateral MoS_2 p-n diodes via selective P implantation and shows clear rectification in the diode I-V characteristics). Furthermore, using a rigorous DFT analysis, the authors confirmed the substitutional p-doping of MoS_2 by incorporation of P atoms at the S atom sites, and found that pre-existing SVs could enhance this doping effect by providing "empty" sites for the P atoms to latch onto (Figure 21e compares the density-of-states plots as a function of energy for pristine MoS_2 as well as P-implanted MoS_2 with and without SVs, showing the shifting of the Fermi level towards the MoS_2 valence band due to the p-doping induced by P atoms). It is worth noting that the low-damage PIII process can also be adapted to substitutionally dope monolayer MoS_2 [427,428].

Recently, another extremely promising result for substitutional doping of MoS_2 using high energy ion implantation was reported by Xu et al. wherein they directly utilized traditional ion-implanters to p-dope few-layer MoS_2 films using P atoms. In their approach, a thin layer of poly(methyl methacrylate) or PMMA resist (200 nm or 1000 nm thick) was spin-coated onto the ultrathin MoS_2 flakes (<10 nm thick) as a "protective masking layer" which helped decelerate the P dopant ions and led to the successful retention of a portion of these ions inside the 2D MoS_2 lattice (Top schematic of Figure 21f illustrates the ion implantation process for MoS_2 utilizing the PMMA mask) [429]. P ions were implanted with implantation energies ranging from 10 keV to 40 keV and a fluence of 5×10^{13} cm^{-2} (with these parameters being more in sync with commercial ion implantation processes as opposed to the low 2 keV energies used by Nipane et al. [427]). Raman, TEM and HRTEM characterization revealed negligible damage to the MoS_2 crystal structure upon removal of the PMMA layer, highlighting an advantage of this PMMA-coated ion implantation technique over the PIII process described above (where the plasma can lead to unintended etching of the 2D MoS_2 layers). The p-doping effect was further confirmed via extensive photoluminescence (PL) measurements as well as electrical characterization of back-gated MoS_2 FETs (Bottom plot of Figure 21f compares the transfer characteristics of a back-gated few-layer MoS_2 FET before and after P ion implantation in the exposed channel region, showing a positive V_{th} shift and decrease in the n-type behavior due to counter p-doping by the implanted P atoms). Although no unintentional etching of the MoS_2 occurred in this process, it was found that the combination of thinner PMMA (200 nm thick) and higher implant energy (40 keV) led to a larger kinetic damage of the MoS_2 lattice, especially in the case of mono- and bilayer films, resulting in the creation of SVs since bombardment of high velocity P atoms can "knock-off" the S atoms from the lattice. Thus, the p-doping effect due to P-implantation can be counterbalanced by the n-doping effect due to creation of these SVs. Further optimization is needed to determine a synergistic relationship between the PMMA mask thickness, MoS$_2$ layer thickness, and the implantation energy to achieve controlled doping profiles without any kinetic damage to the MoS$_2$ lattice [429]. Nonetheless, this ion implantation technique is extremely promising for achieving large-scale, controlled and area-selective doping of MoS$_2$ (both n- and p-type) and is compatible with existing infrastructures in the semiconductor industry.

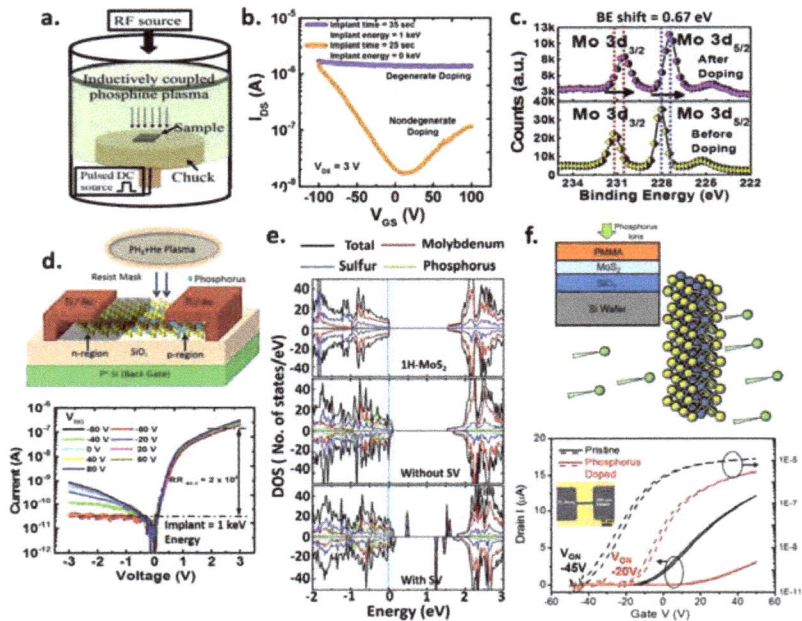

Figure 21. (**a**) Schematic illustration of the P ion implantation process on MoS$_2$ inside the plasma immersion ion implantation (PIII) chamber. (**b**) Transfer characteristics for degenerately (purple curve) and non-degenerately (orange curve) P-doped back-gated MoS$_2$ FETs showing enhanced p-type behavior. Variation of the implant time and energy can help achieve doping controllability in thin MoS$_2$ layers using the PIII process. (**c**) XPS spectra of the Mo 3d core level peaks acquired from the MoS$_2$ film before (yellow) and after (pink) P ion implantation. The peaks show a shift towards lower binding energies after P-doping confirming the downshift of the Fermi level towards the VBE of MoS$_2$. (**d**) **Top schematic**: 3D illustration of the P implantation process used to fabricate lateral MoS$_2$ p-n homojunction diodes. Half the MoS$_2$ channel is masked with a resist to ensure selective p-doping only in the exposed MoS$_2$ regions. **Bottom**: I-V characteristics of a lateral MoS$_2$ p-n homojunction diode at varying back gate biases exhibiting high rectification ratios (up to ~2 × 10^4). (**e**) The density-of-states plots for pristine (**top panel**) and P-implanted MoS$_2$ with (**bottom panel**) and without (**middle panel**) SVs as calculated via DFT. The Fermi level is represented by the vertical line within the faint blue-shaded region. As is evident, the Fermi level shifts towards the MoS$_2$ VBE (overlapping the valence band states) with the incorporation of P atoms at the S atom sites in the MoS$_2$ lattice indicating p-type doping. This shift is even more pronounced in the presence of SVs that serve to enhance the p-doping by providing empty sites for the P atoms. (**a–e**) Adapted with permission from [427]. Copyright 2016 American Chemical Society. (**f**) **Top schematic**: Cross-sectional representation of the PMMA-assisted ion implantation process used for implanting P atoms in the MoS$_2$ lattice. **Bottom**: Comparison of transfer characteristics of a back-gated 5-layer MoS$_2$ FET before (black curves) and after (red curves) P implantation. The inset shows the optical micrograph of the device. After P implantation, the n-type behavior reduces as depicted by the positive shift of the V$_{th}$ due to counter p-doping by the incorporated P atoms. Adapted with permission from [429]. Copyright 2017 IOP Publishing.

16. Conclusions and Future Outlook

Atomically thin semiconducting MoS_2 indeed holds great promise for use as a transistor channel material and can be advantageous for a wide variety of electronic and optoelectronic device applications. The material and device performance projections for MoS_2 certainly seem to give it an edge over conventional bulk semiconductors in ultra-scaled future technology nodes. Moreover, as an ultra-thin, flexible and transparent material, MoS_2 can change the status quo in flexible nanoelectronics and thin-film transistor technologies. However, the promising advantages of MoS_2 can only be utilized to the fullest once several key performance bottlenecks are mitigated. As highlighted in this review, the challenges associated with contact resistance, doping and mobility engineering are of paramount importance and these parameters must be carefully engineered to extract the maximum efficiency from MoS_2-based devices and to make any MoS_2-based technology commercially viable. This review presents a comprehensive overview of a whole host of engineering solutions, reported to date, to mitigate these challenges. Moving forward, the right mix of the most promising and cost-effective techniques must be adopted and further optimized, ensuring their robustness for use on both rigid and flexible platforms. For reducing contact resistance, use of techniques such as phase-engineered 2H/1T metallic contacts, gate-tunable graphene contacts in conjunction with interfacial contact "tunnel" barriers (ultra-thin oxides, 2D hBN etc.), doping via substoichiometric high-κ oxides and 2D/2D Ohmic contacts employing degenerate substitutional doping in the MoS_2 contact regions seem as promising approaches to effectively eliminate the Schottky barrier issue. Concomitantly, the doping selectivity and controllability are extremely crucial to simultaneously achieve both degenerately doped Ohmic S/D contacts as well as non-degenerately doped channel regions with tailored electrical properties (for both n- and p-channel devices) that can effectively be modulated by the gate. In this regard, and to enable efficient MoS_2-based NMOS and PMOS complementary devices, more research effort needs to be devoted to further optimizing the substitutional doping techniques for MoS_2 that have already made a promising start. The layer thickness of MoS_2 and contact architecture scheme can be chosen as per the given application. Few-layer MoS_2, with simultaneous charge injection into all its constituent layers, seems most promising for pure electronic applications as it can afford the maximum carrier mobilities and performance. Optoelectronic applications requiring direct band-gap monolayer MoS_2 pose more stringent challenges due to the extreme susceptibility of the atomically thin MoS_2 body to environmental perturbations. However, these challenges can be met with the right choice of dielectric encapsulation, in conjunction with optimized contact and doping engineering techniques, to ensure maximum performance. Besides all the promising applications MoS_2 can enable all by itself (in the form of ultra-scaled transistors, homojunction devices, etc.), it can also be combined with several other 3D or 2D materials to form various van der Waals heterostructures enabling a wide variety of device applications. Finally, the large-area wafer-scale growth of MoS_2 and its wafer-scale device fabrication techniques must also be co-optimized with special emphasis on producing ultra-clean material and devices, ensuring negligible impurities and defects. Everything considered, the field of 2D atomically thin semiconductors holds great potential for the future and with the current pace of research progress, 2D MoS_2-based commercial applications could soon become a reality.

Author Contributions: A.R. planned and wrote the entire manuscript with contributions from all authors. All authors have given approval to the final version of the manuscript.

Funding: This work was supported by the NASCENT Engineering Research Center (ERC) funded by the National Science Foundation (NSF) (Grant No. EEC-1160494).

Conflicts of Interest: The authors declare no competing financial interest.

References

1. Novoselov, K.S.; Geim, A.K.; Morozov, S.V.; Jiang, D.; Zhang, Y.; Dubonos, S.V.; Grigorieva, I.V.; Firsov, A.A. Electric Field Effect in Atomically Thin Carbon Films. *Science* **2004**, *306*, 666–669. [CrossRef] [PubMed]

2. Moore, G.E. No exponential is forever: But "Forever" can be delayed! [semiconductor industry]. In Proceedings of the 2003 IEEE International Solid-State Circuits Conference, San Francisco, CA, USA, 13 February 2003; Volume 1, pp. 20–23. [CrossRef]

3. Schaller, R.R. Moore's law: Past, present and future. *IEEE Spectr.* **1997**, *34*, 52–59. [CrossRef]

4. Moore, G.E. Cramming more components onto integrated circuits, Reprinted from Electronics, volume 38, number 8, April 19, 1965, pp. 114 ff. *IEEE Solid-State Circuits Soc. Newsl.* **2006**, *11*, 33–35. [CrossRef]

5. Mistry, K. A 45 nm logic technology with high-κ/metal gate transistors, strained silicon, 9 Cu interconnect layers. 193 nm dry patteming, and 100%Pb-free packaging. In Proceedings of the IEEE International Electron Devices Meeting (IEDM), Washington, DC, USA, 10–12 December 2007; pp. 247–250. [CrossRef]

6. Davari, B.; Dennard, R.H.; Shahidi, G.G. CMOS scaling for high performance and low power-the next ten years. *Proc. IEEE* **1995**, *83*, 595–606. [CrossRef]

7. Cartwright, J. Intel enters the third dimension. *Nat. News* **2011**. [CrossRef]

8. Auth, C.; Allen, C.; Blattner, A.; Bergstrom, D.; Brazier, M.; Bost, M.; Buehler, M.; Chikarmane, V.; Ghani, T.; Glassman, T.; et al. A 22 nm high performance and low-power CMOS technology featuring fully-depleted tri-gate transistors, self-aligned contacts and high density MIM capacitors. In Proceedings of the 2012 Symposium on VLSI Technology (VLSIT), Honolulu, HI, USA, 12–14 June 2012; pp. 131–132. [CrossRef]

9. Thompson, S.E.; Parthasarathy, S. Moore's law: The future of Si microelectronics. *Mater. Today* **2006**, *9*, 20–25. [CrossRef]

10. Alamo, J.A. del Nanometre-scale electronics with III–V compound semiconductors. *Nature* **2011**, *479*, 317. [CrossRef] [PubMed]

11. Waldrop, M.M. The chips are down for Moore's law. *Nat. News* **2016**, *530*, 144. [CrossRef] [PubMed]

12. Thompson, S. Mos scaling: Transistor challenges for the 21st century. *Intel Technol. J.* **1998**, *Q3*, 1–19.

13. Haron, N.Z.; Hamdioui, S. Why is CMOS scaling coming to an END? In Proceedings of the 2008 3rd International Design and Test Workshop, Monastir, Tunisia, 20–22 December 2008; pp. 98–103. [CrossRef]

14. Frank, D.J. Power-constrained CMOS scaling limits. *IBM J. Res. Dev.* **2002**, *46*, 235–244. [CrossRef]

15. Yong-Bin, K. Challenges for Nanoscale MOSFETs and Emerging Nanoelectronics. *Trans. Electr. Electron. Mater.* **2010**, *11*, 93–105. [CrossRef]

16. Horowitz, M.; Alon, E.; Patil, D.; Naffziger, S.; Kumar, R.; Bernstein, K. Scaling, power, and the future of CMOS. In Proceedings of the IEEE International Electron Devices Meeting, Washington, DC, USA, 5 December 2005; pp. 7–15. [CrossRef]

17. Kuhn, K.J. Considerations for Ultimate CMOS Scaling. *IEEE Trans. Electron Devices* **2012**, *59*, 1813–1828. [CrossRef]

18. Skotnicki, T.; Hutchby, J.A.; King, T.-J.; Wong, H.S.P.; Boeuf, F. The end of CMOS scaling: Toward the introduction of new materials and structural changes to improve MOSFET performance. *IEEE Circuits Devices Mag.* **2005**, *21*, 16–26. [CrossRef]

19. Cavin, R.K.; Lugli, P.; Zhirnov, V.V. Science and Engineering Beyond Moore's Law. *Proc. IEEE* **2012**, *100*, 1720–1749. [CrossRef]

20. Kuhn, K.J.; Avci, U.; Cappellani, A.; Giles, M.D.; Haverty, M.; Kim, S.; Kotlyar, R.; Manipatruni, S.; Nikonov, D.; Pawashe, C.; et al. The ultimate CMOS device and beyond. In Proceedings of the 2012 International Electron Devices Meeting, San Francisco, CA, USA, 10–13 December 2012; pp. 8.1.1–8.1.4. [CrossRef]

21. Geim, A.K. Graphene: Status and Prospects. *Science* **2009**, *324*, 1530–1534. [CrossRef] [PubMed]

22. Zhu, Y.; Murali, S.; Cai, W.; Li, X.; Suk, J.W.; Potts, J.R.; Ruoff, R.S. Graphene and Graphene Oxide: Synthesis, Properties, and Applications. *Adv. Mater.* **2010**, *22*, 3906–3924. [CrossRef] [PubMed]

23. Novoselov, K.S.; Fal'ko, V.I.; Colombo, L.; Gellert, P.R.; Schwab, M.G.; Kim, K. A roadmap for graphene. *Nature* **2012**, *490*, 192. [CrossRef] [PubMed]

24. Schwierz, F. Graphene transistors. *Nat. Nanotechnol.* **2010**, *5*, 487. [CrossRef] [PubMed]

25. Schwierz, F. Graphene Transistors: Status, Prospects, and Problems. *Proc. IEEE* **2013**, *101*, 1567–1584. [CrossRef]

26. Mas-Ballesté, R.; Gómez-Navarro, C.; Gómez-Herrero, J.; Zamora, F. 2D materials: To graphene and beyond. *Nanoscale* **2011**, *3*, 20–30. [CrossRef] [PubMed]

27. Xu, M.; Liang, T.; Shi, M.; Chen, H. Graphene-Like Two-Dimensional Materials. *Chem. Rev.* **2013**, *113*, 3766–3798. [CrossRef] [PubMed]

28. Gibney, E. The super materials that could trump graphene. *Nat. News* **2015**, *522*, 274. [CrossRef] [PubMed]
29. Chhowalla, M.; Shin, H.S.; Eda, G.; Li, L.-J.; Loh, K.P.; Zhang, H. The chemistry of two-dimensional layered transition metal dichalcogenide nanosheets. *Nat. Chem.* **2013**, *5*, 263. [CrossRef] [PubMed]
30. Wang, Q.H.; Kalantar-Zadeh, K.; Kis, A.; Coleman, J.N.; Strano, M.S. Electronics and optoelectronics of two-dimensional transition metal dichalcogenides. *Nat. Nanotechnol.* **2012**, *7*, 699–712. [CrossRef] [PubMed]
31. Das, S.; Robinson, J.A.; Dubey, M.; Terrones, H.; Terrones, M. Beyond Graphene: Progress in Novel Two-Dimensional Materials and van der Waals Solids. *Annu. Rev. Mater. Res.* **2015**, *45*. [CrossRef]
32. Gong, C.; Zhang, H.; Wang, W.; Colombo, L.; Wallace, R.M.; Cho, K. Band alignment of two-dimensional transition metal dichalcogenides: Application in tunnel field effect transistors. *Appl. Phys. Lett.* **2013**, *103*, 053513. [CrossRef]
33. McDonnell, S.J.; Wallace, R.M. Atomically-thin layered films for device applications based upon 2D TMDC materials. *Thin Solid Films* **2016**, *616*, 482–501. [CrossRef]
34. Kang, J.; Cao, W.; Xie, X.; Sarkar, D.; Liu, W.; Banerjee, K. Graphene and beyond-graphene 2D crystals for next-generation green electronics. In *Micro-and Nanotechnology Sensors, Systems, and Applications VI*; International Society for Optics and Photonics: Baltimore, MD, USA, 2014; Volume 9083, p. 908305. [CrossRef]
35. Chhowalla, M.; Jena, D.; Zhang, H. Two-dimensional semiconductors for transistors. *Nat. Rev. Mater.* **2016**, *1*, 16052. [CrossRef]
36. Fiori, G.; Bonaccorso, F.; Iannaccone, G.; Palacios, T.; Neumaier, D.; Seabaugh, A.; Banerjee, S.K.; Colombo, L. Electronics based on two-dimensional materials. *Nat. Nanotechnol.* **2014**, *9*, 768. [CrossRef] [PubMed]
37. Jariwala, D.; Sangwan, V.K.; Lauhon, L.J.; Marks, T.J.; Hersam, M.C. Emerging Device Applications for Semiconducting Two-Dimensional Transition Metal Dichalcogenides. *ACS Nano* **2014**, *8*, 1102–1120. [CrossRef] [PubMed]
38. Bhimanapati, G.R.; Lin, Z.; Meunier, V.; Jung, Y.; Cha, J.; Das, S.; Xiao, D.; Son, Y.; Strano, M.S.; Cooper, V.R.; et al. Recent Advances in Two-Dimensional Materials beyond Graphene. *ACS Nano* **2015**, *9*, 11509–11539. [CrossRef] [PubMed]
39. Schwierz, F.; Pezoldt, J.; Granzner, R. Two-dimensional materials and their prospects in transistor electronics. *Nanoscale* **2015**, *7*, 8261–8283. [CrossRef] [PubMed]
40. Radisavljevic, B.; Radenovic, A.; Brivio, J.; Giacometti, V.; Kis, A. Single-layer MoS_2 transistors. *Nat. Nanotechnol.* **2011**, *6*, 147–150. [CrossRef] [PubMed]
41. Splendiani, A.; Sun, L.; Zhang, Y.; Li, T.; Kim, J.; Chim, C.-Y.; Galli, G.; Wang, F. Emerging Photoluminescence in Monolayer MoS_2. *Nano Lett.* **2010**, *10*, 1271–1275. [CrossRef] [PubMed]
42. Ellis, J.K.; Lucero, M.J.; Scuseria, G.E. The indirect to direct band gap transition in multilayered MoS_2 as predicted by screened hybrid density functional theory. *Appl. Phys. Lett.* **2011**, *99*, 261908. [CrossRef]
43. Cheiwchanchamnangij, T.; Lambrecht, W.R.L. Quasiparticle band structure calculation of monolayer, bilayer, and bulk MoS_2. *Phys. Rev. B* **2012**, *85*, 205302. [CrossRef]
44. Kuc, A.; Zibouche, N.; Heine, T. Influence of quantum confinement on the electronic structure of the transition metal sulfide TS_2. *Phys. Rev. B* **2011**, *83*, 245213. [CrossRef]
45. Schwierz, F. Nanoelectronics: Flat transistors get off the ground. *Nat. Nanotechnol.* **2011**, *6*, 135. [CrossRef] [PubMed]
46. Lembke, D.; Bertolazzi, S.; Kis, A. Single-Layer MoS_2 Electronics. *Acc. Chem. Res.* **2015**, *48*, 100–110. [CrossRef] [PubMed]
47. Ganatra, R.; Zhang, Q. Few-Layer MoS_2: A Promising Layered Semiconductor. *ACS Nano* **2014**, *8*, 4074–4099. [CrossRef] [PubMed]
48. Venkata Subbaiah, Y.P.; Saji, K.J.; Tiwari, A. Atomically Thin MoS_2: A Versatile Nongraphene 2D Material. *Adv. Funct. Mater.* **2016**, *26*, 2046–2069. [CrossRef]
49. Wang, F.; Wang, Z.; Jiang, C.; Yin, L.; Cheng, R.; Zhan, X.; Xu, K.; Wang, F.; Zhang, Y.; He, J. Progress on Electronic and Optoelectronic Devices of 2D Layered Semiconducting Materials. *Small* **2017**, *13*, 1604298. [CrossRef] [PubMed]
50. Tan, C.; Cao, X.; Wu, X.-J.; He, Q.; Yang, J.; Zhang, X.; Chen, J.; Zhao, W.; Han, S.; Nam, G.-H.; et al. Recent Advances in Ultrathin Two-Dimensional Nanomaterials. *Chem. Rev.* **2017**, *117*, 6225–6331. [CrossRef] [PubMed]

51. Cao, W.; Kang, J.; Liu, W.; Khatami, Y.; Sarkar, D.; Banerjee, K. 2D electronics: Graphene and beyond. In Proceedings of the 2013 Proceedings of the European Solid-State Device Research Conference (ESSDERC), Bucharest, Romania, 16–20 September 2013; pp. 37–44. [CrossRef]

52. Liu, Y.; Weiss, N.O.; Duan, X.; Cheng, H.-C.; Huang, Y.; Duan, X. Van der Waals heterostructures and devices. *Nat. Rev. Mater.* **2016**, *1*, 16042. [CrossRef]

53. Novoselov, K.S.; Neto, A.H.C. Two-dimensional crystals-based heterostructures: Materials with tailored properties. *Phys. Scr.* **2012**, *2012*, 014006. [CrossRef]

54. Geim, A.K.; Grigorieva, I.V. Van der Waals heterostructures. *Nature* **2013**, *499*, 419. [CrossRef] [PubMed]

55. Lotsch, B.V. Vertical 2D Heterostructures. *Annu. Rev. Mater. Res.* **2015**, *45*, 85–109. [CrossRef]

56. Zhang, W.; Wang, Q.; Chen, Y.; Wang, Z.; Wee, A.T.S. Van der Waals stacked 2D layered materials for optoelectronics. *2D Mater.* **2016**, *3*, 022001. [CrossRef]

57. Novoselov, K.S.; Mishchenko, A.; Carvalho, A.; Neto, A.H.C. 2D materials and van der Waals heterostructures. *Science* **2016**, *353*, aac9439. [CrossRef] [PubMed]

58. Hamann, D.M.; Hadland, E.C.; Johnson, D.C. Heterostructures containing dichalcogenides-new materials with predictable nanoarchitectures and novel emergent properties. *Semicond. Sci. Technol.* **2017**, *32*, 093004. [CrossRef]

59. Jariwala, D.; Marks, T.J.; Hersam, M.C. Mixed-dimensional van der Waals heterostructures. *Nat. Mater.* **2017**, *16*, 170. [CrossRef] [PubMed]

60. Liu, H.; Neal, A.T.; Ye, P.D. Channel Length Scaling of MoS$_2$ MOSFETs. *ACS Nano* **2012**, *6*, 8563–8569. [CrossRef] [PubMed]

61. Liu, Y.; Guo, J.; Wu, Y.; Zhu, E.; Weiss, N.O.; He, Q.; Wu, H.; Cheng, H.-C.; Xu, Y.; Shakir, I.; et al. Pushing the Performance Limit of Sub-100 nm Molybdenum Disulfide Transistors. *Nano Lett.* **2016**, *16*, 6337–6342. [CrossRef] [PubMed]

62. Nourbakhsh, A.; Zubair, A.; Sajjad, R.N.; Tavakkoli, A.K.G.; Chen, W.; Fang, S.; Ling, X.; Kong, J.; Dresselhaus, M.S.; Kaxiras, E.; et al. MoS$_2$ Field-Effect Transistor with Sub-10 nm Channel Length. *Nano Lett.* **2016**, *16*, 7798–7806. [CrossRef] [PubMed]

63. Desai, S.B.; Madhvapathy, S.R.; Sachid, A.B.; Llinas, J.P.; Wang, Q.; Ahn, G.H.; Pitner, G.; Kim, M.J.; Bokor, J.; Hu, C.; et al. MoS$_2$ transistors with 1-nanometer gate lengths. *Science* **2016**, *354*, 99–102. [CrossRef] [PubMed]

64. Radisavljevic, B.; Whitwick, M.B.; Kis, A. Integrated Circuits and Logic Operations Based on Single-Layer MoS$_2$. *ACS Nano* **2011**, *5*, 9934–9938. [CrossRef] [PubMed]

65. Wang, H.; Yu, L.; Lee, Y.-H.; Shi, Y.; Hsu, A.; Chin, M.L.; Li, L.-J.; Dubey, M.; Kong, J.; Palacios, T. Integrated Circuits Based on Bilayer MoS$_2$ Transistors. *Nano Lett.* **2012**, *12*, 4674–4680. [CrossRef] [PubMed]

66. Yu, L.; Lee, Y.-H.; Ling, X.; Santos, E.J.G.; Shin, Y.C.; Lin, Y.; Dubey, M.; Kaxiras, E.; Kong, J.; Wang, H.; et al. Graphene/MoS$_2$ Hybrid Technology for Large-Scale Two-Dimensional Electronics. *Nano Lett.* **2014**, *14*, 3055–3063. [CrossRef] [PubMed]

67. Sachid, A.B.; Tosun, M.; Desai, S.B.; Hsu, C.-Y.; Lien, D.-H.; Madhvapathy, S.R.; Chen, Y.-Z.; Hettick, M.; Kang, J.S.; Zeng, Y.; et al. Monolithic 3D CMOS Using Layered Semiconductors. *Adv. Mater.* **2016**, *28*, 2547–2554. [CrossRef] [PubMed]

68. Bertolazzi, S.; Krasnozhon, D.; Kis, A. Nonvolatile Memory Cells Based on MoS$_2$/Graphene Heterostructures. *ACS Nano* **2013**, *7*, 3246–3252. [CrossRef] [PubMed]

69. Kshirsagar, C.U.; Xu, W.; Su, Y.; Robbins, M.C.; Kim, C.H.; Koester, S.J. Dynamic Memory Cells Using MoS$_2$ Field-Effect Transistors Demonstrating Femtoampere Leakage Currents. *ACS Nano* **2016**, *10*, 8457–8464. [CrossRef] [PubMed]

70. Zhang, E.; Wang, W.; Zhang, C.; Jin, Y.; Zhu, G.; Sun, Q.; Zhang, D.W.; Zhou, P.; Xiu, F. Tunable Charge-Trap Memory Based on Few-Layer MoS$_2$. *ACS Nano* **2015**, *9*, 612–619. [CrossRef] [PubMed]

71. Lee, D.; Kim, S.; Kim, Y.; Cho, J.H. One-Transistor–One-Transistor (1T1T) Optoelectronic Nonvolatile MoS$_2$ Memory Cell with Nondestructive Read-Out. *ACS Appl. Mater. Interfaces* **2017**, *9*, 26357–26362. [CrossRef] [PubMed]

72. Radisavljevic, B.; Whitwick, M.B.; Kis, A. Small-signal amplifier based on single-layer MoS$_2$. *Appl. Phys. Lett.* **2012**, *101*, 043103. [CrossRef]

73. Sanne, A.; Ghosh, R.; Rai, A.; Yogeesh, M.N.; Shin, S.H.; Sharma, A.; Jarvis, K.; Mathew, L.; Rao, R.; Akinwande, D.; et al. Radio Frequency Transistors and Circuits Based on CVD MoS$_2$. *Nano Lett.* **2015**, *15*, 5039–5045. [CrossRef] [PubMed]

74. Chang, H.-Y.; Yogeesh, M.N.; Ghosh, R.; Rai, A.; Sanne, A.; Yang, S.; Lu, N.; Banerjee, S.K.; Akinwande, D. Large-Area Monolayer MoS₂ for Flexible Low-Power RF Nanoelectronics in the GHz Regime. *Adv. Mater.* **2016**, *28*, 1818–1823. [CrossRef] [PubMed]

75. Sanne, A.; Park, S.; Ghosh, R.; Yogeesh, M.N.; Liu, C.; Mathew, L.; Rao, R.; Akinwande, D.; Banerjee, S.K. Embedded gate CVD MoS₂ microwave FETs. *Npj 2D Mater. Appl.* **2017**, *1*, 26. [CrossRef]

76. Choi, M.S.; Qu, D.; Lee, D.; Liu, X.; Watanabe, K.; Taniguchi, T.; Yoo, W.J. Lateral MoS₂ p–n Junction Formed by Chemical Doping for Use in High-Performance Optoelectronics. *ACS Nano* **2014**, *8*, 9332–9340. [CrossRef] [PubMed]

77. Deng, Y.; Luo, Z.; Conrad, N.J.; Liu, H.; Gong, Y.; Najmaei, S.; Ajayan, P.M.; Lou, J.; Xu, X.; Ye, P.D. Black Phosphorus–Monolayer MoS₂ van der Waals Heterojunction p–n Diode. *ACS Nano* **2014**, *8*, 8292–8299. [CrossRef] [PubMed]

78. Jariwala, D.; Sangwan, V.K.; Wu, C.-C.; Prabhumirashi, P.L.; Geier, M.L.; Marks, T.J.; Lauhon, L.J.; Hersam, M.C. Gate-tunable carbon nanotube–MoS₂ heterojunction p-n diode. *Proc. Natl. Acad. Sci. USA* **2013**, *110*, 18076–18080. [CrossRef] [PubMed]

79. Lee, E.W.; Lee, C.H.; Paul, P.K.; Ma, L.; McCulloch, W.D.; Krishnamoorthy, S.; Wu, Y.; Arehart, A.R.; Rajan, S. Layer-transferred MoS₂/GaN PN diodes. *Appl. Phys. Lett.* **2015**, *107*, 103505. [CrossRef]

80. Lopez-Sanchez, O.; Lembke, D.; Kayci, M.; Radenovic, A.; Kis, A. Ultrasensitive photodetectors based on monolayer MoS₂. *Nat. Nanotechnol.* **2013**, *8*, 497–501. [CrossRef] [PubMed]

81. Yu, S.H.; Lee, Y.; Jang, S.K.; Kang, J.; Jeon, J.; Lee, C.; Lee, J.Y.; Kim, H.; Hwang, E.; Lee, S.; et al. Dye-Sensitized MoS₂ Photodetector with Enhanced Spectral Photoresponse. *ACS Nano* **2014**, *8*, 8285–8291. [CrossRef] [PubMed]

82. De Fazio, D.; Goykhman, I.; Yoon, D.; Bruna, M.; Eiden, A.; Milana, S.; Sassi, U.; Barbone, M.; Dumcenco, D.; Marinov, K.; et al. High Responsivity, Large-Area Graphene/MoS₂ Flexible Photodetectors. *ACS Nano* **2016**, *10*, 8252–8262. [CrossRef] [PubMed]

83. Kufer, D.; Konstantatos, G. Highly Sensitive, Encapsulated MoS₂ Photodetector with Gate Controllable Gain and Speed. *Nano Lett.* **2015**, *15*, 7307–7313. [CrossRef] [PubMed]

84. Sundaram, R.S.; Engel, M.; Lombardo, A.; Krupke, R.; Ferrari, A.C.; Avouris, P.; Steiner, M. Electroluminescence in Single Layer MoS₂. *Nano Lett.* **2013**, *13*, 1416–1421. [CrossRef] [PubMed]

85. Lopez-Sanchez, O.; Alarcon Llado, E.; Koman, V.; Fontcuberta i Morral, A.; Radenovic, A.; Kis, A. Light Generation and Harvesting in a van der Waals Heterostructure. *ACS Nano* **2014**, *8*, 3042–3048. [CrossRef] [PubMed]

86. Withers, F.; Pozo-Zamudio, O.D.; Mishchenko, A.; Rooney, A.P.; Gholinia, A.; Watanabe, K.; Taniguchi, T.; Haigh, S.J.; Geim, A.K.; Tartakovskii, A.I.; et al. Light-emitting diodes by band-structure engineering in van der Waals heterostructures. *Nat. Mater.* **2015**, *14*, 301. [CrossRef] [PubMed]

87. Cheng, R.; Li, D.; Zhou, H.; Wang, C.; Yin, A.; Jiang, S.; Liu, Y.; Chen, Y.; Huang, Y.; Duan, X. Electroluminescence and Photocurrent Generation from Atomically Sharp WSe₂/MoS₂ Heterojunction p–n Diodes. *Nano Lett.* **2014**, *14*, 5590–5597. [CrossRef] [PubMed]

88. Salehzadeh, O.; Djavid, M.; Tran, N.H.; Shih, I.; Mi, Z. Optically Pumped Two-Dimensional MoS₂ Lasers Operating at Room-Temperature. *Nano Lett.* **2015**, *15*, 5302–5306. [CrossRef] [PubMed]

89. Woodward, R.I.; Howe, R.C.T.; Hu, G.; Torrisi, F.; Zhang, M.; Hasan, T.; Kelleher, E.J.R. Few-layer MoS₂ saturable absorbers for short-pulse laser technology: Current status and future perspectives. *Photonics Res.* **2015**, *3*, A30–A42. [CrossRef]

90. Fontana, M.; Deppe, T.; Boyd, A.K.; Rinzan, M.; Liu, A.Y.; Paranjape, M.; Barbara, P. Electron-hole transport and photovoltaic effect in gated MoS₂ Schottky junctions. *Sci. Rep.* **2013**, *3*. [CrossRef] [PubMed]

91. Tsai, M.-L.; Su, S.-H.; Chang, J.-K.; Tsai, D.-S.; Chen, C.-H.; Wu, C.-I.; Li, L.-J.; Chen, L.-J.; He, J.-H. Monolayer MoS₂ Heterojunction Solar Cells. *ACS Nano* **2014**, *8*, 8317–8322. [CrossRef] [PubMed]

92. Hao, L.Z.; Gao, W.; Liu, Y.J.; Han, Z.D.; Xue, Q.Z.; Guo, W.Y.; Zhu, J.; Li, Y.R. High-performance n-MoS2/i-SiO2/p-Si heterojunction solar cells. *Nanoscale* **2015**, *7*, 8304–8308. [CrossRef] [PubMed]

93. Ur Rehman, A.; Khan, M.F.; Shehzad, M.A.; Hussain, S.; Bhopal, M.F.; Lee, S.H.; Eom, J.; Seo, Y.; Jung, J.; Lee, S.H. n-MoS₂/p-Si Solar Cells with Al2O3 Passivation for Enhanced Photogeneration. *ACS Appl. Mater. Interfaces* **2016**, *8*, 29383–29390. [CrossRef] [PubMed]

94. Perkins, F.K.; Friedman, A.L.; Cobas, E.; Campbell, P.M.; Jernigan, G.G.; Jonker, B.T. Chemical Vapor Sensing with Monolayer MoS₂. *Nano Lett.* **2013**, *13*, 668–673. [CrossRef] [PubMed]

95. Cho, B.; Hahm, M.G.; Choi, M.; Yoon, J.; Kim, A.R.; Lee, Y.-J.; Park, S.-G.; Kwon, J.-D.; Kim, C.S.; Song, M.; et al. Charge-transfer-based Gas Sensing Using Atomic-layer MoS$_2$. *Sci. Rep.* **2015**, *5*, 8052. [CrossRef] [PubMed]

96. Wang, L.; Wang, Y.; Wong, J.I.; Palacios, T.; Kong, J.; Yang, H.Y. Functionalized MoS$_2$ Nanosheet-Based Field-Effect Biosensor for Label-Free Sensitive Detection of Cancer Marker Proteins in Solution. *Small* **2014**, *10*, 1101–1105. [CrossRef] [PubMed]

97. Park, M.; Park, Y.J.; Chen, X.; Park, Y.-K.; Kim, M.-S.; Ahn, J.-H. MoS$_2$-Based Tactile Sensor for Electronic Skin Applications. *Adv. Mater.* **2016**, *28*, 2556–2562. [CrossRef] [PubMed]

98. Roy, T.; Tosun, M.; Cao, X.; Fang, H.; Lien, D.-H.; Zhao, P.; Chen, Y.-Z.; Chueh, Y.-L.; Guo, J.; Javey, A. Dual-Gated MoS$_2$/WSe$_2$ van der Waals Tunnel Diodes and Transistors. *ACS Nano* **2015**, *9*, 2071–2079. [CrossRef] [PubMed]

99. Krishnamoorthy, S.; Lee, E.W.; Lee, C.H.; Zhang, Y.; McCulloch, W.D.; Johnson, J.M.; Hwang, J.; Wu, Y.; Rajan, S. High current density 2D/3D MoS$_2$/GaN Esaki tunnel diodes. *Appl. Phys. Lett.* **2016**, *109*, 183505. [CrossRef]

100. Movva, H.C.P.; Kang, S.; Rai, A.; Kim, K.; Fallahazad, B.; Taniguchi, T.; Watanabe, K.; Tutuc, E.; Banerjee, S.K. Room temperature gate-tunable negative differential resistance in MoS$_2$/hBN/WSe$_2$ heterostructures. In Proceedings of the 2016 74th Device Research Conference (DRC), Newark, DE, USA, 19–22 June 2016; pp. 1–2. [CrossRef]

101. Balaji, Y.; Smets, Q.; de la Rosa, C.J.L.; Lu, A.K.A.; Chiappe, D.; Agarwal, T.; Lin, D.; Huyghebaert, C.; Radu, I.; Mocuta, D.; et al. Tunneling transistors based on MoS$_2$/MoTe$_2$ Van der Waals heterostructures. In Proceedings of the 2017 47th European Solid-State Device Research Conference (ESSDERC), Leuven, Belgium, 11–14 September 2017; pp. 106–109. [CrossRef]

102. Zhu, H.; Wang, Y.; Xiao, J.; Liu, M.; Xiong, S.; Wong, Z.J.; Ye, Z.; Ye, Y.; Yin, X.; Zhang, X. Observation of piezoelectricity in free-standing monolayer MoS$_2$. *Nat. Nanotechnol.* **2015**, *10*, 151. [CrossRef] [PubMed]

103. Wu, W.; Wang, L.; Li, Y.; Zhang, F.; Lin, L.; Niu, S.; Chenet, D.; Zhang, X.; Hao, Y.; Heinz, T.F.; et al. Piezoelectricity of single-atomic-layer MoS$_2$ for energy conversion and piezotronics. *Nature* **2014**, *514*, 470. [CrossRef] [PubMed]

104. Liu, H.F.; Wong, S.L.; Chi, D.Z. CVD Growth of MoS$_2$-based Two-dimensional Materials. *Chem. Vap. Depos.* **2015**, *21*, 241–259. [CrossRef]

105. Walsh, L.A.; Hinkle, C.L. Van der Waals epitaxy: 2D materials and topological insulators. *Appl. Mater. Today* **2017**, *9*, 504–515. [CrossRef]

106. Wong, S.L.; Liu, H.; Chi, D. Recent progress in chemical vapor deposition growth of two-dimensional transition metal dichalcogenides. *Prog. Cryst. Growth Charact. Mater.* **2016**, *62*, 9–28. [CrossRef]

107. Manzeli, S.; Ovchinnikov, D.; Pasquier, D.; Yazyev, O.V.; Kis, A. 2D transition metal dichalcogenides. *Nat. Rev. Mater.* **2017**, *2*, 17033. [CrossRef]

108. Landauer, R. Spatial Variation of Currents and Fields Due to Localized Scatterers in Metallic Conduction. *IBM J. Res. Dev.* **1957**, *1*, 223–231. [CrossRef]

109. Sharvin, Y.V. A Possible Method for Studying Fermi Surfaces. *Sov. J. Exp. Theor. Phys.* **1965**, *21*, 655.

110. Allain, A.; Kang, J.; Banerjee, K.; Kis, A. Electrical contacts to two-dimensional semiconductors. *Nat. Mater.* **2015**, *14*, 1195–1205. [CrossRef] [PubMed]

111. Jena, D.; Banerjee, K.; Xing, G.H. 2D crystal semiconductors: Intimate contacts. *Nat. Mater.* **2014**, *13*, 1076. [CrossRef] [PubMed]

112. ITRS 2.0 Home Page. Available online: http://www.itrs2.net/ (accessed on 18 January 2018).

113. Kaasbjerg, K.; Thygesen, K.S.; Jacobsen, K.W. Phonon-limited mobility in n-type single-layer MoS$_2$ from first principles. *Phys. Rev. B* **2012**, *85*, 115317. [CrossRef]

114. Li, X.; Mullen, J.T.; Jin, Z.; Borysenko, K.M.; Buongiorno Nardelli, M.; Kim, K.W. Intrinsic electrical transport properties of monolayer silicene and MoS$_2$ from first principles. *Phys. Rev. B* **2013**, *87*, 115418. [CrossRef]

115. Jin, Z.; Li, X.; Mullen, J.T.; Kim, K.W. Intrinsic transport properties of electrons and holes in monolayer transition-metal dichalcogenides. *Phys. Rev. B* **2014**, *90*, 045422. [CrossRef]

116. Gunst, T.; Markussen, T.; Stokbro, K.; Brandbyge, M. First-principles method for electron-phonon coupling and electron mobility: Applications to two-dimensional materials. *Phys. Rev. B* **2016**, *93*, 035414. [CrossRef]

117. Cai, Y.; Zhang, G.; Zhang, Y.-W. Polarity-Reversed Robust Carrier Mobility in Monolayer MoS$_2$ Nanoribbons. *J. Am. Chem. Soc.* **2014**, *136*, 6269–6275. [CrossRef] [PubMed]

118. Akinwande, D.; Petrone, N.; Hone, J. Two-dimensional flexible nanoelectronics. *Nat. Commun.* **2014**, *5*, 5678. [CrossRef] [PubMed]

119. Chang, H.-Y.; Yang, S.; Lee, J.; Tao, L.; Hwang, W.-S.; Jena, D.; Lu, N.; Akinwande, D. High-Performance, Highly Bendable MoS$_2$ Transistors with High-κ Dielectrics for Flexible Low-Power Systems. *ACS Nano* **2013**, *7*, 5446–5452. [CrossRef] [PubMed]

120. Pu, J.; Yomogida, Y.; Liu, K.-K.; Li, L.-J.; Iwasa, Y.; Takenobu, T. Highly Flexible MoS$_2$ Thin-Film Transistors with Ion Gel Dielectrics. *Nano Lett.* **2012**, *12*, 4013–4017. [CrossRef] [PubMed]

121. Schwierz, F. (Invited) Performance of Graphene and Beyond Graphene 2D Semiconductor Devices. *ECS Trans.* **2015**, *69*, 231–240. [CrossRef]

122. Yoon, Y.; Ganapathi, K.; Salahuddin, S. How Good Can Monolayer MoS$_2$ Transistors Be? *Nano Lett.* **2011**, *11*, 3768–3773. [CrossRef] [PubMed]

123. Liu, L.; Lu, Y.; Guo, J. On Monolayer MoS$_2$ Field-Effect Transistors at the Scaling Limit. *IEEE Trans. Electron Devices* **2013**, *60*, 4133–4139. [CrossRef]

124. Granzner, R.; Geng, Z.; Kinberger, W.; Schwierz, F. MOSFET scaling: Impact of two-dimensional channel materials. In Proceedings of the 2016 13th IEEE International Conference on Solid-State and Integrated Circuit Technology (ICSICT), Hangzhou, China, 25–28 October 2016; pp. 466–469. [CrossRef]

125. Uchida, K.; Watanabe, H.; Kinoshita, A.; Koga, J.; Numata, T.; Takagi, S. Experimental study on carrier transport mechanism in ultrathin-body SOI nand p-MOSFETs with SOI thickness less than 5 nm. In Proceedings of the International Electron Devices Meeting, San Francisco, CA, USA, 8–11 December 2002; pp. 47–50. [CrossRef]

126. Jena, D. Tunneling Transistors Based on Graphene and 2-D Crystals. *Proc. IEEE* **2013**, *101*, 1585–1602. [CrossRef]

127. Lee, J.; Chang, H.Y.; Ha, T.J.; Li, H.; Ruoff, R.S.; Dodabalapur, A.; Akinwande, D. High-performance flexible nanoelectronics: 2D atomic channel materials for low-power digital and high-frequency analog devices. In Proceedings of the 2013 IEEE International Electron Devices Meeting, Washington, DC, USA, 9–11 December 2013; pp. 19.2.1–19.2.4. [CrossRef]

128. Park, S.; Zhu, W.; Chang, H.Y.; Yogeesh, M.N.; Ghosh, R.; Banerjee, S.K.; Akinwande, D. High-frequency prospects of 2D nanomaterials for flexible nanoelectronics from baseband to sub-THz devices. In Proceedings of the 2015 IEEE International Electron Devices Meeting (IEDM), Washington, DC, USA, 7–9 December 2015; pp. 32.1.1–32.1.4. [CrossRef]

129. Zhu, W.; Park, S.; Yogeesh, M.N.; Akinwande, D. Advancements in 2D flexible nanoelectronics: From material perspectives to RF applications. *Flex. Print. Electron.* **2017**, *2*, 043001. [CrossRef]

130. Cao, W.; Kang, J.; Sarkar, D.; Liu, W.; Banerjee, K. Performance evaluation and design considerations of 2D semiconductor based FETs for sub-10 nm VLSI. In Proceedings of the 2014 IEEE International Electron Devices Meeting, San Francisco, CA, USA, 15–17 December 2014; pp. 30.5.1–30.5.4. [CrossRef]

131. Smithe, K.K.H.; English, C.D.; Suryavanshi, S.V.; Pop, E. Intrinsic electrical transport and performance projections of synthetic monolayer MoS$_2$ devices. *2D Mater.* **2017**, *4*, 011009. [CrossRef]

132. Xu, Y.; Cheng, C.; Du, S.; Yang, J.; Yu, B.; Luo, J.; Yin, W.; Li, E.; Dong, S.; Ye, P.; et al. Contacts between Two- and Three-Dimensional Materials: Ohmic, Schottky, and p–n Heterojunctions. *ACS Nano* **2016**, *10*, 4895–4919. [CrossRef] [PubMed]

133. Giannazzo, F.; Fisichella, G.; Piazza, A.; Di Franco, S.; Greco, G.; Agnello, S.; Roccaforte, F. Impact of contact resistance on the electrical properties of MoS$_2$ transistors at practical operating temperatures. *Beilstein J. Nanotechnol.* **2017**, *8*, 254–263. [CrossRef] [PubMed]

134. Liu, W.; Sarkar, D.; Kang, J.; Cao, W.; Banerjee, K. Impact of Contact on the Operation and Performance of Back-Gated Monolayer MoS$_2$ Field-Effect-Transistors. *ACS Nano* **2015**, *9*, 7904–7912. [CrossRef] [PubMed]

135. Das, S.; Chen, H.-Y.; Penumatcha, A.V.; Appenzeller, J. High Performance Multilayer MoS$_2$ Transistors with Scandium Contacts. *Nano Lett.* **2012**, *13*, 100–105. [CrossRef] [PubMed]

136. Kaushik, N.; Nipane, A.; Basheer, F.; Dubey, S.; Grover, S.; Deshmukh, M.M.; Lodha, S. Schottky barrier heights for Au and Pd contacts to MoS$_2$. *Appl. Phys. Lett.* **2014**, *105*, 113505. [CrossRef]

137. Maurel, C.; Ajustron, F.; Péchou, R.; Seine, G.; Coratger, R. Electrical behavior of the Au/MoS$_2$ interface studied by light emission induced by scanning tunneling microscopy. *Surf. Sci.* **2006**, *600*, 442–447. [CrossRef]

138. Dankert, A.; Langouche, L.; Kamalakar, M.V.; Dash, S.P. High-Performance Molybdenum Disulfide Field-Effect Transistors with Spin Tunnel Contacts. *ACS Nano* **2014**, *8*, 476–482. [CrossRef] [PubMed]

139. Dong, H.; Gong, C.; Addou, R.; McDonnell, S.; Azcatl, A.; Qin, X.; Wang, W.; Wang, W.; Hinkle, C.L.; Wallace, R.M. Schottky Barrier Height of Pd/MoS$_2$ Contact by Large Area Photoemission Spectroscopy. *ACS Appl. Mater. Interfaces* **2017**, *9*, 38977–38983. [CrossRef] [PubMed]

140. Liu, Y.; Stradins, P.; Wei, S.-H. Van der Waals metal-semiconductor junction: Weak Fermi level pinning enables effective tuning of Schottky barrier. *Sci. Adv.* **2016**, *2*, e1600069. [CrossRef] [PubMed]

141. McDonnell, S.; Addou, R.; Buie, C.; Wallace, R.M.; Hinkle, C.L. Defect-Dominated Doping and Contact Resistance in MoS$_2$. *ACS Nano* **2014**, *8*, 2880–2888. [CrossRef] [PubMed]

142. Addou, R.; Colombo, L.; Wallace, R.M. Surface Defects on Natural MoS$_2$. *ACS Appl. Mater. Interfaces* **2015**, *7*, 11921–11929. [CrossRef] [PubMed]

143. Addou, R.; McDonnell, S.; Barrera, D.; Guo, Z.; Azcatl, A.; Wang, J.; Zhu, H.; Hinkle, C.L.; Quevedo-Lopez, M.; Alshareef, H.N.; et al. Impurities and Electronic Property Variations of Natural MoS$_2$ Crystal Surfaces. *ACS Nano* **2015**, *9*, 9124–9133. [CrossRef] [PubMed]

144. Bampoulis, P.; van Bremen, R.; Yao, Q.; Poelsema, B.; Zandvliet, H.J.W.; Sotthewes, K. Defect Dominated Charge Transport and Fermi Level Pinning in MoS$_2$/Metal Contacts. *ACS Appl. Mater. Interfaces* **2017**, *9*, 19278–19286. [CrossRef] [PubMed]

145. KC, S.; Longo, R.C.; Addou, R.; Wallace, R.M.; Cho, K. Impact of intrinsic atomic defects on the electronic structure of MoS$_2$ monolayers. *Nanotechnology* **2014**, *25*, 375703. [CrossRef] [PubMed]

146. Han, Y.; Wu, Z.; Xu, S.; Chen, X.; Wang, L.; Wang, Y.; Xiong, W.; Han, T.; Ye, W.; Lin, J.; et al. Probing Defect-Induced Midgap States in MoS$_2$ Through Graphene–MoS$_2$ Heterostructures. *Adv. Mater. Interfaces* **2015**, *2*, 1500064. [CrossRef]

147. Lu, C.-P.; Li, G.; Mao, J.; Wang, L.-M.; Andrei, E.Y. Bandgap, Mid-Gap States, and Gating Effects in MoS$_2$. *Nano Lett.* **2014**, *14*, 4628–4633. [CrossRef] [PubMed]

148. Kang, J.; Liu, W.; Sarkar, D.; Jena, D.; Banerjee, K. Computational Study of Metal Contacts to Monolayer Transition-Metal Dichalcogenide Semiconductors. *Phys. Rev. X* **2014**, *4*, 031005. [CrossRef]

149. Gong, C.; Colombo, L.; Wallace, R.M.; Cho, K. The Unusual Mechanism of Partial Fermi Level Pinning at Metal–MoS$_2$ Interfaces. *Nano Lett.* **2014**, *14*, 1714–1720. [CrossRef] [PubMed]

150. Farmanbar, M.; Brocks, G. First-principles study of van der Waals interactions and lattice mismatch at MoS$_2$/metal interfaces. *Phys. Rev. B* **2016**, *93*, 085304. [CrossRef]

151. Guo, Y.; Liu, D.; Robertson, J. 3D Behavior of Schottky Barriers of 2D Transition-Metal Dichalcogenides. *ACS Appl. Mater. Interfaces* **2015**, *7*, 25709–25715. [CrossRef] [PubMed]

152. Monch, W. On the physics of metal-semiconductor interfaces. *Rep. Prog. Phys.* **1990**, *53*, 221. [CrossRef]

153. Louie, S.G.; Chelikowsky, J.R.; Cohen, M.L. Ionicity and the theory of Schottky barriers. *Phys. Rev. B* **1977**, *15*, 2154–2162. [CrossRef]

154. Tersoff, J. Schottky Barrier Heights and the Continuum of Gap States. *Phys. Rev. Lett.* **1984**, *52*, 465–468. [CrossRef]

155. Kim, C.; Moon, I.; Lee, D.; Choi, M.S.; Ahmed, F.; Nam, S.; Cho, Y.; Shin, H.-J.; Park, S.; Yoo, W.J. Fermi Level Pinning at Electrical Metal Contacts of Monolayer Molybdenum Dichalcogenides. *ACS Nano* **2017**, *11*, 1588–1596. [CrossRef] [PubMed]

156. Kaushik, N.; Grover, S.; Deshmukh, M.M.; Lodha, S. Metal Contacts to MoS$_2$. In *2D Inorganic Materials beyond Graphene*; World Scientific (Europe): London, UK, 2016; pp. 317–347. ISBN 978-1-78634-269-0.

157. Liu, H.; Si, M.; Deng, Y.; Neal, A.T.; Du, Y.; Najmaei, S.; Ajayan, P.M.; Lou, J.; Ye, P.D. Switching Mechanism in Single-Layer Molybdenum Disulfide Transistors: An Insight into Current Flow across Schottky Barriers. *ACS Nano* **2014**, *8*, 1031–1038. [CrossRef] [PubMed]

158. Sze, S.M.; Ng, K.K. *Physics of Semiconductor Devices*; John Wiley & Sons: Hoboken, NJ, USA, 2006; Online ISBN: 9780470068328. [CrossRef]

159. Ahmed, F.; Sup Choi, M.; Liu, X.; Jong Yoo, W. Carrier transport at the metal–MoS$_2$ interface. *Nanoscale* **2015**, *7*, 9222–9228. [CrossRef] [PubMed]

160. Li, S.-L.; Tsukagoshi, K.; Orgiu, E.; Samorì, P. Charge transport and mobility engineering in two-dimensional transition metal chalcogenide semiconductors. *Chem. Soc. Rev.* **2016**, *45*, 118–151. [CrossRef] [PubMed]

161. Kang, J.; Sarkar, D.; Liu, W.; Jena, D.; Banerjee, K. A computational study of metal-contacts to beyond-graphene 2D semiconductor materials. In Proceedings of the 2012 International Electron Devices Meeting, San Francisco, CA, USA, 10–12 December 2012; pp. 17.4.1–17.4.4. [CrossRef]

162. Deng, J.; Kim, K.; Chuang, C.T.; Wong, H.S.P. The Impact of Device Footprint Scaling on High-Performance CMOS Logic Technology. *IEEE Trans. Electron Devices* **2007**, *54*, 1148–1155. [CrossRef]

163. Wei, L.; Deng, J.; Chang, L.W.; Kim, K.; Chuang, C.T.; Wong, H.S.P. Selective Device Structure Scaling and Parasitics Engineering: A Way to Extend the Technology Roadmap. *IEEE Trans. Electron Devices* **2009**, *56*, 312–320. [CrossRef]

164. English, C.D.; Shine, G.; Dorgan, V.E.; Saraswat, K.C.; Pop, E. Improved Contacts to MoS_2 Transistors by Ultra-High Vacuum Metal Deposition. *Nano Lett.* **2016**, *16*, 3824–3830. [CrossRef] [PubMed]

165. Taur, Y.; Ning, T.H. *Fundamentals of Modern VLSI Devices*; Cambridge University Press: Cambridge, UK, 2013; ISBN 978-1-107-39399-8.

166. Yuan, H.; Cheng, G.; Yu, S.; Walker, A.R.H.; Richter, C.A.; Pan, M.; Li, Q. Field effects of current crowding in metal-MoS_2 contacts. *Appl. Phys. Lett.* **2016**, *108*, 103505. [CrossRef]

167. Guo, Y.; Han, Y.; Li, J.; Xiang, A.; Wei, X.; Gao, S.; Chen, Q. Study on the Resistance Distribution at the Contact between Molybdenum Disulfide and Metals. *ACS Nano* **2014**, *8*, 7771–7779. [CrossRef] [PubMed]

168. Berger, H.H. Models for contacts to planar devices. *Solid-State Electron.* **1972**, *15*, 145–158. [CrossRef]

169. Zhao, Y.; Xu, K.; Pan, F.; Zhou, C.; Zhou, F.; Chai, Y. Doping, Contact and Interface Engineering of Two-Dimensional Layered Transition Metal Dichalcogenides Transistors. *Adv. Funct. Mater.* **2017**, *27*, 1603484. [CrossRef]

170. Porte, L.; de Villeneuve, C.H.; Phaner, M. Scanning tunneling microscopy observation of local damages induced on graphite surface by ion implantation. *J. Vac. Sci. Technol. B Microelectron. Nanometer Struct. Process. Meas. Phenom.* **1991**, *9*, 1064–1067. [CrossRef]

171. Yu, Z.; Ong, Z.-Y.; Li, S.; Xu, J.-B.; Zhang, G.; Zhang, Y.-W.; Shi, Y.; Wang, X. Analyzing the Carrier Mobility in Transition-Metal Dichalcogenide MoS_2 Field-Effect Transistors. *Adv. Funct. Mater.* **2017**, *27*, 1604093. [CrossRef]

172. Radisavljevic, B.; Kis, A. Mobility engineering and a metal–insulator transition in monolayer MoS_2. *Nat. Mater.* **2013**, *12*, 815–820. [CrossRef] [PubMed]

173. Baugher, B.W.H.; Churchill, H.O.H.; Yang, Y.; Jarillo-Herrero, P. Intrinsic Electronic Transport Properties of High-Quality Monolayer and Bilayer MoS_2. *Nano Lett.* **2013**, *13*, 4212–4216. [CrossRef] [PubMed]

174. Jena, D.; Li, M.; Ma, N.; Hwang, W.S.; Esseni, D.; Seabaugh, A.; Xing, H.G. Electron transport in 2D crystal semiconductors and their device applications. In Proceedings of the 2014 Silicon Nanoelectronics Workshop (SNW), Honolulu, HI, USA, 8–9 June 2014; pp. 1–2. Saturation. *Nano Lett.* **2015**, *15*, 5052–5058. [CrossRef]

175. He, G.; Ghosh, K.; Singisetti, U.; Ramamoorthy, H.; Somphonsane, R.; Bohra, G.; Matsunaga, M.; Higuchi, A.; Aoki, N.; Najmaei, S.; et al. Conduction Mechanisms in CVD-Grown Monolayer MoS_2 Transistors: From Variable-Range Hopping to Velocity Saturation. *Nano Lett.* **2015**, *15*, 5052–5058. [CrossRef] [PubMed]

176. Khair, K.; Ahmed, S. Dissipative transport in monolayer MoS_2: Role of remote coulomb scattering. In Proceedings of the 2015 International Workshop on Computational Electronics (IWCE), West Lafayette, IN, USA, 2 September 2015; pp. 1–2. [CrossRef]

177. Mori, T.; Ninomiya, N.; Kubo, T.; Uchida, N.; Watanabe, E.; Tsuya, D.; Moriyama, S.; Tanaka, M.; Ando, A. Characterization of Effective Mobility and Its Degradation Mechanism in MoS_2 MOSFETs. *IEEE Trans. Nanotechnol.* **2016**, *15*, 651–656. [CrossRef]

178. Ahmed, S.; Yi, J. Two-Dimensional Transition Metal Dichalcogenides and Their Charge Carrier Mobilities in Field-Effect Transistors. *Nano-Micro Lett.* **2017**, *9*, 50. [CrossRef]

179. Mirabelli, G.; Gity, F.; Monaghan, S.; Hurley, P.K.; Duffy, R. Impact of impurities, interface traps and contacts on MoS_2 MOSFETs: Modelling and experiments. In Proceedings of the 2017 47th European Solid-State Device Research Conference (ESSDERC), Leuven, Belgium, 11–14 September 2017; pp. 288–291. [CrossRef]

180. Tung, R.T. The physics and chemistry of the Schottky barrier height. *Appl. Phys. Rev.* **2014**, *1*, 011304. [CrossRef]

181. Nishimura, T.; Kita, K.; Toriumi, A. Evidence for strong Fermi-level pinning due to metal-induced gap states at metal/germanium interface. *Appl. Phys. Lett.* **2007**, *91*, 123123. [CrossRef]

182. Hasegawa, H. Fermi Level Pinning and Schottky Barrier Height Control at Metal-Semiconductor Interfaces of InP and Related Materials. *Jpn. J. Appl. Phys.* **1999**, *38*, 1098. [CrossRef]

183. Liu, W.; Kang, J.; Cao, W.; Sarkar, D.; Khatami, Y.; Jena, D.; Banerjee, K. High-performance few-layer-MoS_2 field-effect-transistor with record low contact-resistance. In Proceedings of the 2013 IEEE International Electron Devices Meeting, Washington, DC, USA, 9–11 December 2013; pp. 19.4.1–19.4.4. [CrossRef]

184. Hong, Y.K.; Yoo, G.; Kwon, J.; Hong, S.; Song, W.G.; Liu, N.; Omkaram, I.; Yoo, B.; Ju, S.; Kim, S.; et al. High performance and transparent multilayer MoS$_2$ transistors: Tuning Schottky barrier characteristics. *AIP Adv.* **2016**, *6*, 055026. [CrossRef]

185. Dai, Z.; Wang, Z.; He, X.; Zhang, X.-X.; Alshareef, H.N. Large-Area Chemical Vapor Deposited MoS$_2$ with Transparent Conducting Oxide Contacts toward Fully Transparent 2D Electronics. *Adv. Funct. Mater.* **2017**, *27*, 1703119. [CrossRef]

186. Liu, Y.; Xiao, H.; Goddard, W.A. Schottky-Barrier-Free Contacts with Two-Dimensional Semiconductors by Surface-Engineered MXenes. *J. Am. Chem. Soc.* **2016**, *138*, 15853–15856. [CrossRef] [PubMed]

187. Chuang, S.; Battaglia, C.; Azcatl, A.; McDonnell, S.; Kang, J.S.; Yin, X.; Tosun, M.; Kapadia, R.; Fang, H.; Wallace, R.M.; et al. MoS$_2$ P-type Transistors and Diodes Enabled by High Work Function MoOx Contacts. *Nano Lett.* **2014**, *14*, 1337–1342. [CrossRef] [PubMed]

188. McDonnell, S.; Azcatl, A.; Addou, R.; Gong, C.; Battaglia, C.; Chuang, S.; Cho, K.; Javey, A.; Wallace, R.M. Hole Contacts on Transition Metal Dichalcogenides: Interface Chemistry and Band Alignments. *ACS Nano* **2014**, *8*, 6265–6272. [CrossRef] [PubMed]

189. Musso, T.; Kumar, P.V.; Foster, A.S.; Grossman, J.C. Graphene Oxide as a Promising Hole Injection Layer for MoS$_2$-Based Electronic Devices. *ACS Nano* **2014**, *8*, 11432–11439. [CrossRef] [PubMed]

190. Chuang, H.-J.; Chamlagain, B.; Koehler, M.; Perera, M.M.; Yan, J.; Mandrus, D.; Tománek, D.; Zhou, Z. Low-Resistance 2D/2D Ohmic Contacts: A Universal Approach to High-Performance WSe$_2$, MoS$_2$, and MoSe$_2$ Transistors. *Nano Lett.* **2016**, *16*, 1896–1902. [CrossRef] [PubMed]

191. Kim, Y.; Kim, A.R.; Yang, J.H.; Chang, K.E.; Kwon, J.-D.; Choi, S.Y.; Park, J.; Lee, K.E.; Kim, D.-H.; Choi, S.M.; et al. Alloyed 2D Metal–Semiconductor Heterojunctions: Origin of Interface States Reduction and Schottky Barrier Lowering. *Nano Lett.* **2016**, *16*, 5928–5933. [CrossRef] [PubMed]

192. Giannazzo, F.; Fisichella, G.; Piazza, A.; Agnello, S.; Roccaforte, F. Nanoscale inhomogeneity of the Schottky barrier and resistivity in MoS$_2$ multilayers. *Phys. Rev. B* **2015**, *92*, 081307. [CrossRef]

193. Yuan, H.; Cheng, G.; You, L.; Li, H.; Zhu, H.; Li, W.; Kopanski, J.J.; Obeng, Y.S.; Hight Walker, A.R.; Gundlach, D.J.; et al. Influence of Metal–MoS$_2$ Interface on MoS$_2$ Transistor Performance: Comparison of Ag and Ti Contacts. *ACS Appl. Mater. Interfaces* **2015**, *7*, 1180–1187. [CrossRef] [PubMed]

194. Kim, T.-Y.; Amani, M.; Ahn, G.H.; Song, Y.; Javey, A.; Chung, S.; Lee, T. Electrical Properties of Synthesized Large-Area MoS$_2$ Field-Effect Transistors Fabricated with Inkjet-Printed Contacts. *ACS Nano* **2016**, *10*, 2819–2826. [CrossRef] [PubMed]

195. English, C.D.; Shine, G.; Dorgan, V.E.; Saraswat, K.C.; Pop, E. Improving contact resistance in MoS$_2$ field effect transistors. In Proceedings of the 72nd Device Research Conference, Santa Barbara, CA, USA, 2014, 22–25 June 2014; pp. 193–194. [CrossRef]

196. McDonnell, S.; Smyth, C.; Hinkle, C.L.; Wallace, R.M. MoS$_2$–Titanium Contact Interface Reactions. *ACS Appl. Mater. Interfaces* **2016**, *8*, 8289–8294. [CrossRef] [PubMed]

197. Smyth, C.M.; Addou, R.; McDonnell, S.; Hinkle, C.L.; Wallace, R.M. Contact Metal–MoS$_2$ Interfacial Reactions and Potential Implications on MoS$_2$-Based Device Performance. *J. Phys. Chem. C* **2016**, *120*, 14719–14729. [CrossRef]

198. Zhang, Y.; Ye, J.; Matsuhashi, Y.; Iwasa, Y. Ambipolar MoS$_2$ Thin Flake Transistors. *Nano Lett.* **2012**, *12*, 1136–1140. [CrossRef] [PubMed]

199. Lin, M.-W.; Liu, L.; Lan, Q.; Tan, X.; Dhindsa, K.S.; Zeng, P.; Naik, V.M.; Cheng, M.M.-C.; Zhou, Z. Mobility enhancement and highly efficient gating of monolayer MoS$_2$ transistors with polymer electrolyte. *J. Phys. Appl. Phys.* **2012**, *45*, 345102. [CrossRef]

200. Perera, M.M.; Lin, M.-W.; Chuang, H.-J.; Chamlagain, B.P.; Wang, C.; Tan, X.; Cheng, M.M.-C.; Tománek, D.; Zhou, Z. Improved Carrier Mobility in Few-Layer MoS$_2$ Field-Effect Transistors with Ionic-Liquid Gating. *ACS Nano* **2013**, *7*, 4449–4458. [CrossRef] [PubMed]

201. Wang, F.; Stepanov, P.; Gray, M.; Lau, C.N.; Itkis, M.E.; Haddon, R.C. Ionic Liquid Gating of Suspended MoS$_2$ Field Effect Transistor Devices. *Nano Lett.* **2015**, *15*, 5284–5288. [CrossRef] [PubMed]

202. Zhang, Y.J.; Ye, J.T.; Yomogida, Y.; Takenobu, T.; Iwasa, Y. Formation of a Stable p–n Junction in a Liquid-Gated MoS$_2$ Ambipolar Transistor. *Nano Lett.* **2013**, *13*, 3023–3028. [CrossRef] [PubMed]

203. Jiang, J.; Kuroda, M.A.; Ahyi, A.C.; Isaacs-Smith, T.; Mirkhani, V.; Park, M.; Dhar, S. Chitosan solid electrolyte as electric double layer in multilayer MoS$_2$ transistor for low-voltage operation. *Phys. Status Solidi A* **2015**, *212*, 2219–2225. [CrossRef]

204. Fathipour, S.; Li, H.M.; Remškar, M.; Yeh, L.; Tsai, W.; Lin, Y.; Fullerton-Shirey, S.; Seabaugh, A. Record high current density and low contact resistance in MoS$_2$ FETs by ion doping. In Proceedings of the 2016 International Symposium on VLSI Technology, Systems and Application (VLSI-TSA), Hsinchu, Taiwan, 25–27 April 2016; pp. 1–2. [CrossRef]

205. Liang, J.; Xu, K.; Fullerton, S. MoS$_2$ Field-Effect Transistors Gated with a Two-Dimensional Electrolyte. In *Meeting Abstracts 2017*; The Electrochemical Society: New Orleans, LA, USA, 2017; p. 1362. [CrossRef]

206. Stavitski, N.; van Dal, M.J.H.; Lauwers, A.; Vrancken, C.; Kovalgin, A.Y.; Wolters, R.A.M. Systematic TLM Measurements of NiSi and PtSi Specific Contact Resistance to n- and p-Type Si in a Broad Doping Range. *IEEE Electron Device Lett.* **2008**, *29*, 378–381. [CrossRef]

207. Murakoshi, A.; Iwase, M.; Niiyama, H.; Koike, M.; Suguro, K. Ultralow Contact Resistivity for a Metal/p-Type Silicon Interface by High-Concentration Germanium and Boron Doping Combined with Low-Temperature Annealing. *Jpn. J. Appl. Phys.* **2013**, *52*, 075802. [CrossRef]

208. Yu, H.; Schaekers, M.; Demuynck, S.; Barla, K.; Mocuta, A.; Horiguchi, N.; Collaert, N.; Thean, A.V.Y.; Meyer, K.D. MIS or MS? Source/drain contact scheme evaluation for 7nm Si CMOS technology and beyond. In Proceedings of the 2016 16th International Workshop on Junction Technology (IWJT), Shanghai, China, 9–10 May 2016; pp. 19–24. [CrossRef]

209. Du, Y.; Liu, H.; Neal, A.T.; Si, M.; Ye, P.D. Molecular Doping of Multilayer Field-Effect Transistors: Reduction in Sheet and Contact Resistances. *IEEE Electron Device Lett.* **2013**, *34*, 1328–1330. [CrossRef]

210. Fang, H.; Tosun, M.; Seol, G.; Chang, T.C.; Takei, K.; Guo, J.; Javey, A. Degenerate n-Doping of Few-Layer Transition Metal Dichalcogenides by Potassium. *Nano Lett.* **2013**, *13*, 1991–1995. [CrossRef] [PubMed]

211. Kiriya, D.; Tosun, M.; Zhao, P.; Kang, J.S.; Javey, A. Air-Stable Surface Charge Transfer Doping of MoS$_2$ by Benzyl Viologen. *J. Am. Chem. Soc.* **2014**, *136*, 7853–7856. [CrossRef] [PubMed]

212. Rai, A.; Valsaraj, A.; Movva, H.C.P.; Roy, A.; Ghosh, R.; Sonde, S.; Kang, S.; Chang, J.; Trivedi, T.; Dey, R.; et al. Air Stable Doping and Intrinsic Mobility Enhancement in Monolayer Molybdenum Disulfide by Amorphous Titanium Suboxide Encapsulation. *Nano Lett.* **2015**, *15*, 4329–4336. [CrossRef] [PubMed]

213. Rai, A.; Valsaraj, A.; Movva, H.C.P.; Roy, A.; Tutuc, E.; Register, L.F.; Banerjee, S.K. Interfacial-oxygen-vacancy mediated doping of MoS$_2$ by high-κ dielectrics. In Proceedings of the 2015 73rd Annual Device Research Conference (DRC), Columbus, OH, USA, 21–24 June 2015; pp. 189–190. [CrossRef]

214. Valsaraj, A.; Chang, J.; Rai, A.; Register, L.F.; Banerjee, S.K. Theoretical and experimental investigation of vacancy-based doping of monolayer MoS$_2$ on oxide. *2D Mater.* **2015**, *2*, 045009. [CrossRef]

215. Alharbi, A.; Shahrjerdi, D. Contact engineering of monolayer CVD MoS$_2$ transistors. In Proceedings of the 2017 75th Annual Device Research Conference (DRC), South Bend, IN, USA, 25–28 June 2017; pp. 1–2. [CrossRef]

216. McClellan, C.J.; Yalon, E.; Smithe, K.K.H.; Suryavanshi, S.V.; Pop, E. Effective n-type doping of monolayer MoS$_2$ by AlO$_x$. In Proceedings of the 2017 75th Annual Device Research Conference (DRC), South Bend, IN, USA, 25–28 June 2017; pp. 1–2. [CrossRef]

217. De la Rosa, C.J.L.; Nourbakhsh, A.; Heyne, M.; Asselberghs, I.; Huyghebaert, C.; Radu, I.; Heyns, M.; Gendt, S.D. Highly efficient and stable MoS$_2$ FETs with reversible n-doping using a dehydrated poly(vinyl-alcohol) coating. *Nanoscale* **2016**, *9*, 258–265. [CrossRef] [PubMed]

218. Lim, D.; Kannan, E.S.; Lee, I.; Rathi, S.; Li, L.; Lee, Y.; Khan, M.A.; Kang, M.; Park, J.; Kim, G.-H. High performance MoS$_2$-based field-effect transistor enabled by hydrazine doping. *Nanotechnology* **2016**, *27*, 225201. [CrossRef] [PubMed]

219. Andleeb, S.; Singh, A.K.; Eom, J. Chemical doping of MoS$_2$ multilayer by p-toluene sulfonic acid. *Sci. Technol. Adv. Mater.* **2015**, *16*, 035009. [CrossRef] [PubMed]

220. Wang, W.; Niu, X.; Qian, H.; Guan, L.; Zhao, M.; Ding, X.; Zhang, S.; Wang, Y.; Sha, J. Surface charge transfer doping of monolayer molybdenum disulfide by black phosphorus quantum dots. *Nanotechnology* **2016**, *27*, 505204. [CrossRef] [PubMed]

221. Lockhart de la Rosa, C.J.; Phillipson, R.; Teyssandier, J.; Adisoejoso, J.; Balaji, Y.; Huyghebaert, C.; Radu, I.; Heyns, M.; De Feyter, S.; De Gendt, S. Molecular doping of MoS$_2$ transistors by self-assembled oleylamine networks. *Appl. Phys. Lett.* **2016**, *109*, 253112. [CrossRef]

222. Liu, X.; Qu, D.; Ryu, J.; Ahmed, F.; Yang, Z.; Lee, D.; Yoo, W.J. P-Type Polar Transition of Chemically Doped Multilayer MoS$_2$ Transistor. *Adv. Mater.* **2016**, *28*, 2345–2351. [CrossRef] [PubMed]

223. Tarasov, A.; Zhang, S.; Tsai, M.-Y.; Campbell, P.M.; Graham, S.; Barlow, S.; Marder, S.R.; Vogel, E.M. Controlled Doping of Large-Area Trilayer MoS$_2$ with Molecular Reductants and Oxidants. *Adv. Mater.* **2015**, *27*, 1175–1181. [CrossRef] [PubMed]

224. Sim, D.M.; Kim, M.; Yim, S.; Choi, M.-J.; Choi, J.; Yoo, S.; Jung, Y.S. Controlled Doping of Vacancy-Containing Few-Layer MoS$_2$ via Highly Stable Thiol-Based Molecular Chemisorption. *ACS Nano* **2015**, *9*, 12115–12123. [CrossRef] [PubMed]

225. Min, S.-W.; Yoon, M.; Yang, S.J.; Ko, K.R.; Im, S. Charge Transfer-Induced P-type Channel in MoS$_2$ Flake Field Effect Transistor. *ACS Appl. Mater. Interfaces* **2018**. [CrossRef] [PubMed]

226. Connelly, D.; Faulkner, C.; Clifton, P.A.; Grupp, D.E. Fermi-level depinning for low-barrier Schottky source/drain transistors. *Appl. Phys. Lett.* **2006**, *88*, 012105. [CrossRef]

227. Kobayashi, M.; Kinoshita, A.; Saraswat, K.; Wong, H.-S.P.; Nishi, Y. Fermi level depinning in metal/Ge Schottky junction for metal source/drain Ge metal-oxide-semiconductor field-effect-transistor application. *J. Appl. Phys.* **2009**, *105*, 023702. [CrossRef]

228. Paramahans Manik, P.; Kesh Mishra, R.; Pavan Kishore, V.; Ray, P.; Nainani, A.; Huang, Y.-C.; Abraham, M.C.; Ganguly, U.; Lodha, S. Fermi-level unpinning and low resistivity in contacts to n-type Ge with a thin ZnO interfacial layer. *Appl. Phys. Lett.* **2012**, *101*, 182105. [CrossRef]

229. Agrawal, A.; Shukla, N.; Ahmed, K.; Datta, S. A unified model for insulator selection to form ultra-low resistivity metal-insulator-semiconductor contacts to n-Si, n-Ge, and n-InGaAs. *Appl. Phys. Lett.* **2012**, *101*, 042108. [CrossRef]

230. Gupta, S.; Paramahans Manik, P.; Kesh Mishra, R.; Nainani, A.; Abraham, M.C.; Lodha, S. Contact resistivity reduction through interfacial layer doping in metal-interfacial layer-semiconductor contacts. *J. Appl. Phys.* **2013**, *113*, 234505. [CrossRef]

231. Lee, D.; Raghunathan, S.; Wilson, R.J.; Nikonov, D.E.; Saraswat, K.; Wang, S.X. The influence of Fermi level pinning/depinning on the Schottky barrier height and contact resistance in Ge/CoFeB and Ge/MgO/CoFeB structures. *Appl. Phys. Lett.* **2010**, *96*, 052514. [CrossRef]

232. Roy, A.M.; Lin, J.Y.J.; Saraswat, K.C. Specific Contact Resistivity of Tunnel Barrier Contacts Used for Fermi Level Depinning. *IEEE Electron Device Lett.* **2010**, *31*, 1077–1079. [CrossRef]

233. Chen, J.-R.; Odenthal, P.M.; Swartz, A.G.; Floyd, G.C.; Wen, H.; Luo, K.Y.; Kawakami, R.K. Control of Schottky Barriers in Single Layer MoS$_2$ Transistors with Ferromagnetic Contacts. *Nano Lett.* **2013**, *13*, 3106–3110. [CrossRef] [PubMed]

234. Park, W.; Kim, Y.; Lee, S.K.; Jung, U.; Yang, J.H.; Cho, C.; Kim, Y.J.; Lim, S.K.; Hwang, I.S.; Lee, H.B.R.; et al. Contact resistance reduction using Fermi level de-pinning layer for MoS$_2$ FETs. In Proceedings of the 2014 IEEE International Electron Devices Meeting, San Francisco, CA, USA, 15–17 December 2014; pp. 5.1.1–5.1.4. [CrossRef]

235. Kaushik, N.; Nipane, A.; Karande, S.; Lodha, S. Contact resistance reduction in MoS$_2$ FETs using ultra-thin TiO$_2$ interfacial layers. In Proceedings of the 2015 73rd Annual Device Research Conference (DRC), Columbus, OH, USA, 21–24 June 2015; pp. 211–212. [CrossRef]

236. Kaushik, N.; Karmakar, D.; Nipane, A.; Karande, S.; Lodha, S. Interfacial n-Doping Using an Ultrathin TiO$_2$ Layer for Contact Resistance Reduction in MoS$_2$. *ACS Appl. Mater. Interfaces* **2016**, *8*, 256–263. [CrossRef] [PubMed]

237. Lee, S.; Tang, A.; Aloni, S.; Philip Wong, H.-S. Statistical Study on the Schottky Barrier Reduction of Tunneling Contacts to CVD Synthesized MoS$_2$. *Nano Lett.* **2016**, *16*, 276–281. [CrossRef] [PubMed]

238. Farmanbar, M.; Brocks, G. Controlling the Schottky barrier at MoS$_2$/metal contacts by inserting a BN monolayer. *Phys. Rev. B* **2015**, *91*, 161304. [CrossRef]

239. Su, J.; Feng, L.; Zeng, W.; Liu, Z. Designing high performance metal–mMoS$_2$ interfaces by two-dimensional insertions with suitable thickness. *Phys. Chem. Chem. Phys.* **2016**, *18*, 31092–31100. [CrossRef] [PubMed]

240. Wang, J.; Yao, Q.; Huang, C.-W.; Zou, X.; Liao, L.; Chen, S.; Fan, Z.; Zhang, K.; Wu, W.; Xiao, X.; et al. High Mobility MoS$_2$ Transistor with Low Schottky Barrier Contact by Using Atomic Thick h-BN as a Tunneling Layer. *Adv. Mater.* **2016**, *28*, 8302–8308. [CrossRef] [PubMed]

241. Cui, X.; Shih, E.-M.; Jauregui, L.A.; Chae, S.H.; Kim, Y.D.; Li, B.; Seo, D.; Pistunova, K.; Yin, J.; Park, J.-H.; et al. Low-Temperature Ohmic Contact to Monolayer MoS$_2$ by van der Waals Bonded Co/h-BN Electrodes. *Nano Lett.* **2017**, *17*, 4781–4786. [CrossRef] [PubMed]

242. Chai, J.W.; Yang, M.; Callsen, M.; Zhou, J.; Yang, T.; Zhang, Z.; Pan, J.S.; Chi, D.Z.; Feng, Y.P.; Wang, S.J. Tuning Contact Barrier Height between Metals and MoS$_2$ Monolayer through Interface Engineering. *Adv. Mater. Interfaces* **2017**, *4*, 1700035. [CrossRef]

243. Farmanbar, M.; Brocks, G. Ohmic Contacts to 2D Semiconductors through van der Waals Bonding. *Adv. Electron. Mater.* **2016**, *2*, 1500405. [CrossRef]

244. Musso, T.; Kumar, P.V.; Grossman, J.C.; Foster, A.S. Engineering Efficient p-Type TMD/Metal Contacts Using Fluorographene as a Buffer Layer. *Adv. Electron. Mater.* **2017**, *3*, 1600318. [CrossRef]

245. Su, J.; Feng, L.; Zheng, X.; Hu, C.; Lu, H.; Liu, Z. Promising Approach for High-Performance MoS$_2$ Nanodevice: Doping the BN Buffer Layer to Eliminate the Schottky Barriers. *ACS Appl. Mater. Interfaces* **2017**, *9*, 40940–40948. [CrossRef] [PubMed]

246. Geim, A.K.; Novoselov, K.S. The rise of graphene. *Nat. Mater.* **2007**, *6*, 183. [CrossRef] [PubMed]

247. Choi, W.; Lahiri, I.; Seelaboyina, R.; Kang, Y.S. Synthesis of Graphene and Its Applications: A Review. *Crit. Rev. Solid State Mater. Sci.* **2010**, *35*, 52–71. [CrossRef]

248. Castro Neto, A.H.; Guinea, F.; Peres, N.M.R.; Novoselov, K.S.; Geim, A.K. The electronic properties of graphene. *Rev. Mod. Phys.* **2009**, *81*, 109–162. [CrossRef]

249. Partoens, B.; Peeters, F.M. From graphene to graphite: Electronic structure around the K point. *Phys. Rev. B* **2006**, *74*, 075404. [CrossRef]

250. Novoselov, K.S.; Geim, A.K.; Morozov, S.V.; Jiang, D.; Katsnelson, M.I.; Grigorieva, I.V.; Dubonos, S.V.; Firsov, A.A. Two-dimensional gas of massless Dirac fermions in graphene. *Nature* **2005**, *438*, 197. [CrossRef] [PubMed]

251. Bolotin, K.I.; Sikes, K.J.; Jiang, Z.; Klima, M.; Fudenberg, G.; Hone, J.; Kim, P.; Stormer, H.L. Ultrahigh electron mobility in suspended graphene. *Solid State Commun.* **2008**, *146*, 351–355. [CrossRef]

252. Lemme, M.C.; Echtermeyer, T.J.; Baus, M.; Kurz, H. A Graphene Field-Effect Device. *IEEE Electron Device Lett.* **2007**, *28*, 282–284. [CrossRef]

253. Yu, Y.-J.; Zhao, Y.; Ryu, S.; Brus, L.E.; Kim, K.S.; Kim, P. Tuning the Graphene Work Function by Electric Field Effect. *Nano Lett.* **2009**, *9*, 3430–3434. [CrossRef] [PubMed]

254. Du, Y.; Yang, L.; Zhang, J.; Liu, H.; Majumdar, K.; Kirsch, P.D.; Ye, P.D. Physical understanding of graphene/metal hetero-contacts to enhance MoS$_2$ field-effect transistors performance. In Proceedings of the 72nd Device Research Conference, Santa Barbara, CA, USA, 22–25 June 2014; pp. 147–148. [CrossRef]

255. Du, Y.; Yang, L.; Zhang, J.; Liu, H.; Majumdar, K.; Kirsch, P.D.; Ye, P.D. MoS$_2$ Field-Effect Transistors with Graphene/Metal Heterocontacts. *IEEE Electron Device Lett.* **2014**, *35*, 599–601. [CrossRef]

256. Leong, W.S.; Luo, X.; Li, Y.; Khoo, K.H.; Quek, S.Y.; Thong, J.T.L. Low Resistance Metal Contacts to MoS$_2$ Devices with Nickel-Etched-Graphene Electrodes. *ACS Nano* **2015**, *9*, 869–877. [CrossRef] [PubMed]

257. Liu, Y.; Wu, H.; Cheng, H.-C.; Yang, S.; Zhu, E.; He, Q.; Ding, M.; Li, D.; Guo, J.; Weiss, N.O.; et al. Toward Barrier Free Contact to Molybdenum Disulfide Using Graphene Electrodes. *Nano Lett.* **2015**, *15*, 3030–3034. [CrossRef] [PubMed]

258. Singh, A.K.; Hwang, C.; Eom, J. Low-Voltage and High-Performance Multilayer MoS$_2$ Field-Effect Transistors with Graphene Electrodes. *ACS Appl. Mater. Interfaces* **2016**, *8*, 34699–34705. [CrossRef] [PubMed]

259. Cui, X.; Lee, G.-H.; Kim, Y.D.; Arefe, G.; Huang, P.Y.; Lee, C.-H.; Chenet, D.A.; Zhang, X.; Wang, L.; Ye, F.; et al. Multi-terminal transport measurements of MoS$_2$ using a van der Waals heterostructure device platform. *Nat. Nanotechnol.* **2015**, *10*, 534–540. [CrossRef] [PubMed]

260. Lee, G.-H.; Cui, X.; Kim, Y.D.; Arefe, G.; Zhang, X.; Lee, C.-H.; Ye, F.; Watanabe, K.; Taniguchi, T.; Kim, P.; et al. Highly Stable, Dual-Gated MoS$_2$ Transistors Encapsulated by Hexagonal Boron Nitride with Gate-Controllable Contact, Resistance, and Threshold Voltage. *ACS Nano* **2015**, *9*, 7019–7026. [CrossRef] [PubMed]

261. Guimarães, M.H.D.; Gao, H.; Han, Y.; Kang, K.; Xie, S.; Kim, C.-J.; Muller, D.A.; Ralph, D.C.; Park, J. Atomically Thin Ohmic Edge Contacts Between Two-Dimensional Materials. *ACS Nano* **2016**, *10*, 6392–6399. [CrossRef] [PubMed]

262. Li, S.-L.; Komatsu, K.; Nakaharai, S.; Lin, Y.-F.; Yamamoto, M.; Duan, X.; Tsukagoshi, K. Thickness Scaling Effect on Interfacial Barrier and Electrical Contact to Two-Dimensional MoS$_2$ Layers. *ACS Nano* **2014**, *8*, 12836–12842. [CrossRef] [PubMed]

263. Kwon, J.; Lee, J.-Y.; Yu, Y.-J.; Lee, C.-H.; Cui, X.; Hone, J.; Lee, G.-H. Thickness-dependent Schottky barrier height of MoS$_2$ field-effect transistors. *Nanoscale* **2017**, *9*, 6151–6157. [CrossRef] [PubMed]

264. Das, S.; Appenzeller, J. Screening and interlayer coupling in multilayer MoS$_2$. *Phys. Status Solidi RRL Rapid Res. Lett.* **2013**, *7*, 268–273. [CrossRef]

265. Das, S.; Appenzeller, J. Where Does the Current Flow in Two-Dimensional Layered Systems? *Nano Lett.* **2013**, *13*, 3396–3402. [CrossRef] [PubMed]

266. Na, J.; Shin, M.; Joo, M.-K.; Huh, J.; Jeong Kim, Y.; Jong Choi, H.; Hyung Shim, J.; Kim, G.-T. Separation of interlayer resistance in multilayer MoS$_2$ field-effect transistors. *Appl. Phys. Lett.* **2014**, *104*, 233502. [CrossRef]

267. Bhattacharjee, S.; Ganapathi, K.L.; Nath, D.N.; Bhat, N. Intrinsic Limit for Contact Resistance in Exfoliated Multilayered MoS$_2$ FET. *IEEE Electron Device Lett.* **2016**, *37*, 119–122. [CrossRef]

268. Schmidt, H.; Giustiniano, F.; Eda, G. Electronic transport properties of transition metal dichalcogenide field-effect devices: Surface and interface effects. *Chem. Soc. Rev.* **2015**, *44*, 7715–7736. [CrossRef] [PubMed]

269. Li, S.-L.; Wakabayashi, K.; Xu, Y.; Nakaharai, S.; Komatsu, K.; Li, W.-W.; Lin, Y.-F.; Aparecido-Ferreira, A.; Tsukagoshi, K. Thickness-Dependent Interfacial Coulomb Scattering in Atomically Thin Field-Effect Transistors. *Nano Lett.* **2013**, *13*, 3546–3552. [CrossRef] [PubMed]

270. Lin, M.-W.; Kravchenko, I.I.; Fowlkes, J.; Li, X.; Puretzky, A.A.; Rouleau, C.M.; Geohegan, D.B.; Xiao, K. Thickness-dependent charge transport in few-layer MoS$_2$ field-effect transistors. *Nanotechnology* **2016**, *27*, 165203. [CrossRef] [PubMed]

271. Lockhart de la Rosa, C.J.; Arutchelvan, G.; Leonhardt, A.; Huyghebaert, C.; Radu, I.; Heyns, M.; De Gendt, S. Relation between film thickness and surface doping of MoS$_2$ based field effect transistors. *APL Mater.* **2018**, *6*, 058301. [CrossRef]

272. Yun, W.S.; Han, S.W.; Hong, S.C.; Kim, I.G.; Lee, J.D. Thickness and strain effects on electronic structures of transition metal dichalcogenides: 2H-MX$_2$ semiconductors (M = Mo, W; X = S, Se, Te). *Phys. Rev. B* **2012**, *85*, 033305. [CrossRef]

273. Wang, L.; Meric, I.; Huang, P.Y.; Gao, Q.; Gao, Y.; Tran, H.; Taniguchi, T.; Watanabe, K.; Campos, L.M.; Muller, D.A.; et al. One-Dimensional Electrical Contact to a Two-Dimensional Material. *Science* **2013**, *342*, 614–617. [CrossRef] [PubMed]

274. Min Song, S.; Yong Kim, T.; Jae Sul, O.; Cheol Shin, W.; Jin Cho, B. Improvement of graphene–metal contact resistance by introducing edge contacts at graphene under metal. *Appl. Phys. Lett.* **2014**, *104*, 183506. [CrossRef]

275. Park, H.-Y.; Jung, W.-S.; Kang, D.-H.; Jeon, J.; Yoo, G.; Park, Y.; Lee, J.; Jang, Y.H.; Lee, J.; Park, S.; et al. Extremely Low Contact Resistance on Graphene through n-Type Doping and Edge Contact Design. *Adv. Mater.* **2016**, *28*, 864–870. [CrossRef] [PubMed]

276. Khatami, Y.; Li, H.; Xu, C.; Banerjee, K. Metal-to-Multilayer-Graphene Contact—Part I: Contact Resistance Modeling. *IEEE Trans. Electron Devices* **2012**, *59*, 2444–2452. [CrossRef]

277. Khatami, Y.; Li, H.; Xu, C.; Banerjee, K. Metal-to-Multilayer-Graphene Contact—Part II: Analysis of Contact Resistance. *IEEE Trans. Electron Devices* **2012**, *59*, 2453–2460. [CrossRef]

278. Yoo, G.; Lee, S.; Yoo, B.; Han, C.; Kim, S.; Oh, M.S. Electrical Contact Analysis of Multilayer MoS$_2$ Transistor with Molybdenum Source/Drain Electrodes. *IEEE Electron Device Lett.* **2015**, *36*, 1215–1218. [CrossRef]

279. Chai, Y.; Ionescu, R.; Su, S.; Lake, R.; Ozkan, M.; Ozkan, C.S. Making one-dimensional electrical contacts to molybdenum disulfide-based heterostructures through plasma etching. *Phys. Status Solidi A* **2016**, *213*, 1358–1364. [CrossRef]

280. Zheng, J.; Yan, X.; Lu, Z.; Qiu, H.; Xu, G.; Zhou, X.; Wang, P.; Pan, X.; Liu, K.; Jiao, L. High-Mobility Multilayered MoS$_2$ Flakes with Low Contact Resistance Grown by Chemical Vapor Deposition. *Adv. Mater.* **2017**, *29*, 1604540. [CrossRef] [PubMed]

281. Movva, H.C.P.; Rai, A.; Kang, S.; Kim, K.; Fallahazad, B.; Taniguchi, T.; Watanabe, K.; Tutuc, E.; Banerjee, S.K. High-Mobility Holes in Dual-Gated WSe$_2$ Field-Effect Transistors. *ACS Nano* **2015**, *9*, 10402–10410. [CrossRef] [PubMed]

282. Popov, I.; Seifert, G.; Tománek, D. Designing Electrical Contacts to MoS$_2$ Monolayers: A Computational Study. *Phys. Rev. Lett.* **2012**, *108*, 156802. [CrossRef] [PubMed]

283. Stokbro, K.; Engelund, M.; Blom, A. Atomic-scale model for the contact resistance of the nickel-graphene interface. *Phys. Rev. B* **2012**, *85*, 165442. [CrossRef]

284. Kang, J.; Liu, W.; Banerjee, K. High-performance MoS$_2$ transistors with low-resistance molybdenum contacts. *Appl. Phys. Lett.* **2014**, *104*, 093106. [CrossRef]

285. Song, X.; Zan, W.; Xu, H.; Ding, S.; Zhou, P.; Bao, W.; Zhang, D.W. A novel synthesis method for large-area MoS$_2$ film with improved electrical contact. *2D Mater.* **2017**, *4*, 025051. [CrossRef]

286. Abraham, M.; Mohney, S.E. Annealed Ag contacts to MoS$_2$ field-effect transistors. *J. Appl. Phys.* **2017**, *122*, 115306. [CrossRef]

287. Kappera, R.; Voiry, D.; Yalcin, S.E.; Branch, B.; Gupta, G.; Mohite, A.D.; Chhowalla, M. Phase-engineered low-resistance contacts for ultrathin MoS$_2$ transistors. *Nat. Mater.* **2014**, *13*, 1128–1134. [CrossRef] [PubMed]

288. Lin, Y.-C.; Dumcenco, D.O.; Huang, Y.-S.; Suenaga, K. Atomic mechanism of the semiconducting-to-metallic phase transition in single-layered MoS$_2$. *Nat. Nanotechnol.* **2014**, *9*, 391–396. [CrossRef] [PubMed]

289. Kappera, R.; Voiry, D.; Yalcin, S.E.; Jen, W.; Acerce, M.; Torrel, S.; Branch, B.; Lei, S.; Chen, W.; Najmaei, S.; et al. Metallic 1T phase source/drain electrodes for field effect transistors from chemical vapor deposited MoS$_2$. *APL Mater.* **2014**, *2*, 092516. [CrossRef]

290. Ma, N.; Jena, D. Charge Scattering and Mobility in Atomically Thin Semiconductors. *Phys. Rev. X* **2014**, *4*, 011043. [CrossRef]

291. Ghatak, S.; Pal, A.N.; Ghosh, A. Nature of Electronic States in Atomically Thin MoS$_2$ Field-Effect Transistors. *ACS Nano* **2011**, *5*, 7707–7712. [CrossRef] [PubMed]

292. Komsa, H.-P.; Kurasch, S.; Lehtinen, O.; Kaiser, U.; Krasheninnikov, A.V. From point to extended defects in two-dimensional MoS$_2$: Evolution of atomic structure under electron irradiation. *Phys. Rev. B* **2013**, *88*, 035301. [CrossRef]

293. Qiu, H.; Xu, T.; Wang, Z.; Ren, W.; Nan, H.; Ni, Z.; Chen, Q.; Yuan, S.; Miao, F.; Song, F.; et al. Hopping transport through defect-induced localized states in molybdenum disulphide. *Nat. Commun.* **2013**, *4*, 2642. [CrossRef] [PubMed]

294. Ghatak, S.; Mukherjee, S.; Jain, M.; Sarma, D.D.; Ghosh, A. Microscopic Origin of Charged Impurity Scattering and Flicker Noise in MoS$_2$ field-effect Transistors. *arXiv* **2014**, arXiv:1403.3333. [CrossRef]

295. Liu, D.; Guo, Y.; Fang, L.; Robertson, J. Sulfur vacancies in monolayer MoS$_2$ and its electrical contacts. *Appl. Phys. Lett.* **2013**, *103*, 183113. [CrossRef]

296. Yu, Z.; Pan, Y.; Shen, Y.; Wang, Z.; Ong, Z.-Y.; Xu, T.; Xin, R.; Pan, L.; Wang, B.; Sun, L.; et al. Towards intrinsic charge transport in monolayer molybdenum disulfide by defect and interface engineering. *Nat. Commun.* **2014**, *5*, 5290. [CrossRef] [PubMed]

297. Amani, M.; Lien, D.-H.; Kiriya, D.; Xiao, J.; Azcatl, A.; Noh, J.; Madhvapathy, S.R.; Addou, R.; Kc, S.; Dubey, M.; et al. Near-unity photoluminescence quantum yield in MoS$_2$. *Science* **2015**, *350*, 1065–1068. [CrossRef] [PubMed]

298. Najmaei, S.; Zou, X.; Er, D.; Li, J.; Jin, Z.; Gao, W.; Zhang, Q.; Park, S.; Ge, L.; Lei, S.; et al. Tailoring the Physical Properties of Molybdenum Disulfide Monolayers by Control of Interfacial Chemistry. *Nano Lett.* **2014**, *14*, 1354–1361. [CrossRef] [PubMed]

299. Giannazzo, F.; Fisichella, G.; Piazza, A.; Franco, S.D.; Greco, G.; Agnello, S.; Roccaforte, F. Effect of temperature–bias annealing on the hysteresis and subthreshold behavior of multilayer MoS$_2$ transistors. *Phys. Status Solidi RRL Rapid Res. Lett.* **2016**, *11*, 797–801. [CrossRef]

300. Bolshakov, P.; Zhao, P.; Azcatl, A.; Hurley, P.K.; Wallace, R.M.; Young, C.D. Electrical characterization of top-gated molybdenum disulfide field-effect-transistors with high-κ dielectrics. *Microelectron. Eng.* **2017**, *178*, 190–193. [CrossRef]

301. Bolshakov, P.; Zhao, P.; Azcatl, A.; Hurley, P.K.; Wallace, R.M.; Young, C.D. Improvement in top-gate MoS$_2$ transistor performance due to high quality backside Al$_2$O$_3$ layer. *Appl. Phys. Lett.* **2017**, *111*, 032110. [CrossRef]

302. Young, C.D.; Zhao, P.; Bolshakov-Barrett, P.; Azcatl, A.; Hurley, P.K.; Gomeniuk, Y.Y.; Schmidt, M.; Hinkle, C.L.; Wallace, R.M. Evaluation of Few-Layer MoS$_2$ Transistors with a Top Gate and HfO$_2$ Dielectric. *ECS Trans.* **2016**, *75*, 153–162. [CrossRef]

303. Bhattacharjee, S.; Ganapathi, K.L.; Nath, D.N.; Bhat, N. Surface State Engineering of Metal/MoS$_2$ Contacts Using Sulfur Treatment for Reduced Contact Resistance and Variability. *IEEE Trans. Electron Devices* **2016**, *63*, 2556–2562. [CrossRef]

304. Bhattacharjee, S.; Ganapathi, K.L.; Mohan, S.; Bhat, N. (Invited) Interface Engineering of High-κ Dielectrics and Metal Contacts for High Performance Top-Gated MoS$_2$ FETs. *ECS Trans.* **2017**, *80*, 101–107. [CrossRef]

305. Nan, H.; Wu, Z.; Jiang, J.; Zafar, A.; You, Y.; Ni, Z. Improving the electrical performance of MoS$_2$ by mild oxygen plasma treatment. *J. Phys. Appl. Phys.* **2017**, *50*, 154001. [CrossRef]

306. Guo, J.; Yang, B.; Zheng, Z.; Jiang, J. Observation of abnormal mobility enhancement in multilayer MoS$_2$ transistor by synergy of ultraviolet illumination and ozone plasma treatment. *Phys. E Low-Dimens. Syst. Nanostructures* **2017**, *87*, 150–154. [CrossRef]

307. Leonhardt, A.; Chiappe, D.; Asselberghs, I.; Huyghebaert, C.; Radu, I.; Gendt, S.D. Improving MOCVD MoS$_2$ Electrical Performance: Impact of Minimized Water and Air Exposure Conditions. *IEEE Electron Device Lett.* **2017**, *38*, 1606–1609. [CrossRef]

308. Cheng, Z.; Cardenas, J.A.; McGuire, F.; Franklin, A.D. Using Ar Ion beam exposure to improve contact resistance in MoS$_2$ FETs. In Proceedings of the 2016 74th Annual Device Research Conference (DRC), Newark, DE, USA, 19–22 June 2016; pp. 1–2. [CrossRef]

309. Ho, Y.T.; Chu, Y.C.; Jong, C.A.; Chen, H.Y.; Lin, M.W.; Zhang, M.; Chien, P.Y.; Tu, Y.Y.; Woo, J.; Chang, E.Y. Contact resistance reduction on layered MoS$_2$ by Ar plasma pre-treatment. In Proceedings of the 2016 IEEE Silicon Nanoelectronics Workshop (SNW), Honolulu, HI, USA, 12–13 June 2016; pp. 52–53. [CrossRef]

310. Liu, H. Recent Progress in Atomic Layer Deposition of Multifunctional Oxides and Two-Dimensional Transition Metal Dichalcogenides. *J. Mol. Eng. Mater.* **2016**, *4*, 1640010. [CrossRef]

311. Brent, J.R.; Savjani, N.; O'Brien, P. Synthetic approaches to two-dimensional transition metal dichalcogenide nanosheets. *Prog. Mater. Sci.* **2017**, *89*, 411–478. [CrossRef]

312. Zhou, W.; Zou, X.; Najmaei, S.; Liu, Z.; Shi, Y.; Kong, J.; Lou, J.; Ajayan, P.M.; Yakobson, B.I.; Idrobo, J.-C. Intrinsic Structural Defects in Monolayer Molybdenum Disulfide. *Nano Lett.* **2013**, *13*, 2615–2622. [CrossRef] [PubMed]

313. Najmaei, S.; Yuan, J.; Zhang, J.; Ajayan, P.; Lou, J. Synthesis and Defect Investigation of Two-Dimensional Molybdenum Disulfide Atomic Layers. *Acc. Chem. Res.* **2015**, *48*, 31–40. [CrossRef] [PubMed]

314. Roy, A.; Ghosh, R.; Rai, A.; Sanne, A.; Kim, K.; Movva, H.C.P.; Dey, R.; Pramanik, T.; Chowdhury, S.; Tutuc, E.; et al. Intra-domain periodic defects in monolayer MoS$_2$. *Appl. Phys. Lett.* **2017**, *110*, 201905. [CrossRef]

315. Young, C.D.; Bolshakov, P.; Zhao, P.; Smyth, C.; Khosravi, A.; Hurley, P.K.; Hinkle, C.L.; Wallace, R.M. (Invited) Investigation of Critical Interfaces in Few-Layer MoS$_2$ Field Effect Transistors with High-κ Dielectrics. *ECS Trans.* **2017**, *80*, 219–225 [CrossRef]

316. Ye, M.; Zhang, D.; Yap, Y.K. Recent Advances in Electronic and Optoelectronic Devices Based on Two-Dimensional Transition Metal Dichalcogenides. *Electronics* **2017**, *6*, 43. [CrossRef]

317. Price, K.M.; Franklin, A.D. Integration of 3.4 nm HfO$_2$ into the gate stack of MoS$_2$ and WSe$_2$ top-gate field-effect transistors. In Proceedings of the 2017 75th Annual Device Research Conference (DRC), South Bend, IN, USA, 25–28 June 2017; pp. 1–2. [CrossRef]

318. Casademont, H.; Fillaud, L.; Lefèvre, X.; Jousselme, B.; Derycke, V. Electrografted Fluorinated Organic Ultrathin Film as Efficient Gate Dielectric in MoS$_2$ Transistors. *J. Phys. Chem. C* **2016**, *120*, 9506–9510. [CrossRef]

319. Zhao, P.; Azcatl, A.; Bolshakov, P.; Moon, J.; Hinkle, C.L.; Hurley, P.K.; Wallace, R.M.; Young, C.D. Effects of annealing on top-gated MoS$_2$ transistors with HfO$_2$ dielectric. *J. Vac. Sci. Technol. B Nanotechnol. Microelectron. Mater. Process. Meas. Phenom.* **2017**, *35*, 01A118. [CrossRef]

320. Yang, W.; Sun, Q.-Q.; Geng, Y.; Chen, L.; Zhou, P.; Ding, S.-J.; Zhang, D.W. The Integration of Sub-10 nm Gate Oxide on MoS$_2$ with Ultra Low Leakage and Enhanced Mobility. *Sci. Rep.* **2015**, *5*. [CrossRef] [PubMed]

321. Wang, J.; Li, S.; Zou, X.; Ho, J.; Liao, L.; Xiao, X.; Jiang, C.; Hu, W.; Wang, J.; Li, J. Integration of High-κ Oxide on MoS$_2$ by Using Ozone Pretreatment for High-Performance MoS$_2$ Top-Gated Transistor with Thickness-Dependent Carrier Scattering Investigation. *Small* **2015**, *11*, 5932–5938. [CrossRef] [PubMed]

322. Zou, X.; Wang, J.; Chiu, C.-H.; Wu, Y.; Xiao, X.; Jiang, C.; Wu, W.-W.; Mai, L.; Chen, T.; Li, J.; et al. Interface Engineering for High-Performance Top-Gated MoS$_2$ Field-Effect Transistors. *Adv. Mater.* **2014**, *26*, 6255–6261. [CrossRef] [PubMed]

323. Qian, Q.; Zhang, Z.; Hua, M.; Tang, G.; Lei, J.; Lan, F.; Xu, Y.; Yan, R.; Chen, K.J. Enhanced dielectric deposition on single-layer MoS$_2$ with low damage using remote N$_2$ plasma treatment. *Nanotechnology* **2017**, *28*, 175202. [CrossRef] [PubMed]

324. Price, K.M.; Schauble, K.E.; McGuire, F.A.; Farmer, D.B.; Franklin, A.D. Uniform Growth of Sub-5-Nanometer High-κ Dielectrics on MoS$_2$ Using Plasma-Enhanced Atomic Layer Deposition. *ACS Appl. Mater. Interfaces* **2017**, *9*, 23072–23080. [CrossRef] [PubMed]

325. Wang, X.; Zhang, T.-B.; Yang, W.; Zhu, H.; Chen, L.; Sun, Q.-Q.; Zhang, D.W. Improved integration of ultra-thin high-κ dielectrics in few-layer MoS$_2$ FET by remote forming gas plasma pretreatment. *Appl. Phys. Lett.* **2017**, *110*, 053110. [CrossRef]

326. Son, S.; Yu, S.; Choi, M.; Kim, D.; Choi, C. Improved high temperature integration of Al$_2$O$_3$ on MoS$_2$ by using a metal oxide buffer layer. *Appl. Phys. Lett.* **2015**, *106*, 021601. [CrossRef]

327. Qian, Q.; Li, B.; Hua, M.; Zhang, Z.; Lan, F.; Xu, Y.; Yan, R.; Chen, K.J. Improved Gate Dielectric Deposition and Enhanced Electrical Stability for Single-Layer MoS$_2$ MOSFET with an AlN Interfacial Layer. *Sci. Rep.* **2016**, *6*, 27676. [CrossRef] [PubMed]

328. Azcatl, A.; McDonnell, S.; Santosh, K.C.; Peng, X.; Dong, H.; Qin, X.; Addou, R.; Mordi, G.I.; Lu, N.; Kim, J.; et al. MoS$_2$ functionalization for ultra-thin atomic layer deposited dielectrics. *Appl. Phys. Lett.* **2014**, *104*, 111601. [CrossRef]

329. Azcatl, A.; KC, S.; Peng, X.; Lu, N.; McDonnell, S.; Qin, X.; de Dios, F.; Addou, R.; Kim, J.; Kim, M.J.; et al. HfO$_2$ on UV–O$_3$ exposed transition metal dichalcogenides: Interfacial reactions study. *2D Mater.* **2015**, *2*, 014004. [CrossRef]

330. Cheng, L.; Qin, X.; Lucero, A.T.; Azcatl, A.; Huang, J.; Wallace, R.M.; Cho, K.; Kim, J. Atomic Layer Deposition of a High-κ Dielectric on MoS$_2$ Using Trimethylaluminum and Ozone. *ACS Appl. Mater. Interfaces* **2014**, *6*, 11834–11838. [CrossRef] [PubMed]

331. McDonnell, S.; Brennan, B.; Azcatl, A.; Lu, N.; Dong, H.; Buie, C.; Kim, J.; Hinkle, C.L.; Kim, M.J.; Wallace, R.M. HfO$_2$ on MoS$_2$ by Atomic Layer Deposition: Adsorption Mechanisms and Thickness Scalability. *ACS Nano* **2013**, *7*, 10354–10361. [CrossRef] [PubMed]

332. Zhang, H.; Chiappe, D.; Meersschaut, J.; Conard, T.; Franquet, A.; Nuytten, T.; Mannarino, M.; Radu, I.; Vandervorst, W.; Delabie, A. Nucleation and growth mechanisms of Al$_2$O$_3$ atomic layer deposition on synthetic polycrystalline MoS$_2$. *J. Chem. Phys.* **2016**, *146*, 052810. [CrossRef] [PubMed]

333. Park, J.H.; Movva, H.C.P.; Chagarov, E.; Sardashti, K.; Chou, H.; Kwak, I.; Hu, K.-T.; Fullerton-Shirey, S.K.; Choudhury, P.; Banerjee, S.K.; et al. In Situ Observation of Initial Stage in Dielectric Growth and Deposition of Ultrahigh Nucleation Density Dielectric on Two-Dimensional Surfaces. *Nano Lett.* **2015**, *15*, 6626–6633. [CrossRef] [PubMed]

334. Li, Y.; Xu, C.-Y.; Hu, P.; Zhen, L. Carrier Control of MoS$_2$ Nanoflakes by Functional Self-Assembled Monolayers. *ACS Nano* **2013**, *7*, 7795–7804. [CrossRef] [PubMed]

335. Lee, W.H.; Park, Y.D. Tuning Electrical Properties of 2D Materials by Self-Assembled Monolayers. *Adv. Mater. Interfaces* **2018**, *5*, 1700316. [CrossRef]

336. Park, H.-Y.; Lim, M.-H.; Jeon, J.; Yoo, G.; Kang, D.-H.; Jang, S.K.; Jeon, M.H.; Lee, Y.; Cho, J.H.; Yeom, G.Y.; et al. Wide-Range Controllable n-Doping of Molybdenum Disulfide (MoS$_2$) through Thermal and Optical Activation. *ACS Nano* **2015**, *9*, 2368–2376. [CrossRef] [PubMed]

337. Behura, S.; Berry, V. Interfacial Nondegenerate Doping of MoS$_2$ and Other Two-Dimensional Semiconductors. *ACS Nano* **2015**, *9*, 2227–2230. [CrossRef] [PubMed]

338. Joo, M.-K.; Moon, B.H.; Ji, H.; Han, G.H.; Kim, H.; Lee, G.; Lim, S.C.; Suh, D.; Lee, Y.H. Electron Excess Doping and Effective Schottky Barrier Reduction on the MoS$_2$/h-BN Heterostructure. *Nano Lett.* **2016**, *16*, 6383–6389. [CrossRef] [PubMed]

339. Ong, Z.-Y.; Fischetti, M.V. Mobility enhancement and temperature dependence in top-gated single-layer MoS$_2$. *Phys. Rev. B* **2013**, *88*, 165316. [CrossRef]

340. Jena, D.; Konar, A. Enhancement of Carrier Mobility in Semiconductor Nanostructures by Dielectric Engineering. *Phys. Rev. Lett.* **2007**, *98*, 136805. [CrossRef] [PubMed]

341. Yu, Z.; Ong, Z.Y.; Pan, Y.; Xu, T.; Wang, Z.; Sun, L.; Wang, J.; Zhang, G.; Zhang, Y.W.; Shi, Y.; et al. Electron transport and device physics in monolayer transition-metal dichalcogenides. In Proceedings of the 2016 IEEE International Nanoelectronics Conference (INEC), Chengdu, China, 9–11 May 2016; pp. 1–2. [CrossRef]

342. Sun, Y.; Wang, R.; Liu, K. Substrate induced changes in atomically thin 2-dimensional semiconductors: Fundamentals, engineering, and applications. *Appl. Phys. Rev.* **2017**, *4*, 011301. [CrossRef]

343. Li, X.; Xiong, X.; Li, T.; Li, S.; Zhang, Z.; Wu, Y. Effect of Dielectric Interface on the Performance of MoS$_2$ Transistors. *ACS Appl. Mater. Interfaces* **2017**. [CrossRef] [PubMed]

344. Xu, J.; Chen, L.; Dai, Y.-W.; Cao, Q.; Sun, Q.-Q.; Ding, S.-J.; Zhu, H.; Zhang, D.W. A two-dimensional semiconductor transistor with boosted gate control and sensing ability. *Sci. Adv.* **2017**, *3*, e1602246. [CrossRef] [PubMed]

345. Yu, Z.; Ong, Z.-Y.; Pan, Y.; Cui, Y.; Xin, R.; Shi, Y.; Wang, B.; Wu, Y.; Chen, T.; Zhang, Y.-W.; et al. Realization of Room-Temperature Phonon-Limited Carrier Transport in Monolayer MoS$_2$ by Dielectric and Carrier Screening. *Adv. Mater.* **2016**, *28*, 547–552. [CrossRef] [PubMed]

346. Ganapathi, K.L.; Bhattacharjee, S.; Mohan, S.; Bhat, N. High-Performance HfO$_2$ Back Gated Multilayer MoS$_2$ Transistors. *IEEE Electron Device Lett.* **2016**, *37*, 797–800. [CrossRef]

347. Chamlagain, B.; Cui, Q.; Paudel, S.; Cheng, M.M.-C.; Chen, P.-Y.; Zhou, Z. Thermally oxidized 2D TaS$_2$ as a high-κ gate dielectric for MoS$_2$ field-effect transistors. *2D Mater.* **2017**, *4*, 031002. [CrossRef]

348. Kwon, H.-J.; Jang, J.; Grigoropoulos, C.P. Laser Direct Writing Process for Making Electrodes and High-κ Sol–Gel ZrO$_2$ for Boosting Performances of MoS$_2$ Transistors. *ACS Appl. Mater. Interfaces* **2016**, *8*, 9314–9318. [CrossRef] [PubMed]

349. Lee, H.S.; Baik, S.S.; Lee, K.; Min, S.-W.; Jeon, P.J.; Kim, J.S.; Choi, K.; Choi, H.J.; Kim, J.H.; Im, S. Metal Semiconductor Field-Effect Transistor with MoS$_2$/Conducting NiO$_x$ van der Waals Schottky Interface for Intrinsic High Mobility and Photoswitching Speed. *ACS Nano* **2015**, *9*, 8312–8320. [CrossRef] [PubMed]

350. Zeng, L.; Xin, Z.; Chen, S.; Du, G.; Kang, J.; Liu, X. Remote phonon and impurity screening effect of substrate and gate dielectric on electron dynamics in single layer MoS$_2$. *Appl. Phys. Lett.* **2013**, *103*, 113505. [CrossRef]

351. Konar, A.; Fang, T.; Jena, D. Effect of high-κ gate dielectrics on charge transport in graphene-based field effect transistors. *Phys. Rev. B* **2010**, *82*, 115452. [CrossRef]

352. Fischetti, M.V.; Neumayer, D.A.; Cartier, E.A. Effective electron mobility in Si inversion layers in metal–oxide–semiconductor systems with a high-κ insulator: The role of remote phonon scattering. *J. Appl. Phys.* **2001**, *90*, 4587–4608. [CrossRef]

353. Perebeinos, V.; Avouris, P. Inelastic scattering and current saturation in graphene. *Phys. Rev. B* **2010**, *81*, 195442. [CrossRef]

354. NSM Archive—Physical Properties of Semiconductors. Available online: http://www.ioffe.ru/SVA/NSM/Semicond/ (accessed on 22 January 2018).

355. Dean, C.R.; Young, A.F.; Meric, I.; Lee, C.; Wang, L.; Sorgenfrei, S.; Watanabe, K.; Taniguchi, T.; Kim, P.; Shepard, K.L.; et al. Boron nitride substrates for high-quality graphene electronics. *Nat. Nanotechnol.* **2010**, *5*, 722. [CrossRef] [PubMed]

356. Petrone, N.; Cui, X.; Hone, J.; Chari, T.; Shepard, K. Flexible 2D FETs using hBN dielectrics. In Proceedings of the 2015 IEEE International Electron Devices Meeting (IEDM), Washington, DC, USA, 7–9 December 2015; pp. 19.8.1–19.8.4. [CrossRef]

357. Fallahazad, B.; Movva, H.C.P.; Kim, K.; Larentis, S.; Taniguchi, T.; Watanabe, K.; Banerjee, S.K.; Tutuc, E. Shubnikov-de Haas Oscillations of High-Mobility Holes in Monolayer and Bilayer WSe$_2$: Landau Level Degeneracy, Effective Mass, and Negative Compressibility. *Phys. Rev. Lett.* **2016**, *116*, 086601. [CrossRef] [PubMed]

358. Xu, S.; Wu, Z.; Lu, H.; Han, Y.; Long, G.; Chen, X.; Han, T.; Ye, W.; Wu, Y.; Lin, J.; et al. Universal low-temperature Ohmic contacts for quantum transport in transition metal dichalcogenides. *2D Mater.* **2016**, *3*, 021007. [CrossRef]

359. Roy, T.; Tosun, M.; Kang, J.S.; Sachid, A.B.; Desai, S.B.; Hettick, M.; Hu, C.C.; Javey, A. Field-Effect Transistors Built from All Two-Dimensional Material Components. *ACS Nano* **2014**, *8*, 6259–6264. [CrossRef] [PubMed]

360. Bhattacharjee, S.; Ganapathi, K.L.; Chandrasekar, H.; Paul, T.; Mohan, S.; Ghosh, A.; Raghavan, S.; Bhat, N. Nitride Dielectric Environments to Suppress Surface Optical Phonon Dominated Scattering in High-Performance Multilayer MoS$_2$ FETs. *Adv. Electron. Mater.* **2017**, *3*, 1600358. [CrossRef]

361. Zhu, W.; Neumayer, D.; Perebeinos, V.; Avouris, P. Silicon Nitride Gate Dielectrics and Band Gap Engineering in Graphene Layers. *Nano Lett.* **2010**, *10*, 3572–3576. [CrossRef] [PubMed]

362. Sanne, A.; Ghosh, R.; Rai, A.; Movva, H.C.P.; Sharma, A.; Rao, R.; Mathew, L.; Banerjee, S.K. Top-gated chemical vapor deposited MoS$_2$ field-effect transistors on Si$_3$N$_4$ substrates. *Appl. Phys. Lett.* **2015**, *106*, 062101. [CrossRef]

363. Pelletier, J.; Anders, A. Plasma-based ion implantation and deposition: A review of physics, technology, and applications. *IEEE Trans. Plasma Sci.* **2005**, *33*, 1944–1959. [CrossRef]

364. Williams, J.S. Ion implantation of semiconductors. *Mater. Sci. Eng. A* **1998**, *253*, 8–15. [CrossRef]

365. Jin, S.W.; Cha, J.C.; Lee, H.S.; Son, S.H.; Kim, B.G.; Jung, Y.S. Implant and Anneal Technologies for Memory and CMOS Devices. In Proceedings of the 2016 21st International Conference on Ion Implantation Technology (IIT), Tainan, Taiwan, 26–30 September 2016; pp. 1–5. [CrossRef]

366. Ziegler, J.F. High energy ion implantation. *Nucl. Instrum. Methods Phys. Res. Sect. B Beam Interact. Mater. At.* **1985**, *6*, 270–282. [CrossRef]

367. Pearton, S.J. Ion implantation in iii–v semiconductor technology. *Int. J. Mod. Phys. B* **1993**, *7*, 4687–4761. [CrossRef]

368. Current, M.I. Ion implantation of advanced silicon devices: Past, present and future. *Mater. Sci. Semicond. Process.* **2017**, *62*, 13–22. [CrossRef]

369. Ziegler, J.F. *Ion Implantation Science and Technology*; Elsevier: Amsterdam, The Netherlands, 2012; ISBN 978-0-323-14401-8. [CrossRef]

370. Cho, H.J.; Oh, H.S.; Nam, K.J.; Kim, Y.H.; Yeo, K.H.; Kim, W.D.; Chung, Y.S.; Nam, Y.S.; Kim, S.M.; Kwon, W.H.; et al. Si FinFET based 10nm technology with multi Vt gate stack for low power and high performance applications. In Proceedings of the 2016 IEEE Symposium on VLSI Technology, Honolulu, HI, USA, 14–16 June 2016; pp. 1–2. [CrossRef]

371. Hsu, W.; Kim, T.; Chou, H.; Rai, A.; Banerjee, S.K. Novel BF+ Implantation for High Performance Ge pMOSFETs. *IEEE Electron Device Lett.* **2016**, *37*, 954–957. [CrossRef]

372. Hsu, W.; Kim, T.; Benítez-Lara, A.; Chou, H.; Dolocan, A.; Rai, A.; Josefina Arellano-Jiménez, M.; Palard, M.; José-Yacamán, M.; Banerjee, S.K. Diffusion and recrystallization of B implanted in crystalline and pre-amorphized Ge in the presence of F. *J. Appl. Phys.* **2016**, *120*, 015701. [CrossRef]

373. Hsu, W.; Rai, A.; Wang, X.; Wang, Y.; Kim, T.; Banerjee, S.K. Impact of Junction Depth and Abruptness on the Activation and the Leakage Current in Germanium n$^+$/p Junctions. *arXiv* **2017**, arXiv170506733. [CrossRef]

374. Cheung, N.W. Plasma immersion ion implantation for ULSI processing. *Nucl. Instrum. Methods Phys. Res. Sect. B Beam Interact. Mater. At.* **1991**, *55*, 811–820. [CrossRef]

375. Gibbons, J.F. Ion implantation in semiconductors—Part I: Range distribution theory and experiments. *Proc. IEEE* **1968**, *56*, 295–319. [CrossRef]

376. Borland, J.O. Low temperature activation of ion implanted dopants: A review. In Proceedings of the Extended Abstracts of the Third International Workshop on Junction Technology, Tokyo, Japan, 2–3 December 2002; pp. 85–88. [CrossRef]

377. Holland, O.W.; Appleton, B.R.; Narayan, J. Ion implantation damage and annealing in germanium. *J. Appl. Phys.* **1983**, *54*, 2295–2301. [CrossRef]

378. Gibbons, J.F. Ion implantation in semiconductors—Part II: Damage production and annealing. *Proc. IEEE* **1972**, *60*, 1062–1096. [CrossRef]

379. Pearton, S.J.; Von Neida, A.R.; Brown, J.M.; Short, K.T.; Oster, L.J.; Chakrabarti, U.K. Ion implantation damage and annealing in InAs, GaSb, and GaP. *J. Appl. Phys.* **1988**, *64*, 629–636. [CrossRef]

380. Narayan, J.; Holland, O.W. Characteristics of Ion-Implantation Damage and Annealing Phenomena in Semiconductors. *J. Electrochem. Soc.* **1984**, *131*, 2651–2662. [CrossRef]

381. Hummel, R.E.; Xi, W.; Holloway, P.H.; Jones, K.A. Optical investigations of ion implant damage in silicon. *J. Appl. Phys.* **1988**, *63*, 2591–2594. [CrossRef]

382. Jones, K.S.; Perry, S.; Murray, R.; Hynes, K.; Zhao, X. (Invited) An Overview of Doping Studies in MoS$_2$. *Meet. Abstr.* **2016**, *MA2016-01*, 1298. [CrossRef]

383. Yoon, A.; Lee, Z. Synthesis and Properties of Two Dimensional Doped Transition Metal Dichalcogenides. *Appl. Microsc.* **2017**, *47*, 19–28. [CrossRef]

384. Tong, L.; Liu, T.; Liang, R.; Wang, S.; Chen, J.; Dai, J.; Ye, L.; Tong, L.; Liu, T.; Liang, R.; et al. Growth of Transition Metal Dichalcogenides and Directly Modulating Their Properties by Chemical Vapor Deposition, Growth of Transition Metal Dichalcogenides and Directly Modulating Their Properties by Chemical Vapor Deposition. *Gen. Chem.* **2017**, *3*. [CrossRef]

385. Sun, Y.; Fujisawa, K.; Lin, Z.; Lei, Y.; Mondschein, J.S.; Terrones, M.; Schaak, R.E. Low-Temperature Solution Synthesis of Transition Metal Dichalcogenide Alloys with Tunable Optical Properties. *J. Am. Chem. Soc.* **2017**, *139*, 11096–11105. [CrossRef] [PubMed]

386. Lei, Y.; Pakhira, S.; Fujisawa, K.; Wang, X.; Iyiola, O.O.; Perea López, N.; Laura Elías, A.; Pulickal Rajukumar, L.; Zhou, C.; Kabius, B.; et al. Low-temperature Synthesis of Heterostructures of Transition Metal Dichalcogenide Alloys (W$_x$Mo$_{1-x}$S$_2$) and Graphene with Superior Catalytic Performance for Hydrogen Evolution. *ACS Nano* **2017**, *11*, 5103–5112. [CrossRef] [PubMed]

387. Zheng, Z.; Yao, J.; Yang, G. Centimeter-Scale Deposition of Mo$_{0.5}$W$_{0.5}$Se$_2$ Alloy Film for High-Performance Photodetectors on Versatile Substrates. *ACS Appl. Mater. Interfaces* **2017**, *9*, 14920–14928. [CrossRef] [PubMed]

388. Feng, Q.; Mao, N.; Wu, J.; Xu, H.; Wang, C.; Zhang, J.; Xie, L. Growth of MoS$_{2(1-x)}$Se$_{2x}$ (x = 0.41–1.00) Monolayer Alloys with Controlled Morphology by Physical Vapor Deposition. *ACS Nano* **2015**, *9*, 7450–7455. [CrossRef] [PubMed]

389. Li, H.; Duan, X.; Wu, X.; Zhuang, X.; Zhou, H.; Zhang, Q.; Zhu, X.; Hu, W.; Ren, P.; Guo, P.; et al. Growth of Alloy MoS$_{2x}$Se$_{2(1-x)}$ Nanosheets with Fully Tunable Chemical Compositions and Optical Properties. *J. Am. Chem. Soc.* **2014**, *136*, 3756–3759. [CrossRef] [PubMed]

390. Mann, J.; Ma, Q.; Odenthal, P.M.; Isarraraz, M.; Le, D.; Preciado, E.; Barroso, D.; Yamaguchi, K.; von Son Palacio, G.; Nguyen, A.; et al. 2-Dimensional Transition Metal Dichalcogenides with Tunable Direct Band Gaps: $MoS_{2(1-x)}Se_{2x}$ Monolayers. *Adv. Mater.* **2014**, *26*, 1399–1404. [CrossRef] [PubMed]

391. Chen, Y.; Wen, W.; Zhu, Y.; Mao, N.; Feng, Q.; Zhang, M.; Hsu, H.-P.; Zhang, J.; Huang, Y.-S.; Xie, L. Temperature-dependent photoluminescence emission and Raman scattering from $Mo_{1-x}W_xS_2$ monolayers. *Nanotechnology* **2016**, *27*, 445705. [CrossRef] [PubMed]

392. Kobayashi, Y.; Mori, S.; Maniwa, Y.; Miyata, Y. Bandgap-tunable lateral and vertical heterostructures based on monolayer $Mo_{1-x}W_xS_2$ alloys. *Nano Res.* **2015**, *8*, 3261–3271. [CrossRef]

393. Wang, Z.; Liu, P.; Ito, Y.; Ning, S.; Tan, Y.; Fujita, T.; Hirata, A.; Chen, M. Chemical Vapor Deposition of Monolayer $Mo_{1-x}W_xS_2$ Crystals with Tunable Band Gaps. *Sci. Rep.* **2016**, *6*, 21536. [CrossRef] [PubMed]

394. Chen, Y.; Xi, J.; Dumcenco, D.O.; Liu, Z.; Suenaga, K.; Wang, D.; Shuai, Z.; Huang, Y.-S.; Xie, L. Tunable Band Gap Photoluminescence from Atomically Thin Transition-Metal Dichalcogenide Alloys. *ACS Nano* **2013**, *7*, 4610–4616. [CrossRef] [PubMed]

395. Gong, Y.; Liu, Z.; Lupini, A.R.; Shi, G.; Lin, J.; Najmaei, S.; Lin, Z.; Elías, A.L.; Berkdemir, A.; You, G.; et al. Band Gap Engineering and Layer-by-Layer Mapping of Selenium-Doped Molybdenum Disulfide. *Nano Lett.* **2014**, *14*, 442–449. [CrossRef] [PubMed]

396. Kutana, A.; Penev, E.S.; Yakobson, B.I. Engineering electronic properties of layered transition-metal dichalcogenide compounds through alloying. *Nanoscale* **2014**, *6*, 5820–5825. [CrossRef] [PubMed]

397. Komsa, H.-P.; Krasheninnikov, A.V. Two-Dimensional Transition Metal Dichalcogenide Alloys: Stability and Electronic Properties. *J. Phys. Chem. Lett.* **2012**, *3*, 3652–3656. [CrossRef] [PubMed]

398. Xi, J.; Zhao, T.; Wang, D.; Shuai, Z. Tunable Electronic Properties of Two-Dimensional Transition Metal Dichalcogenide Alloys: A First-Principles Prediction. *J. Phys. Chem. Lett.* **2014**, *5*, 285–291. [CrossRef] [PubMed]

399. Susarla, S.; Kutana, A.; Hachtel, J.A.; Kochat, V.; Apte, A.; Vajtai, R.; Idrobo, J.C.; Yakobson, B.I.; Tiwary, C.S.; Ajayan, P.M. Quaternary 2D Transition Metal Dichalcogenides (TMDs) with Tunable Bandgap. *Adv. Mater.* **2017**, *29*, 1702457. [CrossRef] [PubMed]

400. Moon, R.L.; Antypas, G.A.; James, L.W. Bandgap and lattice constant of GaInAsP as a function of alloy composition. *J. Electron. Mater.* **1974**, *3*, 635–644. [CrossRef]

401. Williams, C.K.; Glisson, T.H.; Hauser, J.R.; Littlejohn, M.A. Energy bandgap and lattice constant contours of iii-v quaternary alloys of the form $A_xB_yC_zD$ or $AB_xC_yD_z$. *J. Electron. Mater.* **1978**, *7*, 639–646. [CrossRef]

402. Glisson, T.H.; Hauser, J.R.; Littlejohn, M.A.; Williams, C.K. Energy bandgap and lattice constant contours of iii–v quaternary alloys. *J. Electron. Mater.* **1978**, *7*, 1–16. [CrossRef]

403. Hill, R. Energy-gap variations in semiconductor alloys. *J. Phys. C Solid State Phys.* **1974**, *7*, 521. [CrossRef]

404. Kim, J.-S.; Ahmad, R.; Pandey, T.; Rai, A.; Feng, S.; Yang, J.; Lin, Z.; Terrones, M.; Banerjee, S.K.; Singh, A.K.; et al. Towards band structure and band offset engineering of monolayer $Mo_{(1-x)}W_{(x)}S_2$ via Strain. *2D Mater.* **2018**, *5*, 015008. [CrossRef]

405. Laskar, M.R.; Nath, D.N.; Ma, L.; Ii, E.W.L.; Lee, C.H.; Kent, T.; Yang, Z.; Mishra, R.; Roldan, M.A.; Idrobo, J.-C.; et al. p-type doping of MoS_2 thin films using Nb. *Appl. Phys. Lett.* **2014**, *104*, 092104. [CrossRef]

406. Suh, J.; Park, T.-E.; Lin, D.-Y.; Fu, D.; Park, J.; Jung, H.J.; Chen, Y.; Ko, C.; Jang, C.; Sun, Y.; et al. Doping against the Native Propensity of MoS_2: Degenerate Hole Doping by Cation Substitution. *Nano Lett.* **2014**, *14*, 6976–6982. [CrossRef] [PubMed]

407. Das, S.; Demarteau, M.; Roelofs, A. Nb-doped single crystalline MoS_2 field effect transistor. *Appl. Phys. Lett.* **2015**, *106*, 173506. [CrossRef]

408. Mirabelli, G.; Schmidt, M.; Sheehan, B.; Cherkaoui, K.; Monaghan, S.; Povey, I.; McCarthy, M.; Bell, A.P.; Nagle, R.; Crupi, F.; et al. Back-gated Nb-doped MoS_2 junctionless field-effect-transistors. *AIP Adv.* **2016**, *6*, 025323. [CrossRef]

409. Xu, E.Z.; Liu, H.M.; Park, K.; Li, Z.; Losovyj, Y.; Starr, M.; Werbianskyj, M.; Fertig, H.A.; Zhang, S.X. p-Type transition-metal doping of large-area MoS_2 thin films grown by chemical vapor deposition. *Nanoscale* **2017**, *9*, 3576–3584. [CrossRef] [PubMed]

410. Gao, J.; Kim, Y.D.; Liang, L.; Idrobo, J.C.; Chow, P.; Tan, J.; Li, B.; Li, L.; Sumpter, B.G.; Lu, T.-M.; et al. Transition-Metal Substitution Doping in Synthetic Atomically Thin Semiconductors. *Adv. Mater.* **2016**, *28*, 9735–9743. [CrossRef] [PubMed]

411. Hallam, T.; Monaghan, S.; Gity, F.; Ansari, L.; Schmidt, M.; Downing, C.; Cullen, C.P.; Nicolosi, V.; Hurley, P.K.; Duesberg, G.S. Rhenium-doped MoS_2 films. *Appl. Phys. Lett.* **2017**, *111*, 203101. [CrossRef]

412. Zhang, K.; Feng, S.; Wang, J.; Azcatl, A.; Lu, N.; Addou, R.; Wang, N.; Zhou, C.; Lerach, J.; Bojan, V.; et al. Manganese Doping of Monolayer MoS_2: The Substrate Is Critical. *Nano Lett.* **2015**, *15*, 6586–6591. [CrossRef] [PubMed]

413. Mishra, R.; Zhou, W.; Pennycook, S.J.; Pantelides, S.T.; Idrobo, J.-C. Long-range ferromagnetic ordering in manganese-doped two-dimensional dichalcogenides. *Phys. Rev. B* **2013**, *88*, 144409. [CrossRef]

414. Ramasubramaniam, A.; Naveh, D. Mn-doped monolayer MoS_2: An atomically thin dilute magnetic semiconductor. *Phys. Rev. B* **2013**, *87*, 195201. [CrossRef]

415. Dietl, T. A ten-year perspective on dilute magnetic semiconductors and oxides. *Nat. Mater.* **2010**, *9*, 965. [CrossRef] [PubMed]

416. Furdyna, J.K. Diluted magnetic semiconductors: An interface of semiconductor physics and magnetism (invited). *J. Appl. Phys.* **1982**, *53*, 7637–7643. [CrossRef]

417. Sato, K.; Bergqvist, L.; Kudrnovský, J.; Dederichs, P.H.; Eriksson, O.; Turek, I.; Sanyal, B.; Bouzerar, G.; Katayama-Yoshida, H.; Dinh, V.A.; et al. First-principles theory of dilute magnetic semiconductors. *Rev. Mod. Phys.* **2010**, *82*, 1633–1690. [CrossRef]

418. Huang, C.; Jin, Y.; Wang, W.; Tang, L.; Song, C.; Xiu, F. Manganese and chromium doping in atomically thin MoS_2. *J. Semicond.* **2017**, *38*, 033004. [CrossRef]

419. Wang, S.Y.; Ko, T.S.; Huang, C.C.; Lin, D.Y.; Huang, Y.S. Optical and electrical properties of MoS_2 and Fe-doped MoS_2. *Jpn. J. Appl. Phys.* **2014**, *53*, 04EH07. [CrossRef]

420. Li, B.; Huang, L.; Zhong, M.; Huo, N.; Li, Y.; Yang, S.; Fan, C.; Yang, J.; Hu, W.; Wei, Z.; et al. Synthesis and Transport Properties of Large-Scale Alloy $Co_{0.16}Mo_{0.84}S_2$ Bilayer Nanosheets. *ACS Nano* **2015**, *9*, 1257–1262. [CrossRef] [PubMed]

421. Yang, L.; Majumdar, K.; Du, Y.; Liu, H.; Wu, H.; Hatzistergos, M.; Hung, P.Y.; Tieckelmann, R.; Tsai, W.; Hobbs, C.; et al. High-performance MoS_2 field-effect transistors enabled by chloride doping: Record low contact resistance (0.5 k$\Omega\cdot\mu$m) and record high drain current (460 $\mu A/\mu$m). In Proceedings of the 2014 Symposium on VLSI Technology (VLSI-Technology), Honolulu, HI, USA, 9–12 June 2014; pp. 1–2. [CrossRef]

422. Yang, L.; Majumdar, K.; Liu, H.; Du, Y.; Wu, H.; Hatzistergos, M.; Hung, P.Y.; Tieckelmann, R.; Tsai, W.; Hobbs, C.; et al. Chloride Molecular Doping Technique on 2D Materials: WS_2 and MoS_2. *Nano Lett.* **2014**, *14*, 6275–6280. [CrossRef] [PubMed]

423. Qin, S.; Lei, W.; Liu, D.; Chen, Y. In-situ and tunable nitrogen-doping of MoS_2 nanosheets. *Sci. Rep.* **2014**, *4*. [CrossRef] [PubMed]

424. Azcatl, A.; Qin, X.; Prakash, A.; Zhang, C.; Cheng, L.; Wang, Q.; Lu, N.; Kim, M.J.; Kim, J.; Cho, K.; et al. Covalent Nitrogen Doping and Compressive Strain in MoS_2 by Remote N_2 Plasma Exposure. *Nano Lett.* **2016**, *16*, 5437–5443. [CrossRef] [PubMed]

425. Neal, A.T.; Pachter, R.; Mou, S. P-type conduction in two-dimensional MoS_2 via oxygen incorporation. *Appl. Phys. Lett.* **2017**, *110*, 193103. [CrossRef]

426. Giannazzo, F.; Fisichella, G.; Greco, G.; Di Franco, S.; Deretzis, I.; La Magna, A.; Bongiorno, C.; Nicotra, G.; Spinella, C.; Scopelliti, M.; et al. Ambipolar MoS_2 Transistors by Nanoscale Tailoring of Schottky Barrier Using Oxygen Plasma Functionalization. *ACS Appl. Mater. Interfaces* **2017**, *9*, 23164–23174. [CrossRef] [PubMed]

427. Nipane, A.; Karmakar, D.; Kaushik, N.; Karande, S.; Lodha, S. Few-Layer MoS_2 p-Type Devices Enabled by Selective Doping Using Low Energy Phosphorus Implantation. *ACS Nano* **2016**, *10*, 2128–2137. [CrossRef] [PubMed]

428. Nipane, A.; Kaushik, N.; Karande, S.; Karmakar, D.; Lodha, S. P-type doping of MoS_2 with phosphorus using a plasma immersion ion implantation (PIII) process. In Proceedings of the 2015 73rd Annual Device Research Conference (DRC), Columbus, OH, USA, 21–24 June 2015; pp. 191–192. [CrossRef]

429. Xu, K.; Zhao, Y.; Lin, Z.; Long, Y.; Wang, Y.; Chan, M.; Chai, Y. Doping of two-dimensional MoS_2 by high energy ion implantation. *Semicond. Sci. Technol.* **2017**, *32*, 124002. [CrossRef]

crystals

MDPI

Review

Vertical Transistors Based on 2D Materials: Status and Prospects

Filippo Giannazzo [1,*] , Giuseppe Greco [1], Fabrizio Roccaforte [1] and Sushant S. Sonde [2,3,*]

1   Consiglio Nazionale delle Ricerche—Institute for Microelectronics and Microsystems (CNR-IMM),
    Strada VIII, 5 I-95121 Catania, Italy; giuseppe.greco@imm.cnr.it (G.G.); fabrizio.roccaforte@imm.cnr.it (F.R.)
2   Institute for Molecular Engineering, The University of Chicago, Eckhardt Research Center,
    5640 South Ellis Avenue, Chicago, IL 60637, USA
3   Center for Nanoscale Materials, Argonne National Laboratory, 9700 Cass Ave, Lemont, IL 60439, USA
*   Correspondence: filippo.giannazzo@imm.cnr.it (F.G.); sushantsonde@uchicago.edu (S.S.S.)

Received: 15 December 2017; Accepted: 29 January 2018; Published: 31 January 2018

Abstract: Two-dimensional (2D) materials, such as graphene (Gr), transition metal dichalcogenides (TMDs) and hexagonal boron nitride (h-BN), offer interesting opportunities for the implementation of vertical transistors for digital and high-frequency electronics. This paper reviews recent developments in this field, presenting the main vertical device architectures based on 2D/2D or 2D/3D material heterostructures proposed so far. For each of them, the working principles and the targeted application field are discussed. In particular, tunneling field effect transistors (TFETs) for beyond-CMOS low power digital applications are presented, including resonant tunneling transistors based on Gr/h-BN/Gr stacks and band-to-band tunneling transistors based on heterojunctions of different semiconductor layered materials. Furthermore, recent experimental work on the implementation of the hot electron transistor (HET) with the Gr base is reviewed, due to the predicted potential of this device for ultra-high frequency operation in the THz range. Finally, the material sciences issues and the open challenges for the realization of 2D material-based vertical transistors at a large scale for future industrial applications are discussed.

Keywords: graphene; 2D materials; van der Waals heterostructures; vertical field effect transistors; hot electron transistors

1. Introduction

In 2004, the pioneering works on the field effect in atomically thin carbon films [1], from then on named graphene (Gr), gave birth to an entirely new research branch of solid-state electronics, focused on the use of two-dimensional (2D) materials and their heterostructures for electronics/optoelectronics devices with unconventional or improved performances compared to traditional semiconductor devices. Due to its excellent carrier mobility (up to ~10^5 cm^2 V^{-1} s^{-1}) [2,3] and micrometer electron mean free path [4–7], Gr has been considered since the first studies as the channel material for fast field effect transistors (FETs) [8]. Essentially, the Gr field effect transistor (GFET) resembles the classical metal-oxide-semiconductor FET architecture, where lateral transport in the channel is modulated by a gate electrode separated from Gr by a thin insulator. As a matter of fact, the lack of a bandgap in the electronics band structure of Gr results in a poor ON/OFF current ratio in GFETs, making them unsuitable for digital (logic) or switching applications [9]. On the other hand, GFETs can be of interest for radio frequency (RF) applications, where fast current modulation of the device operated in the on-state is required and switch-off is not necessarily needed [9,10]. To date, RF GFETs allowing current amplification at very high frequencies (>400 GHz) have been demonstrated [11]. However, the same devices suffer from limited performances in terms of voltage and power amplification, mainly due to the high output conductance resulting from the lack of a bandgap. On the other hand, the peculiar symmetric ambipolar conduction of GFETs has been exploited to demonstrate novel device concepts, such as the RF mixer [12].

To overcome the Gr fundamental limitations arising from the missing bandgap both in digital and RF electronics, new solutions have been explored inside the wide family of 2D materials. In particular, two routes have been followed by scientists working in this field.

The first route was replacing Gr as the channel material in lateral FETs with semiconducting 2D materials, such as some members of the transition metal dichalcogenides (TMDs) family (MoS_2, WS_2, $MoSe_2$, WSe_2) [13] or phosphorene (a 2D lattice composed of phosphorus atoms) [14]. The second route has been to introduce novel device architectures based on van der Waals (vdW) heterostructures obtained by the stacking of 2D materials (such as Gr, hexagonal boron nitride (h-BN), TMDs) [15,16] or by 2D material heterojunctions with thin conventional 3D (i.e., bulk) semiconductors [17]. These devices rely on quite different working principles than traditional lateral FETs and mainly exploit vertical current transport across the interfaces of these materials. They include the tunneling field effect transistors (TFETs) [18,19], the band-to-band tunneling transistor [20], the transistor based on the field effect modulation of the Gr/semiconductor Schottky barrier (barristor) [21] and the hot electron transistor (HET) with the base made with a 2D material [22–25]. These devices typically show high ON/OFF current ratios, not reachable by conventional lateral GFETs, that make them suitable for logic and switching applications. Furthermore, some of these device concepts, like the HET with a Gr base, are especially targeted to operate at ultra-high frequencies up to THz.

This paper reviews recent developments in 2D material-based vertical transistors. Section 2 discusses the open issues of Gr and TMD lateral FETs, and it serves to introduce some of the potential advantages of vertical architectures. Therefore, the main vertical device structures considered so far are presented in the Section 3, where the working principles and the targeted application fields for each structure are discussed. In particular, recent implementations of TFET based on 2D materials are overviewed due to their interest in beyond-CMOS digital electronics. Furthermore, recent experimental activity on the HET with Gr base is presented, considering the predicted potential of these devices in ultra-high frequency electronics. In Section 4, the materials science issues and the open challenges for the realization of 2D material vertical transistors at a large scale are illustrated. Finally, the last section includes a summary and some prospects for future industrial applications of the discussed device structures.

2. Lateral Field Effect Transistors

Due to the proper bandgap (in the range from 1–2 eV), combined with a good stability under ambient conditions, semiconductor TMDs are very promising candidates as channel materials for digital electronics [13]. As an example, MoS_2-channel FETs have been fabricated with large I_{on}/I_{off} ratios (>10^4) and small subthreshold swings (SS < 80 mV/decade) [26], approaching the desired requirements of FETs for CMOS digital circuits. The energy gap of TMDs comes, however, at the cost of a relatively low mobility. As an example, the upper theoretical limit for the mobility of monolayer MoS_2 at room temperature is about 400 cm^2 V^{-1} s^{-1} [27], whereas the experimental values reported so far are in the range of a few tens of cm^2 V^{-1} s^{-1} [28,29].

Besides TMDs, also phosphorene has recently attracted interest as a semiconducting channel material for FETs. A mobility of 286 cm^2 V^{-1} s^{-1} has been reported for few-layer phosphorene [14], whereas values up to ~1000 cm^2 V^{-1} s^{-1} have been shown for multilayer phosphorene with an ~10 nm thickness [30]. However, the main disadvantage of phosphorene (as compared to TMDs) is its chemical reactivity under ambient conditions, which can represent a serious concern for practical applications.

The ultimate thin body of TMDs can be very beneficial for the scaling prospects of lateral FETs for CMOS applications, as discussed in many simulation works [31–34]. As an example, Liu et al. [33] predicted that MoS_2 FETs can meet the requirements of the International Technology Roadmap for Semiconductors (ITRS) [35] down to a minimum channel length of 8 nm. For such aggressively reduced geometries, the low mobility of MoS_2 is not a real issue, because transport can be considered as almost ballistic for channel lengths below 10 nm. Although most of these predictions are based on simulations, some experimental work has been also reported, where the challenges of channel length scaling in TMD lateral transistors down to the nanometric limit started to be addressed [36].

Concerning high frequency applications, the first Gr-based FET capable of RF operation, fabricated using exfoliated Gr from graphite, was reported in 2008 [37]. Later on, the development of advanced synthesis methods, such as epitaxial growth of Gr on SiC by controlled high temperature graphitization [38–41] or chemical vapor deposition (CVD) on catalytic metals [42], provided high quality Gr of a large area for device fabrication. Ultra-scaled transistors with interesting RF performances have been demonstrated both using transferred CVD Gr [43] (see, e.g., Figure 1a) and epitaxial Gr on SiC (see, e.g., Figure 1b) [43,44].

RF transistors are typically used for the amplification of a high frequency input signal (current or voltage), and the amplifier gain decreases with increasing frequency. Hence, the two main figures of merit for RF transistors are the cut-off frequency f_T (i.e., the frequency for which current gain is reduced to unity) and the maximum oscillation frequency f_{MAX} (i.e., the frequency for which power gain is reduced to unity). As in the case of more conventional RF transistors, the f_T of GFETs was found to increase with reducing the channel length L (see, as an example, Figure 1c). A record value of $f_T = 427$ GHz has been reported for scaled devices obtained with a self-aligned fabrication process [11].

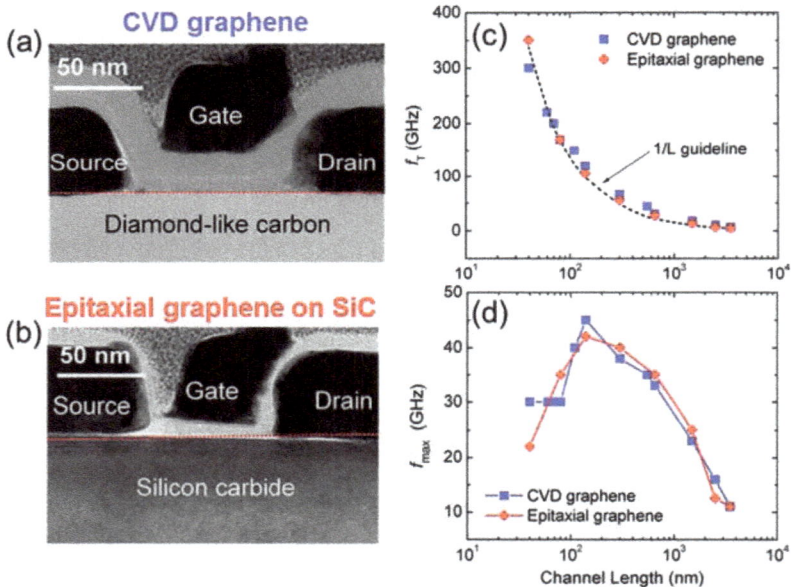

Figure 1. Cross-sectional TEM micrographs of scaled RF graphene (Gr) field effect transistors (GFETs) fabricated with transferred CVD Gr (**a**) and with epitaxial Gr grown on SiC(0001) (**b**); behavior of the cut-off frequency f_T (**c**) and of the maximum oscillation frequency f_{MAX} (**d**) as a function of channel length. Figures adapted with permission from [43].

These values of f_T are comparable with those achieved by the state of the art InP high electron mobility transistors (HEMTs) with similar channel lengths. However, for most RF applications, both high f_T and high f_{MAX} are required, and unfortunately, f_{MAX} values measured for GFETs are significantly smaller than f_T values, as illustrated in Figure 1d [43]. Furthermore, contrary to common expectations for RF FETs, f_{MAX} shows a non-monotonic behavior with the channel length L. In fact, it reaches a peak value around 150 nm and decreases for lower L values.

The reason for these poorer power gain performances can be argued by comparing the theoretical expressions of f_T and f_{MAX} for RF transistors:

$$f_T = \frac{g_m}{2\pi\left(C_{gs} + C_{gd}\right)} \left[\frac{1}{1 + g_d\left(R_s + R_d\right) + g_m\frac{C_{gd}\left(R_s + R_d\right)}{C_{gs} + C_{gd}}}\right] \tag{1}$$

$$f_{MAX} = \frac{f_T}{2\sqrt{g_d\left(R_s + R_{gs}\right) + 2\pi f_T R_{gs} C_{gd}}} \tag{2}$$

Here, C_{gs} and C_{gd} are the gate-source and gate-drain coupling capacitances, R_s and R_d are the source and drain series resistances, R_{gs} is the gate-source resistance, $g_m = dI_D/dV_G$ is the transconductance and $g_d = dI_D/dV_D$ is the output conductance.

The drain current I_D of an FET is proportional to the saturated velocity v_s and to the sheet concentration n_s of the carriers in the channel. Hence, the two prerequisites to achieve a high transconductance are a high v_s and a high dn_s/dV_G, i.e., an effective modulation of the carrier density with the gate bias.

If the R_s and R_d resistance contributions in Equation (1) are properly minimized, the expression of f_T can be approximated as $f_T \approx g_m/[2\pi(C_{gs} + C_{gd})]$, and it results in being independent of the output conductance g_d. For this reason the high carrier mobility and saturation velocity of Gr results in a high transconductance g_m and, hence, in a high f_T. On the other hand, from Equation (2), it is evident that, even minimizing R_s, the output conductance g_d still plays a role in the expression of f_{MAX}. As a matter of fact, the output characteristics of Gr channel FETs exhibit a poor saturation behavior, i.e., a large g_d. This overcompensates the effect of a large g_m, ultimately resulting in degradation of f_{MAX}. The non-monotonic behavior of f_{MAX} in Figure 1d has been also ascribed to the competing contributions from f_T, g_d and R_{gs} as L decreases [43].

The poor saturation of the output characteristics is mainly a consequence of the missing bandgap in the Gr band structure. Hence, this peculiar physical property of Gr not only hinders its application in digital electronics, but severely limits also the high frequency performances of GFETs in terms of power amplification and f_{MAX}. Finally, the high off-state current (I_{off}) of GFETs results in a high power dissipation and represents a significant concern in terms of energy efficiency.

Besides Gr, single and multiple layers of MoS_2 have been also investigated as channel materials in lateral FETs for RF applications. Thanks to an electron saturation velocity $v_s > 3 \times 10^6$ cm/s [45] and to a high bandgap (resulting in a high ratio $g_m/g_d > 30$), MoS_2 FETs can achieve, in principle, both current and power amplification [46–49]. However, quite low values of f_T and f_{MAX} have been reported to date. Initial work on exfoliated monolayer MoS_2 RF FETs yielded $f_T = 2$ GHz and $f_{MAX} = 2.2$ GHz at a gate length of 240 nm [48]. Figure 2 shows a cross-sectional TEM micrograph (a) and the DC output (b) and transfer (c) characteristics of an FET fabricated with multilayer MoS_2 flakes. For these devices, the scaling behavior of f_T and f_{MAX} with the channel length is illustrated in Figure 2d,e, showing how $f_T = 42$ GHz and $f_{MAX} = 50$ GHz are achieved at a gate length of 68 nm [47]. More recently, RF FETs fabricated with monolayer MoS_2 deposited by CVD showed $f_T = 6.7$ GHz and $f_{MAX} = 5.3$ GHz at a gate length of 250 nm [49].

Several issues still need to be addressed to evaluate the real potentialities of TMDs both in digital and RF electronics. Besides the issues related to the lattice defects (such as chalcogen vacancies) [50–52] and impurities [53] commonly present in these compound materials, some critical processing steps need to be developed. These include the fabrication of low resistance source/drain contacts [54,55] and doping [56–58].

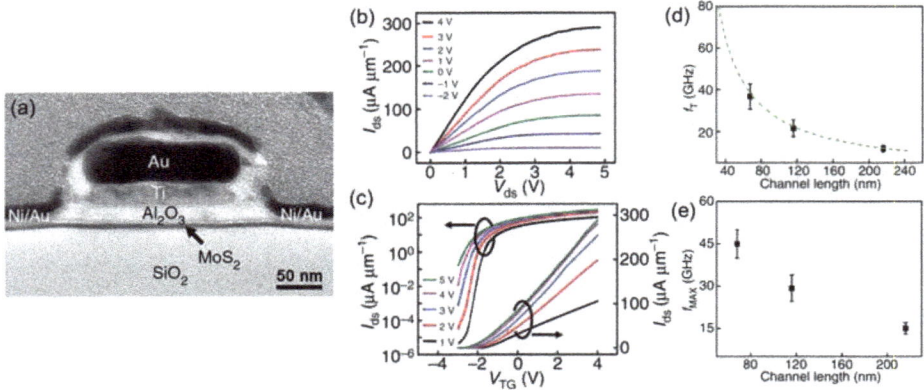

Figure 2. Cross-sectional TEM micrograph (**a**); DC output (**b**) and transfer (**c**) characteristics of an FET fabricated with multilayer MoS$_2$. Scaling behavior of f_T (**d**) and f_{MAX} (**e**) with the channel length. Figures adapted with permission from [47].

As an example, MoS$_2$ thin films are typically unintentionally n-type doped. Furthermore, most of the elementary metals exhibit a Fermi level pinning close to the MoS$_2$ conduction band, resulting in a small (but not negligible) Schottky barrier height (SBH) for electrons' injection and a high SBH for holes' injection. The origin of this Fermi level pinning is still a matter of debate, although some nanoscale electrical investigations highlighted the possible role of the defects present at the MoS$_2$ surface [59,60]. As a matter of fact, this Schottky barrier results in a significant source/drain contact resistance [54,61], which degrades the intrinsic performances of the transistor. Furthermore, the high injection barrier for holes makes it difficult to achieve p-type or ambipolar transport in MoS$_2$ FETs [62,63]. On the other hand, ambipolar transistors can be useful not only for logic (CMOS) applications, but also for some RF circuits. To date, p-type MoS$_2$ transistors have been fabricated by using high work function MoOx contacts [64]. Recently, multilayer MoS$_2$ transistors with ambipolar behavior have been demonstrated by selective-area p-type doping in the source/drain regions with O$_2$ plasma [65]. Besides MoS$_2$, other TMDs have been also considered for FETs' fabrication. As an example, WSe$_2$ is a slightly p-doped semiconductor, allowing the fabrication of both p-type and n-type FETs by proper selection of the metal contacts [66–68]. However, the problem of the contact resistance still holds also in the case of WSe$_2$.

Most of the TMD-based FETs have been fabricated using metals as source-drain contacts and high permittivity (high-k) dielectrics as gate insulators. Recently, some attempts at fabricating FETs with all components formed by 2D materials have been reported. As an example, Roy and co-workers have demonstrated an FET including MoS$_2$ as the channel material, h-BN as the dielectric layer for top electrode isolation and Gr for semi-metal contacts (see the schematic and the optical microscopy in Figure 3a,b) [69]. An optimal Ohmic contact between Gr and MoS$_2$ is possible by tuning the Gr Fermi level by the SiO$_2$/Si back-gate bias. Due to its large bandgap, h-BN acts as a top-gate dielectric, allowing current modulation over several orders of magnitude (see Figure 3c). Furthermore, thanks to its atomically-smooth surface without charge trapping, h-BN forms an ideal interface with the MoS$_2$ channel [3]. One of the most relevant advantages of this smooth interface is that the channel mobility of this device remains constant at high gate bias values (see Figure 3d), different from that in common FETs, where a decrease of mobility is observed at high fields due to the effect of the interface roughness.

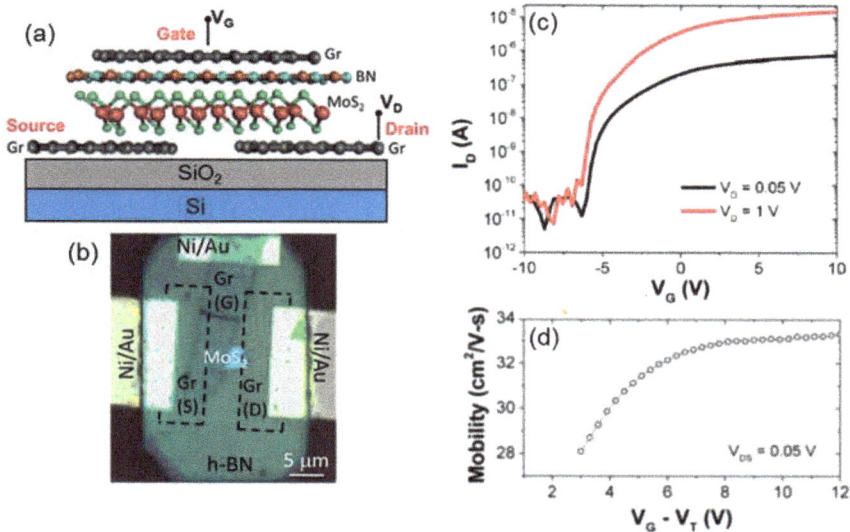

Figure 3. Schematic (**a**) and optical microscopy (**b**) of an FET with all components formed by 2D materials, where MoS$_2$ works as the channel, hexagonal boron nitride (h-BN) as the dielectric layer for top electrode isolation and Gr for semimetal source/drain contacts. Transfer characteristics (**c**) and mobility vs. gate bias behavior (**d**) of this transistor. Figures adapted with permission from [69]

3. Vertical Transistors

3.1. Tunneling Field Effect Transistors

One of the main issues for modern digital electronics is the dramatic increase of power consumption with the increase in the device integration density. As a matter of fact, the minimum supply voltage to switch an MOSFET from the OFF to the ON state is determined by the thermionic emission mechanism of carrier injection over the energy barrier at the source. This mechanism results in a theoretical limit of 60 mV/decade for the minimum subthreshold swing (SS) for MOSFET devices. On the other hand, tunneling field effect transistors (TFETs), based on quantum mechanical tunneling across an energy barrier, have the potential of reduced supply voltage, since a pure tunneling process is not thermally activated. Numerous studies have been conducted in the last decade to implement this device concept using bulk (3D) semiconductors, such as Ge, III-V and Si [70,71], and sub-thermionic SS values have been reported with TFETs based on these materials. However, the demonstrated performances have not been good enough for practical applications. In particular, one of the main limitations is that SS < 60 mV/decade (at room temperature) is typically obtained only at low drain currents, whereas it would be desirable to get such a behavior over a current range of several orders of magnitude. One of the reasons for the difficulty in obtaining sharp switching over a wide current range is the presence of band-tail states in bulk semiconductors [72]. Under this point of view, 2D materials with sharp band edges even at a thickness of a monolayer can represent an interesting platform to implement TFETs.

Britnell and co-workers first reported a TFET with the vertically-stacked heterostructure composed of Gr and thin h-BN [18]. Figure 4a schematically shows the layer stacking (1); the energy band diagram under equilibrium (2); under the effect of the back-gate bias V$_g$ (3); and under the effect of V$_g$ and of a bias V$_b$ between the two Gr layers (4). The basic principle of this vertical transistor is the quantum tunneling between the two Gr electrodes separated by the thin h-BN barrier. Current tunneling was controlled by tuning the density of states in Gr and the associated barrier by the external gate voltage. Figure 4b shows the tunneling current density J vs. the bias V_b for different values of V_g. Figure 4c shows the conductance dJ/dV_b (at $V_b = 0$) as a function of V_g, from which an ON/OFF ratio of ~50 was deduced.

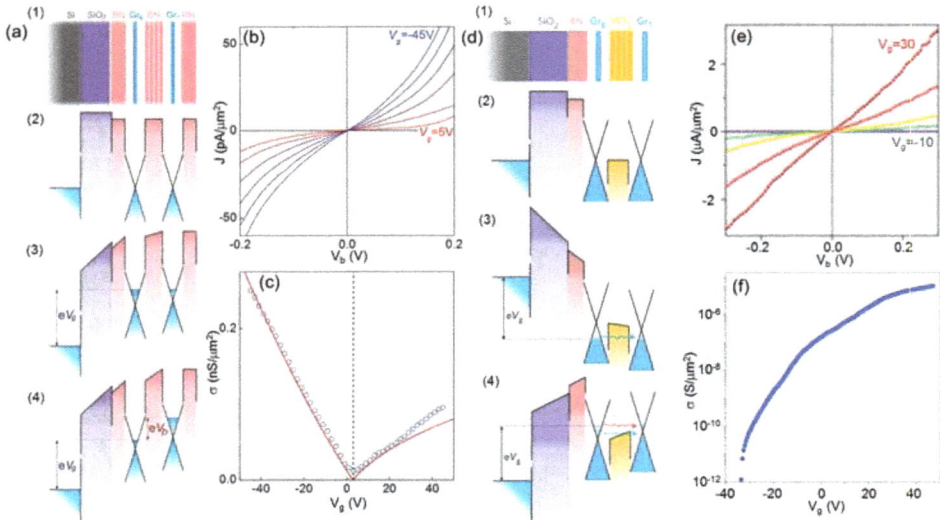

Figure 4. (**a**) Schematic illustration of a Gr/h-BN/Gr field effect tunneling transistor: (1) layer stacking; (2) energy band diagrams under equilibrium; (3) under the effect of the back-gate bias V_g; and (4) under the effect of V_g and of a bias V_b between the two Gr layers; (**b**) tunneling current density J vs. V_b for different values of V_g; (**c**) conductance $\sigma = dJ/dV_b$ as a function of V_g; (**d**) schematic illustration of a Gr/WS$_2$/Gr field effect tunneling transistor: (1) layer stacking; (2) energy band diagrams under equilibrium; (3) under the effect of $V_g < 0$; and (4) under the effect of $V_g > 0$; (**e**) tunneling current density J vs. V_b for different values of V_g; (**f**) conductance $\sigma = dJ/dV_b$ as a function of V_g. Figures adapted with permission from [18,19].

As the transit time associated with tunneling is very low, the field effect tunneling transistor can be potentially suitable for high speed operation. On the other hand, as a result of the direct tunneling mechanism, this device suffers from very low current density (in the order of 10–100 pA/μm^2), making it not useful for practical applications. Starting from the same idea, Georgiou and co-workers reported a Gr vertical FET with WS$_2$ layers as the barrier [19]. Figure 4d schematically shows the layer stacking (1); and the energy band diagrams under equilibrium (2); under the effect of $V_g < 0$ (3); and under the effect of $V_g > 0$ (4). Due to the smaller band gap of WS$_2$, current transport between the two Gr layers occurs by direct tunneling for $V_g < 0$ and by tunneling or thermionic emission for $V_g > 0$. Figure 4e shows the tunneling current density J vs. the bias V_b for different values of V_g, whereas Figure 4f shows the conductance as a function of V_g. As compared to the Gr/h-BN/Gr prototype, this device exhibits much higher ON current (in the order of 1 $\mu A/\mu m^2$) and better current modulation, with an ON/OFF current ratio up to 10^6.

3.1.1. Resonant Interlayer Tunneling Transistors

Progress in the alignment and transfer techniques of 2D materials permitted the demonstration of resonant tunneling phenomena in TFETs. Devices showing gate-tunable negative differential resistance (NDR) of the output characteristics due to resonant tunneling were first obtained by Gr/h-BN/Gr stacks with a precise rotational crystallographic alignment between the two Gr monolayers [73,74]. Since carriers in a Gr monolayer are populated near the *K*-point on the periphery of its Brillouin zone, conservation of both energy and momentum in the tunneling from one layer to the other is allowed only in the presence of a rotational alignment in momentum space.

Following the first demonstrations with double monolayer Gr, resonant TFET using bilayer Gr as the top and bottom electrodes and h-BN as the interlayer tunnel barrier were also demonstrated [75,76]. Due to the more complex band structure of bilayer Gr (with two sub-bands both in the conduction and valence bands at the *K*-point), additional NDR peaks occur at higher interlayer bias [75]. In fact, when the first sub-band of one bilayer energetically aligns with the second sub-band of the opposite bilayer, a second resonant tunneling condition is established. More recently, experimental results for resonant TFETs with multilayer Gr electrodes separated by an h-BN tunnel barrier have been also reported [77]. With an increase in the Gr electrode layer thickness, from bilayer to pentalayer Gr, the resonance peaks have been shown to become narrower in width and stronger in intensity, mainly due to the increase in the density of states with the increase in the Gr thickness. On the other hand, due to the increased complexity in the band structure with multiple sub-bands for thicker Gr, multiple resonance conditions arise in the output characteristics.

h-BN has been widely used as the interlayer in resonant TFETs with symmetric Gr electrodes, due to its good insulating properties and chemically-inert and atomically-flat surface. However, its wide energy bandgap (~5.8 eV) severely limits the peak current at resonance in Gr/h-BN/Gr TFETs [77]. In this context, using TMDs with a smaller bandgap as tunnel barriers may enhance the peak-to-valley ratio of the resonances in the electrical characteristics. Recently, Burg et al. [78] demonstrated gate-tunable resonant tunneling and NDR between two rotationally-aligned bilayer Gr sheets separated by a bilayer WSe_2. Remarkable large interlayer current densities of 2 $\mu A/\mu m^2$ and NDR peak-to-valley ratios of ~4 were observed at room temperature in these device structures.

Recently, the possibility of realizing resonant TFETs using TMD electrodes instead of Gr has been also considered [79,80]. Theoretical reports indicate that vertical heterostructures consisting of two identical monolayer MoS_2 electrodes separated by an h-BN barrier can result in a peak-to-valley ratio several orders of magnitude higher than the best that can be achieved using Gr electrodes [79]. However, practical implementation of resonant tunneling TFETs with identical electrodes (different than Gr) proved to be difficult.

On the other hand, many vertical transistor demonstrators have been implemented with differing bottom and top electrode layers, exploiting the principle of band-to-band tunneling, as discussed in the following.

3.1.2. Band-To-Band Tunneling Vertical Transistor

Roy et al. [81] first experimentally demonstrated interlayer band-to-band tunneling in vertical MoS_2/WSe_2 vdW heterostructures using a dual-gate device architecture. The electric potential and carrier concentration of the MoS_2 and WSe_2 layers were independently controlled by the two symmetric gates. Depending on the gate bias, the device behaves as either an Esaki diode with NDR, a backward diode with large reverse bias tunneling current or a forward rectifying diode with low reverse bias current. Notably, the weak electrostatic screening by the atomically thin MoS_2 and WSe_2 layers resulted in a high gate coupling efficiency for tuning the interlayer band alignments. Later on, Nourbakhsh et al. [82] further investigated band-to-band tunneling in the transverse and lateral directions of the MoS_2/WSe_2 heterojunctions. The room-temperature NDR in a heterojunction diode formed by few-layer WSe_2 stacked on multilayer MoS_2 was attributed to the lateral band-to-band tunneling at the edge of this heterojunction.

A band-to-band tunneling vertical transistor has been demonstrated also using the vdW heterojunction between differently-doped layered semiconductors as WSe_2 and $SnSe_2$, where WSe_2 worked as the back-gate-controlled *p*-layer and $SnSe_2$ was the degenerately *n*-type-doped layer [83].

Yan et al. [84] demonstrated room temperature Esaki tunnel diodes using a vdW heterostructure made of two layered semiconductors with a broken-gap energy band offset: black phosphorus (BP) and tin diselenide ($SnSe_2$). The presence of a thin insulating barrier between BP and $SnSe_2$ enabled the observation of a prominent NDR region in the forward-bias current voltage characteristics, with a peak to valley ratio of 1.8 at 300 K and a weak temperature dependence, indicating electron tunneling as the dominant transport mechanism.

Another recently demonstrated very interesting device concept is based on the field effect modulation of current transport across the *p-n* heterojunction between 3D and 2D semiconductors. The 3D semiconductor component of the heterojunction is heavily doped in equilibrium by substitutional dopants, whereas the doping level of the 2D semiconductor component can be tuned by the field effect. Therefore, gate-tunable 2D–3D *p-n* heterojunctions provide a unique opportunity to realize band-to-band tunneling devices. As an example, Sarkar and co-workers have recently demonstrated a band-to-band tunnel FET with a vertical heterojunction between a *p*-type Ge and an *n*-type bilayer MoS$_2$ [20] (see the schematic representation in Figure 5a). The band diagrams for the device in the OFF and ON states are illustrated in Figure 5b and the transfer characteristics in Figure 5c. By gating the MoS$_2$ into the high *n*-type doping regime, direct tunneling occurs from the Ge valence band to the MoS$_2$ conduction band. Figure 5d illustrates the values of the SS at room temperature as a function of the drain current for this bilayer MoS$_2$/*p*-Ge TFET and for a conventional MOSFET fabricated with a bilayer MoS$_2$ channel. Different from the conventional FET, this TFET exhibits an SS lower than the thermionic limit of 60 mV/decade in the considered drain current range (from 10^{-13} to 10^{-9} A). On the other hand, for larger drain currents, significantly larger SS values are obtained.

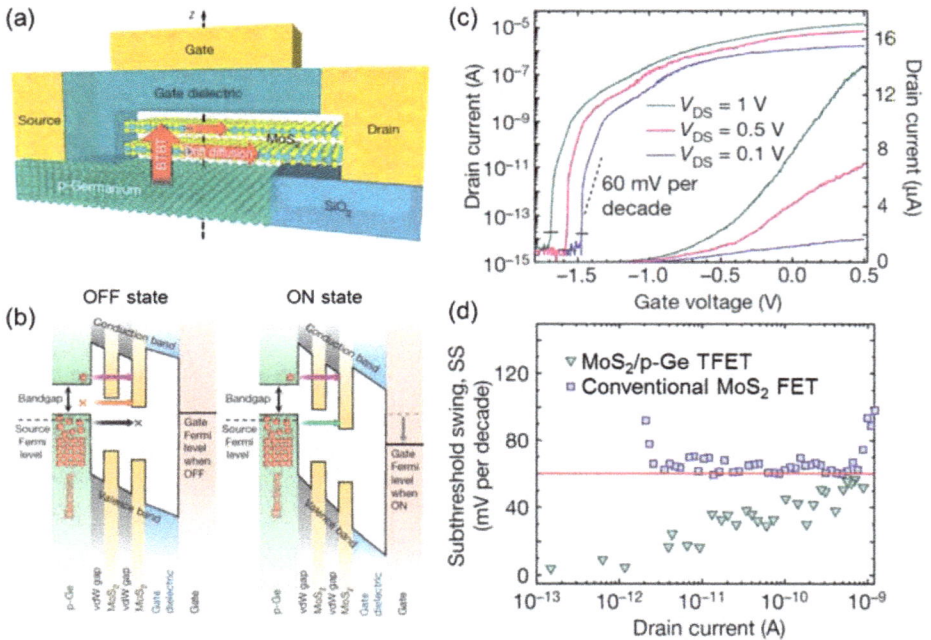

Figure 5. (a) Schematic cross-section of a gate-modulated bilayer MoS$_2$/*p*-Ge junction; (b) band diagrams of the device in the OFF and ON state; (c) transfer characteristics of the device for different V$_{DS}$; (d) comparison of the subthreshold swing (SS) of this bilayer MoS$_2$/*p*-Ge tunneling field effect transistor (TFET) with that of a conventional FET with a bilayer MoS$_2$ channel. Figures adapted with permission from [20].

3.2. Gate Modulated Schottky Barrier Transistor (Barristor)

The Barristor device concept is based on the tunability of the Schottky barrier height of a Gr contact with a semiconductor by an external electric field. Clearly, a nearly ideal interface between Gr and the semiconductor, without interface states responsible for Fermi level pinning, is required to achieve an efficient field effect modulation of the Schottky barrier height. The first Barristor was demonstrated by transferring CVD graphene onto hydrogen-passivated Si, thus obtaining a nearly ideal Schottky diode behavior both with *n*- and *p*-type Si [21]. Figure 6a illustrates a cross-sectional schematic of a Gr/Si barristor, and Figure 6b shows the band diagram for the Gr/*n*-Si Schottky junction for $V_g > 0$ and $V_g < 0$. The modulation of the Gr/*n*-Si Schottky barrier height with the gate bias is shown in Figure 6c, and the resulting output characteristics of the device (for different V_g values) are reported in Figure 6d. A current ON/OFF ratio of ~10^5 under forward bias was achieved, which is suitable for digital logic applications.

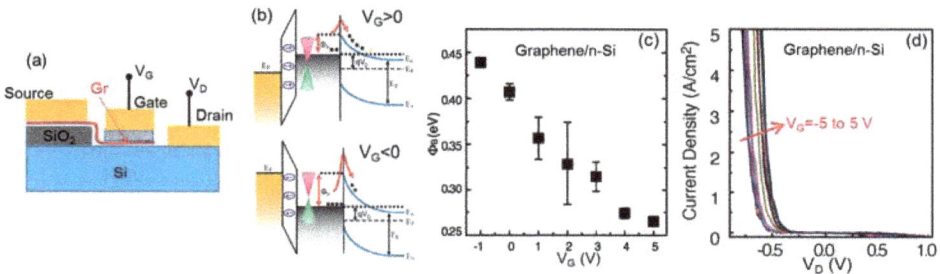

Figure 6. (**a**) Schematic cross-section of a Gr/Si Barristor; (**b**) band-diagrams of the Gr/*n*-Si device for $V_g > 0$ (**a**) and $V_g < 0$; (**c**) behavior of the Gr/*n*-Si Schottky barrier height vs. the gate voltage V_g; (**d**) Current density vs. V_D for different V_g from -5 to 5 V. Figures adapted with permission from [21].

The early demonstration of the Barristor was based on the vdW heterostructure between a 2D material (i.e., Gr) and a 3D material (i.e., Si). More recently, a vertical transistor working on the same principle has been demonstrated using the 2D/2D vdW heterostructure between Gr and a TMD. Also in this case, current modulation was obtained by electric-field tuning of the Schottky barrier between Gr and the TMD, whereas a proper metal layer provides an Ohmic contact with the TMD. Yu and coworkers have demonstrated such a device with a Gr/few-layer MoS_2/metal heterostructure. This FET showed an ON/OFF ratio larger than 100 and a high current density of 5000 A/cm^2 [85]. Moriya and co-workers further improved the current modulation to >10^5 and current density up to 10^4 A/cm^2 with a similar structure, but better interface fabrication [86]. The advantages of this type of vertical transistor are the large current density and the small device scale, providing high potential for future high density integration circuits.

3.3. Hot Electron Transistor

The hot electron transistor (HET) is a three-terminal (i.e., emitter, base and collector) heterostructure device where the ultra-thin base layer is sandwiched between two thin insulating barriers (i.e., the emitter-base and base-collector barriers), as schematically illustrated in Figure 7a. For a sufficiently high forward bias V_{BE} applied between the base and the emitter, electrons are injected into the base by Fowler–Nordheim (FN) tunneling through the barrier or by thermionic emission above the barrier, depending on the barrier height and thickness. A key aspect for the HET operation is that the injected electrons (hot electrons) have a higher energy compared to the Fermi energy of the electrons' thermal population (cold electrons) in the base. Ideally, for a base thickness lower than the scattering mean free path of hot electrons, a large fraction of the injected electrons can traverse the base ballistically, i.e., without losing energy, and finally reach the edge of the base-collector barrier.

This barrier is aimed to act as an energy filter, which allows the hot electrons to reach the collector and reflects back the electrons with insufficient energy. These reflected electrons eventually become part of the cold electrons' population in the base and contribute to base current (I_B), whereas the hot electrons reaching the collector give rise to the collector current (I_C). Besides transmitting hot electrons, the base-collector barrier must be thick and high enough to block the leakage current I_{BCleak} of cold electrons from the base to the collector.

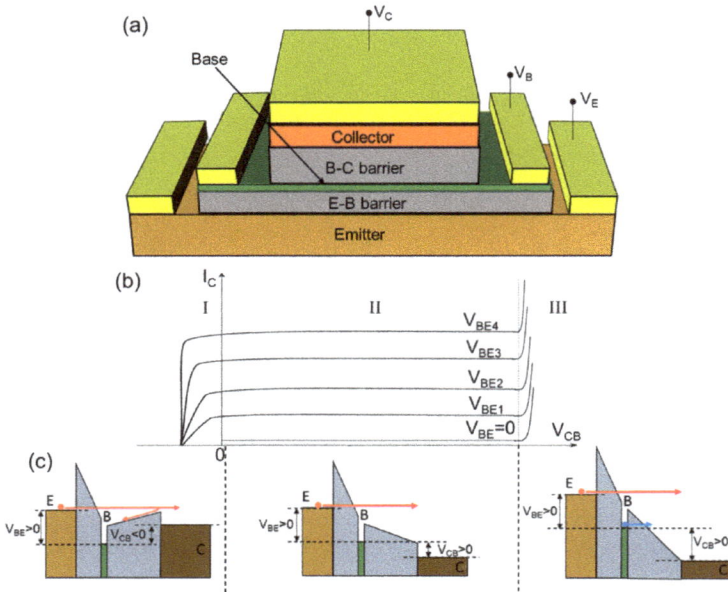

Figure 7. (**a**) Schematic illustration of a hot electron transistor (HET); (**b**) ideal output characteristics I_C-V_{CB} for the transistor biased in the common-base configuration (V_B = 0 and V_{BE} = V_B − V_E > 0) for different V_{BE} values; (**c**) energy band diagrams for different V_{CB} biasing regimes.

Figure 7b,c illustrates the DC electrical characteristics and the band diagrams for an HET biased in the common-base configuration (i.e., with V_B = 0 and V_{BE} = V_B − V_E > 0). Depending on the values of the potential difference V_{CB} = V_C − V_B, three current transport regions can be observed in the output characteristics I_C-V_{CB} (Figure 7b). For V_{CB} > 0 (Region II), I_C is almost independent of V_{CB}, i.e., all the injected hot electrons are transmitted above the B-C barrier (current saturation regime of the transistor). For V_{CB} < 0 (Region I), the collector edge of the B-C barrier is raised up, and part of the hot electrons is reflected back in the base, resulting in a decrease of I_C with increasing negative values of V_{CB}, up to device switch-off. For large positive values of V_{CB} (Region III), the leakage current (I_{BCleak}) contribution of cold electrons injected by FN tunneling through the B-C barrier becomes large, and this leads to a rapid increase of I_C as a function of V_{CB}.

The main figures of merits for DC operation of an HET are the common-base current transfer ratio α = I_C/I_E and the common-emitter current gain β = I_C/I_B. For good DC performances, $\alpha \approx 1$ and β as large as possible are needed.

In the case of an HET, the high-frequency figures of merit, i.e., the cutoff frequency f_T and the maximum oscillation frequency f_{MAX}, can be expressed as follows:

$$f_T = \frac{1}{2\pi\left(\tau_d + \frac{C_{EB}+C_{BC}}{g_m}\right)} \tag{3}$$

$$f_{MAX} = \sqrt{\frac{f_T}{2\pi R_B C_{BC}}} \tag{4}$$

where τ_d is the total delay time associated with electrons' transit in the E-B barrier layer, in the base and in the B-C filtering layer, C_{EB} and C_{BC} are the capacitances of the two barriers, $g_m = dJ_C/dV_{BE}$ is the transconductance and R_B is the base resistance. Clearly, the most effective way to maximize f_T is the increase of g_m. In fact, a reduction of the barrier layer capacitances would imply an increase of the E-B and B-C barrier thicknesses, with a consequent impact on the transit delay times across these barriers. Under saturation conditions, when all the hot electrons injected from the emitter reach the collector ($J_C \approx J_E$), the transconductance $g_m \approx dJ_E/dV_{BE}$. As the emitter current is injected over a barrier, it exhibits an exponential dependence on V_{BE}, i.e., $J_E \propto \exp(qV_{BE}/kT)$. As a result, $g_m \propto qJ_E/kT$. This means that a high injection current density is one of the main requirements to achieve a high cut-off frequency f_T. The R_B term in Equation (4) is the resistance associated with "lateral" current transport in the base layer from the device active area to the base contact. Hence, R_B is the sum of different contributions, i.e., the "intrinsic" base resistance $R_{B\_int} \propto \rho/d_B$ (with ρ the base resistivity and d_B the base thickness), the resistance of the Ohmic metal contact with the base and the access resistance from this contact to the device active area. All these contributions should be minimized to achieve a low R_B. Of course, the most challenging issue to obtain high f_{MAX} is to fabricate an ultra-thin base (allowing ballistic transport of hot electrons in the vertical direction) while maintaining low enough intrinsic and extrinsic base resistances. However, for most of the bulk materials, reducing the film thickness to the nanometer or sub-nanometer range implies an increase of the resistivity, due to the dominance of surface roughness and/or grain boundaries' scattering, as well as to the presence of pinholes and other structural defects in the film.

Indeed, the HET device concept was introduced more than 50 years ago by Mead [87]. Since then, several material systems have been considered for HET implementation, including metal thin films [87–90], complex oxides [91], superconducting materials [92], III-V and III-nitride semiconductor heterostructures [93–97]. However, the successful demonstration of high-performance HETs has been limited by the difficulty to scale the base thickness below the electron mean free path of the carriers. In this context, 2D materials, in particular Gr and TMDs, can represent ideal candidates to fabricate the base of HETs, since they maintain excellent conduction properties and structural integrity down to single atomic layer thickness, allowing one to overcome the base scalability issue.

Theoretical studies have predicted that, with an optimized structure, f_T and f_{MAX} up to several terahertz [98], I_{on}/I_{off} over 10^5, high current and voltage gains can be achieved with a Gr-based HET (GBHET). The first experimental prototypes of GBHETs were reported by Vaziri et al. [22] and by Zeng et al. [23] in 2013. Those demonstrators were fabricated on Si wafers using a CMOS-compatible technology and were based on metal/insulator/Gr/SiO$_2$/n$^+$-Si stacks, where n$^+$-doped Si substrate worked as the emitter, a few nm thick SiO$_2$ as the E-B barrier, a thicker high-k insulator (Al$_2$O$_3$ or HfO$_2$) as the B-C barrier and the topmost metal layer as the collector. Figure 8a shows a schematic of the device structure, while Figure 8b illustrates the band diagrams in the OFF and ON states. The measured common-base output characteristics of this device are reported in Figure 8c, showing a collector current I_C nearly independent of V_{CB} and strongly dependent on V_{BE}.

In spite of the wide modulation of I_C as a function of V_{BE}, these first prototypes suffered from a high threshold voltage and a very poor injected current density (in the order of $\mu A/cm^2$) due to the high Si/SiO$_2$ barrier, hindering their application at high frequencies.

In order to improve the current injection efficiency, other materials have been investigated as E-B barrier layers in replacement of SiO$_2$ [99]. As an example, using a 6 nm-thick HfO$_2$ (including a 0.5-nm interfacial SiO$_2$) deposited by atomic layer deposition results in an improved threshold voltage and a higher injected current density. Further improvements have been obtained using a TmSiO/TiO$_2$ (1 nm/5 nm) bilayer, where the thin TmSiO layer (with low electron affinity) in contact with the Si emitter allows high current injection by step tunneling, while the thicker TiO$_2$ layer (with higher electron affinity) serves to block the leakage current from the Si valence band. For this GBHET with a TmSiO/TiO$_2$ E-B barrier, a collector current density J$_C \approx 4$ A/cm^2 (more than five orders of magnitude higher than in the first prototypes) was obtained at V$_{BE}$ = 5 V and for V$_{BC}$ = 0. However, the device still suffers from low values of $\alpha \approx 0.28$ and $\beta \approx 0.4$, which can be due to the insufficient quality of the interface between Gr and the deposited B-C barrier.

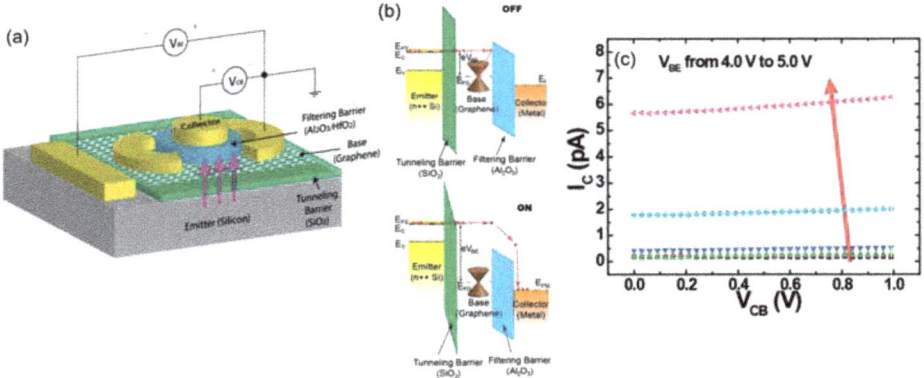

Figure 8. (**a**) Schematic illustration of a Si/SiO$_2$/Gr/Al$_2$O$_3$/Ti Gr-based hot electron transistor (HET) (GBHET) biased in the common base configuration; (**b**) band diagrams in the OFF and in the ON state; and (**c**) output characteristics I$_C$-V$_{CB}$ of the device. Figures adapted from [23].

Besides Gr, monolayer MoS$_2$ has been also considered as the base material. As an example, Torres et al. [25] demonstrated an HET device based on a stack of ITO/HfO$_2$/MoS$_2$/SiO$_2$/n$^+$-Si, where the n^+-doped Si substrate worked as the emitter, thermally-grown SiO$_2$ (3 nm) as the E-B tunneling barrier, a monolayer of CVD-grown MoS$_2$ as the base, the HfO$_2$ layer (55 nm thick) as the B-C barrier and the topmost ITO as the collector electrode. This device showed an improved value of $\alpha \approx 0.95$ with respect to the previously described Gr-base HET prototypes, mainly due to the lower conduction band offset between MoS$_2$ and HfO$_2$ (1.52 eV), with respect to the cases of Gr/HfO$_2$ (2.05 eV) and Gr/Al$_2$O$_3$ (3.3 eV). In spite of this, the collector current density of these devices was still poor (in the order of μA/cm^2), due to the high E-B barrier between Si and SiO$_2$.

The above discussed attempts to implement the GBHET device using Si as the emitter material have been mainly motivated by the perspective of integrating this new technology with the state-of-the-art CMOS fabrication platform. More recently, the possibility of demonstrating GBHETs by the integration of Gr with nitride semiconductors has been investigated. GaN/AlGaN or GaN/AlN heterostructures are excellent systems to be used as emitter/emitter-base barriers, due to the presence of high density 2DEG at the interface and to the high structural quality of the barrier layer. Thermionic emission has been demonstrated as the main current transport mechanisms in GaN/AlGaN/Gr systems with a thick (~20 nm) AlGaN barrier layer [100–102]. Very efficient current injection by FN tunneling has been recently shown in the case of GaN/AlN/Gr heterojunctions with an ultra-thin (3 nm) AlN barrier [103].

Figure 9a,b illustrates a cross-sectional schematic and the band diagram of a recently-demonstrated GBHET based on a GaN/AlN/Gr/WSe$_2$/Au stack [103]. The 3-nm AlN tunneling barrier was grown on top of a bulk GaN substrate (n$^+$-doped), working as the emitter. In order to circumvent the problems related to the poor interface quality between Gr and conventional insulators or semiconductors deposited on top of it, an exfoliated WSe$_2$ layer (forming a vdW heterojunction with Gr) was adopted as the B-C barrier layer. The resulting Gr/WSe$_2$ Schottky junction is characterized by a low barrier height due the small band offset (~0.54 eV) between Gr and WSe$_2$. Figure 9c shows the common-base output characteristics (I$_C$-V$_{CB}$) for different values of the emitter injection current I$_E$ in the case of a GBHET with a 2.6 nm-thick WSe$_2$ barrier. Furthermore, Figure 9d plots the common-base current transfer ratio

$\alpha = I_C/I_E$ as a function of V_{CB} in the same bias range. Three current transport regimes can be identified in Figure 9c,d. At intermediate V_{CB} bias (Region II), I_C is almost independent of the V_{CB} and $\alpha \approx 1$, indicating that almost all the injected hot electrons are able to overcome the Gr/WSe$_2$ Schottky barrier and reach the collector. For $V_{CB} < 0$ (Region I), the injected electrons from the emitter are reflected back by the elevated B-C potential barrier, resulting in a reduced I_C and $\alpha < 1$. Finally, at higher positive V_{CB}, current starts to increase due to the increasing contribution of cold electrons' leakage current from the base. Although this device showed excellent DC characteristics in terms of α, its operating V_{BC} window was very limited (~0.3 V), due to the poor blocking capability of the B-C junction with an ultrathin WSe$_2$ barrier. Increasing the WSe$_2$ thickness improved the blocking capability of the B-C barrier, but resulted in a reduced value of α. As an example, $\alpha = 0.75$ was evaluated for a GaN/AlN/Gr/WSe$_2$/Au GBHET with a 10 nm-thick WSe$_2$ barrier [103].

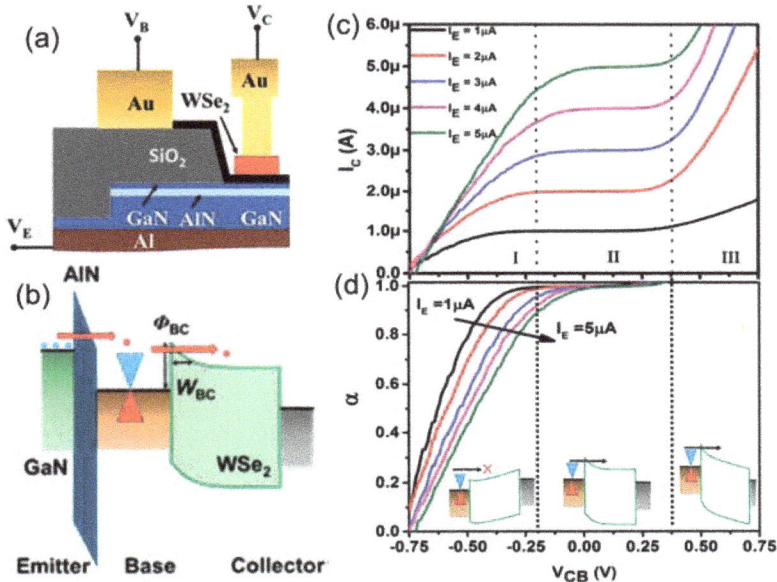

Figure 9. (a) Cross-section schematic and (b) band diagram of a GBHET based on the GaN/AlN/Gr/WSe$_2$/Au stack; (c) common-base output characteristics (I_C-V_{CB}) for different values of the emitter injection current I_E in the case of a GBHET with a 2.6 nm-thick WSe$_2$ barrier; (d) plots of $\alpha = I_C/I_E$ as a function of V_{CB} in the same bias range. Images adapted with permission from [103].

Table 1 reports a comparison of the main DC electrical parameters (i.e., the collector current density J_C, the common-base current transfer ratio α and the common-emitter current gain β) for the HETs with a Gr or MoS$_2$ base reported in the literature. Some examples of HETs fully based on nitride-semiconductors with a sub-10-nm base thickness are reported for comparison. In spite of the theoretically-predicted superior performances (related to ballistic transport in the atomically-thin Gr base), GBHETs still suffer from reduced values of J_C, α and β with respect to HETs fabricated by bandgap engineering of III-N semiconductors (even with a thicker GaN base). This reduced GBHET performance can be due to the non-ideal quality of Gr interfaces with the emitter and collector barriers, indicating that further work will be necessary in this direction.

Table 1. Comparison of J_C, α and β for the-state-of-the-art HETs with a Gr or MoS_2 base and for nitride semiconductor-based HETs with a sub-10-nm base thickness

Emitter/Emitter-Base Barrier	Base (Thickness)	Base-Collector Barrier	J_C (A/cm^2)	α	β	Reference
Si/SiO_2	Gr (0.35 nm)	Al_2O_3	$\sim 1 \times 10^{-5}$	~0.06	~0.06	[22]
Si/SiO_2	Gr (0.35 nm)	Al_2O_3, HfO_2	$\sim 5 \times 10^{-5}$	~0.44	~0.78	[23]
$Si/TmSiO/TiO_2$	Gr (0.35 nm)	Si	~4	~0.28	~0.4	[99]
GaN/AlN	Gr (0.35 nm)	WSe_2 (10 nm)	~50	~0.75	4–6	[103]
Si/SiO_2	MoS_2 (0.7 nm)	HfO_2	$\sim 1 \times 10^{-6}$	~0.95	~4	[25]
$GaN/Al_{0.24}Ga_{0.76}N$	GaN (10 nm)	$Al_{0.08}Ga_{0.92}N$	$\sim 5 \times 10^3$	~0.97		[95]
GaN/AlN	GaN/InGaN (7 nm)	GaN	$\sim 2.5 \times 10^3$	>0.5	>1	[97]
GaN/AlN	GaN (8 nm)	AlGaN/GaN	$\sim 46 \times 10^3$	~0.93	~14.5	[96]

4. Materials Science Issues and Challenges

The device structures reviewed in this paper are based on vdW heterostructures of 2D materials (Gr, TMDs, h-BN) [15] or on mixed-dimensional vdW heterostructures [17] formed by the integration of 2D materials with 3D semiconductors and insulators (bulk or thin films). In many cases, proof of concept devices have been fabricated by the transfer of the individual 2D components, obtained by mechanical or chemical exfoliation of flakes from layered bulk crystals. As a matter of fact, the large area growth method of electronic quality 2D materials and heterostructures are mandatory to move from proof-of-concept devices to industrial applications.

Nowadays, high quality Gr can be grown on a large area by CVD on catalytic metals, such as copper [42], followed by transfer to arbitrary substrates [104]. Although this is a very versatile and widely-used method, it suffers from some drawbacks related to Gr damage and polymer contaminations during the transfer procedure, as well as of possible adhesion problems between Gr and the substrate. Furthermore, it typically introduces undesired metal (Cu, Fe) contaminations [105] originating from the growth substrate and the typically used Cu etchants. An intense research activity is still in progress to optimize Gr transfer procedures to minimize Gr defectivity and contaminations associated with Gr manipulation [106–109].

Notwithstanding the above-mentioned issues, large area (cm^2) Gr heterojunctions with semiconductors are currently fabricated by optimized transfer of CVD-grown Gr. These have been used for the fabrication of device arrays using semiconductor fab-compatible approaches. As an example, Gr junctions with AlGaN/GaN heterostructures showing excellent lateral uniformity have been reported [100,110] and are currently investigated as building blocks for HET devices. Figure 10a,b reports two representative morphologies of the AlGaN surface without (a) and with (b) a single-layer Gr membrane on top. Figure 10c,d shows two arrays of local current-voltage characteristics measured by conductive atomic force microscopy (CAFM) at the different positions on bare AlGaN- and Gr-coated AlGaN, respectively. In both cases, all the I-V curves exhibit a rectifying behavior, with a lower Schottky barrier height for the Gr/AlGaN junction. Noteworthy, a very narrow spread between different curves is observed for the Gr/AlGaN junction, indicating an excellent lateral homogeneity of the Gr/AlGaN Schottky contact.

Under many respects, the direct growth/deposition of Gr on the target substrate would be highly desirable. However, to date, high quality Gr growth has been demonstrated only on a few semiconducting or semi-insulating materials, such as silicon-carbide [38–40,111,112] and, more recently, germanium [113]. Single or few layers of Gr can be obtained on the Si face (0001) of hexagonal SiC, either by controlled sublimation of Si at high temperatures (typically > 1650 °C) in Ar at atmospheric pressure or by direct CVD deposition at lower temperatures (~1450 °C) using an external carbon source (such as C_3H_8) with H_2 or H_2/Ar carrier gases [111]. Gr grown on SiC(0001), commonly named epitaxial graphene (EG), generally exhibits a precise epitaxial orientation with respect to the substrate, which originates from the peculiar nature of the interface, i.e., the presence of a carbon buffer layer with mixed sp^2/sp^3 hybridization sharing covalent bonds with the Si face of SiC [114,115]. This buffer layer has a strong impact both on the lateral (i.e., in plane) current transport in EG, causing a reduced carrier mobility, and on the vertical current transport at the EG/SiC interface [116,117].

Hydrogen intercalation at the interface between the buffer layer and Si face has been demonstrated to be efficient in increasing Gr carrier mobility and tuning the Schottky barrier and, hence, the vertical current transport across the Gr/SiC interface [118]. Recently, CVD growth of Gr from carbon precursors on nitride semiconductor (AlN) substrates/templates has been also investigated. Gr deposition on these non-catalytic surfaces represents a challenging task, as it requires significantly higher temperatures as compared to conventional deposition on metals. The first experimental works addressing this issue showed the possibility of depositing a few layers of Gr both on bulk AlN (Al and N face) and on AlN templates grown on different substrates, such as Si(111) and SiC, at temperatures >1250 °C using propane (C_3H_8) as the carbon source, without significantly degrading the morphology of AlN substrates/templates [119,120]. In spite of the very promising results of these experiments, further work will be required to evaluate the feasibility and the effects of CVD Gr growth onto AlN/GaN or AlGaN/GaN heterostructures. Moreover, the possibility of integrating these high temperature processes in the fabrication flow of GBHETs with the GaN/AlN (or GaN/AlGaN) emitter needs to be investigated.

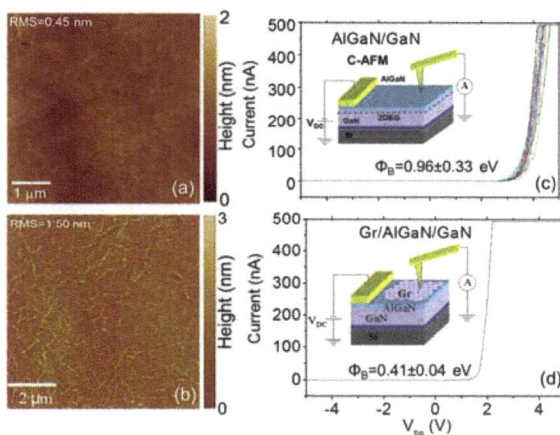

Figure 10. AFM morphologies of an AlGaN/GaN heterostructure (**a**) and of Gr transferred onto AlGaN/GaN (**b**). Current-voltage characteristics measured by conductive atomic force microscopy (CAFM) on an array of different positions on the bare AlGaN surface (**c**) and on the Gr-coated AlGaN surface (**d**). Figures adapted with permission from [100].

Although most of TMD-based devices are still fabricated using exfoliated flakes, much progress has been made in the last few years in the CVD deposition of MoS_2 and other TMDs, both on insulating substrates, such as SiO_2 [121,122] and sapphire [123], and on semiconductors, such as GaN [124]. Noteworthy, an epitaxial registry with the substrate has been observed for CVD-grown MoS_2 on sapphire and on GaN.

High quality thin insulating layers are key components for most of the above-discussed lateral and vertical devices based on Gr and TMDs. In this context, due to the layer-by-layer deposition mechanism, atomic layer deposition (ALD) has been considered as method of choice to grow thin high-k dielectrics (such as Al_2O_3 and HfO_2) on Gr and TMDs [125]. The main challenge related to ALD on the chemically inert and dangling-bonds' free surface of 2D materials is the activation of nucleation sites from which the growth can initiate. Several approaches have been explored so far to this aim, like ex situ deposition of metal or metal-oxide seed layers [126] or pre-functionalization of Gr [127]. Recently, highly uniform Al_2O_3 films with very low leakage current and a high breakdown field have been deposited on Gr by a two-step thermal ALD process, resulting in minimal degradation of the Gr electronic/structural properties [128]. As a matter of fact, the interface between Gr and TMDs with common insulators is not atomically flat. Furthermore, interface or near-interface defects are typically present in the oxide and act as trapping states for electrons/holes.

In this respect, due to its atomically-flat crystal surface, excellent insulating properties and chemical inertness, h-BN represents an ideal ultra-flat substrate and interlayer dielectric for Gr (to which it is closely lattice matched, within 1.6%) and, by extension, other 2D semiconductor materials [3]. Although single and multilayer h-BN used for device demonstration are still mainly obtained by exfoliation from the bulk crystal, significant progress has been also made towards the controlled synthesis of large-area, high-quality h-BN films by CVD on metal catalysts, such as Ni [129], Cu [130] and Ni/Cu alloys [131]. Recently, Sonde et al. reported a detailed study clarifying the mechanisms of CVD h-BN growth on Ni and Co thin films on SiO_2/Si substrates [132], which could lead to large area (up to wafer scale) growth of h-BN thin films on arbitrary substrates in a transfer-free manner. As schematically illustrated in Figure 11a, after exposure to ammonia (NH_3) and diborane (B_2H_6) precursors at high temperature (~1050 °C), diffusion of boron (B) and nitrogen (N) in Ni occurs, followed by segregation/precipitation of h-BN multilayers both on the upper and the buried face of the Ni film. These h-BN films showed excellent insulating properties, with a breakdown field of $9.34\ MV\cdot cm^{-1}$, as determined from current-voltage characteristics measured by CAFM (see Figure 11b). Finally, the quality of h-BN as a substrate for back-gated Gr field effect transistors has been evaluated. The Gr resistance measured under forward and backward gate bias sweep is reported in Figure 11c, showing minimal hysteresis associated with the absence of charge trapping at the Gr/h-BN interface.

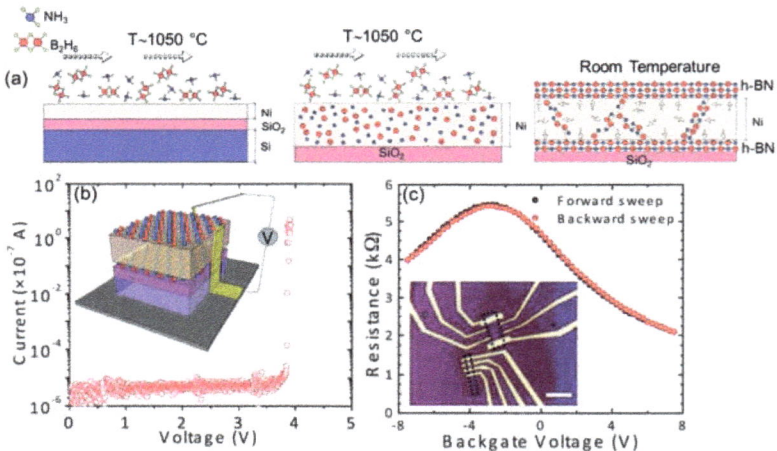

Figure 11. (**a**) Schematic illustration of the mechanism of h-BN CVD growth on Ni thin films by ammonia (NH_3) and diborane (B_2H_6) precursors at high temperature (~1050 °C); (**b**) estimation of the breakdown field ($9.34\ MV\cdot cm^{-1}$) of the multilayer h-BN (10–11 layers) by current-voltage measurements with CAFM; (**c**) resistance of a Gr field effect transistor with an h-BN back-gate, showing minimal hysteresis between forward and backward gate bias sweep. Figures adapted with permission from [132].

PMMA-assisted transfer is the simplest way to construct arbitrary 2D heterostructures. However, the quality of the interface can be affected by the trapping of polymer, solvents or chemicals used for transfer. This represents a major issue, especially for large-area 2D heterostructures. In this respect, the direct synthesis of vertically-stacked 2D heterojunctions, obtained by CVD growth of one 2D material on another, would be highly desirable, as the direct growth would result, in principle, in clean heterojunction interfaces. The early van der Waals epitaxy experiments started from Gr and h-BN, which share a similar lattice constant. Yang et al. reported a plasma-assisted deposition method for the growth of single domain Gr on the h-BN substrate [133]. Gr grows with a preferred orientation with respect to the h-BN lattice, and the size of the domain is only restricted by the area of underlying h-BN. Furthermore, Shi et al. have obtained a vertically-stacked MoS_2/Gr heterostructure via thermal decomposition of ammonium thiomolybdate precursors on Gr surfaces [134]. In spite of the 28% mismatch between MoS_2 and Gr lattice constants, Gr is still a good growth platform for MoS_2, as the growth of MoS_2 on Gr involves strain to accommodate the lattice mismatch. Lin et al. demonstrated the direct growth of MoS_2, WSe_2 and h-BN on epitaxial Gr on SiC through CVD methods [135], showing how the morphology of the underlying Gr strongly affects the growth and the properties of top heterostructures. In particular, strain, wrinkling and defects on the surface of Gr provide the nucleation centers for the upper layer material growth.

5. Summary and Outlook

We have reviewed the state-of-the-art of 2D material-based vertical transistors for logic and high frequency electronics.

Regarding logic applications, vdW heterostructures obtained by 2D/2D or 2D/3D material stacking have been explored by several research groups as a platform to implement tunneling field effect transistors (TFETs). Resonant TFETs have been demonstrated by rotationally-aligned Gr monolayers or bilayers separated by a tunnel barrier (h-BN or TMD). However, although these prototypes permitted exploring interesting physical phenomena, the possibility of realizing vdW heterostructures with a precise crystallographic alignment on a large area represents a big challenge, making real applications of resonant TFETs in the near future difficult. On the other hand, TFETs relying on the band-to-band-tunneling across the interface of different semiconducting layered materials could have more realistic prospects of practical applications, once further progress in van der Waals epitaxy of TMDs is achieved.

Regarding high frequency applications of 2D materials, lateral Gr FETs with a very high cut-off frequency ($f_T > 400$ GHz) have been demonstrated, exploiting the high Gr channel mobility. However, these devices suffer from a lower maximum oscillation frequency f_{MAX}, due to the poor saturation of the output characteristics mainly originating from the missing bandgap of Gr. Vertical transistors based on 2D/2D or 2D/3D material heterostructures can represent an alternative to lateral Gr FETs to realize RF functions. In particular, the hot electrons transistor (HET) has been theoretically predicted to be suitable for ultra-high-frequency applications, with f_T and f_{MAX} values in the THz range. The main requirement for the implementation of this device concept, i.e., an ultrathin base allowing both ballistic transport in the vertical direction and low base resistance in the lateral direction, can be fulfilled by 2D materials, in particular Gr. However, although much progress has been made in the last few years in the fabrication of Gr-based HETs, the electrical performances of these demonstrators are still lower than those of previously-reported HET devices fabricated with III-V heterostructures (even with a thicker base) and far from the state-of-the-art RF HEMTs (which represent the benchmark for any competing RF device concept). Further improvements in the emitter-base and base-collector barrier layers and interfaces are still required to achieve the theoretical DC and RF performances of HETs.

Generally speaking, the perspective of industrial applications of 2D material-based devices is strongly related to the possibility of growing individual 2D layers and, possibly, vdW heterostructures on a large area.

Acknowledgments: The authors want to acknowledge all of the following colleagues for useful discussions and participation in the experiments: Gabriele Fisichella., Emanuela Schilirò, Salvatore Di Franco, Patrick Fiorenza, Raffaella Lo Nigro, Ioannis Deretzis, Antonino La Magna, Giuseppe Nicotra, Corrado Bongiorno, Corrado Spinella (CNR—Institute for Microelectronics and Microsystems (CNR-IMM), Catania, Italy); Sebastiano Ravesi, Stella Lo Verso, Ferdinando Iucolano (STMicroelectronics, Catania, Italy); Eric Frayssinet, Roy Dagher, Adrien Michon, Yvon Cordier (CNRS—Centre de Recherche sur l'Hétéro-Epitaxie et ses Applications (CNRS-CRHEA), Valbonne, France); Mike Leszczynski, Piotr Kruszewski, Pawel Prystawko (TopGaN, Warsaw, Poland); Rositza Yakimova, Anelia Kakanakova (Linkoping University, Linkoping, Sweden); Bela Pecz (Institute for Technical Physics and Materials Science Research, Centre for Energy Research, Hungarian Academy of Science,

Budapest, Hungary); and Luigi Colombo (Texas Instruments, Texas, TX, USA). This paper has been supported, in part, by the Flag-ERA project "GraNitE: Graphene heterostructures with Nitrides for high frequency Electronics".

Conflicts of Interest: The authors declare no conflict of interest.

References

1. Novoselov, K.S.; Geim, A.K.; Morozov, S.V.; Jiang, D.; Zhang, Y.; Dubonos, S.V.; Grigorieva, I.V.; Firsov, A.A. Electric Field Effect in Atomically Thin Carbon Films. *Science* **2004**, *306*, 666–669. [CrossRef] [PubMed]

2. Bolotin, K.I.; Sikes, K.J.; Hone, J.H.; Stormer, L.; Kim, P. Temperature-Dependent Transport in Suspended Graphene. *Phys. Rev. Lett.* **2008**, *101*, 096802. [CrossRef] [PubMed]

3. Dean, C.R.; Young, A.F.; Meric, I.; Lee, C.; Wang, L.; Sorgenfrei, S.; Watanabe, K.; Taniguchi, T.; Kim, P.; Shepard, K.L.; et al. Boron nitride substrates for high-quality graphene electronics. *Nat. Nanotechnol.* **2010**, *5*, 722–726. [CrossRef] [PubMed]

4. Mayorov, A.S.; Gorbachev, R.V.; Morozov, S.V.; Britnell, L.; Jalil, R.; Ponomarenko, L.A.; Blake, P.; Novoselov, K.S.; Watanabe, K.; Taniguchi, T.; et al. Micrometer-scale ballistic transport in encapsulated graphene at room temperature. *Nano Lett.* **2011**, *11*, 2396–2399. [CrossRef] [PubMed]

5. Giannazzo, F.; Raineri, V. Graphene: Synthesis and nanoscale characterization of electronic properties. *Rivista del Nuovo Cimento* **2012**, *35*, 267–304.

6. Sonde, S.; Giannazzo, F.; Vecchio, C.; Yakimova, R.; Rimini, E.; Raineri, V. Role of graphene/substrate interface on the local transport properties of the two-dimensional electron gas. *Appl. Phys. Lett.* **2010**, *97*, 132101. [CrossRef]

7. Giannazzo, F.; Sonde, S.; Lo Nigro, R.; Rimini, E.; Raineri, V. Mapping the Density of Scattering Centers Limiting the Electron Mean Free Path in Graphene. *Nano Lett.* **2011**, *11*, 4612–4618. [CrossRef] [PubMed]

8. Lemme, M.C.; Echtermeyer, T.J.; Baus, M.; Kurz, H. A Graphene Field-Effect Device. *IEEE Electron Device Lett.* **2007**, *28*, 282–284. [CrossRef]

9. Schwierz, F. Graphene transistors. *Nat. Nanotechnol.* **2010**, *5*, 487–496. [CrossRef] [PubMed]

10. Liao, L.; Duan, X. Graphene for radio frequency electronics. *Mater. Today* **2012**, *15*, 328–338. [CrossRef]

11. Cheng, R.; Bai, J.; Liao, L.; Zhou, H.; Chen, Y.; Liu, L.; Lin, Y.-C.; Jiang, S.; Huang, Y.; Duan, X. High-frequency self-aligned graphene transistors with transferred gate stacks. *Proc. Natl. Acad. Sci. USA* **2012**, *109*, 11588–11592. [CrossRef] [PubMed]

12. Wang, H.; Hsu, A.; Wu, J.; Kong, J.; Palacios, T. Graphene-Based Ambipolar RF Mixers. *IEEE Electron Device Lett.* **2010**, *31*, 906–908. [CrossRef]

13. Wang, Q.H.; Zadeh, K.K.; Kis, A.; Coleman, J.N.; Strano, M.S. Electronics and optoelectronics of two-dimensional transition metal dichalcogenides. *Nat. Nanotechnol.* **2012**, *7*, 699–712. [CrossRef] [PubMed]

14. Liu, H.; Neal, A.T.; Zhu, Z.; Luo, Z.; Xu, X.; Tománek, D.; Ye, P.D. Phosphorene: An Unexplored 2D Semiconductor with a High Hole Mobility. *ACS Nano* **2014**, *8*, 4033–4041. [CrossRef] [PubMed]

15. Geim, A.K.; Grigorieva, I.V. Van der Waals heterostructures. *Nature* **2013**, *499*, 419–425. [CrossRef] [PubMed]

16. Wang, H.; Liu, F.; Fu, W.; Fang, Z.; Zhoue, W.; Liu, Z. Two-dimensional heterostructures: Fabrication, characterization, and application. *Nanoscale* **2014**, *6*, 12250–12272. [CrossRef] [PubMed]

17. Jariwala, D.; Marks, T.J.; Hersam, M.C. Mixed-dimensional van der Waals heterostructures. *Nat. Mater.* **2017**, *16*, 170–181. [CrossRef] [PubMed]

18. Britnell, L.; Gorbachev, R.V.; Jalil, R.; Belle, B.D.; Schedin, F.; Mishchenko, A.; Georgiou, T.; Katsnelson, M.I.; Eaves, L.; Morozov, S.V.; et al. Field-Effect Tunneling Transistor Based on Vertical Graphene Heterostructures. *Science* **2012**, *335*, 947–950. [CrossRef] [PubMed]

19. Georgiou, T.; Jalil, R.; Belle, B.D.; Britnell, L.; Gorbachev, R.V.; Morozov, S.V.; Kim, Y.-J.; Gholinia, A.; Haigh, S.J.; Makarovsky, O.; et al. Vertical field-effect transistor based on graphene-WS$_2$ heterostructures for flexible and transparent electronics. *Nat. Nanotechnol.* **2013**, *8*, 100–103. [CrossRef] [PubMed]

20. Sarkar, D.; Xie, X.; Liu, W.; Cao, W.; Kang, J.; Gong, Y.; Kraemer, S.; Ajayan, P.M.; Banerjee, K. A subthermionic tunnel field-effect transistor with an atomically thin channel. *Nature* **2015**, *526*, 91–95. [CrossRef] [PubMed]

21. Yang, H.; Heo, J.; Park, S.; Song, H.J.; Seo, D.H.; Byun, K.-E.; Kim, P.; Yoo, I.; Chung, H.-J.; Kim, K. Graphene barristor, a triode device with a gate-controlled Schottky barrier. *Science* **2012**, *336*, 1140–1143. [CrossRef] [PubMed]

22. Vaziri, S.; Lupina, G.; Henkel, C.; Smith, A.D.; Ostling, M.; Dabrowski, J.; Lippert, G.; Mehr, W.; Lemme, M.C. A graphene-based hot electron transistor. *Nano Lett.* **2013**, *13*, 1435–1439. [CrossRef] [PubMed]

23. Zeng, C.; Song, E.B.; Wang, M.; Lee, S.; Torres, C.M.; Tang, J.; Weiller, B.H.; Wang, K.L. Vertical graphene-base hot electron transistor. *Nano Lett.* **2013**, *13*, 2370–2375. [CrossRef] [PubMed]

24. Vaziri, S.; Smith, A.D.; Östling, M.; Lupina, G.; Dabrowski, J.; Lippert, G.; Mehr, W.; Driussi, F.; Venica, S.; Di Lecce, V.; et al. Going ballistic: Graphene hot electron transistors. *Solid State Commun.* **2015**, *224*, 64–75. [CrossRef]

25. Torres, C.M.; Lan, Y.W.; Zeng, C.; Chen, J.H.; Kou, X.; Navabi, A.; Tang, J.; Montazeri, M.; Adleman, J.R.; Lerner, M.B.; et al. High-Current Gain Two-Dimensional MoS$_2$-Base Hot-Electron Transistors. *Nano Lett.* **2015**, *15*, 7905–7912. [CrossRef] [PubMed]

26. Radisavljevic, B.; Radenovic, A.; Brivio, J.; Giacometti, V.; Kis, A. Single-layer MoS$_2$ transistors. *Nat. Nanotechnol.* **2011**, *6*, 147–150. [CrossRef] [PubMed]

27. Kaasbjerg, K.; Thygesen, K.S.; Jacobsen, K.W. Phonon-limited mobility in n-type single-layer MoS$_2$ from first principles. *Phys. Rev. B* **2012**, *85*, 115317. [CrossRef]

28. Radisavljevic, B.; Kis, A. Mobility engineering and a metal–insulator transition in monolayer MoS$_2$. *Nat. Mater.* **2013**, *12*, 815–820. [CrossRef] [PubMed]

29. Fuhrer, M.S.; Hone, J. Measurement of mobility in dual-gate MoS$_2$ transistor. *Nat. Nanotechnol.* **2013**, *8*, 146–147. [CrossRef] [PubMed]

30. Li, L.; Yu, Y.; Ye, G.J.; Ge, Q.; Ou, X.; Wu, H.; Feng, D.; Chen, X.H.; Zhang, Y. Black phosphorus field-effect transistors. *Nat. Nanotechnol.* **2014**, *9*, 372–377. [CrossRef] [PubMed]

31. Yoon, Y.; Ganapathi, K.; Salahuddin, S. How good can monolayer MoS$_2$ transistors be? *Nano Lett.* **2011**, *11*, 3768–3773. [CrossRef] [PubMed]

32. Alam, K.; Lake, R. Monolayer MoS$_2$ transistors beyond the technology road map. *IEEE Trans. Electron Devices* **2012**, *59*, 3250–3254. [CrossRef]

33. Liu, L.; Lu, Y.; Guo, J. On monolayer MoS$_2$ field-effect transistors at the scaling limit. *IEEE Trans. Electron Devices* **2013**, *60*, 4133–4139. [CrossRef]

34. Majumdar, K.; Hobbs, C.; Kirsch, P.D. Benchmarking transition metal dichalcogenide MOSFET in the ultimate physical scaling limit. *IEEE Electron Device Lett.* **2014**, *35*, 402–404. [CrossRef]

35. International Technology Roadmap for Semiconductors (ITRS, 2013). Available online: http://www.itrs2.net/ (accessed on 1 December 2017).

36. Desai, S.B.; Madhvapathy, S.R.; Sachid, A.B.; Llinas, J.P.; Wang, Q.; Ahn, G.H.; Pitner, G.; Kim, M.J.; Bokor, J.; Hu, C.; et al. MoS$_2$ transistors with 1-nanometer gate lengths. *Science* **2016**, *354*, 99–102. [CrossRef] [PubMed]

37. Meric, I.; Baklitskaya, N.; Kim, P.; Shepard, K.L. RF performance of to p-gated, zero-bandgap graphene field-effect transistors. In Proceedings of the IEEE International Electron Devices Meeting (IEDM 2008), San Francisco, CA, USA, 15–17 December 2008; pp. 1–4.

38. Berger, C.; Song, Z.; Li, X.; Wu, X.; Brown, N.; Naud, C.; Mayou, D.; Li, T.; Hass, J.; Marchenkov, A.N.; et al. Electronic confinement and coherence in patterned epitaxial graphene. *Science* **2006**, *312*, 1191–1196. [CrossRef] [PubMed]

39. Emtsev, K.V.; Bostwick, A.; Horn, K.; Jobst, J.; Kellogg, G.L.; Ley, L.; McChesney, J.L.; Ohta, T.; Reshanov, S.A.; Röhrl, J.; et al. Towards wafer-size graphene layers by atmospheric pressure graphitization of silicon carbide. *Nat. Mater.* **2009**, *8*, 203–207. [CrossRef] [PubMed]

40. Virojanadara, C.; Syvajarvi, M.; Yakimova, R.; Johansson, L.I.; Zakharov, A.A.; Balasubramanian, T. Homogeneous large-area graphene layer growth on 6H-SiC(0001). *Phys. Rev. B* **2008**, *78*, 245403. [CrossRef]

41. Vecchio, C.; Sonde, S.; Bongiorno, C.; Rambach, M.; Yakimova, R.; Rimini, E.; Raineri, V.; Giannazzo, F. Nanoscale structural characterization of epitaxial graphene grown on off-axis 4H-SiC (0001). *Nanoscale Res. Lett.* **2011**, *6*, 269. [CrossRef] [PubMed]

42. Li, X.; Cai, W.; An, J.; Kim, S.; Nah, J.; Yang, D.; Piner, R.; Velamakanni, A.; Jung, I.; Tutuc, E.; et al. Large-area synthesis of high-quality and uniform graphene films on copper foils. *Science* **2009**, *324*, 1312–1314. [CrossRef] [PubMed]

43. Wu, Y.; Jenkins, K.A.; Valdes-Garcia, A.; Farmer, D.B.; Zhu, Y.; Bol, A.A.; Dimitrakopoulos, C.; Zhu, W.; Xia, F.; Avouris, P.; et al. State-of-the-Art Graphene High-Frequency Electronics. *Nano Lett.* **2012**, *12*, 3062–3067. [CrossRef] [PubMed]

44. Lin, Y.M.; Dimitrakopoulos, C.; Jenkins, K.A.; Farmer, D.B.; Chiu, H.Y.; Grill, A.; Avouris, P. 100-GHz transistors from wafer-scale epitaxial graphene. *Science* **2010**, *327*, 662. [CrossRef] [PubMed]

45. Li, X.; Mullen, J.T.; Jin, Z.; Borysenko, K.M.; Nardelli, M.B.; Kim, K.W. Intrinsic electrical transport properties of monolayer silicene and MoS$_2$ from first principles. *Phys. Rev. B* **2013**, *87*, 115418. [CrossRef]

46. Akinwande, D.; Petrone, N.; Hone, J. Two-dimensional flexible nanoelectronics. *Nat. Commun.* **2014**, *5*, 5678. [CrossRef] [PubMed]

47. Cheng, R.; Jian, S.; Chen, Y.; Weiss, N.; Cheng, H.C.; Wu, H.; Huang, Y.; Duan, X. Few-layer molybdenum disulfide transistors and circuits for high-speed flexible electronics. *Nat. Commun.* **2014**, *5*, 5143. [CrossRef] [PubMed]

48. Krasnozhon, D.; Lembke, D.; Nyffeler, C.; Leblebici, Y.; Kis, A. MoS$_2$ Transistors Operating at Gigahertz Frequencies. *Nano Lett.* **2014**, *14*, 5905–5911. [CrossRef] [PubMed]

49. Sanne, A.; Ghosh, R.; Rai, A.; Yogeesh, M.N.; Shin, S.H.; Sharma, A.; Jarvis, K.; Mathew, L.; Rao, R.; Akinwande, D.; et al. Radio Frequency Transistors and Circuits Based on CVD MoS$_2$. *Nano Lett.* **2015**, *15*, 5039–5045. [CrossRef] [PubMed]

50. Hong, J.; Hu, Z.; Probert, M.; Li, K.; Lv, D.; Yang, X.; Gu, L.; Mao, N.; Feng, Q.; Xie, L.; et al. Exploring atomic defects in molybdenum disulphide monolayers. *Nat. Commun.* **2015**, *6*, 6293. [CrossRef] [PubMed]

51. Yu, Z.; Pan, Y.; Shen, Y.; Wang, Z.; Ong, Z.-Y.; Xu, T.; Xin, R.; Pan, L.; Wang, B.; Sun, L.; et al. Towards intrinsic charge transport in monolayer molybdenum disulfide by defect and interface engineering. *Nat. Commun.* **2014**, *5*, 5290. [CrossRef] [PubMed]

52. Guo, Y.; Liu, D.; Robertson, J. Chalcogen vacancies in monolayer transition metal dichalcogenides and Fermi level pinning at contacts. *Appl. Phys. Lett.* **2015**, *106*, 173106. [CrossRef]

53. Addou, R.; McDonnell, S.; Barrera, D.; Guo, Z.; Azcatl, A.; Wang, J.; Zhu, H.; Hinkle, C.L.; Quevedo-Lopez, M.; Alshareef, H.N.; et al. Impurities and Electronic Property Variations of Natural MoS$_2$ Crystal Surfaces. *ACS Nano* **2015**, *9*, 9124–9133. [CrossRef] [PubMed]

54. Das, S.; Chen, H.-Y.; Penumatcha, A.V.; Appenzeller, J. High Performance Multi-layer MoS$_2$ Transistors with Scandium Contacts. *Nano Lett.* **2013**, *13*, 100–105. [CrossRef] [PubMed]

55. Kappera, R.; Voiry, D.; Yalcin, S.E.; Branch, B.; Gupta, G.; Mohite, A.D.; Chhowalla, M. Phase-Engineered Low-Resistance Contacts for Ultrathin MoS$_2$ Transistors. *Nat. Mater.* **2014**, *13*, 1128–1134. [CrossRef] [PubMed]

56. Suh, J.; Park, T.-E.; Lin, D.-Y.; Fu, D.; Park, J.; Jung, H.J.; Chen, Y.; Ko, C.; Jang, C.; Sun, Y.; et al. Doping against the Native Propensity of MoS$_2$: Degenerate Hole Doping by Cation Substitution. *Nano Lett.* **2014**, *14*, 6976–6982. [CrossRef] [PubMed]

57. Fang, H.; Tosun, M.; Seol, G.; Chang, T.C.; Takei, K.; Guo, J.; Javey, A. Degenerate n-Doping of Few-Layer Transition Metal Dichalcogenides by Potassium. *Nano Lett.* **2013**, *13*, 1991–1995. [CrossRef] [PubMed]

58. Nipane, A.; Karmakar, D.; Kaushik, N.; Karande, S.; Lodha, S. Few-Layer MoS$_2$ p-type Devices Enabled by Selective Doping Using Low Energy Phosphorus Implantation. *ACS Nano* **2016**, *10*, 2128–2137. [CrossRef] [PubMed]

59. McDonnell, S.; Addou, R.; Buie, C.; Wallace, R.M.; Hinkle, C.L. Defect-Dominated Doping and Contact Resistance in MoS$_2$. *ACS Nano* **2014**, *8*, 2880–2888. [CrossRef] [PubMed]

60. Giannazzo, F.; Fisichella, G.; Piazza, A.; Agnello, S.; Roccaforte, F. Nanoscale inhomogeneity of the Schottky barrier and resistivity in MoS$_2$ multilayers. *Phys. Rev. B* **2015**, *92*, 081307. [CrossRef]

61. Giannazzo, F.; Fisichella, G.; Piazza, A.; Di Franco, S.; Greco, G.; Agnello, S.; Roccaforte, F. Impact of Contact Resistance on the Electrical Properties of MoS$_2$ Transistors at Practical Operating Temperatures. *Beilstein J. Nanotechnol.* **2017**, *8*, 254–263. [CrossRef] [PubMed]

62. Das, S.; Prakash, A.; Salazar, R.; Appenzeller, J. Toward Low-Power Electronics: Tunneling Phenomena in Transition Metal Dichalcogenides. *ACS Nano* **2014**, *8*, 1681–1689. [CrossRef] [PubMed]

63. Giannazzo, F.; Fisichella, G.; Piazza, A.; Di Franco, S.; Greco, G.; Agnello, S.; Roccaforte, F. Effect of Temperature–Bias Annealing on the Hysteresis and Subthreshold Behavior of Multilayer MoS$_2$ Transistors. *Phys. Status Solidi RRL* **2016**, *10*, 797–801. [CrossRef]

64. Chuang, S.; Battaglia, C.; Azcatl, A.; McDonnell, S.; Kang, J.S.; Yin, X.; Tosun, M.; Kapadia, R.; Fang, H.; Wallace, R.M.; et al. MoS$_2$ P-Type Transistors and Diodes Enabled by High Work Function MoOx Contacts. *Nano Lett.* **2014**, *14*, 1337–1342. [CrossRef] [PubMed]

65. Giannazzo, F.; Fisichella, G.; Greco, G.; Di Franco, S.; Deretzis, I.; La Magna, A.; Bongiorno, C.; Nicotra, G.; Spinella, C.; Scopelliti, M.; et al. Ambipolar MoS$_2$ Transistors by Nanoscale Tailoring of Schottky Barrier Using Oxygen Plasma Functionalization. *ACS Appl. Mater. Interfaces* **2017**, *9*, 23164–23174. [CrossRef] [PubMed]

66. Fang, H.; Chuang, S.; Chang, T.C.; Takei, K.; Takahashi, T.; Javey, A. High-Performance Single Layered WSe$_2$ p-FETs with Chemically Doped Contacts. *Nano Lett.* **2012**, *12*, 3788–3792. [CrossRef] [PubMed]

67. Tosun, M.; Chuang, S.; Fang, H.; Sachid, A.B.; Hettick, M.; Lin, Y.; Zeng, Y.; Javey, A. High-Gain Inverters Based on WSe$_2$ Complementary Field-Effect Transistors. *ACS Nano* **2014**, *8*, 4948–4953. [CrossRef] [PubMed]

68. Yu, L.; Zubair, A.; Santos, E.J.G.; Zhang, X.; Lin, Y.; Zhang, Y.; Palacios, T. High-Performance WSe$_2$ Complementary Metal Oxide Semiconductor Technology and Integrated Circuits. *Nano Lett.* **2015**, *15*, 4928–4934. [CrossRef] [PubMed]

69. Roy, T.; Tosun, M.; Kang, J.S.; Sachid, A.B.; Desai, S.B.; Hettick, M.; Hu, C.C.; Javey, A. Field-Effect Transistors Built from All Two-Dimensional Material Components. *ACS Nano* **2014**, *8*, 6259–6264. [CrossRef] [PubMed]

70. Esaki, L. New phenomenon in narrow germanium p-n junctions. *Phys. Rev.* **1958**, *109*, 603. [CrossRef]

71. Chang, L.L. Resonant tunneling in semiconductor double barriers. *Appl. Phys. Lett.* **1974**, *24*, 593. [CrossRef]

72. Agarwal, S.; Yablonovitch, E. Band-Edge Steepness Obtained from Esaki Backward Diode Current–Voltage Characteristics. *IEEE Trans. Electron Devices* **2014**, *61*, 1488–1493. [CrossRef]

73. Britnell, L.; Gorbachev, R.V.; Geim, A.K.; Ponomarenko, L.A.; Mishchenko, A.; Greenaway, M.T.; Fromhold, T.M.; Novoselov, K.S.; Eaves, L. Resonant tunnelling and negative differential conductance in graphene transistors. *Nat. Commun.* **2013**, *4*, 1794. [PubMed]

74. Mishchenko, A.; Tu, J.S.; Cao, Y.; Gorbachev, R.V.; Wallbank, J.R.; Greenaway, M.T.; Morozov, V.E.; Morozov, S.V.; Zhu, M.J.; Wong, S.L.; et al. Twist-controlled resonant tunnelling in graphene/boron nitride/graphene heterostructures. *Nat. Nanotechnol.* **2014**, *9*, 808–813. [PubMed]

75. Kang, S.; Fallahazad, B.; Lee, K.; Movva, H.; Kim, K.; Corbet, C.M.; Taniguchi, T.; Watanabe, K.; Colombo, L.; Register, L.F.; et al. Bilayer Graphene–Hexagonal Boron Nitride Heterostructure Negative Differential Resistance Interlayer Tunnel FET. *IEEE Electron Device Lett.* **2015**, *36*, 405–407.

76. Fallahazad, B.; Lee, K.; Kang, S.; Xue, J.; Larentis, S.; Corbet, C.; Kim, K.; Movva, H.C.P.; Taniguchi, T.; Watanabe, K.; et al. Gate-Tunable Resonant Tunneling in Double Bilayer Graphene Heterostructures. *Nano Lett.* **2015**, *15*, 428–433. [PubMed]

77. Kang, S.; Prasad, N.; Movva, H.C.P.; Rai, A.; Kim, K.; Mou, X.; Taniguchi, T.; Watanabe, K.; Register, L.F.; Tutuc, E.; et al. Effects of electrode layer band structure on the performance of multilayer graphene–hBN–graphene interlayer tunnel field effect transistors. *Nano Lett.* **2016**, *16*, 4975–4981. [PubMed]

78. Burg, G.W.; Prasad, N.; Fallahazad, B.; Valsaraj, A.; Kim, K.; Taniguchi, T.; Watanabe, K.; Wang, Q.; Kim, M.J.; Register, L.F.; et al. Coherent interlayer tunneling and negative differential resistance with high current density in double bilayer graphene–WSe$_2$ heterostructures. *Nano Lett.* **2017**, *17*, 3919–3925. [CrossRef] [PubMed]

79. Campbell, P.M.; Tarasov, A.; Joiner, C.A.; Ready, W.J.; Vogel, E.M. Enhanced Resonant Tunneling in Symmetric 2D Semiconductor Vertical Heterostructure Transistors. *ACS Nano* **2015**, *9*, 5000–5008. [PubMed]

80. Srivastava, A.; Fahad, M.S. Vertical MoS$_2$/hBN/MoS$_2$ interlayer tunneling field effect transistor. *Solid-State Electron.* **2016**, *126*, 96–103.

81. Roy, T.; Tosun, M.; Cao, X.; Fang, H.; Lien, D.-H.; Zhao, P.; Chen, Y.-Z.; Chueh, Y.-L.; Guo, J.; Javey, A. Dual-Gated MoS$_2$/WSe$_2$ van der Waals Tunnel Diodes and Transistors. *ACS Nano* **2015**, *9*, 2071–2079. [PubMed]

82. Nourbakhsh, A.; Zubair, A.; Dresselhaus, M.S.; Palacios, T. Transport properties of a MoS$_2$/WSe$_2$ heterojunction transistor and its potential for application. *Nano Lett.* **2016**, *16*, 1359–1366. [PubMed]

83. Roy, T.; Tosun, M.; Hettick, M.; Ahn, G.H.; Hu, C.; Javey, A. 2D–2D tunneling field-effect transistors using WSe$_2$/SnSe$_2$ heterostructures. *Appl. Phys. Lett.* **2016**, *108*, 83111. [CrossRef]

84. Yan, R.; Fathipour, S.; Han, Y.; Song, B.; Xiao, S.; Li, M.; Ma, N.; Protasenko, V.; Muller, D.A.; Jena, D.; et al. Esaki diodes in van der Waals heterojunctions with broken-gap energy band alignment. *Nano Lett.* **2015**, *15*, 5791–5798. [CrossRef] [PubMed]

85. Yu, W.J.; Li, Z.; Zhou, H.; Chen, Y.; Wang, Y.; Huang, Y.; Duan, X. Vertically stacked multi-heterostructures of layered materials for logic transistors and complementary inverters. *Nat. Mater.* **2013**, *12*, 246–252. [CrossRef] [PubMed]

86. Moriya, R.; Yamaguchi, T.; Inoue, Y.; Morikawa, S.; Sata, Y.; Masubuchi, S.; Machida, T. Large current modulation in exfoliated-graphene/MoS$_2$/metal vertical heterostructures. *Appl. Phys. Lett.* **2014**, *105*, 083119. [CrossRef]

87. Mead, C.A. Operation of Tunnel-Emission Devices. *J. Appl. Phys.* **1961**, *32*, 646–652. [CrossRef]

88. Atalla, M.M.; Soshea, R.W. Hot-carrier triodes with thin-film metal base. *Solid-State Electron.* **1963**, *6*, 245–250. [CrossRef]

89. Hensel, J.C.; Levi, A.F.J.; Tung, R.T.; Gibson, J.M. Transistor action in Si/CoSi$_2$/Si heterostructures. *Appl. Phys. Lett.* **1985**, *47*, 151–153. [CrossRef]

90. Rosencher, E.; Badoz, P.A.; Pfister, J.C.; Arnaud d'Avitaya, F.; Vincent, G.; Delage, S. Study of ballistic transport in Si-CoSi$_2$-Si metal base transistors. *Appl. Phys. Lett.* **1986**, *49*, 271–273. [CrossRef]

91. Yajima, T.; Hikita, Y.; Hwang, H.Y. A heteroepitaxial perovskite metal-base transistor. *Nat. Mater.* **2011**, *10*, 198–201. [CrossRef] [PubMed]

92. Tonouchi, M.; Sakai, H.; Kobayashi, T.; Fujisawa, K. A novel hot-electron transistor employing superconductor base. *IEEE Trans. Magn.* **1987**, *23*, 1674–1677. [CrossRef]

93. Heiblum, M.; Thomas, D.C.; Knoedler, C.M.; Nathan, M.I. Tunneling hot-electron transfer amplifier: A hot-electron GaAs device with current gain. *Appl. Phys. Lett.* **1985**, *47*, 1105–1107. [CrossRef]

94. Shur, M.S.; Bykhovski, A.D.; Gaska, R.; Asif Khan, M.; Yang, J.W. AlGaN–GaN–AlInGaN induced base transistor. *Appl. Phys. Lett.* **2000**, *76*, 3298–3300. [CrossRef]

95. Dasgupta, S.; Raman, N.A.; Speck, J.S.; Mishra, U.K. Experimental demonstration of III-nitride hot-electron transistor with GaN base. *IEEE Electron Device Lett.* **2011**, *32*, 1212–1214. [CrossRef]

96. Yang, Z.; Zhang, Y.; Nath, D.N.; Khurgin, J.B.; Rajan, S. Current gain in sub-10nm base GaN tunneling hot electron transistors with AlN emitter barrier. *Appl. Phys. Lett.* **2015**, *106*, 032101. [CrossRef]

97. Gupta, G.; Ahmadi, E.; Hestroffer, K.; Acuna, E.; Mishra, U.K. Common Emitter Current Gain >1 in III-N Hot Electron Transistors with 7-nm GaN/InGaN Base. *IEEE Electron Device Lett.* **2015**, *36*, 439–441. [CrossRef]

98. Kong, B.D.; Jin, Z.; Kim, K.W. Hot-Electron Transistors for Terahertz Operation Based on Two-Dimensional Crystal Heterostructures. *Phys. Rev. Appl.* **2014**, *2*, 054006. [CrossRef]

99. Vaziri, S.; Belete, M.; Dentoni Litta, E.; Smith, A.D.; Lupina, G.; Lemme, M.C.; Östlinga, M. Bilayer insulator tunnel barriers for graphene-based vertical hot-electron transistors. *Nanoscale* **2015**, *7*, 13096–13104. [CrossRef] [PubMed]

100. Fisichella, G.; Greco, G.; Roccaforte, F.; Giannazzo, F. Current transport in graphene/AlGaN/GaN vertical heterostructures probed at nanoscale. *Nanoscale* **2014**, *6*, 8671–8680. [CrossRef] [PubMed]

101. Giannazzo, F.; Fisichella, G.; Greco, G.; Roccaforte, F. Challenges in graphene integration for high-frequency electronics. In *AIP Conference Proceedings*; AIP Publishing: Melville, NY, USA, 2016; Volume 1749, p. 020004.

102. Giannazzo, F.; Fisichella, G.; Greco, G.; La Magna, A.; Roccaforte, F.; Pecz, B.; Yakimova, R.; Dagher, R.; Michon, A.; Cordier, Y. Graphene integration with nitride semiconductors for high power and high frequency electronics. *Phys. Status Solidi A* **2017**, *214*, 1600460. [CrossRef]

103. Zubair, A.; Nourbakhsh, A.; Hong, J.-Y.; Qi, M.; Song, Y.; Jena, D.; Kong, J.; Dresselhaus, M.; Palacios, T. Hot Electron Transistor with van der Waals Base-Collector Heterojunction and High-Performance GaN Emitter. *Nano Lett.* **2017**, *17*, 3089–3096. [CrossRef] [PubMed]

104. Kim, K.S.; Zhao, Y.; Jang, H.; Lee, S.Y.; Kim, J.M.; Kim, K.S.; Ahn, J.-H.; Kim, P.; Choi, J.-Y.; Hong, B.H. Large-scale pattern growth of graphene films for stretchable transparent electrodes. *Nature* **2009**, *457*, 706–710. [CrossRef] [PubMed]

105. Lupina, G.; Kitzmann, J.; Costina, I.; Lukosius, M.; Wenger, C.; Wolff, A.; Vaziri, S.; Östling, M.; Pasternak, I.; Krajewska, A.; et al. Residual Metallic Contamination of Transferred Chemical Vapor Deposited Graphene. *ACS Nano* **2015**, *9*, 4776–4785. [CrossRef] [PubMed]

106. Fisichella, G.; Di Franco, S.; Roccaforte, F.; Ravesi, S.; Giannazzo, F. Microscopic mechanisms of graphene electrolytic delamination from metal substrates. *Appl. Phys. Lett.* **2014**, *104*, 233105. [CrossRef]

107. Hong, J.-Y.; Shin, Y.C.; Zubair, A.; Mao, Y.; Palacios, T.; Dresselhaus, M.S.; Kim, S.H.; Kong, J. A Rational Strategy for Graphene Transfer on Substrates with Rough Features. *Adv. Mater.* **2016**, *28*, 2382–2392. [CrossRef] [PubMed]

108. Choi, J.-K.; Kwak, J.; Park, S.-D.; Yun, H.D.; Kim, S.-Y.; Jung, M.; Kim, S.Y.; Park, K.; Kang, S.; Kim, S.-D.; et al. Growth of Wrinkle-Free Graphene on Texture-Controlled Platinum Films and Themal-Assisted Transfer of Large-Scale Patterned Graphene. *ACS Nano* **2015**, *9*, 679–686. [CrossRef] [PubMed]

109. Kim, H.H.; Lee, S.K.; Lee, S.G.; Lee, E.; Cho, K. Wetting-Assisted Crack- and Wrinkle-Free Transfer of Wafer-Scale Graphene onto Arbitrary Substrates over a Wide Range of Surface Energies. *Adv. Funct. Mater.* **2016**, *26*, 2070–2077. [CrossRef]

110. Giannazzo, F.; Fisichella, G.; Greco, G.; Schilirò, E.; Deretzis, I.; Lo Nigro, R.; La Magna, A.; Roccaforte, F.; Iucolano, F.; Lo Verso, S.; et al. Fabrication and Characterization of Graphene Heterostructures with Nitride Semiconductors for High Frequency Vertical Transistors. *Phys. Status Solidi A* **2017**, 1700653. [CrossRef]

111. Michon, A.; Vézian, S.; Roudon, E.; Lefebvre, D.; Zielinski, M.; Chassagne, T.; Portail, M. Effects of pressure, temperature, and hydrogen during graphene growth on SiC (0001) using propane-hydrogen chemical vapor deposition. *J. Appl. Phys.* **2013**, *113*, 203501. [CrossRef]

112. Bouhafs, C.; Zakharov, A.A.; Ivanov, I.G.; Giannazzo, F.; Eriksson, J.; Stanishev, V.; Kühne, P.; Iakimov, T.; Hofmann, T.; Schubert, M.; et al. Multi-scale investigation of interface properties, stacking order and decoupling of few layer graphene on C-face 4H-SiC. *Carbon* **2017**, *116*, 722–732. [CrossRef]

113. Lee, J.-H.; Lee, E.K.; Joo, W.-J.; Jang, Y.; Kim, B.-S.; Lim, J.Y.; Choi, S.-H.; Ahn, S.J.; Ahn, J.R.; Park, M.-H.; et al. Wafer-Scale Growth of Single-Crystal Monolayer Graphene on Reusable Hydrogen-Terminated Germanium. *Science* **2014**, *344*, 286–288. [CrossRef] [PubMed]

114. Varchon, F.; Feng, R.; Hass, J.; Li, X.; Nguyen, B.N.; Naud, C.; Mallet, P.; Veuillen, J.Y.; Berger, C.; Conrad, E.H.; et al. Electronic Structure of Epitaxial Graphene Layers on SiC: Effect of the Substrate. *Phys. Rev. Lett.* **2007**, *99*, 126805. [CrossRef] [PubMed]

115. Nicotra, G.; Ramasse, Q.M.; Deretzis, I.; La Magna, A.; Spinella, C.; Giannazzo, F. Delaminated Graphene at Silicon Carbide Facets: Atomic Scale Imaging and Spectroscopy. *ACS Nano* **2013**, *7*, 3045–3052. [CrossRef] [PubMed]

116. Giannazzo, F.; Deretzis, I.; La Magna, A.; Roccaforte, F.; Yakimova, R. Electronic transport at monolayer-bilayer junctions in epitaxial graphene on SiC. *Phys. Rev. B* **2012**, *86*, 235422. [CrossRef]

117. Sonde, S.; Giannazzo, F.; Raineri, V.; Yakimova, R.; Huntzinger, J.-R.; Tiberj, A.; Camassel, J. Electrical properties of the graphene/4H-SiC (0001) interface probed by scanning current spectroscopy. *Phys. Rev. B* **2009**, *80*, 241406. [CrossRef]

118. Speck, F.; Jobst, J.; Fromm, F.; Ostler, M.; Waldmann, D.; Hundhausen, M.; Weber, H.; Seyller, T. The quasi-freestanding nature of graphene on H-saturated SiC (0001). *Appl. Phys. Lett.* **2011**, *99*, 122106. [CrossRef]

119. Michon, A.; Tiberj, A.; Vezian, S.; Roudon, E.; Lefebvre, D.; Portail, M.; Zielinski, M.; Chassagne, T.; Camassel, J.; Cordier, Y. Graphene growth on AlN templates on silicon using propane-hydrogen chemical vapor deposition. *Appl. Phys. Lett.* **2014**, *104*, 071912. [CrossRef]

120. Dagher, R.; Matta, S.; Parret, R.; Paillet, M.; Jouault, B.; Nguyen, L.; Portail, M.; Zielinski, M.; Chassagne, T.; Tanaka, S.; et al. High temperature annealing and CVD growth of few-layer graphene on bulk AlN and AlN templates. *Phys. Status Solidi A* **2017**, 1600436. [CrossRef]

121. Van der Zande, A.M.; Huang, P.Y.; Chenet, D.A.; Berkelbach, T.C.; You, Y.; Lee, G.-H.; Heinz, T.F.; Reichman, D.R.; Muller, D.A.; Hone, J.C. Grains and grain boundaries in highly crystalline monolayer molybdenum disulphide. *Nat. Mater.* **2013**, *12*, 554–561. [CrossRef] [PubMed]

122. Kang, K.; Xie, S.; Huang, L.; Han, Y.; Huang, P.Y.; Mak, K.F.; Kim, C.-J.; Muller, D.; Park, J. High-mobility three-atom-thick semiconducting films with wafer-scale homogeneity. *Nature* **2015**, *520*, 656–660. [CrossRef] [PubMed]

123. Dumcenco, D.; Ovchinnikov, D.; Marinov, K.; Lazic, P.; Gibertini, M.; Marzari, N.; Sanchez, O.L.; Kung, Y.-C.; Krasnozhon, D.; Chen, M.-W.; et al. Large-Area Epitaxial Monolayer MoS_2. *ACS Nano* **2015**, *9*, 4611–4620. [CrossRef] [PubMed]

124. Ruzmetov, D.; Zhang, K.; Stan, G.; Kalanyan, B.; Bhimanapati, G.R.; Eichfeld, S.M.; Burke, R.A.; Shah, P.B.; O'Regan, T.P.; Crowne, F.J.; et al. Vertical 2D/3D Semiconductor Heterostructures Based on Epitaxial Molybdenum Disulfide and Gallium Nitride. *ACS Nano* **2016**, *10*, 3580–3588. [CrossRef] [PubMed]

125. Vervuurt, R.H.J.; Kessels, W.M.M.; Bol, A.A. Atomic Layer Deposition for Graphene Device Integration. *Adv. Mater. Interfaces* **2017**, 1700232. [CrossRef]

126. Kim, S.; Nah, J.; Jo, I.; Shahrjerdi, D.; Colombo, L.; Yao, Z.; Tutuc, E.; Banerjee, S.K. Realization of a High Mobility Dual-Gated Graphene Field-Effect Transistor with Al_2O_3 Dielectric. *Appl. Phys. Lett.* **2009**, *94*, 062107. [CrossRef]

127. Jandhyala, S.; Mordi, G.; Lee, B.; Lee, G.; Floresca, C.; Cha, P.-R.; Ahn, J.; Wallace, R.M.; Chabal, Y.J.; Kim, M.J.; et al. Atomic Layer Deposition of Dielectrics on Graphene Using Reversibly Physisorbed Ozone. *ACS Nano* **2012**, *6*, 2722–2730. [CrossRef] [PubMed]

128. Fisichella, G.; Schilirò, E.; Di Franco, S.; Fiorenza, P.; Lo Nigro, R.; Roccaforte, F.; Ravesi, S.; Giannazzo, F. Interface Electrical Properties of Al_2O_3 Thin Films on Graphene Obtained by Atomic Layer Deposition with an in Situ Seedlike Layer. *ACS Appl. Mater. Interfaces* **2017**, *9*, 7761–7771. [CrossRef] [PubMed]

129. Ismach, A.; Chou, H.; Ferrer, D.A.; Wu, Y.; McDonnell, S.; Floresca, H.C.; Covacevich, A.; Pope, C.; Piner, R.; Kim, M.J.; et al. Toward the controlled synthesis of hexagonal boron nitride films. *ACS Nano* **2012**, *6*, 6378–6385. [CrossRef] [PubMed]

130. Tay, R.Y.; Griep, M.H.; Mallick, G.; Tsang, S.H.; Singh, R.S.; Tumlin, T.; Teo, E.H.T.; Karna, S.P. Growth of large single-crystalline two-dimensional boron nitride hexagons on electropolished copper. *Nano Lett.* **2014**, *14*, 839–846. [CrossRef] [PubMed]

131. Lu, G.; Wu, T.; Yuan, Q.; Wang, H.; Wang, H.; Ding, F.; Xie, X.; Jiang, M. Synthesis of large single-crystal hexagonal boron nitride grains on Cu–Ni alloy. *Nat. Commun.* **2015**, *6*, 6160. [CrossRef] [PubMed]

132. Sonde, S.; Dolocan, A.; Lu, N.; Corbet, C.; Kim, J.; Tutuc, E.; Banerjee, S.K.; Colombo, L. Ultrathin wafer-scale hexagonal boron nitride on dielectric surfaces by diffusion and segregation mechanism. *2D Mater.* **2017**, *4*, 025052. [CrossRef]

133. Yang, W.; Chen, G.; Shi, Z.; Liu, C.-C.; Zhang, L.; Xie, G.; Cheng, M.; Wang, D.; Yang, R.; Shi, D.; et al. Epitaxial growth of single-domain graphene on hexagonal boron nitride. *Nat. Mater.* **2013**, *12*, 792–797. [CrossRef] [PubMed]

134. Shi, Y.; Zhou, W.; Lu, A.-Y.; Fang, W.; Lee, Y.-H.; Hsu, A.L.; Kim, S.M.; Kim, K.K.; Yang, H.Y.; Li, L.-J.; et al. van der Waals Epitaxy of MoS_2 Layers Using Graphene as Growth Templates. *Nano Lett.* **2012**, *12*, 2784–2791. [CrossRef] [PubMed]

135. Lin, Y.-C.; Lu, N.; Perea-Lopez, N.; Li, J.; Lin, Z.; Peng, X.; Lee, C.H.; Sun, C.; Calderin, L.; Browning, P.N.; et al. Direct Synthesis of van der Waals Solids. *ACS Nano* **2014**, *8*, 3715–3723. [CrossRef] [PubMed]

crystals

MDPI

Review

Van der Waals Heterostructure Based Field Effect Transistor Application

Jingyu Li [†], Xiaozhang Chen [†], David Wei Zhang and Peng Zhou *

State Key Laboratory of ASIC and System, School of Microelectronics, Fudan University, Shanghai 200433, China; 17212020024@fudan.edu.cn (J.L.); 17112020001@fudan.edu.cn (X.C.); dwzhang@fudan.edu.cn (D.W.Z.)
* Correspondence: pengzhou@fudan.edu.cn
† These authors contributed equally to this work.

Received: 16 November 2017; Accepted: 20 December 2017; Published: 26 December 2017

Abstract: Van der Waals heterostructure is formed by two-dimensional materials, which applications have become hot topics and received intensive exploration for fabricating without lattice mismatch. With the sustained decrease in dimensions of field effect transistors, van der Waals heterostructure plays an important role in improving the performance of devices because of its prominent electronic and optoelectronic behavior. In this review, we discuss the process of assembling van der Waals heterostructures and thoroughly illustrate the applications based on van der Waals heterostructures. We also present recent innovation in field effect transistors and van der Waals stacks, and offer an outlook of the development in improving the performance of devices based on van der Waals heterostructures.

Keywords: van der Waals heterostructure; field effect transistor; two-dimensional material

---

1. Introduction

In the past few decades, graphene, which exhibits extraordinary electronic and optoelectronic properties in exploring low-dimensional materials, has become the most advanced research in the field of two-dimensional layered material (2DLM) science, solid-state physics and engineering [1–3]. Because graphene originally has excellent parameters, such as mechanical stiffness, strength and elasticity, very high electrical and thermal conductivity, graphene-based transistor preforms better than conventional Si-based transistors [1] in fabricating high-frequency transistors with the possible cut-off frequency f_T = 1THz at a channel length of about 100 nm [4] and photodetectors with a wide spectral from ultraviolet to infrared [1]. Therefore, the discovery of graphene creates a new generation of electronic devices with atomically thin geometry and unprecedented combination of speed and flexibility. Except for the advantages and merits of graphene-based transistor mentioned above, there are some shortcomings, such as the unique zero bandgap, hindering the graphene's application in electronic industry [3]. To overcome the defect brought by graphene device, this extensive library of 2DLMs, such as graphene, phosphorene, hexagonal boron nitride and transition metal dichalcogenide mentioned in this article with selected properties opens up the possibility of heterogeneous integration at the atomic scale, creating novel hybrid structures that display totally new physics and enable unique functionality [5]. Because 2DLMs have various bandgap, the different material shows different property. For example, black phosphorus shows the characteristic of conductor; molybdenum disulfide (MoS_2) and tungsten selenide (WSe_2) are semiconductors; and boron nitride (BN) performs as insulator. 2DLMs with various properties make it possible to form special combination of heterogeneous integration at atom scale, such as van der Waals heterostructures (vdWHs).

In general, heterostructures formed by 2DLM monolayers are assembled by covalent bond force within layers and stacked together by van der Waals force between layers. Thus, there is no free dangling bond between 2DLM layers. In contrast to typical nanostructures persecuted by dangling bond and trap state on the surfaces [6], monolayer are free of these disadvantages (Figure 1a,b) and exhibit extraordinary electronic and optoelectronic properties. Additionally, without free dangling bonds, the interface between neighbor 2DLM layers are assembled by van der Waals forces, which is much weaker than chemical bond force. Therefore, the van der Waals heterostructure can be easily isolated by exfoliation with the help of taps [7]. Although some highly disparate materials have a great lattice mismatch in creating heterostructure, those can be assembled together by van der Waals force. This allows diverse 2DLMs to construct various van der Waals heterostructures (vdWHs) with completely novel properties and functions.

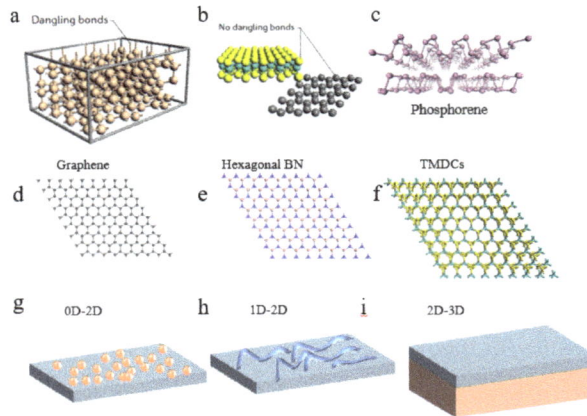

Figure 1. Two-dimensional layered materials and van der Waals heterostructures: (**a**,**b**) materials with and without dangling bond; (**c**) anisotropic transport behavior of phosphorene in different direction; (**d**–**f**) monolayer graphene, hexagonal boron nitride and transition metal dichalcogenides, graphene shows semimetal characteristic, hexagonal boron nitride behaves as insulator and transition metal dichalcogenides act as semiconductor; (**g**–**i**) van der Waals heterostructure formed with free dangling bond materials; (**g**) 0D (particles or quantum dots) and 2D layered materials stack; (**h**) 1D (nanowires) and 2D layered materials stack; and (**i**) 2D layered materials and 3D bulk materials stack. (**a**–**e**) are reproduced with permission from Nature Publishing Group, Ref [6,8].

The early study in vdWHs focus on the combination of 2DLMs with 0D (for example, plasmonic nanoparticles and quantum dots) and 1D (for example, nanowires and nanoribbons) nanostructures [9–12]. These novel 0D–2D (Figure 1f) and 1D–2D (Figure 1g) vdWHs structures have fabricated many highly performed devices, such as photodetectors with highly optical response [10] and transistors with high speed and flexibility [12]. In recent years, most efforts have been taken to form 2D–2D integration vdWHs by vertically stacking various 2DLMs; for example, Dean fabricated a graphene–BN heterostructure transistor which Hall mobility reached ~25,000 $cm^2 \cdot V^{-1} \cdot s^{-1}$ at high density [13], and Withers stacked metallic graphene, insulating hexagonal boron nitride and various semiconducting monolayers into complex but carefully designed sequences, which exhibited an extrinsic quantum efficiency of nearly 10% and the emission can be tuned over a wide range of frequencies by appropriately choosing and combining 2D semiconductors [14]. 2D–2D vdWHs with diverse 2DLMs make it possible to accurately control and modulate the transport of charge carriers, excitons and photons within the atomic surface, and greatly promote the new generation of unique devices. Apart from 2D–2D vdWHs, 2DLMs can also be integrated with 3D bulk materials to form 2D–3D structures [8,15] (Figure 1f–h), which attract some interest. Hot electron transistors based on 2D–3D heterojunctions have been reported [16]. Gate tunable p-n junctions based on 2D–3D heterojunctions have been reported [17,18]. Heterojunction formed with MoS_2 and p type Ge can fabricate high performance tunneling field transistor [18].

Recently, most vdWHs are formed by mechanically stacking different monolayers together. Although this method allows great flexibility, it is slow and inconvenient. Thus, some simple and high-quality techniques are being developed, for example, large-area 2DLMs grown by chemical vapor deposition (CVD) and direct growth of heterostructure by CVD or physical epitaxy, which improves the growth quality of monolayers and the properties of heterostructure exhibited [19–21].

In this review, we mainly concentrate on the integration, properties, application and technology of vdWHs. First, we give a brief introduction of typical materials, which got highly unprecedented attention in the field of materials science and solid-state physics in the past few years, and discuss the electronical properties about them. Then, we focus on the applications of field effect transistors based on vdWHs, including elevating mobility, decreasing the contract resistance and conventional transistors like tunneling field effect transistor. Except for traditional FETs, we investigated some late-model structure, namely the vertical thin film transistor. Finally, we analyze the process of mechanically and chemically stacking van der Waals heterostructures. Meanwhile, we discuss the process of fabricating the field effect transistors (FETs) by van der Waals stacking methods.

2. Two Dimensional Materials

Two-dimensional materials, which constitute van der Waals heterostructures (vdWHs), have been explored to exhibit entirely different characteristics (Figure 2a), including conductors, semimetals, semiconductors, and insulators. Large group of two-dimensional materials provide various band gap and different electronic and optical performance, which offers the theoretical possibility of different device application requirement. With different 2D materials to form heterostructure, not only novel properties can be observed but also highly reliable nano-device can be achieved as well.

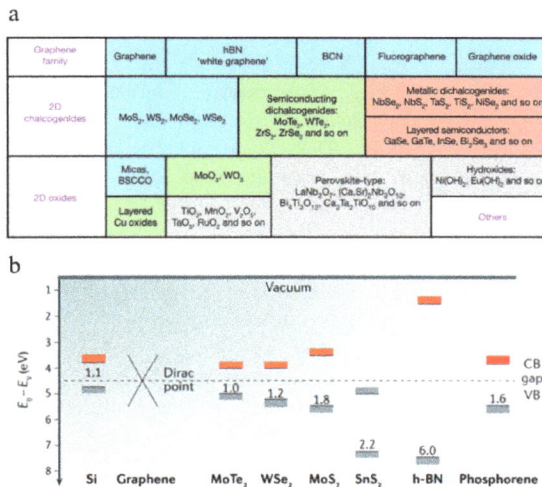

Figure 2. Different 2D layered materials and electronic properties. (**a**) Classified 2D layered materials based on different conductive characteristic. Monolayers proved to be stable in air condition are marked as blue. Those probably stable in air are shaded green. Those monolayers stable in inert atmosphere are shaded pink. Grey shading shows that monolayer can be successfully exfoliated from bulk materials. Reproduced with permission from Nature Publishing Group, Ref [22] (**b**) Band gaps of 2D layered materials compared with Dirac point. Reproduced with the permission from IEEE in Ref [23].

2.1. Graphene

With a honeycomb lattice structure assembled by strong covalent bond force, crystal structure shown in Figure 1d, graphene only has a single atomic carbon layer, which can be manufactured via mechanically exfoliating from bulk graphite. Due to bulk graphite has weak interlayer connection formed by van der Waals force, the large-area monolayer graphene can be easily and successfully exfoliated from bulk material with the help of a scotch tape. Owing to the unique lattice structure and atom combination, graphene has zero band gap which means charge carriers move in a ultrahigh speed about 25,000 $cm^2 \cdot V^{-1} \cdot s^{-1}$ at room temperature [24]. In other words, with high conductivity in 2DLMs family, graphene behaves like metal, which is the significant component of forming vdWHs. Transistors with graphene electrodes will have a lower contact resistance compared to transistors without graphene electrodes [25]. In flash memory devices, graphene also acts as charge trap layers [26]. Due to the single-atom thickness and specific density of state (DOS), it exhibits partial optical and electrostatic transparency under tunable work function. For possessing this characteristic, graphene serves as contact material integrated with other two-dimensional semiconductors in vdHW transistors, which achieves an appropriate contact resistance and admirable electrical performance [27].

Bilayer graphene, which is constituted of two atomic carbon-layers stacked by van der Waals interaction under specific direction, does not exhibit the same properties as monolayer graphene. That is because the interaction between two neighboring layers provides extra van der Waals electric field, which inducts bilayer's bandgap [28] and forms four pseudospin flavors [29]. Although bilayer graphene has a band gap up to 250 mV under gate controlling, it is still smaller than other 2DLM like transition metal dichalcogenide, and even bulk Si (the band gap is about 1.2 eV). Even though monolayer graphene has zero band gap, it can be used in other ways such as graphene electrode in transistor, charge trap layer in flash memory and metal–graphene interfaces which can provide photo-generated carriers in photodetectors [30,31]. Moreover, bilayer graphene transistors also have a degradation of the carrier mobility compared with monolayer graphene [2].

2.2. Transition Metal Dichalcogenide

Transition metal dichalcogenide (TMDC) is the group of two-dimensional materials with chemical form MX_2, where M stands for transition metal (molybdenum and tungsten) and X represents for the chalcogenide (sulfur, selenium or tellurium). The arrangement of atomic layered TMDC is X-M-X form. The crystal structure is shown in Figure 1f. Considering the same reason that adjacent layers of TMDC are weakly held together by van der Waals force, monolayers of TMDC can also be directly split from bulk crystals (regardless of artificiality or natural mineral) by scotch tape, called mechanical exfoliation, and it can also be fabricated by chemical vapor deposition (CVD) processes [32,33]. From bulk structure to two-dimensional layered structure, the most frequently studied TMDCs, such as MoS_2, $MoSe_2$, WS_2 and WSe_2, show various band gaps (Figure 2b) and tunable electronic properties with its volume changing from bulk material to monolayer [16]. At room temperature, as the thickness of TMDC decreases from bulk to monolayer, its band gap ranges from 1 eV to 2 eV and carrier mobility reaches over 100 $cm^2 \cdot V^{-1} \cdot s^{-1}$. Owing to the varied bandgap resulting in tunable electrical characteristics, there are some distinctive properties observed in TMDC, for example, valley Hall effect [34], valley polarization [35], superconductivity [36] and photoelectric characteristic [37,38]. TMDC's lack of dangling bands is another reason that makes it attractive for use in FET channel material. This means the lattice mismatching is out of consideration during the formation of TMDC–TMDC van der Waals interconnection.

2.3. Phosphorene

Phosphorene is a monolayer material constituted of phosphorus, with a vertically interlaced hexagonal lattice that looks like armchairs. Phosphorene's bandgap varies from 0.33 eV to 1.5 eV depending on the number of layers stacked [39]. Notably, an individual property is observed in few layers of phosphorene, i.e., anisotropic transportation in different phosphorene's crystal orientations [40]. Specifically, transporting in the armchair direction (x direction) is much more efficient than in zigzag direction (Figure 1c). In addition, a p-type semiconductor characteristic emerges in phosphorene. When the hole mobility exceeds 1000 $cm^2 \cdot V^{-1} \cdot s^{-1}$ in certain direction [41], the phosphorene becomes the significant 2D material for electronic device. Later report shows that the drain current of FETs based on phosphorene can be modulated to 10^5, which means the I_{on}/I_{off} ratio reaches 10^5 [42]. The tunable bandgap of phosphorene make it possible to react with different spectral regime from visible to infrared, which can be applied in fabricating multispectral photodetector [43].

The main challenge for phosphorene's extensive application is the environmental stability, which means it can be easily oxidized [44]. To prevent the surface oxidization which leads to the decrease of device performance, some solutions have been adopted, for example, using h-BN [45], atomic layer of Al_2O_3 [46,47] and hydrophobic fluoropolymer [48] as encapsulating materials during process of van der Waals stacking.

2.4. Hexagonal Boron Nitride

Hexagonal boron nitride (h-BN), also called white graphite, has the same lattice structure as graphite. (Figure 1e) For boron and nitrogen are combined together by the strong covalent sp^2 hybridized chemical bond, h-BN exists as insulator with a large bandgap of 6 eV [49]. Mechanical exfoliation can provide an atomically smooth surface without dangling bond and carrier traps, so a few layers of h-BN can be used as ultra-flat insulating substrates, the gate dielectrics [50] and ideal encapsulate material [24] in heterostructure transistors [13]. Besides, ultrathin h-BN can be served as great tunneling barrier material in tunneling devices [51,52].

2.5. Other Various 2D Materials for vdWHs

In the past few years, novel 2D TMDCs draw a lot of concentration on creating new types of devices; meanwhile, traditional 2D materials such as MoS_2 and WSe_2 have been considered once more. In virtue of doping technology, high-k amorphous titanium suboxide (ATO) is used as an n-type charge transfer dopant on the monolayer MoS_2, which decreases the contact resistance to 180 $\Omega \cdot \mu m$ [53]. Utilizing AlO_x, another method of n-type doping in MoS_2 is demonstrated, which can also reduce the contact resistance to 480 $\Omega \cdot \mu m$ [54]. After confirming operational parameters of chemical vapor deposition, ultrathin $MoS_{2(1-x)}Se_{2x}$ alloy nano-flakes with atomic or few-layer thickness can be obtained and it have high activity and durability during the hydrogen evolution reaction [55]. These various 2D materials show variable bandgaps, tunable optical and electronic properties [56,57].

3. Techniques

Two-dimensional materials can be assembled or grown directly into heterostructures, where the monolayers are held together by van der Waals forces. Although variety of 2D materials have been found, the assembly techniques currently used allow only certain types to build up the heterojunction. At the same time, the alternative technique, called sequential growth of monolayers, potentially do well in large scale production. Actually, the growth method has some limitations, such as slow growth and hard to control, and is still in early stages. Nevertheless, a great number of experiments with van der Waals heterostructure indicate that 2D materials are versatile and practical for further applications [58].

3.1. Assembly

Direct mechanical assembly is the most commonly used techniques of assembling heterostructures at present. This techniques was been put forward in 2010 by Dean et al., who demonstrated a very high level of performance provided by graphene and h-BN [13].

In early works, the technique is only used in preparing a flake of 2D material on a sacrificial membrane, such as PDMS, aligning and placing it on the substrate, and then remove or dissolve the membrane (Figure 3A–G). Then, the progress is repeated to deposit the other 2D material on the top of the former layer. Because the materials have close contact with the sacrificial layer, which is dissolved finally, the interface between TMDC and sacrificial layer will be contaminated. However, annealing can remove this contamination and achieve high interface quality, so the mobility of graphene device can reach 10^6 cm^2·V^{-1}·s^{-1} [13].

The other substantially cleaner method has the same principle that van der Waals interaction exist between layers. After sticking the first layer onto membrane, the second layer is directly stuck on the first layer, instead of dissolving the membrane (Figure 3G–O). Then, the process is repeated several times to get expected multilayer. This method leads to cleaner interface over large areas, and higher electron mobility.

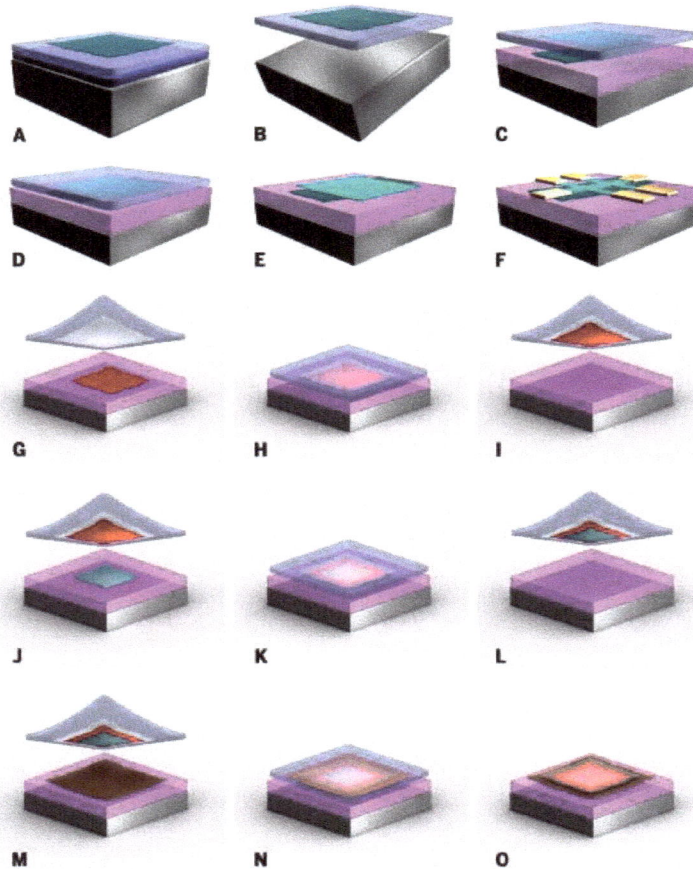

Figure 3. Assemble van der Waals heterostructures techniques: wet-transfer and pick-and-lift. (**A–F**) Wet-transfer technique: (**A**) A 2D crystal prepared on a sacrificial layer. (**B**) Pick up the sacrificial layer with 2D material. The layer is then placed on (**C**) and the second layer is transferred the same way (**D**). Then, it is placed on top of the first layer (**E**). Deposited metal on the proper location (**F**). (**G–O**). Pick-and-lift technique: A 2D material on a membrane is aligned (**G**) and then placed on top of another 2D crystal (**H**). The process is repeated to lift additional crystals (**I–L**). Finally, the whole multilayer is placed on the substrate (**M–O**), and the membraned is dissolved. Reused with the permission from American Association for the Advancement of Science in the Ref [58].

3.2. Growth Methods

Although the assembly methods can stack the layers and achieve a clean interface, it still has some disadvantages, for example, the stacking position cannot be accurately controlled, which affects the device properties [22]. The direct growth of vertical layered heterostructures via chemical vapor deposition (CVD) is much more promising in terms of being controllable and scalable.

The direct growth of heterostructures can be divided into two parts: (1) sequential CVD growth of 2D materials layered; and (2) molecular beam epitaxy (MBE). The detailed information is presented in the following.

3.2.1. CVD

For the CVD system of MoS_2 growth, as depicted in Figure 4, two separated furnaces are used to precisely control temperature applied on both precursors and substrate. Sulfur powder and the substrate are in the same pipe while different furnaces, where the interaction temperature can be independently controlled. Twenty milligrams of MoO_3 powder, which temperature is dominated by furnace 2, is loaded between sulfur powder and the substrate and has an independent mini-pipe (diameter 1 cm). This is to prevent any cross-contamination and reaction between sulfur and MoO_3, resulting in the decrease amount of MoO_3 and unstable precursor supply in the vapor phase. Typical growth temperatures used for sulfur, MoO_3, and the substrate are ~180 °C, ~300 °C, and ~800 °C, respectively. To ensure the sufficient supply of sulfur vapor in the system, it is preheated before increasing the temperature of MoO_3.

This simple but scalable CVD growth approach can realize the direct fabrication of large-area MoS_2/h-BN heterostructures with cleaner interface and better contact, compared with mechanical transfer method or assembly methods [59].

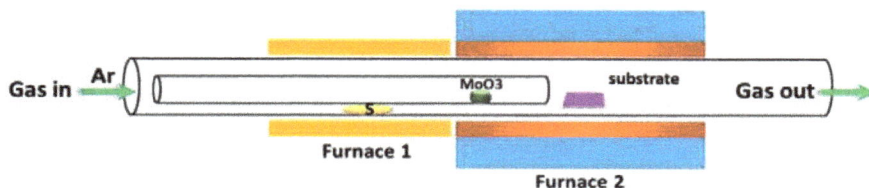

Figure 4. Schematic illustration of CVD growth methods for MoS_2 growth. Reproduced with permission from American Chemical Society Publishing Group in Ref [59].

3.2.2. MBE

As one of the most commonly used growth methods of heterostructures, CVD has successfully accomplished several two-dimensional layered materials [60]. Although it is almost theoretically perfect, the reaction condition of CVD is hard to control at present and the results reported publicly can only form a few well-defined islands over the entire wafer. Considering the defects, MBE is an alternative approach for CVD in some ways [61].

As the name suggests, MBE is a kind of physical deposition method that takes place in ultrahigh vacuum. Figure 5 shows a typical MBE system. The required components are heated into vapor and sprayed on the substrate through a small hole. At the same time, a controlling molecule beam is used to scan the substrate line by line, and the molecules or atoms are arranged layer by layer in lattice structure. The ultra-high vacuum(UHV) environment minimize the contamination of the interface between substrate and growing flake. Because there are few particles in UHV environment, the ionized atoms and molecules traveled in nearly collision-free path until hitting the substrate or the chilled oven walls, where atoms condense immediately, and thus are effectively cleaned up from the reacting system without extra contamination. Because the components are sprayed onto substrate through a small hole, it can be turned off almost instantly and changed into another reactant whenever necessary. Most importantly, MBE growth can be simplistically described as "spray painting" the substrate crystal with layers of atoms, changing the composition or impurity in each layer until a desired structure is obtained [62].

Compared with CVD, MBE can potentially grow products overspread the whole wafer, precisely control the rate of interaction by means of regulating the rate of spraying and easily change the reactant as requirement. In this sense, MBE is almost the ideal method of vdWHs fabrication since the composition can be precisely controlled layer by layer.

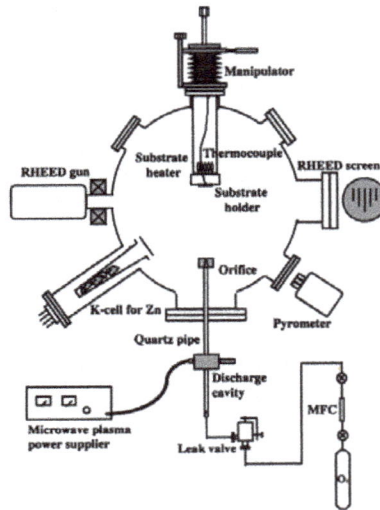

Figure 5. Schematic diagram of the MBE system. Reproduced with the permission from American Institute of Physics in the Ref [63].

4. FETs Based on Heterostructures Applications

Various 2D materials provide diverse electronic properties from conductors to insulators which benefits new devices creation. Van der Waals heterostructures exhibit ultrathin thickness and special properties that give a chance to build transistors at nanoscale size compared to bulk materials. Thus, novel field effect transistors based on van der Waals heterostructures have developed.

4.1. Decrease the Contact Resistance

Two-dimensional semiconductors (2DSCs), especially TMDCs, have sparked immense interest in the electronic devices since the MoS_2-based transistor was first reported [64]. Source and drain are often made of Metal-TMD directly contact, which induce a high contact resistance Rc. Even the contact resistance is larger than Si based transistors. The excessive contact resistance Rc on the source and drain interface is a great challenge for 2DSCs devices.

The high contact resistance, induced by a high Schottky barrier [65], prevent the charge carriers moving through the metal-semiconductor contacts. While the typical 2DLMs contact resistance currently achieved is 100 times higher than Si-based electronics (<20 $\Omega \cdot \mu m$) [66], there is still much work to do to decrease the actual contact resistance (Figure 6a). Some strategies have been used to reduce the contact resistance with the help of graphene/metal heterostructures. Schematics are shown in Figure 6d–f.

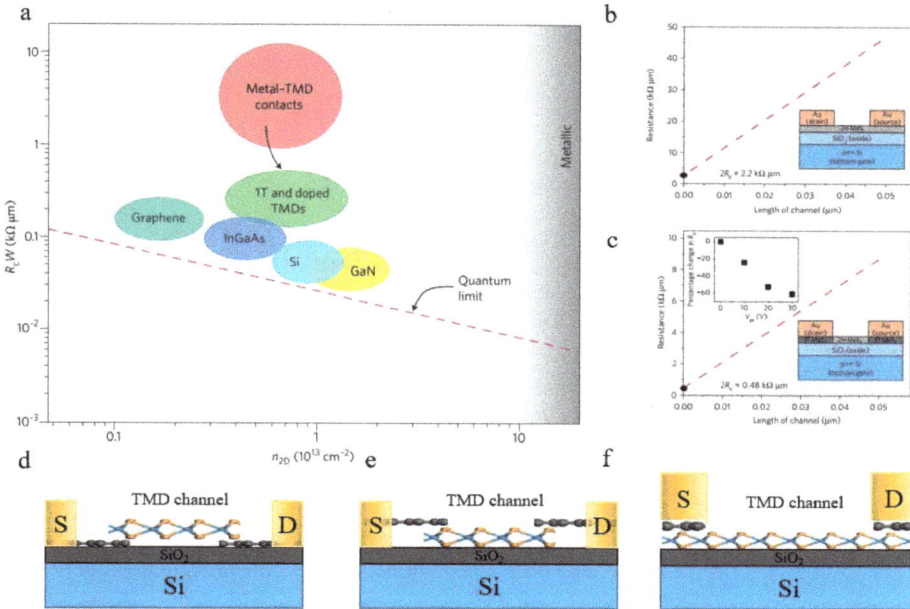

Figure 6. Contact resistance for different two-dimensional materials: (**a**) various contact resistance of materials against quantum limits; Reused with the permission from Nature publishing Group in the Ref [65] (**b**) contact resistance with traditional 2H-MoS$_2$ material; (**c**) contact resistance with 1T-MoS$_2$ and 2H MoS$_2$ as channel material; (**d**) coplanar contacts of graphene electrodes; (**e**) staggered contacts of graphene electrodes; and (**f**) hybrid contacts of graphene electrodes. (**b**,**c**) are reproduced with the permission from Nature publishing Group in the Ref [67]. (**d**–**f**) are adapted with the permission from Nature publishing Group Ref [5].

In silicon-based transistor, doping is conventionally used in traditional process of decreasing contact resistance at the source and drain interface, which is not reliable in TMDC-based transistor. That is because a highly doped silicon-based transistor can provide more carriers to fill valence band [65]. However, injecting other elements into TMDC seriously damages the covalent bonds between the atoms and breaks TMDC's unique lattice structure, thus a great deal of defects are generated into layers resulting in Fermi-level pinning, which causes undesired contact resistance [68,69]. Besides TMDC, graphene is another 2DLM drawing extensive concern. The greatest contribution of graphene is forming van der Waals stacks with metal and 2DLM, which commendably decreases the contact resistance due to its special band structure. Because graphene is single-element 2DLM, it can be doped similar to silicon. Compared with intrinsic graphene, the highly-doped graphene decreases the height of Schottky barrier formed by the connection of graphene and metal, which means the contact resistance is greatly reduced [70]. A typical TMDC-based transistor, which uses monolayer graphene as metal electrodes, 2DLMs such as tungsten diselenide (WSe$_2$) as semiconductor channel, and hexagonal boron nitride (h-BN) as gate dielectric, exhibits very high performance, such as high current on/off ratio, remarkable temperature stability, low contact resistance and low subthreshold slop [71]. Even though the Schottky barriers is still high (0.22 eV) at the graphene–WSe$_2$ contact, the on-state current is dominated by tunneling through the barriers which makes the barriers almost transparent to the charge carriers. After using 3 nm h-BN as gate dielectric and 1.5 nm WSe$_2$ flake as channel material, the tunneling distance is only 1.6 nm [72]. There are three types of interface of graphene–2DSCs contact: coplanar contacts (Figure 6d), staggered contacts (Figure 6e) and hybrid contacts (Figure 6f).

The coplanar structure is that the TMD channel materials and source/drain electrodes are on the same side of graphene. With the bottom graphene contact for 2DSCs transistors, nearly perfect work function matching lead to a barrier-free contact, which can significantly decrease Rc. MoS_2 and WSe_2 transistors using this strategy have been demonstrated [25,70,73,74]. Staggered contacts mean that TMD channel and source/drain electrodes are on different sides of graphene. With this strategy, transistors can be less susceptible to lithographic contamination, because the polymer residue did not touch TMD channel directly. Otherwise the residue will affect the charge transport at the interface. With this method, high performance MoS_2 transistors [75–78] and air stable black phosphorus transistors [45] have been reported. Metal–graphene hybrid structure as a unique contact to reach minimized Rc (300 $\Omega \cdot \mu m$) compared to metal–TMD structure. With graphene inserted as the interlayer buffer, the carrier injection was enhanced at metal–graphene–TMD channel interfaces. For example, nickel–etched–graphene electrodes have achieved a relatively low resistance [79]; Ti–graphene–MoS_2 stacks can ruduce Rc from $12.1 \pm 1.2\ \Omega \cdot mm$ to $3.7 \pm 0.3\ \Omega \cdot mm$ to some degree compared with stacking Ti onto MoS_2 directly [80].

When integrating two-dimensional transition metal dichalcogenides with metal, the valence band shows unusual Fermi level pinning. Compared with chemical doping, which will destroy the covalent bond of TMDs within layers, the stacking of graphene can create a undamaged van der Waals contact and a ultraclean interface, which prevents Fermi level pinning [25,75]. There are some other reports demonstrate a stacking structure formed by MoS_2 and graphene to reduce the connect resistance [73,77,81]. In this structure, graphene and metal are connected together to form electrodes and molybdenum disulfide (MoS_2) is used as channel. The tunability of doping graphene's work function can significantly improve the ohmic contact to MoS_2 [77].

Besides graphene, other 2DLMs such as MoS_2 can also be doped with element to improve the mobility of charge carriers which leads to low contact resistance, for example, physisorbing chloride molecule into WS_2 or MoS_2 [82]. With the method of physisorbed molecules, contact resistance can be obviously decreased [83]. Other efforts of contacting electrode metal materials with TMDs create the low Schottky barrier. For example, the connection between molybdenum (Mo) and MoS_2 forms perfect interface, leading to ultralow Schottky barrier about 0.1 eV [84]. Other methods based on van der Waals heterostructure to decrease Rc have include phase-engineering. With the organolithium chemical method, MoS_2 changes from 1T phase to 2H phase. 2H phase MoS_2 behaves as electronic channel, as shown in Figure 6b. This 1T–2H–1T structure can receive relatively low Rc [67] (Figure 6c). To create stable homojunction contact of $MoTe_2$, some reports use laser to achieve two phase interconversion. Stacking semiconducting hexagonal (2H) and metallic monoclinic (1T) $MoTe_2$ together can fabricate an ohmic heterophase homojunction, which is necessary in lowing contact resistance [67]. Phase engineering, therefore, provide a way of creating lateral heterostructures with low resistance, but additional work is needed to achieve values comparable to state-of-art materials and close to the quantum limit.

4.2. Tunneling Field Effect Transistor and Barrister

The principle of tunneling field effect transistors (TFETs) is band-to-band tunneling [85]. The barriers induce by van der Waals gap narrow because the width of van der Waals gap is below 1 nm, which will lead to charge carriers tunneling [86]. For width of barriers narrowing at the contact interface, the tunneling current between source and drain increases rapidly in silicon-based transistor. However, vertical tunneling transistor based on vdWHs is different from conventional TEFTs. In vertical structure, two separate monolayers of graphene serve as electrodes, and ultrathin dielectric material are stuck between them. Based on gate-voltage tunability of the DOS in graphene, the height of Schottky barrier reduces and realizes the charge carriers' transportation.

Transistors based on that principle use very thin BN (1.4 nm) and MoS_2 as vertical transport barrier, making devices have the switching ratio of ~10,000 at room temperature [87]. Further study of this type of heterostructure is the resonant tunneling and gate voltage-tunable negative differential conductance [88]. Other studies have put forward how to achieve the resonant tunneling with the help of crystallographic orientation alignment of the graphene-BN-graphene heterostructure [89,90].

In addition of graphene, two dimensional TMDs have sharp band edges and ultra-narrow channel length, which has been the suitable candidates for TFETs. By correctly choosing two 2D layers of semiconductors to make up a suitable heterojunction (one layer is n-type and the other layer is p-type), it can achieve both large tunneling current and low subthreshold swing (SS). Subthreshold swing (SS) shows the current transformation with the change of voltage, when the device works in subthreshold state. With the proper heterojunction band offsets between the two 2D semiconductor layers (Figure 7a,b), the tunneling current will be significantly increased because of the band alignments. TFETs with $SnSe_2/WSe_2$ van der Waals heterostructure (Figure 7c) have achieved the SS of 80 mV/dec at room temperature and I_{ON}/I_{OFF} ratio exceeding 10^6 [91] (Figure 7d). In addition, dual-gated MoS_2/WSe_2 van der Waals heterostructures were demonstrated and carrier concentration of MoS_2 and WSe_2 can be independently controlled by up and down symmetric gates, which achieved a high efficiency of 80% due to the weak electrostatic barrier through atomic layers [92].

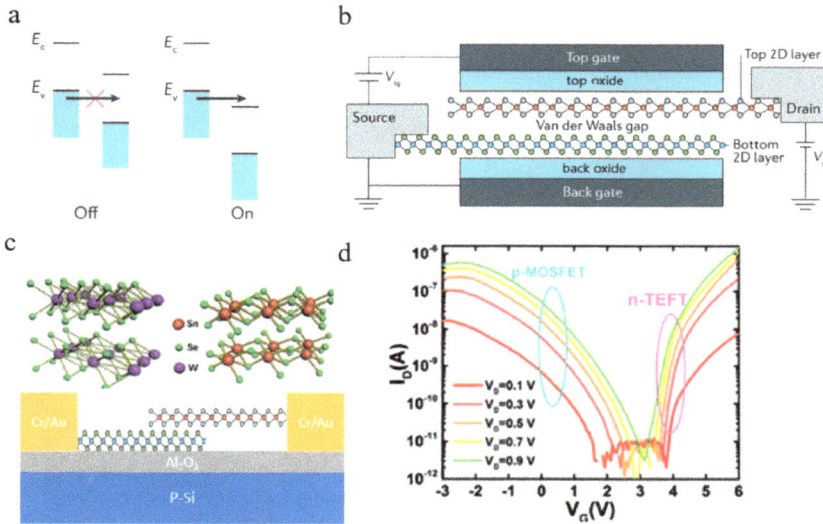

Figure 7. Tunneling field effect transistors structure and transfer curve (**a**) Band offset of p-type semiconductor and n-type semiconductor. (**b**) A schematic illustration of TFETs. The devices consist of ultrathin p-type layer and n-type layer connected to source and drain, respectively. V_{tg} is the gate tunable voltage and V_{ds} is the drain voltage. (**c**) $WSe_2/SnSe_2$ heterostructure based TFET schematic diagram. (**d**) Transfer curve of $WSe_2/SnSe_2$ TFET with the gate voltage of 0.1 V, 0.3 V, 0.5 V, 0.7 V, 0.9 V. (**a,b**) are adapted with the permission of Nature publishing Group in ref [6] (**c,d**)are adapted with the permission of John Wiley & Sons, Inc. in Ref [91].

4.3. Van der Waals Heterostructure Based Vertical Transistors (VFETs)

With ultrathin structure and suitable band structure, 2DLMs can create totally new devices such as vertical transistors. Compared with conventional planar field effect transistors, which consist of a source and a drain plate electrode with a conducting channel located between them, vertical transistors often use graphene and metal as source and drain electrodes, respectively. In planar transistors

(Figure 8a), the gate electrical field (E) controls the channel carriers to influx. Thus, the channel current density (J) is always perpendicular to the gate electrical field (E) (J⊥E). The channel length is decided by the area of 2DSCs. In other word, the lithographic accuracy which decides the distance of source and drain would influence the channel length. In particular, the tunable work function and partial electrostatic transparency make graphene become the appropriate candidate of interconnecting with semiconductor to fabricate vertical transistors, in which graphene acts as the active contact. In this fundamentally different geometry, as shown in Figure 8b, the channel carriers transport direction is parallel to the gate electrical field (J∥E). The channel length is decided by the thickness of 2DSCs which can be prominently shorter than the planar transistors. With effectively modulating the work function of graphene by introducing the gate voltage, the Schottky barrier between graphene and semiconductor changed and result in large current injection, which brings a large on/off ratio that cannot be achieved in planar transistors.

Different kinds of transistors based on vertical vdWHs have been reported according to its special device structure. N-channel vertical FET created by sandwiching multilayered MoS_2 between graphene and metal panel (Figure 8c), has the on/off ratio >10^3 at room temperature [93]. Due to van der Waals interface between MoS_2 and graphene, the Schottky barrier restrains current injection. Introducing an external electric field can strongly modulate the height of Schottky barrier owing to the small density of states of graphene [94]. The band structure of graphene/MoS_2/metal heterostructure is shown in Figure 8d. The source/drain voltage (V_B) drives current follow vertically into multilayered MoS_2 channel. The carriers transport is dominated by thermionic-emission (TE). The application of V_B changes the Fermi level in graphene due to the capacitive coupling through MoS_2. With the increasing V_B, the Fermi level will reach the Dirac point at some point, as shown in the right panel of Figure 8d [95,96]. Other 2DLMs such as WS_2 possessing the same geometrical structure of MoS_2 VFETs have been reported, achieving extremely large on-state current [97].

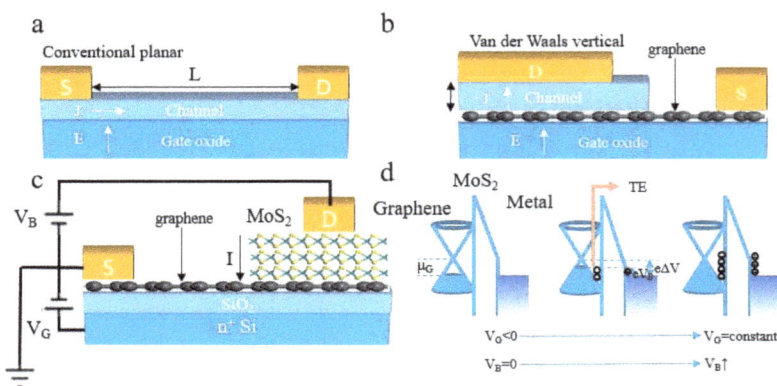

Figure 8. Van der Waals vertical stack structure and devices: (**a**) conventional planar transistor, channel current exists in planar area; (**b**) van der Waals vertical stack and channel current follow in vertical direction; (**a,b**) are adapted with the permission of Nature Publishing Group in Ref [5] (**c**) schematic of metal/MoS_2/graphene vertical stack and measurement circuit; Adapted with the permission of AIP Publishing LLC in Ref [94] (**d**) band alignment of metal/MoS_2/graphene vertical stack with a constant V_G and changing with the increase of V_B. Adapted with the permission of American Institute of Physics Publishing LLC in Ref [95].

As graphene has attracted increasing attention among researchers, other carbon-based materials such as fullerene also draw attention to be assembled with graphene or 2DLMs to create organic vertical field effect transistors (OVFETs). In other words, the work function of fullerene (C_{60}) is closed to graphene. Thus, the graphene/C_{60}/metal VFET with a measured current on/off ratio up to 10^5

has been reported [98] and the logic inverter based on this structure also has been reported yet [99]. Some types of organic or molecular crystals can act as semiconductor. After stacking on graphene to form 2D–3D heterojunction, which is an excellent vertical channel for VFETs, new opportunities in developing novel OVFETs exist. At the same time, organic transistors such as 2D quasi-freestanding molecular based graphene transistors which had the mobility up to 10 cm$^2 \cdot$V$^{-1} \cdot$s^{-1} have been achieved [100].

4.4. Photodetector and Diodes Based on van der Waals Heterostructure

Based on fine optical characteristic of two-dimensional TMDCs, photovoltaic applications about van der Waals heterostructure combined with such materials have been explored. The typical application is the photodetectors made of graphene act as channel material and MoS$_2$ act as light sensitive material [101]. Under the combination of two kinds of 2D materials which have different tunable work functions, photoexcited electrons and holes can be accumulated in conductor layers. Shown in Figure 9a, photoexcited pairs have been observed in the heterostructure of MoS$_2$/WSe$_2$ [102] and MoSe$_2$/WSe$_2$ [103]. Most photovoltaic devices can be created by combining few layers of TMDCs with graphene. Using graphene as electrodes and TMDCs sandwiched between two graphene electrodes, one can create efficient photoexcited carriers inject into graphene electrodes [104,105]. In other words, graphene can form a typical ohmic contacts which will significantly decrease the resistance and serve as a transparent electrode.

If p-type and n-type 2D semiconductors stack together with van der Waals force, then atomic layers of p-n junctions can be acquired. Gate tunable diode based on TMDCs have been abundantly created. Various combinations of 2D semiconductors were used to construct p-n junction for photodetectors (typical p-n structure and band structure shown in Figure 9b) to realize the excellent performance of responsivity and detectivity [106–112]. Heterostructures such as MoS$_2$/WSe$_2$ [107], black phosphorus/MoS$_2$ [109], and MoS$_2$/WS$_2$ [110] p-n junctions are widely used in photodetectors. GaTe/MoS$_2$ heterostructure received a high photovoltaic performance whose photoresponsivity can reach 21.83 AW^{-1} [112].

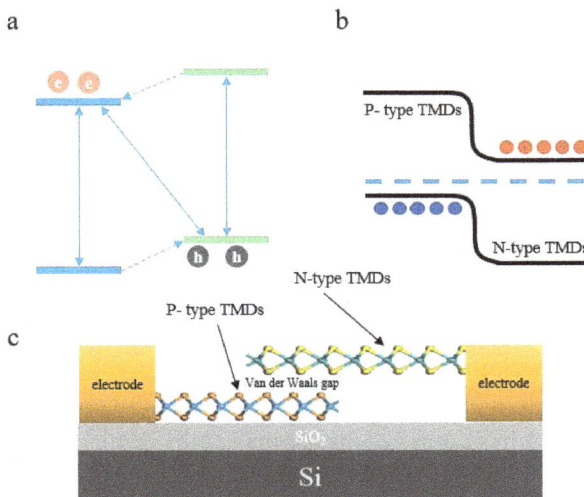

Figure 9. Photodetector and diodes schematic diagram: (**a**) photoexcited pairs generate in MoS$_2$/WSe$_2$ heterostructure; (**b**) band structure of n-MoS$_2$ and p-WSe$_2$ p-n junction; and (**c**) schematic illustration of vertical stack p-n diode based on two-dimensional TMDCs.

4.5. Memory

There are many types of memory devices which include dynamic random access memory (DRAM), static random access memory (SRAM), resistive random access memory (RRAM) and flash memory. The memory cells of DRAM and SRAM are based on the one-transistor and one-capacitance (1T1C) structure and six transistors (6T), respectively. The 1T1C DRAM cell is shown in Figure 10a. The capacitance charging and discharging controlled by the transistor on-state and off-state. When the word line voltage reaches the transistor on-state voltage, the current will follow through transistor and charge the capacitance when the bit line voltage is at high potential. This operation is called the programming process. When the bit line voltage is at low potential, the charged capacitance will discharge and the current will follow through the transistor into bit line, which is called erasing process. When the word line voltage controls transistor at the off-state, the charge carriers will be held in capacitance. After setting bit line at half of the high potential, the voltage will rise if the capacitance is charged and will decline if the capacitance is discharged. This is called the reading operation of DRAM. However, DRAM is a typical volatile memory, which means that the storage performance will disappear under the power-down state.

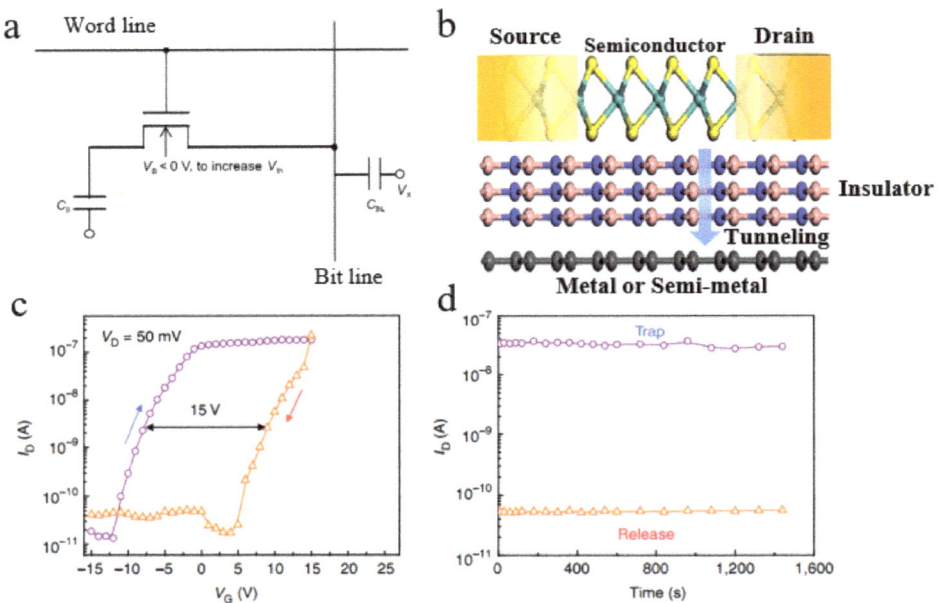

Figure 10. Schematic diagram of different type of memory and performance curve: (**a**) 1T1C structure of dynamic random access memory; (**b**) semiconductor–insulator–metal heterostructure of flash memory; (**c**) gate hysteresis curve of flash memory; (**d**) stability and endurance of flash memory. (**c,d**) are reproduced with permission from Nature Publishing Group, Ref [113].

A new concept of constructing memory is based on the domain wall motion which means that crystalline particles with the same polarization direction as the external electric or magnetic field inside the crystal become larger. This process is done by the movement of domain wall. Current driven domain wall motion and electrically controlling the magnetization switching by spin orbit torque have been proposed for information storage. Decreasing the current density to realize the domain wall motion plays an important role in practical applications [114]. Basically, the external magnetic field is necessary to control the current-induced magnetization switching direction by spin orbit torque [115,116]. Later research has reported that the electric field definitively control the current induced ferromagnet switching at room temperature with the polarized ferroelectric substrate [117]. Piezo voltage solely controlling the magnetization switching at room temperature without external magnetic field has been realized, which can largely decrease the energy consumption compared with the electric current switching magnetization [118]. This concept offers an idea to develop ferroelectric random access memory, magnetic random access memory and phase change random access memory.

Another block of memory is flash, a kind of non-volatile memory. Flash memory is based on 2D material stacks established on a semiconductor–insulator–metal (SIM) structure (Figure 10b). MoS_2/BN/Graphene heterostructure for use in flash memory has been widely explored [113,119]. Graphene and MoS_2 were utilized as charge trapping layers and channel layers and h-BN acted as a tunnel barrier, respectively. Under a considerable voltage positive bias, the tunneling current exponentially increases with the decreasing of BN thickness, which will lead to the increasing charge injection into the trap layer. Thus, the thickness of insulator becomes especially important in SIM structure. The critical thickness of tunneling barrier is 7 nm. Tunneling current cannot be formed in the thick layers (>10 nm), while leakage current will exist in the thinner layers (3.5 nm), which will cause charging process failure [113]. When gate voltage becomes negative in releasing process, charge carriers trapped in the graphene layer will be transferred into MoS_2. Thus, the charge retention and gate hysteresis that contains a memory window can be observed (Figure 10c). Retention time of trapping and releasing current shows the endurance of charge trapping flash device (Figure 10d). Detailed band is structure shown in Figure 11a with the charging and releasing process.

Nonvolatile floating gate flash memory based on black phosphorus/BN/MoS_2 stacks have been reported, where MoS_2 behaves as charge trapping layer and black phosphorus acts as channel [120]. Charge trap memory with high k dielectric materials can greatly improve the storage performance. Few-layer MoS_2 channel and three-dimensional Al_2O_3/HfO_2/Al_2O_3 stack together to form a charge trap flash memory device, exhibiting an unprecedented gate window of 20 V and a program/erase current ratio of 10^4 [121]. Band diagram of program/erase state of device under different top gate bias (V_{tg}) is shown in Figure 11d. Positive V_{tg} carry out the programming process. Electrons tunneling from MoS_2 channel are accumulated in HfO_2 charge-trap layer. Negative V_{tg} carry out the erasing process. Holes tunnel from MoS_2 channel to HfO_2 charge-trap layer. The charge-trap layer holds different charges, causing the threshold shifting. Thus, memory window exists. Besides MoS_2 channel based charge-trap memory, Ambipolar WSe_2 channel based charge-trap memory devices have also been reported [122]. Subsequently, black phosphorus based charge-trap memory devices with a Al_2O_3/HfO_2/Al_2O_3 charge trap stack have been demonstrated, where the long data retention with only 30% charge loss after 10 years were obtained [123]. Beside floating gate memory, novel structure based on vertically stacked 2D materials (WSe_2/h-BN/graphene) has the gate metal stack directly on the graphene, thus we call it semi-floating gate (SFG) flash memory [124]. The structure is shown in Figure 11b,c. When a positive voltage is applied on Si, electrons are accumulated in the WSe_2. These electrons tunnel through BN and begin to be accumulated in graphene due to the different potential between WSe_2 and graphene which is considered as programing process. When the positive bias is removed, electrons are still trapped in graphene due to the potential barriers of h-BN. When applying a small negative voltage bias on Si, the electrons tunnel through BN and are injected into WSe_2 channel as an erasing process. Similar behaviors can be observed for holes when a negative voltage is applied on Si. When treated with plasmas, multilayer MoS_2 transistors can serve as nonvolatile, highly durable flash memory with multibit data storage capability. This novel multibit flash memory is desirable for improving data storage density in order to reach manufacturing process [125].

Memory have made significant progress due to the multiduty van der Waals heterostructures. New type of memory devices structure began to spring up. However, effort still needs to be made to improve the memory performance such as stability, reliability and time of memory operation. 2D van der Waals heterostructures offer a new route of building novel memory.

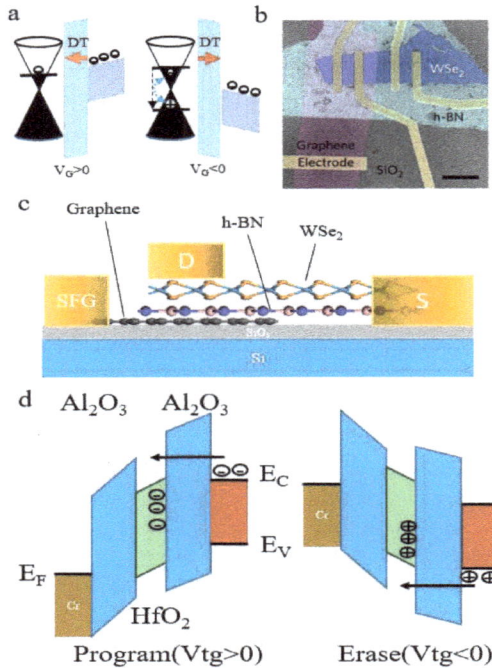

Figure 11. Semi-floating gate flash memory and band structure diagram of SIM heterostructure: (**a**) band structure with different gate voltage and charge carriers following direction; (**b**) SEM image of semi-floating gate memory based on WSe₂/h-BN/Graphene stack; (**c**) schematic diagram of SFG device; and (**d**) band structure of charge trap memory operations. (**a–c**) are reproduced with permission from Nature Publishing Group, Ref [124]. (**d**) is reproduced with permission from American Chemical Society Publishing Group in Ref [121].

5. Perspective

The inherent absolute advantages of 2D materials—exhibiting free of dangling bond and gate tunable electronic properties—make them promising candidates for van der Waals stacks, which are widely used in novel device design. However, the burgeoning van der Waals heterostructures applied to electronic devices remain in their infancy. Technical improvement still needs to be explored in the following effort.

Conventional laboratory-stage process of assembling van der Waals stack is mechanical exfoliation and transfer methods. The shape of exfoliated 2D materials is irregular and the size can only reach micron order. It is hard to acquire monolayer materials and most are few layers. Large area 2D materials are still needed in some cases during transfer process. Although with the utilize of chemical vapor deposition, large-scale synthesizing 2D materials and van der Waals heterostructures is developed, it still cannot be applied in integrated technique and lacks controllability of the relative orientation, size, shape and interlayer spacing with increasing vdWH layers. Furthermore, the electrical performance of chemically synthesized 2DLMs is inferior to their exfoliated counterpart. Thus, there is still much room to improve the process of assembling van der Waals heterostructure.

FETs based on van der Waals heterostructures have largely focused on attaining high performance with certain parameter to measure. However, parameters such as low SS at extensive range of gate voltage biases, which will lead to large on/off ratios with low input voltage biases, have become important criteria for measuring the performance of the devices. Besides, contact resistance is still a big deal for limiting the improvement of devices. There is still plenty of room to lower the contact resistance, as it is still above theoretical minimum, even larger than Si based transistors. The contact resistance must be lower by at least an order of magnitude to <100 $\Omega\cdot\mu m$. In 2D TMDCs, based on phase engineering, changing semiconductor to metal phases is a promising route of creating an ohmic contact which can greatly lower the contact resistance. In addition, the majority of TMDCs are unipolar, n-type or p-type. Controllable and stable doping process is still a major challenge to overcome the achievement of bipolar—both n-type and p-type—semiconductors or van der Waals stack of homogenous material, which will enable the next generation of high speed and low power devices. Direct growth of homogenous 2D semiconductor with stable doping process to form different types of heterostructures and phase engineering to decrease contact resistance are topics for future research. Scalable fabrication of van der Waals stacks also need to be explored. Creating new structure of semiconductor devices based on van der Waals heterostructures still has a long way to go.

Acknowledgments: This work was supported by the National Natural Science Foundation of China (61622401, 61734003, and 61376093) and National Key Research and Development Program (2017YFB0405600 and 2016YFA0203900).

Author Contributions: Jingyu Li conceived and designed the workflow of the paper. Jingyu Li performed the literature search, collected the data meeting the criteria, performed the schematic diagram analysis and figures. Jingyu Li and Xiaozhang Chen wrote the paper. All authors revised the manuscript.

Conflicts of Interest: The authors declare no conflict of interest.

References

1. Novoselov, K.S.; Fal, V.; Colombo, L.; Gellert, P.; Schwab, M.; Kim, K. A roadmap for graphene. *Nature* **2012**, *490*, 192–200. [CrossRef] [PubMed]
2. Schwierz, F. Graphene transistors. *Nat. Nanotechnol.* **2010**, *5*, 487–496. [CrossRef] [PubMed]
3. Weiss, N.O.; Zhou, H.; Liao, L.; Liu, Y.; Jiang, S.; Huang, Y.; Duan, X. Graphene: An emerging electronic material. *Adv. Mater.* **2012**, *24*, 5782–5825. [CrossRef] [PubMed]
4. Liao, L.; Bai, J.; Cheng, R.; Lin, Y.-C.; Jiang, S.; Qu, Y.; Huang, Y.; Duan, X. Sub-100 nm channel length graphene transistors. *Nano Lett.* **2010**, *10*, 3952–3956. [CrossRef] [PubMed]
5. Liu, Y.; Weiss, N.O.; Duan, X.; Cheng, H.-C.; Huang, Y.; Duan, X. Van der Waals heterostructures and devices. *Nat. Rev. Mater.* **2016**, *1*, 16042. [CrossRef]
6. Chhowalla, M.; Jena, D.; Zhang, H. Two-dimensional semiconductors for transistors. *Nat. Rev. Mater.* **2016**, *1*, 16052. [CrossRef]
7. Novoselov, K.; Jiang, D.; Schedin, F.; Booth, T.; Khotkevich, V.; Morozov, S.; Geim, A. Two-dimensional atomic crystals. *Proc. Natl. Acad. Sci. USA* **2005**, *102*, 10451–10453. [CrossRef] [PubMed]
8. Jariwala, D.; Marks, T.J.; Hersam, M.C. Mixed-dimensional van der Waals heterostructures. *Nat. Mater.* **2016**, *16*, 170. [CrossRef] [PubMed]
9. Liu, Y.; Cheng, R.; Liao, L.; Zhou, H.; Bai, J.; Liu, G.; Liu, L.; Huang, Y.; Duan, X. Plasmon resonance enhanced multicolour photodetection by graphene. *Nat. Commun.* **2011**, *2*, 579. [CrossRef] [PubMed]
10. Konstantatos, G.; Badioli, M.; Gaudreau, L.; Osmond, J.; Bernechea, M.; De Arquer, F.P.G.; Gatti, F.; Koppens, F.H. Hybrid graphene-quantum dot phototransistors with ultrahigh gain. *Nat. Nanotechnol.* **2012**, *7*, 363–368. [CrossRef] [PubMed]
11. Liao, L.; Bai, J.; Qu, Y.; Lin, Y.-C.; Li, Y.; Huang, Y.; Duan, X. High-κ oxide nanoribbons as gate dielectrics for high mobility top-gated graphene transistors. *Proc. Natl. Acad. Sci. USA* **2010**, *107*, 6711–6715. [CrossRef] [PubMed]
12. Liao, L.; Lin, Y.-C.; Bao, M.; Cheng, R.; Bai, J.; Liu, Y.; Qu, Y.; Wang, K.L.; Huang, Y.; Duan, X. High speed graphene transistors with a self-aligned nanowire gate. *Nature* **2010**, *467*, 305. [CrossRef] [PubMed]
13. Dean, C.R.; Young, A.F.; Meric, I.; Lee, C.; Wang, L.; Sorgenfrei, S.; Watanabe, K.; Taniguchi, T.; Kim, P.; Shepard, K.L. Boron nitride substrates for high-quality graphene electronics. *Nat. Nanotechnol.* **2010**, *5*, 722–726. [CrossRef] [PubMed]

14. Withers, F.; Del Pozo-Zamudio, O.; Mishchenko, A.; Rooney, A.; Gholinia, A.; Watanabe, K.; Taniguchi, T.; Haigh, S.; Geim, A.; Tartakovskii, A. Light-emitting diodes by band-structure engineering in van der Waals heterostructures. *Nat. Mater.* **2015**, *14*, 301–306. [CrossRef] [PubMed]

15. Yang, H.; Heo, J.; Park, S.; Song, H.J.; Seo, D.H.; Byun, K.-E.; Kim, P.; Yoo, I.; Chung, H.-J.; Kim, K. Graphene barristor, a triode device with a gate-controlled schottky barrier. *Science* **2012**, *336*, 1140–1143. [CrossRef] [PubMed]

16. Vaziri, S.; Smith, A.D.; Östling, M.; Lupina, G.; Dabrowski, J.; Lippert, G.; Mehr, W.; Driussi, F.; Venica, S.; Di Lecce, V. Going ballistic: Graphene hot electron transistors. *Solid State Commun.* **2015**, *224*, 64–75. [CrossRef]

17. Jeong, H.; Bang, S.; Oh, H.M.; Jeong, H.J.; An, S.-J.; Han, G.H.; Kim, H.; Kim, K.K.; Park, J.C.; Lee, Y.H. Semiconductor–insulator–semiconductor diode consisting of monolayer MoS_2, h-BN, and GaN heterostructure. *ACS Nano* **2015**, *9*, 10032–10038. [CrossRef] [PubMed]

18. Sarkar, D.; Xie, X.; Liu, W.; Cao, W.; Kang, J.; Gong, Y.; Kraemer, S.; Ajayan, P.M.; Banerjee, K. A subthermionic tunnel field-effect transistor with an atomically thin channel. *Nature* **2015**, *526*, 91–95. [CrossRef] [PubMed]

19. Zhang, Z.; Chen, P.; Duan, X.; Zang, K.; Luo, J.; Duan, X. Robust epitaxial growth of two-dimensional heterostructures, multiheterostructures, and superlattices. *Science* **2017**, *357*, 788–792. [CrossRef] [PubMed]

20. Duan, X.; Wang, C.; Shaw, J.C.; Cheng, R.; Chen, Y.; Li, H.; Wu, X.; Tang, Y.; Zhang, Q.; Pan, A. Lateral epitaxial growth of two-dimensional layered semiconductor heterojunctions. *Nat. Nanotechnol.* **2014**, *9*, 1024–1030. [CrossRef] [PubMed]

21. Gong, Y.; Lin, J.; Wang, X.; Shi, G.; Lei, S.; Lin, Z.; Zou, X.; Ye, G.; Vajtai, R.; Yakobson, B.I. Vertical and in-plane heterostructures from WS_2/MoS_2 monolayers. *Nat. Mater.* **2014**, *13*, 1135–1142. [CrossRef] [PubMed]

22. Geim, A.K.; Grigorieva, I.V. Van der Waals heterostructures. *Nature* **2013**, *499*, 419–425. [CrossRef] [PubMed]

23. Jena, D. Tunneling transistors based on graphene and 2-D crystals. *Proc. IEEE* **2013**, *101*, 1585–1602. [CrossRef]

24. Mayorov, A.S.; Gorbachev, R.V.; Morozov, S.V.; Britnell, L.; Jalil, R.; Ponomarenko, L.A.; Blake, P.; Novoselov, K.S.; Watanabe, K.; Taniguchi, T. Micrometer-scale ballistic transport in encapsulated graphene at room temperature. *Nano Lett.* **2011**, *11*, 2396–2399. [CrossRef] [PubMed]

25. Liu, Y.; Wu, H.; Cheng, H.-C.; Yang, S.; Zhu, E.; He, Q.; Ding, M.; Li, D.; Guo, J.; Weiss, N.O. Toward barrier free contact to molybdenum disulfide using graphene electrodes. *Nano Lett.* **2015**, *15*, 3030–3034. [CrossRef] [PubMed]

26. Bertolazzi, S.; Krasnozhon, D.; Kis, A. Nonvolatile memory cells based on MoS_2/graphene heterostructures. *ACS Nano* **2013**, *7*, 3246–3252. [CrossRef] [PubMed]

27. Neto, A.C.; Guinea, F.; Peres, N.M.; Novoselov, K.S.; Geim, A.K. The electronic properties of graphene. *Rev. Mod. Phys.* **2009**, *81*, 109. [CrossRef]

28. Zhang, Y.; Tang, T.-T.; Girit, C.; Hao, Z.; Martin, M.C.; Zettl, A.; Crommie, M.F.; Shen, Y.R.; Wang, F. Direct observation of a widely tunable bandgap in bilayer graphene. *Nature* **2009**, *459*, 820. [CrossRef] [PubMed]

29. Novoselov, K.S.; McCann, E.; Morozov, S.V.; Falko, V.I.; Katsnelson, M.I.; Zeitler, U.; Jiang, D.; Schedin, F.; Geim, A.K. Unconventional quantum hall effect and berry's phase of 2π in bilayer graphene. *Nat. Phys.* **2006**, *2*, 177. [CrossRef]

30. Xia, F.; Mueller, T.; Lin, Y.-M.; Valdes-Garcia, A.; Avouris, P. Ultrafast graphene photodetector. *Nat. Nanotechnol.* **2009**, *4*, 839–843. [CrossRef] [PubMed]

31. Mueller, T.; Xia, F.; Avouris, P. Graphene photodetectors for high-speed optical communications. *Nat. Photonics* **2010**, *4*, 297–301. [CrossRef]

32. Wang, X.; Xia, F. Van der Waals heterostructures: Stacked 2D materials shed light. *Nat. Mater.* **2015**, *14*, 264. [CrossRef] [PubMed]

33. Reale, F.; Sharda, K.; Mattevi, C. From bulk crystals to atomically thin layers of group vi-transition metal dichalcogenides vapour phase synthesis. *Appl. Mater. Today* **2016**, *3*, 11–22. [CrossRef]

34. Mak, K.F.; McGill, K.L.; Park, J.; McEuen, P.L. The valley hall effect in MoS_2 transistors. *Science* **2014**, *344*, 1489–1492. [CrossRef] [PubMed]

35. Zeng, H.; Dai, J.; Yao, W.; Xiao, D.; Cui, X. Valley polarization in MoS_2 monolayers by optical pumping. *Nat. Nanotechnol.* **2012**, *7*, 490–493. [CrossRef] [PubMed]

36. Kong, D.; Cui, Y. Opportunities in chemistry and materials science for topological insulators and their nanostructures. *Nat. Chem.* **2011**, *3*, 845–849. [CrossRef] [PubMed]

37. Yin, X.; Ye, Z.; Chenet, D.A.; Ye, Y.; O'Brien, K.; Hone, J.C.; Zhang, X. Edge nonlinear optics on a MoS_2 atomic monolayer. *Science* **2014**, *344*, 488–490. [CrossRef] [PubMed]

38. Carozo, V.; Wang, Y.; Fujisawa, K.; Carvalho, B.R.; McCreary, A.; Feng, S.; Lin, Z.; Zhou, C.; Perea-López, N.; Elías, A.L. Optical identification of sulfur vacancies: Bound excitons at the edges of monolayer tungsten disulfide. *Sci. Adv.* **2017**, *3*, e1602813. [CrossRef] [PubMed]

39. Tran, V.; Soklaski, R.; Liang, Y.; Yang, L. Layer-controlled band gap and anisotropic excitons in few-layer black phosphorus. *Phys. Rev. B* **2014**, *89*, 235319. [CrossRef]

40. Liu, H.; Neal, A.T.; Zhu, Z.; Luo, Z.; Xu, X.; Tománek, D.; Peide, D.Y. Phosphorene: An unexplored 2D semiconductor with a high hole mobility. *ACS Nano* **2014**, *8*, 4033–4041. [CrossRef] [PubMed]

41. Qiao, J.; Kong, X.; Hu, Z.-X.; Yang, F.; Ji, W. High-mobility transport anisotropy and linear dichroism in few-layer black phosphorus. *Nat. Commun.* **2014**, *5*, 4475. [CrossRef] [PubMed]

42. Li, L.; Yu, Y.; Ye, G.J.; Ge, Q.; Ou, X.; Wu, H.; Feng, D.; Chen, X.H.; Zhang, Y. Black phosphorus field-effect transistors. *Nat. Nanotechnol.* **2014**, *9*, 372–377. [CrossRef] [PubMed]

43. Engel, M.; Steiner, M.; Avouris, P. Black phosphorus photodetector for multispectral, high-resolution imaging. *Nano Lett.* **2014**, *14*, 6414–6417. [CrossRef] [PubMed]

44. Du, Y.; Ouyang, C.; Shi, S.; Lei, M. Ab initio studies on atomic and electronic structures of black phosphorus. *J. Appl. Phys.* **2010**, *107*, 093718. [CrossRef]

45. Avsar, A.; Vera-Marun, I.J.; Tan, J.Y.; Watanabe, K.; Taniguchi, T.; Castro Neto, A.H.; Özyilmaz, B. Air-stable transport in graphene-contacted, fully encapsulated ultrathin black phosphorus-based field-effect transistors. *ACS Nano* **2015**, *9*, 4138–4145. [CrossRef] [PubMed]

46. Na, J.; Lee, Y.T.; Lim, J.A.; Hwang, D.K.; Kim, G.-T.; Choi, W.K.; Song, Y.-W. Few-layer black phosphorus field-effect transistors with reduced current fluctuation. *ACS Nano* **2014**, *8*, 11753–11762. [CrossRef] [PubMed]

47. Wood, J.D.; Wells, S.A.; Jariwala, D.; Chen, K.-S.; Cho, E.; Sangwan, V.K.; Liu, X.; Lauhon, L.J.; Marks, T.J.; Hersam, M.C. Effective passivation of exfoliated black phosphorus transistors against ambient degradation. *Nano Lett.* **2014**, *14*, 6964–6970. [CrossRef] [PubMed]

48. Kim, J.-S.; Liu, Y.; Zhu, W.; Kim, S.; Wu, D.; Tao, L.; Dodabalapur, A.; Lai, K.; Akinwande, D. Toward air-stable multilayer phosphorene thin-films and transistors. *Sci. Rep.* **2015**, *5*, 8989. [CrossRef] [PubMed]

49. Gorbachev, R.V.; Riaz, I.; Nair, R.R.; Jalil, R.; Britnell, L.; Belle, B.D.; Hill, E.W.; Novoselov, K.S.; Watanabe, K.; Taniguchi, T. Hunting for monolayer boron nitride: Optical and raman signatures. *Small* **2011**, *7*, 465–468. [CrossRef] [PubMed]

50. Meric, I.; Dean, C.R.; Petrone, N.; Wang, L.; Hone, J.; Kim, P.; Shepard, K.L. Graphene field-effect transistors based on boron–nitride dielectrics. *Proc. IEEE* **2013**, *101*, 1609–1619. [CrossRef]

51. Britnell, L.; Gorbachev, R.V.; Jalil, R.; Belle, B.D.; Schedin, F.; Katsnelson, M.I.; Eaves, L.; Morozov, S.V.; Mayorov, A.S.; Peres, N.M. Electron tunneling through ultrathin boron nitride crystalline barriers. *Nano Lett.* **2012**, *12*, 1707–1710. [CrossRef] [PubMed]

52. Lee, G.-H.; Yu, Y.-J.; Lee, C.; Dean, C.; Shepard, K.L.; Kim, P.; Hone, J. Electron tunneling through atomically flat and ultrathin hexagonal boron nitride. *Appl. Phys. Lett.* **2011**, *99*, 243114. [CrossRef]

53. Rai, A.; Valsaraj, A.; Movva, H.C.P.; Roy, A.; Ghosh, R.; Sonde, S.; Kang, S.; Chang, J.; Trivedi, T.; Dey, R.; et al. Air stable doping and intrinsic mobility enhancement in monolayer molybdenum disulfide by amorphous titanium suboxide encapsulation. *Nano Lett* **2015**, *15*, 4329–4336. [CrossRef] [PubMed]

54. McClellan, C.J.; Yalon, E.; Smithe, K.K.H.; Suryavanshi, S.V.; Pop, E. Effective n-type doping of monolayer MoS$_2$ by AlO$_x$. Presented at the 2017 75th Annual Device Research Conference (DRC), South Bend, IN, USA, 25–28 June 2017; pp. 1–2.

55. Gong, Q.; Cheng, L.; Liu, C.; Zhang, M.; Feng, Q.; Ye, H.; Zeng, M.; Xie, L.; Liu, Z.; Li, Y. Ultrathin MoS$_{2(1-x)}$Se$_{2x}$ alloy nanoflakes for electrocatalytic hydrogen evolution reaction. *ACS Catal.* **2015**, *5*, 2213–2219. [CrossRef]

56. Jiang, Y.; Li, L.; Yang, R.Q.; Gupta, J.A.; Aers, G.C.; Dupont, E.; Baribeau, J.-M.; Wu, X.; Johnson, M.B. Type-I interband cascade lasers near 3.2 µm. *Appl. Phys. Lett.* **2015**, *106*, 041117. [CrossRef]

57. Feng, Q.; Mao, N.; Wu, J.; Xu, H.; Wang, C.; Zhang, J.; Xie, L. Growth of MoS$_{2(1-x)}$Se$_{2x}$ (x = 0.41–1.00) monolayer alloys with controlled morphology by physical vapor deposition. *ACS Nano* **2015**, *9*, 7450–7455. [CrossRef] [PubMed]

58. Novoselov, K.; Mishchenko, A.; Carvalho, A.; Neto, A.C. 2D materials and van der Waals heterostructures. *Science* **2016**, *353*, aac9439. [CrossRef] [PubMed]

59. Wang, S.; Wang, X.; Warner, J.H. All chemical vapor deposition growth of MoS2: H-BN vertical van der Waals heterostructures. *ACS Nano* **2015**, *9*, 5246–5254. [CrossRef] [PubMed]

60. Diaz, H.C.; Chaghi, R.; Ma, Y.; Batzill, M. Molecular beam epitaxy of the van der Waals heterostructure MoTe$_2$ on MoS$_2$: Phase, thermal, and chemical stability. *2D Mater.* **2015**, *2*, 044010. [CrossRef]

61. Yue, R.; Barton, A.T.; Zhu, H.; Azcatl, A.; Pena, L.F.; Wang, J.; Peng, X.; Lu, N.; Cheng, L.; Addou, R. HfSe$_2$ thin films: 2D transition metal dichalcogenides grown by molecular beam epitaxy. *ACS Nano* **2014**, *9*, 474–480. [CrossRef] [PubMed]

62. Cho, A.; Arthur, J. Molecular beam epitaxy. *Prog. Solid State Chem.* **1975**, *10*, 157–191. [CrossRef]

63. Chen, Y.; Bagnall, D.; Koh, H.-T.; Park, K.-T.; Hiraga, K.; Zhu, Z.; Yao, T. Plasma assisted molecular beam epitaxy of zno on c-plane sapphire: Growth and characterization. *J. Appl. Phys.* **1998**, *84*, 3912–3918. [CrossRef]

64. Radisavljevic, B.; Radenovic, A.; Brivio, J.; Giacometti, i.V.; Kis, A. Single-layer MoS$_2$ transistors. *Nat. Nanotechnol.* **2011**, *6*, 147–150. [CrossRef] [PubMed]

65. Jena, D.; Banerjee, K.; Xing, G.H. 2D crystal semiconductors: Intimate contacts. *Nat. Mater.* **2014**, *13*, 1076. [CrossRef] [PubMed]

66. Arden, W.M. The international technology roadmap for semiconductors—perspectives and challenges for the next 15 years. *Curr. Opin. Solid State Mater. Sci.* **2002**, *6*, 371–377. [CrossRef]

67. Kappera, R.; Voiry, D.; Yalcin, S.E.; Branch, B.; Gupta, G.; Mohite, A.D.; Chhowalla, M. Phase-engineered low-resistance contacts for ultrathin MoS$_2$ transistors. *Nat. Mater.* **2014**, *13*, 1128. [CrossRef] [PubMed]

68. Allain, A.; Kang, J.; Banerjee, K.; Kis, A. Electrical contacts to two-dimensional semiconductors. *Nat. Mater.* **2015**, *14*, 1195. [CrossRef] [PubMed]

69. Das, S.; Chen, H.-Y.; Penumatcha, A.V.; Appenzeller, J. High performance multilayer MoS$_2$ transistors with scandium contacts. *Nano Lett.* **2012**, *13*, 100–105. [CrossRef] [PubMed]

70. Chuang, H.-J.; Tan, X.; Ghimire, N.J.; Perera, M.M.; Chamlagain, B.; Cheng, M.M.-C.; Yan, J.; Mandrus, D.; Tománek, D.; Zhou, Z. High mobility WSe$_2$ p-and n-type field-effect transistors contacted by highly doped graphene for low-resistance contacts. *Nano Lett.* **2014**, *14*, 3594–3601. [CrossRef] [PubMed]

71. Das, S.; Gulotty, R.; Sumant, A.V.; Roelofs, A. All two-dimensional, flexible, transparent, and thinnest thin film transistor. *Nano Lett.* **2014**, *14*, 2861–2866. [CrossRef] [PubMed]

72. Das, S.; Prakash, A.; Salazar, R.; Appenzeller, J. Toward low-power electronics: Tunneling phenomena in transition metal dichalcogenides. *ACS Nano* **2014**, *8*, 1681–1689. [CrossRef] [PubMed]

73. Qu, D.; Liu, X.; Ahmed, F.; Lee, D.; Yoo, W.J. Self-screened high performance multi-layer MoS$_2$ transistor formed by using a bottom graphene electrode. *Nanoscale* **2015**, *7*, 19273–19281. [CrossRef] [PubMed]

74. Roy, T.; Tosun, M.; Kang, J.S.; Sachid, A.B.; Desai, S.B.; Hettick, M.; Hu, C.C.; Javey, A. Field-Effect transistors built from all two-dimensional material components. *ACS Nano* **2014**, *8*, 6259–6264. [CrossRef] [PubMed]

75. Cui, X.; Lee, G.-H.; Kim, Y.D.; Arefe, G.; Huang, P.Y.; Lee, C.-H.; Chenet, D.A.; Zhang, X.; Wang, L.; Ye, F. Multi-terminal transport measurements of MoS$_2$ using a van der Waals heterostructure device platform. *Nat. Nanotechnol.* **2015**, *10*, 534–540. [CrossRef] [PubMed]

76. Yoon, J.; Park, W.; Bae, G.Y.; Kim, Y.; Jang, H.S.; Hyun, Y.; Lim, S.K.; Kahng, Y.H.; Hong, W.K.; Lee, B.H. Highly flexible and transparent multilayer MoS$_2$ transistors with graphene electrodes. *Small* **2013**, *9*, 3295–3300. [CrossRef] [PubMed]

77. Yu, L.; Lee, Y.-H.; Ling, X.; Santos, E.J.; Shin, Y.C.; Lin, Y.; Dubey, M.; Kaxiras, E.; Kong, J.; Wang, H. Graphene/MoS$_2$ hybrid technology for large-scale two-dimensional electronics. *Nano Lett.* **2014**, *14*, 3055–3063. [CrossRef] [PubMed]

78. Lee, G.-H.; Cui, X.; Kim, Y.D.; Arefe, G.; Zhang, X.; Lee, C.-H.; Ye, F.; Watanabe, K.; Taniguchi, T.; Kim, P. Highly stable, dual-gated MoS$_2$ transistors encapsulated by hexagonal boron nitride with gate-controllable contact, resistance, and threshold voltage. *ACS Nano* **2015**, *9*, 7019–7026. [CrossRef] [PubMed]

79. Leong, W.S.; Luo, X.; Li, Y.; Khoo, K.H.; Quek, S.Y.; Thong, J.T. Low resistance metal contacts to mos2 devices with nickel-etched-graphene electrodes. *ACS Nano* **2014**, *9*, 869–877. [CrossRef] [PubMed]

80. Du, Y.; Yang, L.; Zhang, J.; Liu, H.; Majumdar, K.; Kirsch, P.D.; Peide, D.Y. MoS$_2$ field-effect transistors with graphene/metal heterocontacts. *IEEE Electron Device Lett.* **2014**, *35*, 599–601.

81. Larentis, S.; Tolsma, J.R.; Fallahazad, B.; Dillen, D.C.; Kim, K.; MacDonald, A.H.; Tutuc, E. Band offset and negative compressibility in graphene- MoS$_2$ heterostructures. *Nano Lett.* **2014**, *14*, 2039–2045. [CrossRef] [PubMed]

82. Yang, L.; Majumdar, K.; Liu, H.; Du, Y.; Wu, H.; Hatzistergos, M.; Hung, P.; Tieckelmann, R.; Tsai, W.; Hobbs, C. Chloride molecular doping technique on 2D materials: WS_2 and MoS_2. *Nano Lett.* **2014**, *14*, 6275–6280. [CrossRef] [PubMed]

83. Fang, H.; Tosun, M.; Seol, G.; Chang, T.C.; Takei, K.; Guo, J.; Javey, A. Degenerate n-doping of few-layer transition metal dichalcogenides by potassium. *Nano Lett.* **2013**, *13*, 1991–1995. [CrossRef] [PubMed]

84. Kang, J.; Liu, W.; Banerjee, K. High-performance MoS_2 transistors with low-resistance molybdenum contacts. *Appl. Phys. Lett.* **2014**, *104*, 093106. [CrossRef]

85. Appenzeller, J.; Lin, Y.-M.; Knoch, J.; Avouris, P. Band-to-band tunneling in carbon nanotube field-effect transistors. *Phys. Rev. Lett.* **2004**, *93*, 196805. [CrossRef] [PubMed]

86. Pierucci, D.; Henck, H.; Avila, J.; Balan, A.; Naylor, C.H.; Patriarche, G.; Dappe, Y.J.; Silly, M.G.; Sirotti, F.; Johnson, A.C. Band alignment and minigaps in monolayer MoS_2-graphene van der Waals heterostructures. *Nano Lett.* **2016**, *16*, 4054–4061. [CrossRef] [PubMed]

87. Britnell, L.; Gorbachev, R.; Jalil, R.; Belle, B.; Schedin, F.; Mishchenko, A.; Georgiou, T.; Katsnelson, M.; Eaves, L.; Morozov, S. Field-effect tunneling transistor based on vertical graphene heterostructures. *Science* **2012**, *335*, 947–950. [CrossRef] [PubMed]

88. Britnell, L.; Gorbachev, R.; Geim, A.; Ponomarenko, L.; Mishchenko, A.; Greenaway, M.; Fromhold, T.; Novoselov, K.; Eaves, L. Resonant tunnelling and negative differential conductance in graphene transistors. *Nat. Commun.* **2013**, *4*, 1794. [CrossRef] [PubMed]

89. Mishchenko, A.; Tu, J.; Cao, Y.; Gorbachev, R.; Wallbank, J.; Greenaway, M.; Morozov, V.; Morozov, S.; Zhu, M.; Wong, S. Twist-controlled resonant tunnelling in graphene/boron nitride/graphene heterostructures. *Nat. Nanotechnol.* **2014**, *9*, 808–813. [CrossRef] [PubMed]

90. Greenaway, M.; Vdovin, E.; Mishchenko, A.; Makarovsky, O.; Patane, A.; Wallbank, J.; Cao, Y.; Kretinin, A.; Zhu, M.; Morozov, S. Resonant tunnelling between the chiral landau states of twisted graphene lattices. *Nat. Phys.* **2015**, *11*, 1057–1062. [CrossRef]

91. Yan, X.; Liu, C.; Li, C.; Bao, W.; Ding, S.; Zhang, D.W.; Zhou, P. Tunable $SnSe_2/WSe_2$ heterostructure tunneling field effect transistor. *Small* **2017**, *13*. [CrossRef] [PubMed]

92. Roy, T.; Tosun, M.; Cao, X.; Fang, H.; Lien, D.-H.; Zhao, P.; Chen, Y.-Z.; Chueh, Y.-L.; Guo, J.; Javey, A. Dual-gated MoS_2/WSe_2 van der Waals tunnel diodes and transistors. *Acs Nano* **2015**, *9*, 2071–2079. [CrossRef] [PubMed]

93. Yu, W.J.; Li, Z.; Zhou, H.; Chen, Y.; Wang, Y.; Huang, Y.; Duan, X. Vertically stacked multi-heterostructures of layered materials for logic transistors and complementary inverters. *Nat. Mater.* **2013**, *12*, 246. [CrossRef] [PubMed]

94. Moriya, R.; Yamaguchi, T.; Inoue, Y.; Morikawa, S.; Sata, Y.; Masubuchi, S.; Machida, T. Large current modulation in exfoliated-graphene/MoS_2/metal vertical heterostructures. *Appl. Phys. Lett.* **2014**, *105*, 083119. [CrossRef]

95. Moriya, R.; Yamaguchi, T.; Inoue, Y.; Sata, Y.; Morikawa, S.; Masubuchi, S.; Machida, T. Influence of the density of states of graphene on the transport properties of graphene/MoS_2/metal vertical field-effect transistors. *Appl. Phys. Lett.* **2015**, *106*, 223103. [CrossRef]

96. Sata, Y.; Moriya, R.; Morikawa, S.; Yabuki, N.; Masubuchi, S.; Machida, T. Electric field modulation of schottky barrier height in graphene/ $MoSe_2$ van der Waals heterointerface. *Appl. Phys. Lett.* **2015**, *107*, 023109. [CrossRef]

97. Georgiou, T.; Jalil, R.; Belle, B.D.; Britnell, L.; Gorbachev, R.V.; Morozov, S.V.; Kim, Y.-J.; Gholinia, A.; Haigh, S.J.; Makarovsky, O. Vertical field-effect transistor based on graphene-WS_2 heterostructures for flexible and transparent electronics. *Nat. Nanotechnol.* **2013**, *8*, 100–103. [CrossRef] [PubMed]

98. Parui, S.; Pietrobon, L.; Ciudad, D.; Vélez, S.; Sun, X.; Casanova, F.; Stoliar, P.; Hueso, L.E. Gate-controlled energy barrier at a graphene/molecular semiconductor junction. *Adv. Funct. Mater.* **2015**, *25*, 2972–2979. [CrossRef]

99. Hlaing, H.; Kim, C.-H.; Carta, F.; Nam, C.-Y.; Barton, R.A.; Petrone, N.; Hone, J.; Kymissis, I. Low-voltage organic electronics based on a gate-tunable injection barrier in vertical graphene-organic semiconductor heterostructures. *Nano Lett.* **2014**, *15*, 69–74. [CrossRef] [PubMed]

100. He, D.; Zhang, Y.; Wu, Q.; Xu, R.; Nan, H.; Liu, J.; Yao, J.; Wang, Z.; Yuan, S.; Li, Y. Two-dimensional quasi-freestanding molecular crystals for high-performance organic field-effect transistors. *Nat. Commun.* **2014**, *5*. [CrossRef] [PubMed]

101. Roy, K.; Padmanabhan, M.; Goswami, S.; Sai, T.P.; Ramalingam, G.; Raghavan, S.; Ghosh, A. Graphene-MoS$_2$ hybrid structures for multifunctional photoresponsive memory devices. *Nat. Nano.* **2013**, *8*, 826–830. [CrossRef] [PubMed]

102. Fang, H.; Battaglia, C.; Carraro, C.; Nemsak, S.; Ozdol, B.; Kang, J.S.; Bechtel, H.A.; Desai, S.B.; Kronast, F.; Unal, A.A.; et al. Strong interlayer coupling in van der Waals heterostructures built from single-layer chalcogenides. *Proc. Natl. Acad. Sci. USA* **2014**, *111*, 6198–6202. [CrossRef] [PubMed]

103. Rivera, P.; Schaibley, J.R.; Jones, A.M.; Ross, J.S.; Wu, S.; Aivazian, G.; Klement, P.; Seyler, K.; Clark, G.; Ghimire, N.J.; et al. Observation of long-lived interlayer excitons in monolayer MoSe$_2$–WSe$_2$ heterostructures. *Nat. Commun.* **2015**, *6*, 6242. [CrossRef] [PubMed]

104. Gan, X.; Shiue, R.-J.; Gao, Y.; Meric, I.; Heinz, T.F.; Shepard, K.; Hone, J.; Assefa, S.; Englund, D. Chip-integrated ultrafast graphene photodetector with high responsivity. *Nat. Photon.* **2013**, *7*, 883–887. [CrossRef]

105. Mudd, G.W.; Svatek, S.A.; Hague, L.; Makarovsky, O.; Kudrynskyi, Z.R.; Mellor, C.J.; Beton, P.H.; Eaves, L.; Novoselov, K.S.; Kovalyuk, Z.D.; et al. High broad-band photoresponsivity of mechanically formed inse–graphene van der Waals heterostructures. *Adv. Mater.* **2015**, *27*, 3760–3766. [CrossRef] [PubMed]

106. Lee, C.-H.; Lee, G.-H.; van der Zande, A.M.; Chen, W.; Li, Y.; Han, M.; Cui, X.; Arefe, G.; Nuckolls, C.; Heinz, T.F.; et al. Atomically thin p–n junctions with van der Waals heterointerfaces. *Nat. Nano.* **2014**, *9*, 676–681. [CrossRef] [PubMed]

107. Furchi, M.M.; Pospischil, A.; Libisch, F.; Burgdörfer, J.; Mueller, T. Photovoltaic effect in an electrically tunable van der Waals heterojunction. *Nano. Lett.* **2014**, *14*, 4785–4791. [CrossRef] [PubMed]

108. Deng, Y.; Luo, Z.; Conrad, N.J.; Liu, H.; Gong, Y.; Najmaei, S.; Ajayan, P.M.; Lou, J.; Xu, X.; Ye, P.D. Black phosphorus–monolayer MoS$_2$ van der Waals heterojunction p–n diode. *ACS Nano* **2014**, *8*, 8292–8299. [CrossRef] [PubMed]

109. Huo, N.; Kang, J.; Wei, Z.; Li, S.-S.; Li, J.; Wei, S.-H. Novel and enhanced optoelectronic performances of multilayer MoS$_2$–WS$_2$ heterostructure transistors. *Adv. Funct. Mater.* **2014**, *24*, 7025–7031. [CrossRef]

110. Choi, M.S.; Qu, D.; Lee, D.; Liu, X.; Watanabe, K.; Taniguchi, T.; Yoo, W.J. Lateral MoS$_2$ p–n junction formed by chemical doping for use in high-performance optoelectronics. *ACS Nano* **2014**, *8*, 9332–9340. [CrossRef] [PubMed]

111. Kiran, M.; Tran, T.; Smillie, L.; Haberl, B.; Subianto, D.; Williams, J.; Bradby, J. Temperature-dependent mechanical deformation of silicon at the nanoscale: Phase transformation versus defect propagation. *J. Appl. Phys.* **2015**, *117*, 205901. [CrossRef]

112. Wang, F.; Wang, Z.; Xu, K.; Wang, F.; Wang, Q.; Huang, Y.; Yin, L.; He, J. Tunable gate- MoS$_2$ van der Waals p–n junctions with novel optoelectronic performance. *Nano Lett.* **2015**, *15*, 7558–7566. [CrossRef] [PubMed]

113. Vu, Q.A.; Shin, Y.S.; Kim, Y.R.; Kang, W.T.; Kim, H.; Luong, D.H.; Lee, I.M.; Lee, K.; Ko, D.-S.; Heo, J. Two-terminal floating-gate memory with van der waals heterostructures for ultrahigh on/off ratio. *Nat. Commun.* **2016**, *7*, 12725. [CrossRef] [PubMed]

114. Wang, K.; Edmonds, K.; Irvine, A.; Tatara, G.; De Ranieri, E.; Wunderlich, J.; Olejnik, K.; Rushforth, A.; Campion, R.; Williams, D. Current-driven domain wall motion across a wide temperature range in a (Ga, Mn)(As, P) device. *Appl. Phys. Lett.* **2010**, *97*, 262102. [CrossRef]

115. Li, Y.; Cao, Y.; Wei, G.; Li, Y.; Ji, Y.; Wang, K.; Edmonds, K.; Campion, R.; Rushforth, A.; Foxon, C. Anisotropic current-controlled magnetization reversal in the ferromagnetic semiconductor (Ga, Mn) as. *Appl. Phys. Lett.* **2013**, *103*, 022401. [CrossRef]

116. Yang, M.; Cai, K.; Ju, H.; Edmonds, K.W.; Yang, G.; Liu, S.; Li, B.; Zhang, B.; Sheng, Y.; Wang, S. Spin-orbit torque in Pt/CoNiCo/Pt symmetric devices. *Sci. Rep.* **2016**, *6*, 20778. [CrossRef] [PubMed]

117. Sheng, Y.; Zhang, B.; Zhang, N.; Liu, S.; Zheng, H.; Wang, K. Electric field control of deterministic current-induced magnetization switching in a hybrid ferromagnetic/ferroelectric structure. *Nat. Mater.* **2017**, *16*, 712–716.

118. Zhang, B.; Meng, K.-K.; Yang, M.-Y.; Edmonds, K.; Zhang, H.; Cai, K.-M.; Sheng, Y.; Zhang, N.; Ji, Y.; Zhao, J.-H. Piezo voltage controlled planar hall effect devices. *Sci. Rep.* **2016**, *6*, 28458. [CrossRef] [PubMed]

119. Choi, M.S.; Lee, G.-H.; Yu, Y.-J.; Lee, D.-Y.; Lee, S.H.; Kim, P.; Hone, J.; Yoo, W.J. Controlled charge trapping by molybdenum disulphide and graphene in ultrathin heterostructured memory devices. *Nat. Commun.* **2013**, *4*, 1624. [CrossRef] [PubMed]

120. Li, D.; Wang, X.; Zhang, Q.; Zou, L.; Xu, X.; Zhang, Z. Nonvolatile floating-gate memories based on stacked black phosphorus–boron nitride–MoS$_2$ heterostructures. *Adv. Funct. Mater.* **2015**, *25*, 7360–7365. [CrossRef]

121. Zhang, E.; Wang, W.; Zhang, C.; Jin, Y.; Zhu, G.; Sun, Q.; Zhang, D.W.; Zhou, P.; Xiu, F. Tunable charge-trap memory based on few-layer MoS$_2$. *ACS Nano* **2014**, *9*, 612–619. [CrossRef] [PubMed]

122. Liu, C.; Yan, X.; Wang, J.; Ding, S.; Zhou, P.; Zhang, D.W. Eliminating overerase behavior by designing energy band in high-speed charge-trap memory based on WSe$_2$. *Small* **2017**, *13*. [CrossRef] [PubMed]

123. Feng, Q.; Yan, F.; Luo, W.; Wang, K. Charge trap memory based on few-layer black phosphorus. *Nanoscale* **2016**, *8*, 2686–2692. [CrossRef] [PubMed]

124. Li, D.; Chen, M.; Sun, Z.; Yu, P.; Liu, Z.; Ajayan, P.M.; Zhang, Z. Two-dimensional non-volatile programmable p-n junctions. *Nat. Nanotechnol.* **2017**, *12*, 901. [CrossRef] [PubMed]

125. Chen, M.; Nam, H.; Wi, S.; Priessnitz, G.; Gunawan, I.M.; Liang, X. Multibit data storage states formed in plasma-treated MoS$_2$ transistors. *ACS Nano* **2014**, *8*, 4023–4032. [CrossRef] [PubMed]

crystals

MDPI

Review
Graphene Coated Nanoprobes: A Review

Fei Hui, Shaochuan Chen, Xianhu Liang, Bin Yuan, Xu Jing, Yuanyuan Shi and Mario Lanza *

Institute of Functional Nano& Soft Materials, Collaborative Innovation Center of Suzhou Nano Science and Technology, Soochow University, Suzhou 215123, China; huifei324@126.com (F.H.); scchen_19@163.com (S.C.); xianhuliang_21@163.com (X.L.); bin_yuan2016@126.com (B.Y.); ntjingxu@163.com (X.J.); syy0909078@126.com (Y.S.)
* Correspondence: mlanza@suda.edu.cn

Academic Editor: Filippo Giannazzo
Received: 25 July 2017; Accepted: 28 August 2017; Published: 8 September 2017

Abstract: Nanoprobes are one of the most important components in several fields of nanoscience to study materials, molecules and particles. In scanning probe microscopes, the nanoprobes consist on silicon tips coated with thin metallic films to provide additional properties, such as conductivity. However, if the experiments involve high currents or lateral frictions, the initial properties of the tips can wear out very fast. One possible solution is the use of hard coatings, such as diamond, or making the entire tip out of a precious material (platinum or diamond). However, this strategy is more expensive and the diamond coatings can damage the samples. In this context, the use of graphene as a protective coating for nanoprobes has attracted considerable interest. Here we review the main literature in this field, and discuss the fabrication, performance and scalability of nanoprobes.

Keywords: graphene; nanoprobes; coatings; atomic force microscopy; wear

---

1. Introduction

Sharp probe tips with an apex radius <100 nm are widely used in many different fields of science including electronics [1], physics [2], chemistry [3], biology [4] and medicine [5], as they allow local characterization and manipulation of materials with a nanometric spatial resolution. Depending on the equipment and application in which they are used, nanoprobes can have a wide variety of shapes, material composition and prizes. One of the main problems of nanoprobes is that they can lose their initial properties (e.g., sharpness, conductivity) after several experiments, leading to false data collection and increases in the cost of the research (i.e., new probes need to be used). Therefore, understanding the degradation process and the speed of nanoprobes during each experiment is essential to ensure high quality, reliable and cheap data collection.

Scanning Probe Microscopes (SPM) are among the most advanced equipments for nanoscale characterization and patterning, as they can achieve high spatial resolution both laterally (<0.1 nm) and vertically (<0.01 nm) [6,7]. The nanoprobes used in SPMs consist of a cantilever with a sharp tip at its end, which is the only part that contacts the sample being tested. The lateral resolution of the data collected with an SPM in most cases depends on the sharpness of the nanoprobe at its apex: the smaller the tip radius, the smaller the tip–sample contact area, and therefore the higher the lateral resolution [6]. When using an SPM to analyze the morphological properties of a material (e.g., topography, adhesion force) normally Si or Si_3N_4 nanoprobes are used, as they can be easily fabricated by standard silicon bulk micromachining technology [8]. The tip radius at the tip apex can be as small as ~2 nm. For electrical modes of SPMs, such as conductive atomic force microscopy (C-AFM), electrical force microscopy (EFM) and Kelvin probe force microscopy (KPFM), the probes need to be conductive, a capability that can be provided by coating the Si or Si_3N_4 nanoprobes with a thin (~20 nm) conductive varnish [9]. The conductive varnish (normally a metal or doped diamond) should be thick enough to provide stable conductivity to the tips under high current densities and frictions, and at the same time thin enough for not increasing the radius at the apex too much (this would result in a loss of lateral resolution). Maintaining the initially high conductivity and sharpness of conductive nanoprobes during several experiments is one of the main problems for the users of electrical modes of SPMs [10]. It is worth noting that the contact area between the tip of an SPM and the sample is typically <100 nm^2, and that the minimum current that this equipment can measure is ~1pA. Therefore, the minimum current that an SPM can detect is 1 A/cm^2. This is already a very high current density, but it can go even higher, up to 10^9 A/cm^2 if the currents detected with the SPM increase into the milliampere regime [11]. According to the International Technology Roadmap of Semiconductors [12], sub-10 nm resolution and a dynamic range of 10^{16}–10^{20} A/cm^2 is required to acquire electrical information in future nanoscale devices [13,14].

In the market place, one can also find conductive nanoprobes for SPMs made of solid metals or doped diamond. However, these nanoprobes show many disadvantages: (i) much higher cost (up to 10-times) due to the use of precious materials and hone techniques; (ii) very few suppliers and a very limited range of spring constants, resonance frequencies and tip radiuses; and (iii) damage to the samples due to high stiffness (for the diamond tips), which not only produces degradation of the sample under test, but also abundant adhesion of particles to the tip apex and subsequent reduction of its sharpness and conductivity. Table 1 summarizes the most used conductive nanoprobes for SPMs, and classifies them into four categories depending on their material structure: metal varnished Si probes, doped diamond varnished Si probes, solid metallic probes and solid doped diamond probes. Despite the wide range of conductive nanoprobes available in the market, currently none of them possess high spatial resolution, high conductivity, long durability and low cost at the same time. Therefore the exploration of novel materials with high conductivity and wear resistance is necessary to promote nanoscale characterization techniques.

Table 1. Specifications of the most used commercial conductive nanoprobes from different manufacturers. The prices represent the cost given by the local distributors in Shanghai.

Type	Model	Tip Coating (nm)	Bulk Materials	Tip Radius (nm)	Spring k (N/m)	Freq (kHz)	Manufacturer	Unit Price ($)
	SCM-PIC	PtIr	n-doped Si	20 + 5	0.2 (0.1–0.4)	13 (10–16)	Bruker	41.9
	OSCM-PT	Pt (20)	Si	15 + 10	2 (0.6–3.5)	70 (50–90)	Bruker	51.2
	SCM-PTSI	Pt/Si	n-doped Si	15 + 10	2.8 (1–5)	75 (50–100)	Bruker	156.7
	SMIM-150	TiW	Si$_3$N$_4$	50 ± 10	8 (7–9)	75 (70–80)	Bruker	139.8
	MESP	Co/Cr	Si	35 + 15	2.8 (1–5)	75 (50–100)	Bruker	116.7
	Arrow CONTPT	Cr/PtIr (5/25)	Si	33 ± 10	0.2 (0.06–0.38)	14 (10–19)	NanoWorld	38
Metal	CONTPT	Cr/PtIr (5/25)	Si	30 ± 10	0.2 (0.07–0.4)	13 (9–17)	NanoWorld	42.98
varnished	ATEC-CONTPT	Cr/PtIr (5/25)	Si	33 ± 10	0.2 (0.02–0.75)	15 (7–25)	Nanosensors	41.39
Si tip	PPP-CONTPT	Cr/PtIr (5/25)	Si	30 ± 10	0.2 (0.02–0.77)	13 (6–21)	Nanosensors	46.11
	PtSi-NCH	Pt	Si	30 ± 10	42 (10–130)	330 (204–497)	Nanosensors	152.08
	ACCESSS-NC-GG	Au	Si	30	113	330	App Nano	53.99
	TiN-ACT	TiN	Si	70	37	300	App Nano	39.5
	AC240TM	Ti/Pt (5/20)	Si	28 ± 10	2 (0.3–4.8)	70 (45–95)	Olympus	35.94
	NSC14/Pt	Pt or Au	Si	<30	5 (1.8–13)	160 (110–220)	μ-Masch	40.3
	Electri Tap 190-G	Cr/Pt	Si	<25	48 (20–100)	190 ± 60	Budgetsensors	37.26
	CDT-FMR	Doped diamond	Si	83 ± 17	2.8 (1.2–5.5)	75 (60–90)	NanoWorld	143.66
	CDT-CONTR	Doped diamond	Si	83 ± 17	0.2 (0.02–0.77)	13 (6–21)	Nanosensors	152.08
Doped	CDT-NCLR	Doped diamond	Si	83 ± 17	48 (21–98)	190 (146–236)	Nanosensors	152.08
diamond	DD-ACCESS-NC	Doped diamond	Si	100–300	93	320	App Nano	154.48
varnished	DDESP-FM	Doped diamond	Si	150 + 50	6.2 (3–11.4)	105 (80–103)	Bruker	132.8
Si tip	AD-0.5-AS	Single crystal diamond	Si	10 ± 5	0.5 (0.1–1)	30 (10–50)	Bruker	186.5
	AD-0.5-SS	Single crystal diamond	Si	<5	0.5 (0.1–1)	30 (10–50)	Bruker	279.6
Solid metal AFM tip	RMN-12PT400B	None	Pt	15 ± 5	0.3 (0.18–0.42)	4.5 (3.15-5.85)	Bruker	74.5

Table 1. *Cont.*

Type	Model	Tip Coating (nm)	Bulk Materials	Tip Radius (nm)	Spring k (N/m)	Freq (kHz)	Manufacturer	Unit Price ($)
Solid doped diamond AFM tip	SSRM-DIA	None	Diamond	5–20	3/11/27	-	Bruker (IMEC)	372.2
	P-CT1T2S	None	Diamond	-	0.71	50	Advanced Creative Solution Technology	1050
	P-CTCR1S	None	Diamond	<10 nm	0.35	35	Advanced Creative Solution Technology	950

Researchers have been working on designing prototypes of nanoprobes coated with specific materials. For example, Dai et al. [15] modified conventional SPM probe tips by attaching multiple walled carbon nanotubes (diameter of 5–20 nm and length of 1 μm) to their apex using epoxy and manual manipulation under optical microscope. Afterwards, the attachment method was improved with the assistance of DC current flow [16]. Tay et al. [17] attached single metallic (tungsten or cobalt) nanowires to commercial AFM tips, and successfully used them to profile a steep sidewall structure with high resolution due to their tiny radius of curvature (1–2 nm) and high aspect ratio (length of ~100 nm, height of ~1.5 μm). Bakhti et al. [18] grew a gold nano-filament with radius of <3 nm and length of 10–100 nm on the apex of conductive SPM nanoprobes, and the resulting nanoprobe was shown to be chemically inert with improved lateral resolution (observable from the topographic maps). The emergence of two dimensional (2D) materials with superior properties (mechanical strength, flexibility, transparency, thermal conductivity, chemical stability, among others) [19–22], has also been attractive in the field of conductive nanoprobes engineering. Several works [23–29] have demonstrated that graphene would be an ideal coating material to enhance the lifetime of a conductive nanoprobe, as it can provide high conductivity and mechanical robustness without increasing the tip radius (it is just one atom thick) and/or modifying the spring constant of the cantilever (its mass is negligible, while hard coatings like diamond have a considerable mass that bend the cantilevers and alter their mechanical and dynamic properties). Furthermore, graphene could be used to functionalize the surface of the probes, providing additional properties like hydrophobicity and piezoelectricity. In this review, we present a detailed summary of all the graphene coated nanoprobes developed, and describe several characteristics including the fabrication technologies and performances.

2. Graphene-Coated AFM Probes Production

2.1. Direct Chemical Vapor Deposition of Graphene on AFM Nanoprobes

Chemical vapor deposition (CVD) is a widespread methodology used to produce high quality, large area and continuous graphene films on the surface of different metal catalysts (e.g., Cu [30], Fe [31], Ni [32]). To do so, the metallic substrate (typically a foil) is introduced in a tube CVD furnace and heated at high temperatures (>850 °C) while the graphene precursor (typically CH_4 or C_2H_5OH) is introduced in the tube with the assistance of a carrier gas (typically H_2/Ar). Using this approach, the precursor seeds (carbon-containing molecules) can precipitate at random locations on the surface of the metallic sample, and they grow until merging into each other, forming a homogeneous (but polycrystalline) graphene film—intuitively this process is similar to placing several ice cubes on a flat table and waiting until they melt and form a homogeneous water film. By controlling the amount of precursor in the chamber, the temperature and growth time, the properties of the graphene sheets can be tuned (e.g., domain size, graphene thickness). As the growth takes place at all locations of the sample simultaneously, the growth process is scalable.

Wen et al. [23] attempted to grow graphene on the surface of Au-varnished Si nanoprobes via CVD. To do so, they inserted the nanoprobes (from Mikromash model CSC38/Cr-Au) in a tube furnace and ran the CVD process (Figure 1a). C_2H_5OH was used to supply the carbon source and delivered to the Au-coated nanoprobe by the mixed gases of Ar/H_2 for 5 min, under a flow rate of 100 sccm at 750–850 °C. The authors claim that, as the tips are varnished with Au (which can serve as a metallic catalyst), the graphene would form on the surface of the Au and eventually cover the entire surface of the nanoprobe. In order to demonstrate the correct growth of graphene the authors compare the Raman spectra of the graphene films grown on Au electrodes and Au-varnished tips, and both showed peaks at 1597 cm^{-1} (Figure 1b). However, the scanning electron microscopy (SEM) images of the tip apex shown by the authors reveal the material deposited on the tip apex appears to be very un-homogenous and thick (Figure 1c), meaning that this carbon-rich material may not hold the genuine properties of 2D graphene sheets.

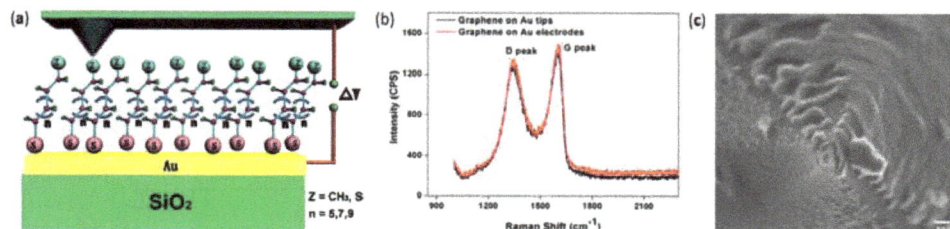

Figure 1. Direct chemical vapor deposition of graphene on metal-varnished AFM tips. (**a**) Schematic representation of an Au-varnished AFM tip coated with graphene that is being used as top electrode in a molecular junction; (**b**) Raman spectrum of the graphene formed on the Au electrode (red) and Au-coated AFM tip (black line); (**c**) SEM image of the tip apex after the CVD process. The surface appears to be covered by a discontinuous and thick layer, rather than atomically-thin continuous graphene. Reproduced with permission from [23], copyright Wiley-VCH (2012).

It should be highlighted that the growth of 2D materials on metallic substrates has been normally performed using metallic foils, not metal-varnished surfaces. The reason is that large amounts of metal need to be available on the surface of the sample to avoid massive diffusion and de-wetting at high temperatures >850 °C. Very few authors successfully achieved the CVD growth of graphene or any other 2D material on metal-varnished samples (300 nm SiO_2/Si) [33–35], and in all cases the thickness of the metal was >500 nm. In Ref. [23], the authors used standard Au-varnished Si nanoprobes for SPMs, on which the Au-varnish is <70 nm thick. They recognized discontinuous graphene growth due to uncontrollable Au melting. Despite the authors showing that their graphene coated nanoprobes achieved enhanced performance (90% yield) as a molecular junction, to the best of our knowledge this approach has never been reproduced by these or other authors, and the graphene grown by this method contains a large number of defects, which is indicated by the strong D peak. Therefore, this is not an ideal methodology for graphene coating on the AFM tips.

2.2. Transfer of CVD-Grown Graphene onto AFM Probes

As the temperatures required for the CVD growth of graphene are very high, avoiding the use of this method directly on the tips is necessary. One of the main advantages of CVD-grown 2D materials is that they can be prepared on a substrate that ensures very high quality (e.g., Cu foils) and is then transferred onto the target device. Lanza et al. [24] coated commercially available AFM probes with a sheet of graphene previously grown on a Cu foil. During the fabrication process two samples were prepared independently and merged (see Figure 2a). Single layer graphene was synthesized via CVD approach on a 25 μm thick Cu foil, using 20 sccm methane gas mixed with 10 sccm hydrogen in a tube furnace working at a growth temperature of 1000 °C for 15 min. After that, the furnace was cooled down to room temperature under the flow rate of 10 sccm hydrogen. During this process, carbon-containing sources precipitated and eventually graphene films formed on both sides of the Cu substrate. After the growth, the graphene/Cu/graphene stack was introduced to an oxygen plasma furnace to remove one side of the graphene layers. The resulting graphene/Cu stack was fixed in a spinner and a drop of PMMA was deposited on top of the graphene sheet and spun at 1000 rpm for 1 min. Then, the sample was backed on a hot plate at 170 °C for 5 min until the PMMA became solid, resulting in an average thickness of ~200 nm. The PMMA/graphene/Cu stack was deposited on the surface of a $FeCl_3$ solution for etching the Cu substrate. After that, the PMMA/graphene stack was fished and washed, first in a 2% HCl solution and later in deionized water. Meanwhile, a commercial PtIr-varnished silicon AFM tip was fully wrapped and glued on a cleaned piece of Si wafer. To do so, the surface of the Si was first covered with a ~200 nm layer of PMMA (spun at 1000 rpm for 1 min and baked at 170 °C for 5 min). Then, the tip was fixed manually simply by pushing and partially immersing the chip containing the AFM tip in the soft PMMA substrate. Then the resulting sample was covered again with PMMA. During this process, the cantilever of the AFM nanoprobe was bent due to the weight of PMMA. The probe tip used was the CONTPt from Nanosensors, which has the following main properties: tip radius = 10 nm, cantilever length = 450 μm, spring constant = 0.2 N/m, and resonance frequency = 13 kHz; the thickness of the PtIr varnish was 20 nm.

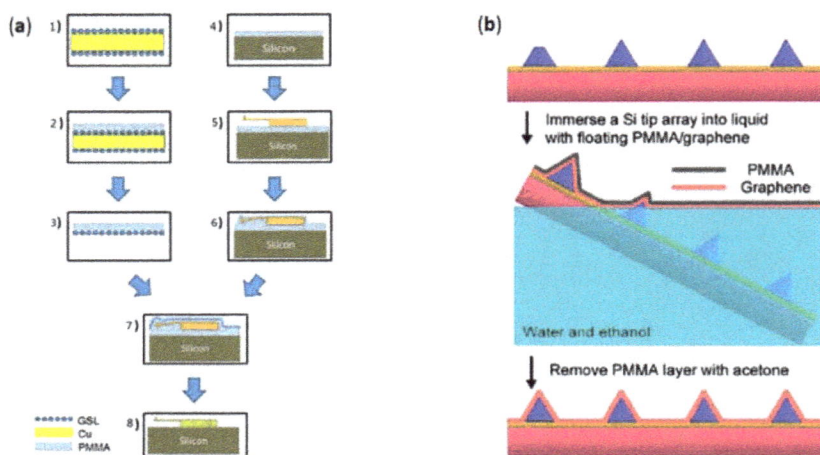

Figure 2. Transfer of CVD-grown graphene onto AFM probe tips. (**a**) Step-by-step schematic of the process followed to transfer graphene (grown on Cu foils) onto AFM tips. Reproduced with permission from [24], copyright Wiley-VCH (2013): (**b**) Experimental protocol used to fabricate graphene coated tip arrays. Reproduced with permission from [25], copyright AAAS.

Once the tip is immobilized on the surface of the Si substrate by the bottom and top PMMA layers, it is used as target substrate to pick up the PMMA/graphene sample prepared previously (see Figure 2a), and the entire structure was dried at room temperature. Finally, all the PMMA was removed via acetone vapor. This is the most critical step of this technique, as highlighted by the authors [24]. Basically, a very ingenious method was proposed to remove the PMMA in this work, that is, instead of liquid acetone, a large amount of PMMA was sufficiently removed by boiling acetone vapor (~68 °C) for 30 min. A designed set up with two glass containers was used to keep the AFM tip continuously exposed to the vaporized acetone, which effectively reduced the residual of contamination on the AFM tip. As a result, the geometric shape of the tip was fully wrapped by flexible graphene film and finally achieved the conformal graphene coated AFM tip.

This kind of transfer method was also used in the work reported by Shim et al. [25], in which a 10–20 layer thick graphene stack grown on a 4-inch Ni/Si wafer (Graphene Laboratories Inc., Calverton, NY, USA) was used as graphene coating for scanning probe arrays. As usual, a ~70 nm thick PMMA layer was spin-coated on as-grown graphene/Ni/Si at 500 rpm for 10 s with a speed of 100 rpm, followed by 5000 rpm for 60 s with a speed of 1000 rpm. The PMMA/graphene/Ni/Si stack was preserved at room temperature for 24 h and dried naturally. Afterwards, small pieces (1×1 cm^2) of wafer were immersed into aqueous FeCl$_3$ solution (1 M) for etching away the Ni film, which produced the separation of the PMMA/graphene film from the substrate. Similarly, the PMMA/graphene film was rinsed with deionized water and then fished with the target substrate. In this case the authors designed a target substrate consisting on a transparent glass with 100 nm SiO$_2$ with more than 4489 pyramidal tips (without cantilever, see Figure 2b) arranged in a matrix distribution. This target substrate was used to fish the PMMA/graphene stack, a process that was done by keeping a constant angle of 40° with respect to the liquid surface. The sample was exposed 48 h to air atmosphere at room temperature for natural drying, and the PMMA/graphene coated HSL tip array was immersed in acetone for 2 h to remove the PMMA. Finally, the sample was cleaned again in ethanol.

2.3. Mold-Assisted Transfer of CVD-Grown Graphene onto AFM Probes

Martin-Olmos et al. [26] developed a different transfer method based on the use of a Cu mold, on which the graphene was grown before being filled with SU-8 photoresist. The entire fabrication process is displayed in Figure 3a.

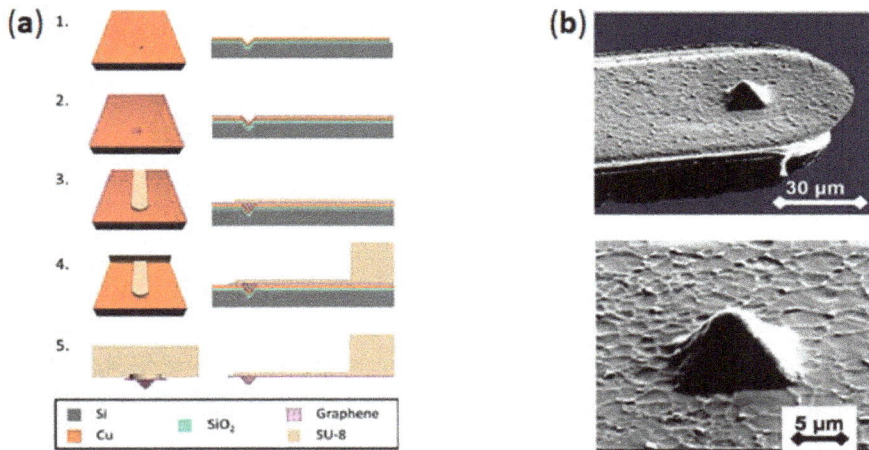

Figure 3. Schematic of graphene coated SU-8 AFM probes obtained by CVD graphene transfer using a mold. (**a**) Fabrication procedure of graphene coated SU-8 AFM probes; (**b**) SEM images of released graphene coated SU-8 AFM probes (top) and zoom in probe apex (bottom). Reproduced with permission from [26], copyright ACS Nano (2013).

First, a 100 nm SiO$_2$/Si wafer was patterned with circles of different sizes (a few micrometers) via lithography. The sample was then immersed in potassium hydroxide (KOH) etchant to generate the inverted pyramids with different sizes, which were used as the molds for AFM tips. Thermal thin silicon dioxide (which served as a sacrificial layer) was grown on the patterned wafers, followed by the deposition of a 500 nm thick high-purity copper film, which acted as a catalytic metal for the CVD growth of graphene. In this case, the authors used a suitable metallic thickness that might be able to withstand the thermal heating during the 2D material growth. The molded substrates were heated in a tube furnace up to 800 °C under hydrogen gas flow of 5 sccm. Then, 35 sccm of methane (CH$_4$) was introduced into the chamber, which produced the decomposition of methane and formation of monolayer graphene. The resulting graphene coated Cu mold was filled with SU-8 photoresist (10 μm) by using a spinner. Finally, the SU-8 photoresist was patterned by lithography and the substrate was etched: first the etching of SiO_2 using KOH and then the Cu using $FeCl_3$ (more experimental details can be found in Ref. [26]). The SEM images of the released SU-8 graphene coated nanoprobes (Figure 3b) show interesting and correct graphene coating on the SU-8 photoresist, which can be easily distinguished by the characteristic wrinkles.

2.4. Direct Graphite-Like Thin Film Deposition on AFM Nanoprobes

Graphene can be also successfully deposited on the apex of standard nanoprobes without the use of any CVD steps. Pacios et al. [27] reported the fabrication of an ultrathin graphite-coated AFM tip using a sputtering deposition followed by in situ annealing. First, a 30 nm thick amorphous carbon film was deposited on the as-received spherical Si/SiO_2 or rounded Si bulk AFM tips via radio frequency (RF) sputtering. Then, a 100 nm thick platinum (Pt) catalyst film was directly deposited on the carbon film, also via RF sputtering. The process is highlighted in Figure 4a. The tip is then annealed in a quartz tube furnace under an argon atmosphere for 30 min at a temperature of 800 °C. This step produced the spreading of carbon from the graphite film into metal catalyst due to the high solubility of carbon in the Pt. When cooling down, the carbon separates and graphitic flakes are generated on the surface of the Pt film (see Figure 4b). Cross-section TEM images exhibited the successful growth of layered graphite film on the Pt-coated tips (Figure 4c,d). The thickness of the ultrathin graphite film obtained by this method is around 20 nm (see Figure 4d). For this experiment, the authors used AFM tips with different tip radiuses (ranging from 2 μm to 90 nm), heights (typically 10–15 μm) and force constants (ranging from 48 N/m to 0.2 N/m).

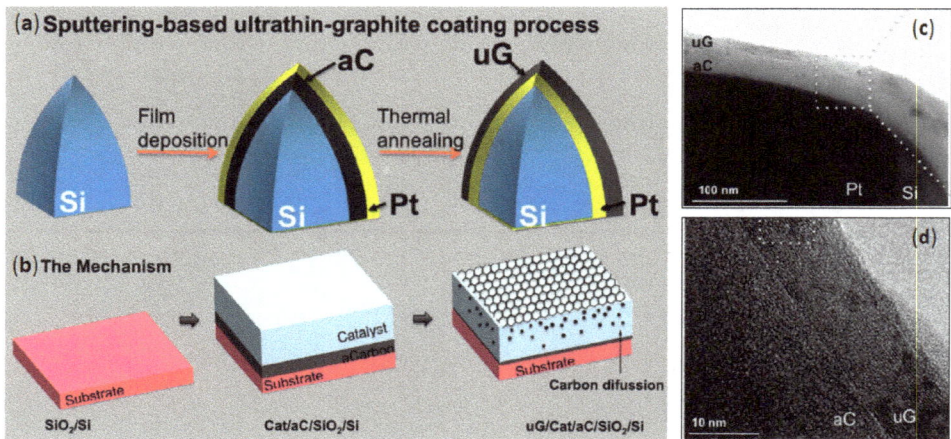

Figure 4. Ultrathin graphite growth on high aspect ratio features. (**a**) Schematic of the growth process on a 3D shaped AFM nanoprobe; (**b**) Growth mechanism of thin graphite by thin film deposition technology and thermal annealing; Cross-sectional TEM image of the resulting graphite coated AFM tip in low (**c**) and high (**d**) magnification. Reproduced with permission from [27], copyright Springer Nature Publishing Group (2016).

The advantages of this work are: (i) unlike thermal evaporation, in which carbon and Pt layers are evaporated in two separated chambers, here sputtering can produce both layers in the same run and the quality of the materials can be precisely controlled by the bias and pressure during the deposition process; (ii) the changeable directionality of sputtering helps to achieve full coverage for the irregular shapes of AFM tips; (iii) as a metal catalyst, platinum can provide better surface morphology and homogeneity for the growth of high quality 2D materials due to its higher melting temperature and lower thermal expansion coefficient; and (iv) Pt cannot be easily oxidized, in contrast with what happens to other metals (such as Ni or Cu), and its chemically inert property contributes to decreasing surface irregularities and keeping its catalytic ability. Compared to traditional CVD technology, the authors claim that this method is a clean, simple, less hazardous and reproducible technique, which makes it possible for large-scale manufacturing. In Figure 4c,d a cross-section of the graphitic film grown on the surface of the tip apex can be seen. Layered structure can be successfully obtained, as observed from the high resolution TEM images.

2.5. Liquid Phase Graphene Flakes Coated AFM Probes

Although AFM probes coated with graphene can be achieved by using some of the above-mentioned methodologies, complicated procedures are not advisable for the mass production needed in industry. Recently, high quality solution-processed graphene was used to coat different kinds of nanoprobes [28]. The graphene sheets were synthesized from graphite powder. Based on a series of redox reactions, graphite powder (2 g) was firstly oxidized by the Hummers–Offeman method with the assistance of H_2SO_4 (12 mL), $K_2S_2O_8$ (3.0 g) and P_2O_5 (3.0 g) at 80 °C for 5 h. Then, H_2SO_4 (150 mL), $KMnO_4$ (25 g) and 30 mL H_2O_2 were added to the resulting product in order to help to re-oxidize it, and thin flakes of graphene oxide were successfully obtained after: (i) washing (in 1:10 HCl and pure water), (ii) drying naturally, (iii) purifying (by dialysis for 1 week), and (iv) an ultrasonic bath. The 50 mL graphene oxide solution (0.1 mg/mL) was reduced by mixing it with hydration hydrazine (5 mL). The mix was then stirred for 24 h at 80 °C, filtered and dried, which resulted in a black powder graphene. Finally, the graphene solution was prepared by mixing the black graphene powder (5 mg) and pure water (1 mL), and sonicating at 50 W for 10 min. The resulting solution contained large amounts of graphene sheets with an average thickness of 0.7 nm and size of <1 μm. Currently there are several suppliers that provide these types of graphene solutions at a very low price. However, when selecting a graphene solution supplier one needs to verify that the solution really contains large area atomically thin graphene sheets, not just thick graphite particles. One recent study analyzed the size and thickness of the sheets in liquid phase exfoliated graphene from more than 20 different companies, and it was concluded that only two were able to provide micrometer scale sheets with thicknesses below 10 layers [36]. Also, several companies add polymers to the solution in order to stabilize it. It is important to select graphene solutions that do not use these polymers, as they may fall between the tip apex and the graphene flake, and result in a thick and non-conductive coating.

Once the solution is ready, coating the graphene tip is very simple, cheap and fast. In Hui's work [28], commercial AFM nanoprobes with a conductive coating of 20 nm Pt or PtIr from different manufacturers (Olympus and Bruker) were immersed in the graphene solution for less than one minute (see Figure 5). By swinging the probe, the graphene sheets readily attached to the sharp AFM tip by van der Waals forces. Van der Waals forces are much higher at very sharp morphologies [37], which means that all the flakes tend to attach there (i.e., many other locations of the nanoprobes remain uncoated, but very good conformal coating and high reproducibility is achieved at the tip apex). After that, the graphene coated nanoprobe was left to dry naturally. Alternatively, the nanoprobes coated by this method can also be dried using a N_2 gun at a very low gas flow. N_2 blowing also enhances the adhesion between the graphene and the tip apex. Additionally, the amount of graphene sheets attached to the tip apex can be tuned by adjusting the concentration of the graphene solution. This cost-efficient methodology could be used to achieve high-yields of graphene coated nanoprobes and facilitate their industrial production.

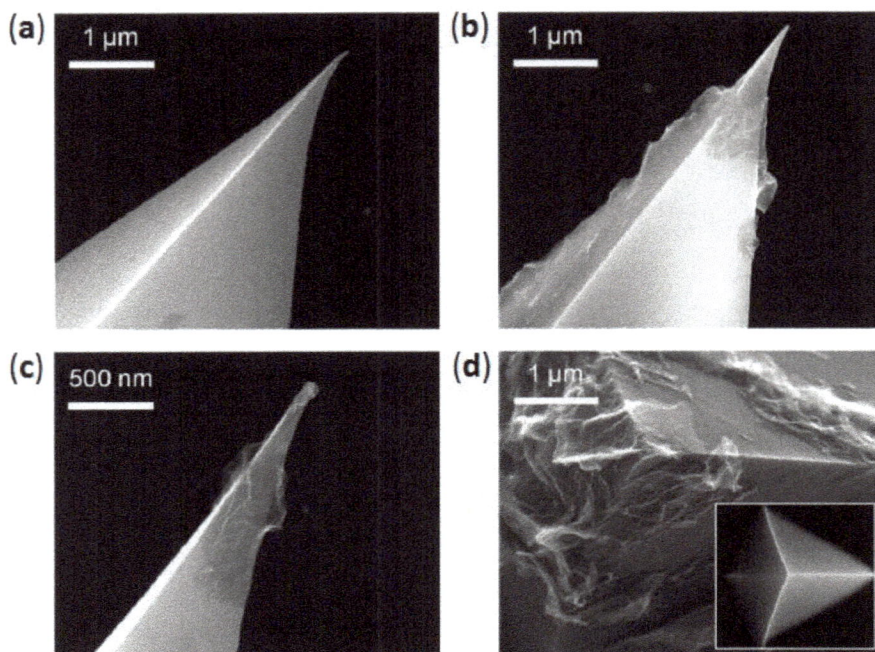

Figure 5. Different commercial Pt-varnished AFM probes coated by solution processed graphene flakes. SEM images of as-received OMCL-AC240 nanoprobe before (**a**) and after coating (**b**); (**c,d**) are OSCM-PT and SCM-PIC AFM nanoprobes coated with low density of graphene sheets (respectively). Reproduced with permission from [28], copyright The Royal Society of Chemistry (2016).

3. Perspectives on the Fabrication of Graphene Coated AFM Probes

Based on the above discussions, liquid-phase exfoliated graphene coating seems to be the most promising methodology due to its low cost, fast coating process and excellent compatibility with industry. Here, the possible manufacturing procedure based on the current technologies being used by AFM tips manufacturers (such as Nanoworld) is proposed as follows [38]. The basic fabrication process of standard tips is described in Figure 6. A piece of silicon wafer is oxidized on both sides (Figure 6a). Then, a layer of photo resist is spin-coated onto the back side of the $SiO_2/Si/SiO_2$ wafer, and subsequently exposed to UV light using a mask (Figure 6b–c). Similarly, the front side is patterned by the same method but using a different mask (see Figure 6d–e). The silicon dioxide on the front side is removed by isotropic wet etching methodology (Figure 6f) and then the photoresist on both sides is dissolved (Figure 6g). The formation of probes can be observed after the anisotropic wet etching of silicon by KOH in the following steps, until the oxide shields falls off (as shown in image Figure 6h,i). A silicon nitride layer is deposited to protect the tip side of the probe from the damage through further wet etching of silicon, which defines the thickness of cantilever (Figure 6j,k). Finally, silicon bulk probes are isolated by removing the silicon nitride layer. In order to obtain functional coating probes, metallic or magnetic materials are required to be deposited on top of Si bulk probes by specific methods, such as sputtering, atomic layer deposition (ALD) and/or E-beam evaporation. This process is normally conducted on six-inch wafers (see Figure 7a), and after the process the chips are patterned within the silicon wafer (Figure 7b). Then, they need to be removed and placed in sticky boxes (using vacuum tweezers to avoiding scratches) for commercialization. The photograph of a wafer patterned with AFM tips, as well as the zoomed-in SEM images of the chips containing the cantilevers and sharp tips are shown in Figure 7a,b, respectively.

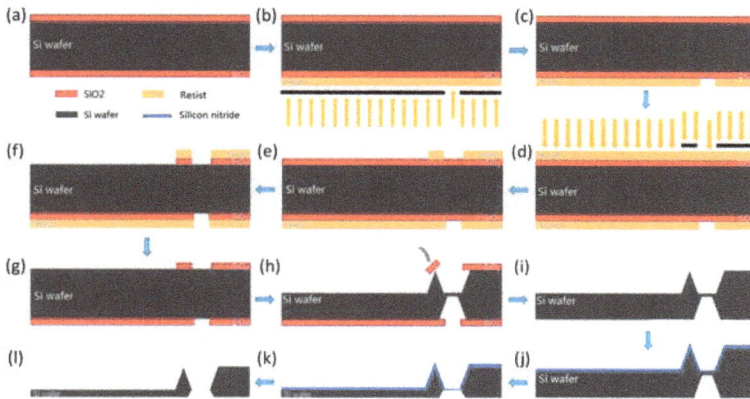

Figure 6. Schematic illustration of standard AFM probes manufacturing by Nanoworld. (**a**) Silicon wafer with both sides of silicon dioxides; (**b,d**) exposure of the back and front sides to photoresist through a mask; (**c,e**) development of the exposed photoresist; (**f**) isotropic wet etching of silicon oxide; (**g**) dissolution of photo resist; (**h,i**) anisotropic wet etching of silicon by KOH and silicon oxide; (**j–l**) deposition and isotropic wet etching of silicon and silicon nitride, respectively. Reproduced with permission from [38], copyright Nanoworld 2008.

The coating process via liquid phase exfoliation presented in Ref. [28] has been carried out by manipulating probe tips one-by-one from commercial sticky boxes using standard tweezers, and immersing them into a tube containing the graphene solution. Despite the success and reproducibility of the experiments, this method should be optimized for the coating of several tips in parallel and avoid tip-induced scratch in the chip that contains the cantilever and tips (see Figure 7c). The ideal is that AFM tip manufacturers incorporate the graphene coating process at the end of the production chain of the AFM tips. To do so, the as-fabricated wafers containing the tips should be immersed in a container filled with graphene solution for a certain period of time. The concentration of the graphene solution, the immersion time and the use of agitation and/or voltage may be tuned to improve the percentage of tips coated and the quality of the coating.

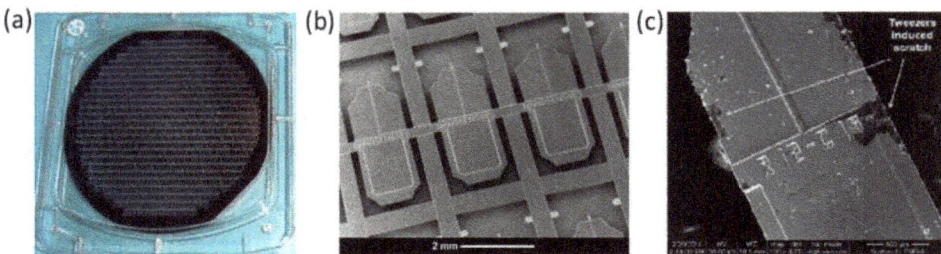

Figure 7. As-fabricated AFM probes by Nanoworld manufacturer. (**a**) Wafer scale of AFM probes; SEM images of (**b**) as-fabricated AFM probes. Reproduced with permission from Ref. [38], copyright Nanoworld GmbH 2008; (**c**) Individual probe scratched by tweezers when picking it up. Reproduced with permission from [28], copyright The Royal Society of Chemistry (2016).

4. Functionalities of Graphene Coated AFM Probes

Graphene coated AFM probes have been fabricated successfully by some of the methodologies described above. After fabrication, these devices show enhanced performance in several types of experiments such as, topographic and current maps, current-voltage curves, force-distance curves, as well as statistical analysis of variability and durability. Graphene coated nanoprobes have been used in different fields of science, including electronics, mechanics, and physics [23–29].

4.1. High Wear Resistance in Lateral Scans

The graphene coated SU-8 nanoprobes fabricated by mold-assisted graphene transfer [26] exhibited mechanical properties and conductivity different to those of uncoated (graphene-free) SU-8 probes. In experiments both tapping mode and contact mode images are collected on a calibration sample, which consists of 800 nm diameter and 40 nm thick Ag pillars on a SiO_2 substrate, to characterize the polymer AFM probes. As shown in Figure 8a, under tapping mode, both probes can resolve the topographic information, and the scanned image using a graphene coated SU-8 probe shows similar resolution (10% loss) as the uncoated probe. This may a result of the slightly increased radius of the tip apex caused by the graphene layer (this indicates that the layer is not as thin as believed) and/or non-uniform surface of the graphene film. The interesting point came when scanning the sample in contact mode. The images scanned using uncoated (graphene-free) SU-8 probes show much lower lateral and vertical resolution than those collected using graphene coated ones, and the images appear blurry and with the profiles repeated. The characteristic doubled (repeated) features in the topographic scans are a clear indication that a tip with two apexes has been formed. This is related to the removal of some material at the tip apex; the volume loss in the tip can produce irregular shapes, making it possible that more than one location of the tip touches the sample, which results in repeated profiles along the scanned area.

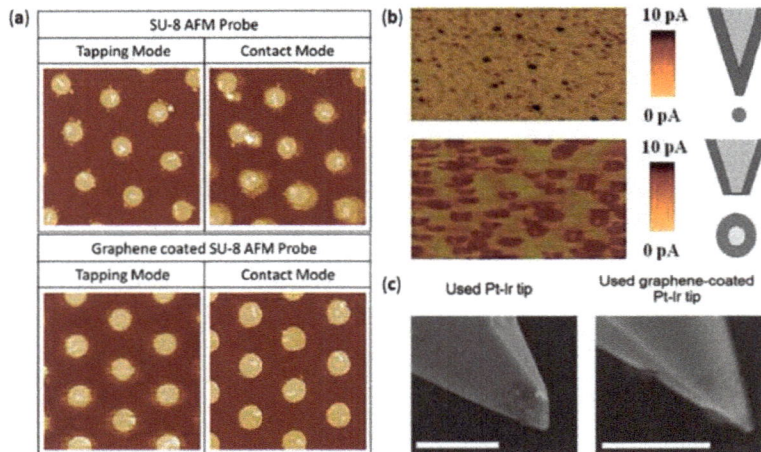

Figure 8. Mechanical exploration of graphene coated AFM compared with uncoated ones. (a) Topographic images collected by both tapping mode and contact mode on the sample of Ag/SiO₂. Reproduced with permission from [26], copyright American Chemical Society (2013); (b) 400 nm × 250 nm current maps of HfO_2/SiO_2 stack using a new (top) and worn-out commercial PtIr AFM conductive tip (bottom), the schematic in the right side represents the degree of wearing of tips; (c) SEM images indicate the degree of damage of the PtIr tip with and without graphene coating after use. The scale bars are 5 μm (left) and 3 μm (right), respectively. Reproduced with permission from [24], copyright Wiley-VCH (2013).

The reason why the tip loses resolution so fast is the low hardness of the SU-8 material (~0.43 GPa [39], compared to Si ~11.9 GPa [40]), leading to the fast wearing of (volume removal) the apex after scanning the sample consecutively. This phenomenon has been confirmed by collecting current maps of 4 nm HfO_2/1 nm SiO_2 stacks (under a constant bias), using a new (top) and worn out (bottom) commercial conductive AFM (see Figure 8b). When the tip is new (top map) small conductive spots can be detected, indicating the flow of tunneling current across the ultra-thin bilayer insulator. On the contrary, the worn tip shows abundant ring-like conducting features (instead of spots, see bottom current map) which is related to the removal of the tip apex (see schematics in Figure 8b). Fortunately, the superb mechanical properties of graphene can effectively protect the apex from wearing even after several scans (see Figure 8c).

4.2. Avoiding Water Perturbations at the Tip–Sample Junction

The interaction between the tips and a silicon substrate has been compared before and after graphene coating. Force-distance (F–Z) curves have been collected at several locations of the sample. The F–Z curves represent the vertical force that the tip applies to the surface of the sample, which varies with the distance between them. The force vertically applied is proportional to the deflection of the cantilever, and for this reason deflection-force (D–Z) curves can also give relevant information (the Y-axis in Figure 9a is deflection, not force, but this is also acceptable if relative variations want to be studied). The D–Z collected with the graphene-free Pt-varnished Si tip on the surface of a Si substrate shows the typical shape: (i) the force stays at zero when the tip is far away from the sample surface of the sample (Z << 0); (ii) starts to increase when the tip is in contact with the sample. In theory that should happen for Z > 0, but in Figure 9a the deflection starts to increase at Z ~−25 nm; the reason is that the initial position of the tip is not at Z > 0, but a bit (~25 nm) above the surface of the sample in order to avoid damage of the tip); (iii) during tip retraction, a high negative peak can be observed, which is related to the adhesion (anti-repulsive) force between the tip and the sample and (iv) once the tip is beyond the limited distance this type of force suddenly disappears (a jump in the D–Z curve can be distinguished).

As shown in Figure 9a, the adhesion force detected in the D–Z curve performed with the Pt-coated AFM tip is 1680 nN. Interestingly, when the same experiment is carried out using a graphene coated tip (obtained with the graphite thin film deposition method explained in Section 2.4) the shape of the D–Z curve is similar, but the adhesion force is much smaller. While different materials at the tip apex may produce different tip-sample interactions, a feasible explanation for such a huge difference in the adhesion peak is as follows. The force that plays the major role in tip–sample adhesion in AFM experiments is due to capillary effects [41]. Basically, when an AFM tip is placed in contact with the sample in normal atmospheric conditions, a water layer related to the relative humidity is formed between the tip and a sample's surface, which produces a water meniscus at the junction. The depth of the water layers and the size of the water meniscus increases with the relative humidity (see Figure 9d) [42]. Stukalov et al. [43] demonstrated that the contact force between a Si AFM tip and a mica substrate can increase greatly depending on the relative humidity (see Figure 9c). Taking into account that the reduction observed in Figure 9a,b is one order of magnitude, this indicates that graphene is an excellent material for avoiding tip-sample capillary effects, which are very annoying in several types of SPM experiments [44–46]. This observation can also be related to the hydrophobic nature of the graphene coating [47]. On the contrary, the Pt coating of the graphene-free tip is highly hydrophilic.

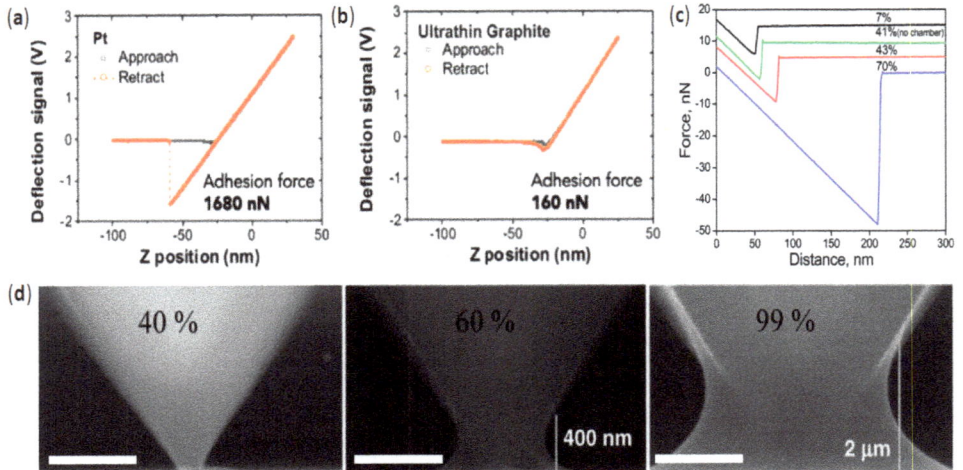

Figure 9. Tip-sample interaction. Force-distance curves collected on Silicon substrate with a Pt coated AFM tip (**a**) and ultrathin graphite coated tip (**b**). Reproduced with permission from [27], copyright Springer Nature Publishing Group (2016); (**c**) Representative force-distance curves obtained by retracting a Si cantilever from a freshly-cleaved mica surface at different RH values that are listed next to each curve. Reproduced with permission from [43], copyright American Institute of Physics (2006); (**d**) Sequence of images collected at various relative humidity. The scale bars are 1 μm. Reproduced with permission from [42], copyright American Chemical Society (2005).

The effect of water meniscus minimization/removal at the tip-sample junction due to the graphene coating have very positive impact in several different types of experiments, especially in electronic measurements. Hui et al. [28] collected forward and backward current-voltage (I–V) curves on the surface of an n-type Si wafer using both standard PtIr varnished AFM tips with and without graphene coating (solution processed liquid-phase exfoliation graphene). The results are displayed in Figure 10a.

The standard tip shows a large onset potential (V_{ON}, defined as the minimum voltage that shows currents above the noise level) followed by a sharp current increase in the forward curve. Then, the backward tip shows a clear shift towards lower potentials, indicating that the tip-sample junction has become more conductive; moreover, the shape of this I–V curve is less sharp, and more similar to the typical conduction of silicon. On the contrary, the I–V curves collected with graphene coated AFM tips show currents similar to the backward curve collected with graphene-free tips. It should be highlighted that the expected conduction mechanism between the PtIr coating (in fact it is 95% Pt) and the n-type Si substrate (which has no native oxide, as it was removed via hydrofluoric acid) is Schottky conduction, and the observations during the forward I–V curve collected with the graphene-free PtIr-varnished tip are unexpected.

Nevertheless, this behavior can be explained by the presence of a water meniscus and a nanometric water layer between the graphene-free PtIr-varnished AFM tip and the sample. The water layer between the tip and the sample increases the resistivity at the junction, which is why V_{ON} is larger. When the electrical field is large enough current starts to flow and the tip physically contacts the n-type Si sample, which produces a shift in the I–V curve towards lower potentials. After that, the shape observed in the I–V curve is typical for Schottky conduction. To demonstrate this hypothesis, several I–V curves have been collected at different locations of the sample using the standard (graphene-free) tips, and the forward and backward I–V curves have been fitted to the conduction across a $Pt/H_2O/Si$ heterojunction (similar to the tip-sample junction) tuning the different H_2O thicknesses. The fittings were done with the software MDLab [48]. As displayed in Figure 10b, the calculations indicate that the forward I–V curves (blue) can only be fitted using a H_2O thickness of 10 Å, while the backward I–V curves can be fitted with an almost negligible H_2O thickness of 1 Å, respectively. In contrast, graphene coated AFM tips do not show this problem (see Figure 10c,d), and the currents registered are real (correct) from the initial (forward) I–V curve. Again, the hydrophobic nature of the graphene coating had a positive influence on the measurements.

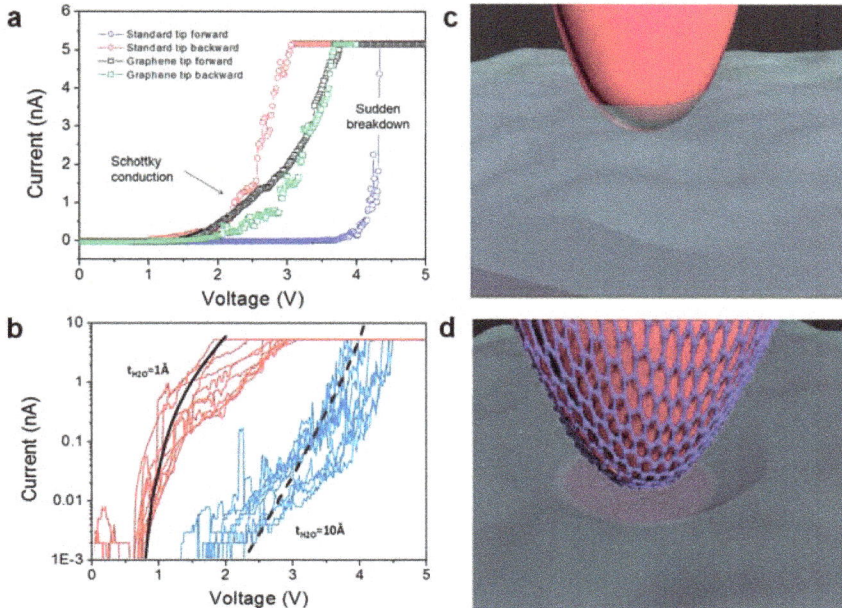

Figure 10. (**a**) Typical forward and backward I–V curves collected with standard and graphene nanoprobes on a piece of n-type silicon; (**b**) Fitting of the forward and backward I–V curves collected with the standard tip to the charge transport model; 3D schematics of (**c**) standard and (**d**) graphene nanoprobe, which shows the water resistance of graphene. Reproduced with permission from [28], copyright The Royal Society of Chemistry (2016).

Lanza et al. [29] has proved this by studying the electrical behaviors of HfO_2/Si stacks using as-received AFM tips (Figure 11a–c) and graphene coated AFM tips (Figure 11d–f), respectively. As shown in Figure 11, topographic and current maps were collected simultaneously under a contact force of 0.1 nN and voltage of 2 V in a high vacuum environment (10^{-7} torr). Comparing these two current maps (white represents 0 pA), it is obvious that the dark spots, which represent the local currents through HfO_2, are smaller in Figure 11e than the current collected by the uncoated AFM tip (Figure 11b). Also, the statistical analysis shown in Figure 11c,f corroborate the smaller size of the conductive spots collected using graphene coated tips and an obvious shift to higher currents compared to uncoated ones (insets in Figure 11c,f), which further shows a lower tip-sample contact area for the graphene-coated tips due to keeping good conductivity. In general, the smoother topographic surface and individual conductive spots observed in Figure 11 demonstrate that the graphene-coated AFM tips possess not only high lateral resolution, but also superior electrical performance.

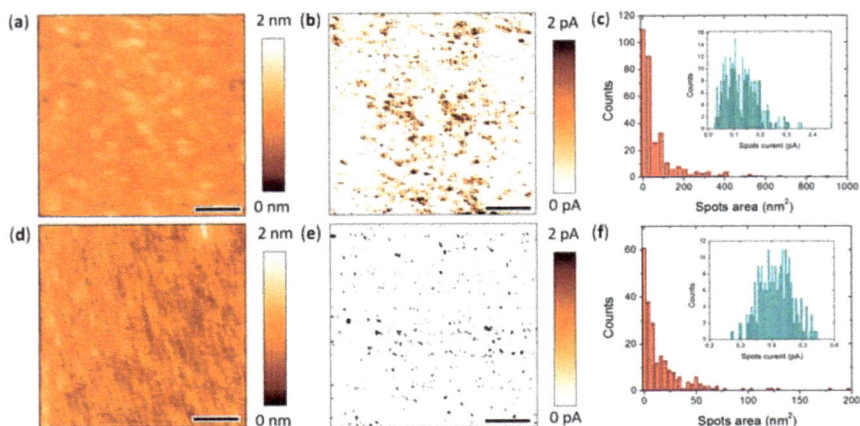

Figure 11. Topographic (**a,d**) and current maps (**b,e**) recorded for as-received (top panel) and graphene-coated AFM tips (bottom panel) on 2.5 nm HfO_2/Si stack; (**c,f**) Statistical analysis of the conductive spots are measured when scanning the surface of HfO_2/SiO_2/Si stack. Reproduced with permission from [29], copyright The Royal Society of Chemistry (2013).

4.3. Lower Data Variability

Before analyzing the variability of graphene coated nanoprobes, the intrinsic variability of standard (graphene-free) probes should be discussed. The variability of standard nanoprobes is mainly affected by two factors: (i) the variability of the parameters of the tips (tip radius and spring constant). It is important to note that the manufacturers of AFM probes allow deviations of the tip radius up to 100% (e.g., from 100 nm to 200 nm [49]), and the typical ranges for spring constant deviations can be as high as one order of magnitude (e.g., from 1.5 N/m to 18.3 N/m [49]); and (ii) the presence of a water meniscus at the tip-sample junction. Standard AFM probes varnished with metal are hydrophilic (which can form a thick water meniscus at the junction, see Figure 9d), while graphene-coated probes are hydrophobic (which repulses the water from the tip and produces a clean junction). Different amounts of water at the junction can have a huge effect in experiments, especially when measuring adhesion force (see Figure 9) or currents (see Figure 10 and Refs. [50,51]). In this section, the effect of the graphene coating on these two sources of variability is analyzed.

When discussing the variability of the tip radius and spring constant induced by the graphene coating, the observations are clear: the effect of the graphene coating on these two parameters is much lower than the intrinsic variability provided by the manufacturer. First, the typical thickness of the graphene sheets used to coat the tips in the liquid-phase exfoliated method (which is the most promising) varies from one to five layers, which equals 0.46 nm–2.3 nm. Second, the mass of the graphene nanosheets used in the liquid-phase exfoliated method is negligible, which does not modify the spring constant of the cantilever [28]. The effect of the graphene coating may be much more relevant if its thickness and area is increased (for example, using the graphite-like film deposition method described in Section 2.4).

Regarding the variability induced by the water meniscus, the use of graphene coating is beneficial because it reduces (if not completely removes) the presence of water at the junction. Therefore, this should result in even lower variability compared to its graphene-free counterparts. Wen et al. [23] statistically collected several groups of I–V curves in octanemonothiol-based molecular junctions using both graphene-free and graphene coated Au-varnished nanoprobes (fabricated via direct CVD-growth on the surface of the Au varnish), leading to two types of molecular junctions, Au/octanemonothiol/Au (Figure 12a) and Au/octanemonothiol/graphene/Au (Figure 12b). In total, more than 1000 curves are collected using 18 tips, and each line in the plots represents the average of 100 I–V curves. Comparing them, larger tip-to-tip variance in current up to 3 orders of magnitude has been detected in the molecular junctions without graphene. Additionally, a relatively small variation of less than 1 order of magnitude has been detected using graphene coated nanoprobes as molecular junction.

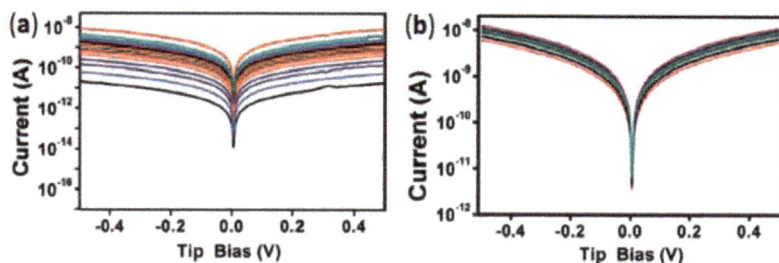

Figure 12. Variability and stability test. (a) Average I-V data for Au/octanemonothiol/Au junctions with 18 different Au tips; (b) Average I-V data for Au/octanemonothiol/graphene/Au junctions with 18 different graphene tips. Reproduced with permission from [23], copyright Wiley-VCH (2012).

Statistical analyses of the experimental data using nanoprobes coated with graphene via liquid-phase exfoliation method have been also carried out [28,52]. More than 100 I–V curves collected on a native oxide free n-type Si substrate have been collected (see Figure 13a,b) for each type of tip. In order to avoid big differences related to the presence of water between the tip and the sample, only backward curves have been considered (see also explanations related to Figure 10). Despite using only backward curves the results still show that both the onset potential and the variability is smaller when using graphene coated nanoprobes. Finally, groups of I–V curves have been measured with the same types of tips on a metallic substrate. The plots show saturation of the currents at ±5 nA, with a linear shape going from −5 nA to +5 nA. The saturation voltage (voltage at which the I–V curves reach 5 nA) has been evaluated statistically. Figure 13c shows that the variability of the data is not only smaller when using graphene coated AFM tips, but also the variability when comparing different devices is much smaller. Hui et al. [28] advise that the smaller onset voltage of the graphene coated probes may be affected by the different work functions between Pt and graphene tip electrodes. Indeed, the work function of the probes varies from ~6.1 eV (Pt) to ~4.8 eV (Pt/graphene) [53], as well as the absence of water perturbations, together contributing to the smaller onset voltage and the variability of I–V curves.

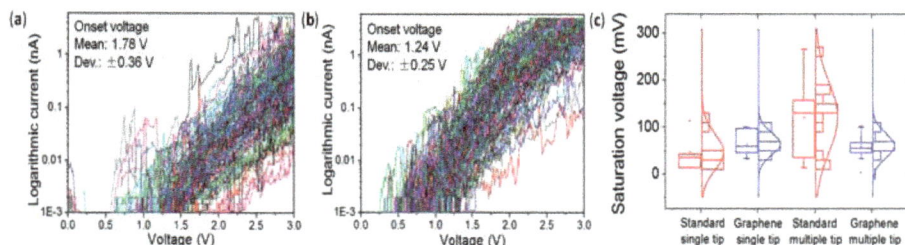

Figure 13. 100 backward I–V curves collected with both standard (a) and graphene coated nanoprobes; (b) the onset voltage and the deviation of the graphene nanoprobes is much smaller. Reproduced with permission from [28], copyright The Royal Society of Chemistry (2016); (c) Variance statistics analysis for single and multiple tip measurements at the saturation voltage (I = −5 nA) for fresh probes (in red) or graphene coated (in blue). Reproduced with permission from [52], copyright Elsevier (2016).

4.4. High Stability vs. High Currents and Mechanical Strains: Enhanced Lifetime

This is probably one of the properties that can bring most benefit to the users of graphene coated nanoprobes. Graphene coated nanoprobes fabricated by various methods have been demonstrated to have superior stability when performing different types of experiments (see Figure 14). Lanza et al. [24] performed an experiment that consisted of collecting several I-V curves at different (randomly selected) locations of a 3 nm thick HfO_2 stack (in order to monitor the tunneling currents and dielectric breakdown) using standard and graphene coated AFM tips (fabricated via CVD graphene transfer). The main characteristic was that a source meter was connected directly to the tip of the CAFM, so that high voltages (>10 V) and high currents (up to mA) could be registered [54]. While the graphene coated AFM tip is able to distinguish the dielectric breakdown at several different locations (Figure 14d), the standard tip cannot withstand the high currents (the current is 10^{-4} A, therefore, the current density is 10^8 A/cm^2 because the typical effective area at the tip-sample junction for this sample is 100 nm^2), and shows obvious current reduction with total conductivity degradation after 5 I–V curves.

In a different experiment by Hui et al. [28], the durability of the graphene-free and graphene coated tips (using a liquid-phase exfoliation method) was analyzed via current maps collected on highly conductive samples (graphene/Cu foils) under a constant bias of 1 V and a deflection setpoint of 4 V. As shown in Figure 14b,e, while the standard CAFM tip lost its conductivity in just 13 scans (Figure 14b), the graphene-coated tip kept good conductivity even after 92 scans (Figure 14e). In the first experiment (Figure 14a,d) the tip was held static at one position on the sample; therefore, the lateral friction was minimal and the wearing was related to the high currents. On the contrary (Figure 14b,e), the most harmful thing for the tips was the friction, as the currents did not increase above the nanometer scale. In any case, graphene coated AFM tips showed good stability vs. these harmful tests. In Figure 14c, an AFM tip degraded in a static experiment driving large currents can be observed via SEM; the metallic varnishes that provide high conductivity melted. In Figure 14f an SEM image of a graphene coated tip after a sequence of current maps can be observed. No degradation is detected, other than small particles attached to the tip during the scans.

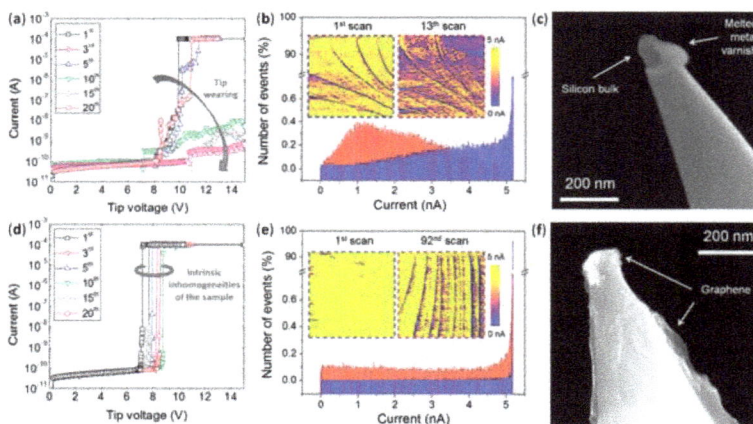

Figure 14. Reliability and durability test. Current vs. Voltage (I–V) curves performed on different positions of HfO_2/Si stack by (**a**) uncoated AFM tip and (**d**) graphene coated tips. Reproduced with permission from [24], copyright Wiley-VCH (2013). Current spectra and the insets in (**b**) are the 1st (blue) and 13th (red) CAFM current maps collected on graphene/Cu foil with a standard metal-varnished nanoprobe. Current spectra and the insets in (**e**) are the 1st (blue) and 92th (red) CAFM current maps collected with a graphene coated nanoprobe. SEM images of AFM tips without (**c**) and with graphene coating (**f**) after testing. Reproduced with permission from [28], copyright The Royal Society of Chemistry (2015).

Apart from all the above benefits, more recently graphene coated colloidal probes, i.e., SiO_2 microspheres, were developed to achieve an ultra-low and robust friction coefficient of 0.003 under high contact pressure, in which the super-lubricity of graphite and graphene plays a significant role [55]. Moreover, other properties of graphene, such as, flexibility, transparency, and low toxicity, enable the graphene coated probes to be competitive in the development of flexible and transparency microelectromechanical systems, as well as medical diagnosis.

5. Conclusions

In conclusion, different approaches using CVD growth, CVD transfer, sputtering and liquid phase exfoliated graphene have been used to coat nanoprobes with graphene. Direct CVD growth is highly questionable because the high temperatures used can damage the tip, and CVD transfer is too expensive due to the involvement of human labor. Direct sputtering shows good performance but the thermal budget is also high. So far, the liquid phase exfoliated approach has proved to be the cheapest and most industry-compatible method. All graphene coated nanoprobes show lifetimes much longer than their uncoated counterparts when used in different experiments. Specifically, they possess high wear resistance, stability and mechanical strength, as well as lower variability, which enables the application of the graphene-coated nanoprobes in the fields of electronics, mechanics and biology in the near future. In order to truly realize the manufacturing of graphene coated probes in industry, systematic techniques need to be further undertaken and optimized for mass production.

Acknowledgments: This work has been supported by the Young 1000 Global Talent Recruitment Program of the Ministry of Education of China, the National Natural Science Foundation of China (grants no. 61502326, 41550110223, 11661131002), the Jiangsu Government (grant no. BK20150343), the Ministry of Finance of China (grant no. SX21400213) and the Young 973 National Program of the Chinese Ministry of Science and Technology (grant no. 2015CB932700). The Collaborative Innovation Center of Suzhou Nano Science & Technology, the Jiangsu Key Laboratory for Carbon-Based Functional Materials & Devices, the Priority Academic Program Development of Jiangsu Higher Education Institutions, the 111 Project from the State Administration of Foreign Experts Affairs, and the Opening Project of Key Laboratory of Microelectronic Devices & Integrated Technology (Institute of Microelectronics, Chinese Academy of Sciences) are also acknowledged.

Conflicts of Interest: The authors declare no conflict of interest.

References

1. Ji, Y.; Pan, C.; Zhang, M.; Long, S.; Lian, X.; Miao, F.; Hui, F.; Shi, Y.; Larcher, L.; Wu, E.; et al. Boron nitride as two dimensional dielectric: reliability and dielectric breakdown. *Appl. Phys. Lett.* **2016**, *108*, 012905. [CrossRef]

2. Song, X.; Hui, F.; Gilmore, K.; Wang, B.; Jing, G.; Fan, Z.; Grustan-Gutierrez, E.; Shi, Y.; Lombardi, L.; Hodge, S.A.; et al. Enhanced piezoelectric effect at the edges of stepped molybdenum disulfide. *Nanoscale* **2017**, *9*, 6237–6245. [CrossRef] [PubMed]

3. Paxton, W.F.; Spruell, J.M.; Stoddart, J.F. Heterogeneous catalysis of a copper-coated atomic force microscopy tip for direct-write click chemistry. *J. Am. Chem. Soc.* **2009**, *131*, 6692–6694. [CrossRef] [PubMed]

4. Park, K.D.; Raschke, M.B.; Jang, M.J.; Kim, J.H.; O, B.H.; Park, S.G.; Lee, E.H.; Lee, S.G. Near-field imaging of cell membranes in liquid enabled by active scanning probe mechanical resonance control. *J. Phys. Chem. C.* **2016**, *120*, 21138–21144. [CrossRef]

5. Pan, N.; Rao, W.; Standke, S.J.; Yang, Z. Using dicationic ion-pairing compounds to enhance the single cell mass spectrometry analysis using the single-probe: A microscale sampling and ionization device. *Anal. Chem.* **2016**, *88*, 6812–6819. [CrossRef] [PubMed]

6. Bining, G.; Quate, C.F.; Gerber, C. Atomic Force Microscope. *Phys. Rev. Lett.* **1986**, *56*, 930–933. [CrossRef] [PubMed]

7. Kawahara, K.; Shirasawa, T.; Lin, C.L.; Nagao, R.; Tsukahara, N.; Takahashi, T.; Arafune, R.; Kawai, M.; Takagi, N. Atomic structure of "multilayer silicene" grown on Ag(111): Dynamical low energy electron diffraction analysis. *Surf. Sci.* **2016**, *651*, 70–75. [CrossRef]

8. Hafner, J.H.; Cheung, C.L.; Woolley, A.T.; Lieber, C.M. Structural and functional imaging with carbon nanotube AFM probes. *Prog. Biophys. Mol. Biol.* **2001**, *77*, 73–110. [CrossRef]

9. Yapici, M.K.; Lee, H.; Zou, J.; Liang, H. Gold-coated scanning probes for direct "write" of sub-micron metallic structures. *Micro Nano Lett.* **2008**, *3*, 90–94. [CrossRef]

10. Boggild, P.; Hansen, T.M.; Kuhn, O.; Grey, F. Scanning nanoscale multiprobes for conductivity measurements. *Rev. Sci. Instrum.* **2000**, *7*, 2781–2783. [CrossRef]

11. Lanza, M. *Conductive Atomic Force Microscopy: Applications in Nanomaterials*; Wiley-VCH Berlin: Berlin, Germany, 2017; ISBN 978-3-527-34091-0.

12. Hoefflinger, B. *The International Technology Roadmap for Semiconductors (ITRS)*, 2001 ed.; Springer: Berlin, Germany, 2001.

13. Nourbakhsh, A.; Zubair, A.; Sajjad, R.N.; Amir, T.K.G.; Chen, W.; Fang, S.; Ling, X.; Kong, J.; Dresselhaus, M.S.; Kaxiras, E.; et al. MoS2 field-effect transistor with sub-10 nm channel length. *Nano Lett.* **2016**, *16*, 7798–7806. [CrossRef] [PubMed]

14. Xu, K.; Chen, D.; Yang, F.; Wang, Z.; Yin, L.; Wang, F.; Cheng, R.; Liu, K.; Xiong, J.; Liu, Q.; et al. Sub-10 nm nanopattern architecture for 2D material field-effect transistors. *Nano Lett.* **2017**, *17*, 1065–1070. [CrossRef] [PubMed]

15. Dai, H.J.; Hafner, J.H.; Rinzler, A.G.; Colbert, D.T.; Smalley, R.E. Nanotubes as nanoprobes in scanning probe microscopy. *Nature* **1996**, *384*, 147–150. [CrossRef]

16. Xu, J.; Shingaya, Y.; Zhao, Y.; Nakayama, T. In situ, controlled and reproducible attachment of carbon nanotubes onto conductive AFM tip. *Appl. Surf. Sci.* **2015**, *335*, 11–16. [CrossRef]

17. Tay, A.B.H.; Thong, J.T.L. Fabrication of super-sharp nanowire atomic force microscope probes using a field emission induced growth technique. *Rev. Sci. Instrum.* **2004**, *75*, 3248–3255. [CrossRef]

18. Bakhti, S.; Destouches, N.; Hubert, C.; Reynaud, S.; Vocanson, F.; Ondarcuhu, T.; Epicier, T. Growth of single gold nanofilaments at the apex of conductive atomic force microscope tips. *Nanoscale* **2016**, *8*, 7496–7500. [CrossRef] [PubMed]

19. Neto, A.H.C.; Guinea, F.; Peres, N.M.R.; Novoselov, K.S.; Geim, A.K. The electronic properties of graphene. *Rev. Mod. Phys.* **2009**, *81*, 109. [CrossRef]

20. Chen, S.; Brown, L.; Levendorf, M.; Cai, W.; Ju, S.Y.; Edgeworth, J.; Li, X.; Magnuson, C.W.; Velamakanni, A.; Piner, R.D.; et al. Oxidation resistance of graphene-coated Cu and Cu/Ni alloy. *ACS Nano* **2011**, *5*, 1321. [CrossRef] [PubMed]

21. Lee, C.; Wei, X.; Kysar, J.W.; Hone, J. Measurement of the elastic properties and intrinsic strength of monolayer graphene. *Science* **2008**, *321*, 385. [CrossRef] [PubMed]

22. Balandin, A.; Ghosh, S.; Bao, W.; Calizo, I.; Teweldebrhan, D.; Miao, F.; Lau, C.N. Superior thermal conductivity of single-layer graphene. *Nano Lett.* **2008**, *8*, 902. [CrossRef] [PubMed]

23. Wen, Y.; Chen, J.Y.; Guo, Y.; Wu, B.; Yu, G.; Liu, Y. Multilayer graphene-coated atomic force microscopy tips for molecular junctions. *Adv. Mater.* **2012**, *24*, 3482–3485. [CrossRef] [PubMed]

24. Lanza, M.; Bayerl, A.; Gao, T.; Porti, M.; Nafria, M.; Jing, G.Y.; Zhang, Y.F.; Liu, Z.F.; Duan, H.L. Graphene-coated atomic force microscope tips for reliable nanoscale electrical characterization. *Adv. Mater.* **2013**, *25*, 1440–1444. [CrossRef] [PubMed]

25. Shim, W.; Brown, K.A.; Zhou, X.Z.; Rasin, B.; Liao, X.; Mirkin, C.A. Multifunctional cantilever-free scanning probe arrays coated with multilayer graphene. *Proc. Natl. Acad. Sci. USA* **2012**, *109*, 18311–18317. [CrossRef] [PubMed]

26. Martin-Olmos, C.; Rasool, H.I.; Weiller, B.H.; Gimzewski, J.K. Graphene MEMS: AFM probe performance improvement. *ACS Nano* **2013**, *7*, 4164–4170. [CrossRef] [PubMed]

27. Pacios, M.; Hosseini, P.; Fan, Y.; He, Z.; Krause, O.; Hutchison, J.; Warner, J.H.; Bhasskaran, H. Direct manufacturing of ultrathin graphite on three-dimensional nanoscale features. *Sci. Rep.* **2016**, *6*, 22700. [CrossRef] [PubMed]

28. Hui, F.; Vajha, P.; Shi, Y.; Ji, Y.; Duan, H.; Padovani, A.; Larcher, L.; Li, X.R.; Xu, J.J.; Lanza, M. Moving graphene devices from lab to market: Advanced graphene-coated nanoprobes. *Nanoscale* **2016**, *8*, 8466–8473. [CrossRef] [PubMed]

29. Lanza, M.; Gao, T.; Yin, Z.; Zhang, Y.; Liu, Z.; Tong, Y.; Shen, Z.; Duan, H. Nanogap based graphene coated AFM tips with high spatial resolution, conductivity and durability. *Nanoscale* **2015**, *5*, 10816–10823. [CrossRef] [PubMed]

30. Kim, K.S.; Zhao, Y.; Jang, H.; Lee, S.Y.; Kime, J.M.; Kim, K.S.; Ahn, J.H.; Kim, P.; Choi, J.Y.; Hong, B.H. Large-scale pattern growth of graphene films for stretchable transparent electrodes. *Nature* **2009**, *457*, 706. [CrossRef] [PubMed]

31. Xue, Y.Z.; Wu, B.; Guo, Y.L.; Huang, L.P.; Jiang, L.; Chen, J.Y.; Geng, D.C.; Liu, Y.Q.; Hu, W.P.; Yu, G. Synthesis of large-area, few-layer graphene on iron foil by chemical vapor deposition. *Nano Res.* **2011**, *4*, 1208. [CrossRef]

32. Reina, A.; Thiele, S.; Jia, X.; Bhaviripudi, S.; Dresselhaus, M.S.; Schaefer, J.A.; Kong, J. Growth of large-area single- and bi-layer graphene by controlled carbon precipitation on polycrystalline Ni surfaces. *Nano Res.* **2009**, *2*, 509–516. [CrossRef]

33. Nam, J.; Kim, D.C.; Yun, H.; Shin, D.H.; Nam, S.; Lee, W.K.; Hwang, J.Y.; Lee, S.W.; Weman, H.; Kim, K.S. Chemical vapor deposition of graphene on platinum: Growth and substrate interaction. *Carbon* **2017**, *111*, 733–740. [CrossRef]

34. Li, X.; Cai, W.; Colombo, L.; Ruoff, R.S. Evolution of graphene growth on Ni and Cu by carbon isotope labeling. *Nano Lett.* **2009**, *9*, 4268–4272. [CrossRef] [PubMed]

35. Uchida, Y.; Iwaizako, T.; Mizuno, S.; Tsuji, M.; Ago, H. Epitaxial chemical vapor deposition growth of monolayer hexagonal boron nitride on Cu(111)/sapphire substrate. *Phys. Chem. Chem. Phys.* **2017**, *19*, 8230–8235. [CrossRef] [PubMed]

36. Castro-Neto, A.H. Plenary talk in Graphchina 2016, Qingdao (China). 22–24 September.

37. Yang, L.; Hu, J.H.; Qin, J. The van der waals force between arbitrary-shaped particle and a plane surface connected by a liquid bridge in humidity environment. *Granul. Matter* **2014**, *16*, 903–909. [CrossRef]

38. Russell, P. AFM Probe Manufacturing. AFM TIP Webinar. Available online: https://www. agilent.com/cs/library/slidepresentation/Public/AFM%20Probe%20ManufacturingNanoworld_tip_technologyPRussell07.pdf (accessed on 10 November 2008).

39. Al-Halhouji, A.T.; Kampen, I.; Krah, T.; Büttgenbach, S. Nanoindentation Testing of SU-8 Photoresist Mechanical Properties. *Microelectron. Eng.* **2008**, *85*, 942–944. [CrossRef]

40. Bhushan, B.; Li, X.D. Micromechanical and Tribological Characterization of Doped Single-Crystal Silicon and Polysilicon Films for Microelectromechanical Systems Devices. *J. Mater. Res.* **1997**, *12*, 54–63. [CrossRef]

41. Cappella, B.; Dietler, G. Force-distance curves by atomic force microscopy. *Surf. Sci. Rep.* **1999**, *34*, 1–104. [CrossRef]

42. Weeks, B.L.; Vaughn, M.W. Direct imaging of meniscus formation in atomic force microscopy using environmental scanning electron microscopy. *Langmuir* **2015**, *21*, 8096–8098. [CrossRef] [PubMed]

43. Stukalov, O.; Murray, C.A.; Jacina, A.; Dutcher, J.R. Relative humidity control for atomic force microscopes. *Rev. Sci. Instrum.* **2006**, *77*, 033704. [CrossRef]

44. Yang, G.; Vesenka, J.P.; Bustamante, C.J. Effects of tip-sample forces and humidity on the imaging of DNA with a scanning force microscope. *Scanning* **1996**, *18*, 344–350. [CrossRef]

45. Ebenstein, Y.; Nahum, E.; Banin, U. Tapping mode atomic force microscopy for nanoparticle sizing: Tip-sample interaction effects. *Nano Lett.* **2002**, *2*, 945–950. [CrossRef]

46. Xiao, X.; Qian, L. Investigation of humidity-dependent capillary force. *Langmuir* **2000**, *16*, 8153–8158. [CrossRef]

47. Li, D.; Muller, M.B.; Gilje, S.; Kaner, R.B.; Wallace, G.G. Processable aqueous dispersions of graphene nanosheets. *Nat. Nanotech.* **2008**, *3*, 101–105. [CrossRef] [PubMed]

48. Puglisi, F.M.; Larcher, L.; Pan, C.; Xiao, N.; Shi, Y.; Hui, F.; Lanza, M. 2D h-BN based RRAM devices. Proceedings of 2016 IEEE International Electron Devices Meeting (IEDM), San Francisco, CA, USA, 3–7 December 2016.

49. Conductive Diamond Coated Tip-Force Modulation Mode-Reflex Coating. Available online: http://www.nanosensors.com/Conductive-Diamond-Coated-Tip-Force-Modulation-Mode-Reflex-Coating-afm-tip-CDT-FMR (accessed on 2 August 2017).

50. Lanza, M.; Porti, M.; Nafría, M.; Aymerich, X.; Wittaker, E.; Hamilton, B. Electrical resolution during Conductive AFM measurements under different environmental conditions and contact forces. *Rev. Sci. Instrum.* **2010**, *81*, 106110. [CrossRef] [PubMed]

51. Lanza, M.; Porti, M.; Nafría, M.; Aymerich, X.; Wittaker, E.; Hamilton, B. UHV CAFM characterization of high-k dielectrics: effect of the technique resolution on the pre- and post-breakdown electrical measurements. *Microelectron. Reliab.* **2010**, *50*, 1312–1315. [CrossRef]

52. Hui, F.; Vajha, P.; Ji, Y.; Pan, C.; Grustan-Gutierrez, E.; Duan, H.; He, P.; Ding, G.; Shi, Y.; Lanza, M. Variability of graphene devices fabricated using graphene inks: atomic force microscope tips. *Surf. Coat. Tech.* **2017**, *320*, 391–395. [CrossRef]

53. Khomyakov, P.A.; Giovannetti, G.; Rusu, P.C.; Brocks, G.; Brink, H.; Kelly, P.J. First-principles study of the interaction and charge transfer between graphene and metals. *Phys. Rev.* **2009**, *79*, 195425. [CrossRef]

54. Blasco, X.; Nafria, M.; Aymerich, X. Enhanced electrical performance for conductive atomic force microscopy. *Rev. Sci. Instrum.* **2005**, *76*, 016105. [CrossRef]

55. Liu, S.W.; Wang, H.P.; Xu, Q.; Ma, T.B.; Yu, G.; Zhang, C.; Geng, D.; Yu, Z.; Zhang, S.; Wang, W.; et al. Robust microscale superlubricity under high contact pressure enabled by graphene-coated microsphere. *Nat. Commun.* **2017**, *8*, 14029. [CrossRef] [PubMed]

crystals

MDPI

Review

A Review on Metal Nanoparticles Nucleation and Growth on/in Graphene

Francesco Ruffino [1],* and Filippo Giannazzo [2] (ORCID)

[1]   Dipartimento di Fisica ed Astronomia-Università di Catania and MATIS IMM-CNR, via S. Sofia 64, 95123 Catania, Italy
[2]   Consiglio Nazionale delle Ricerche—Institute for Microelectronics and Microsystems (CNR-IMM), Strada VIII, 5 I-95121 Catania, Italy; filippo.giannazzo@imm.cnr.it
*   Correspondence: francesco.ruffino@ct.infn.it; Tel.: +39-09-5378-5466

Received: 8 June 2017; Accepted: 11 July 2017; Published: 13 July 2017

Abstract: In this review, the fundamental aspects (with particular focus to the microscopic thermodynamics and kinetics mechanisms) concerning the fabrication of graphene-metal nanoparticles composites are discussed. In particular, the attention is devoted to those fabrication methods involving vapor-phase depositions of metals on/in graphene-based materials. Graphene-metal nanoparticles composites are, nowadays, widely investigated both from a basic scientific and from several technological point of views. In fact, these graphene-based systems present wide-range tunable and functional electrical, optical, and mechanical properties which can be exploited for the design and production of innovative and high-efficiency devices. This research field is, so, a wide and multidisciplinary section in the nanotechnology field of study. So, this review aims to discuss, in a synthetic and systematic framework, the basic microscopic mechanisms and processes involved in metal nanoparticles formation on graphene sheets by physical vapor deposition methods and on their evolution by post-deposition processes. This is made by putting at the basis of the discussions some specific examples to draw insights on the common general physical and chemical properties and parameters involved in the synergistic interaction processes between graphene and metals.

Keywords: graphene; metal nanoparticles; nanocomposites; physical vapor deposition; kinetics

1. Introduction: Metal-Based Graphene Nanocomposites in the Nanotechnology Revolution

Free-standing graphene, also known as one layer graphite, was firstly obtained in 2004 [1,2]. Then, the scientific and technological research has seen an exceptional continuing grow of the interest in graphene and graphene-based materials since the properties of these materials can drastically revolutionize the modern-day technology. Graphene, in fact, presents several disruptive properties as compared to the standard semiconducting materials which were, until now, at the basis of the technological development. As examples, graphene is characterized by extraordinary carrier mobility (200,000 cm^2 V^{-1} s^{-1} [3]), thermal conductivity (~5000 Wm^{-1} K^{-1} [4–6]), white light transmittance (~97.3% [7]), and specific surface area (~2630 m^2 g^{-1} [8]). These properties make graphene the key material in the current nanotechnology revolution and the ideal material for the fabrication of functional devices finding applications in electronics, energy generation and storage (batteries, fuel cells and solar cells), plasmonics, sensors, supercapacitors and other nano-devices [1,2,9–16]. In view of such applications, the synergistic interaction of graphene with other nano-sized materials can offer the pathway to produce novel graphene-based composites with artificial and tunable properties arising from the exotic combination of the properties of the single materials forming the composites [17–30]. Recently, various materials (polymers [17–21], carbon nanotubes [17–19,22], semiconducting materials [17–19,23], insulating materials [17–19,24–30], etc.) have been used to produce graphene-based composites. As specific examples: (1) graphene has drawn a great attention as a filler material in polymer nanocomposites due to its very low resistivity, thermal stability, and superior mechanical strength. In addition, it presents very high dispersibility in many polymers even it can show synergistic properties with polymer matrices [17]. The resulting flexible nanocomposites can show enhanced electrical, thermal and mechanical properties with respect to the pure polymeric matrix and these properties can be successful exploited for renewable energy sources applications (supercapacitors, polymer based solar cells, etc.); (2) graphene combined with TiO_2, ZnO, Fe_3O_4, MnO_2, SiO_2 micro- and nano-structures can be used in photocatalysis, photovoltaics, optoelectronics, supercapacitor, Li ion battery and magnetic drug carrier applications [17–19]. Nowadays, the range of graphene-based nanocomposites is extremely wide: a multitude of organic and inorganic materials are, currently, used in combination to graphene and the physical and chemical properties of the resulting composites are studied in view of cutting-edge applications. In particular, the framework regarding graphene-metal nanoparticles (NPs) composites acquired a great relevance [17–19,31–43]. Systems fabricated by anchoring Au, Ag, Pd, Pt, Ni, Cu, and more other NPs on graphene sheets are, today, largely studied due to the broad range of application exploiting the specific optical, electrical, mechanical, magnetic properties arising from the microscopic interactions between the NPs and the graphene. Depending on the nature of the metal NPs, the graphene–metal NPs composites find applications in areas such as Surface Enhanced Raman Scattering (SERS), nanoelectronics, photovoltaics, catalysis, electrochemical sensing, hydrogen storage, etc. [17–19,31–71].

In general, composites fabricated combining graphene and metal NPs attract great attention since they result versatile hybrid materials presenting unconventional properties arising from the atomic-scale mixing of the properties of graphene and NPs. In fact, in addition to graphene, metal NPs are another main character in the nanotechnology field of study. Due to electron confinement and surface effects metal NPs present size-dependent electrical, optical, mechanical properties different from the corresponding bulk counterparts and these properties are routinely exploited in plasmonic, sensing, electrical, catalytic applications [72–74]. The notable aspect is that these size-dependent properties of metal NPs can be coupled to the properties of graphene to obtain a composite artificial material presenting un-precedent characteristics and performances arising from the (controlled) mixing of the properties of the component elements. For example, nanocomposite materials obtained by metal NPs and thin metal nano-grained films deposited on graphene sheets were successful employed in transistors, optical and electrochemical sensors, solar cells, batteries [17–19,31–43].

A deep understanding and control of the electrical properties of metal/graphene interface is crucial for future applications of this material in electronics and optoelectronics. Current injection at the junction between a three dimensional metal contact and two-dimensional graphene with very different densities of states is an interesting physical problem. Furthermore, the specific contact resistance at the metal/graphene junction [75–78] currently represents one of the main limiting factors for the performances of lateral and vertical graphene transistors both on rigid and flexible substrates [79–85]. Several solutions have been investigated to minimize this resistive contribution [86–88].

In this context, the key point of study is the interaction occurring at the graphene-metal interface [89–113]. In fact, for example, the electronic properties of graphene are dramatically influenced by interaction with metallic atoms [101–104]. So, this interaction crucially affects the electronic transport properties of graphene based transistors [105–113]. In this context, the detailed description and comprehension of the metal-graphene interactions is the key step toward the control of processes and properties of the graphene-metal NPs composite materials and, so, to develop effective applications [59]. The graphene–metal NPs composites can be prepared by several methods such as chemical reduction, photochemical synthesis, microwave assisted synthesis, electroless metallization, and physical vapor deposition processes [17–19]. In particular, this paper reviews the basic aspects of physical vapor based synthesis methods of graphene-metal NPs composites. Physical vapor deposition processes, such as thermal evaporation or sputtering, are traditional methods to produce metal NPs and films on substrates [114–122]. These methods are acquiring large interest for the production of metal

NPs-graphene composites with specific physico-chemical properties exploitable in specific applications (SERS, catalysis, nanoelectronics) [54–67]. In fact, they are simple, versatile and high-throughput and the general microscopic thermodynamics and kinetics mechanisms involved in the nucleation and growth processes of atoms on surfaces are well-known. In this sense, physical vapor deposition processes are a convenient way of depositing a range of metallic materials onto graphene sheets. Atoms deposited on a substrate undergo competing kinetic and thermodynamic processes which establish the final NPs or film structure [114–116]. The adsorbed atoms (or adatoms) transport process involves random hopping phenomena on the surface dictated by the surface diffusivity D (which determines the diffusion length) obeying an Arrhenius law [114–116]. So, these adatoms, randomly diffuse across the substrate surface until they can join together forming a nucleus or they can stop at some particular surface defect or they can re-evaporate from the surface. This situation is largely influenced by the adatom-substrate interaction and process parameters (substrate temperature, arrival flux, etc.). Materials deposited by physical vapor processes can adopt a variety of morphologies which are tunable by the control of the deposition process parameters. In addition, post-deposition processes can allow a further control of the NPs or films morphology ad structure by inducing further specific thermodynamics and kinetics driven self-organization phenomena. It is evident, so, the key importance assumed by the understanding of the growth kinetics of the metal NPs and films on graphene sheets to infer how the interaction with the graphene and the process parameters influences the metal film morphology and, as a consequence, the overall nanocomposite properties.

On the basis of these considerations, the review is organized as follows:

The first part (Section 2) is devoted to adsorption and diffusion of metal atoms on/in graphene and on the influence of these parameters on the metal NPs nucleation and growth processes. This section describes the fundamental microscopic thermodynamics and kinetics processes occurring during vapor-phase depositions (i.e., evaporation or sputtering) of metals on graphene sheets and resulting in the formation of metal NPs or films; particular attention is devoted to theoretical (Section 2.1) and experimental (Section 2.2) studies focused on the interaction, after adsorption, of metals atoms with graphene and on how this interaction influences the adatoms diffusivity and the final metal NPs structure and morphology. Critical discussions on the specific involved microscopic phenomena (adsorption, diffusion, nucleation, ripening, coalescence, etc.) and on the corresponding parameters (surface energies, diffusivity, activation energy, etc.) are presented.

The second part (Section 3) is devoted to the review of data concerning the production of metal NPs arrays on graphene exploiting the dewetting process of deposited metal films. The dewetting process of a metal film deposited on a substrate is the clustering phenomenon of the continuous metal layer driven by the lowering of the total surface free energy. Nowadays, the controlled dewetting of thin metal films on functional substrates is widely used as a low-cost, versatile, high-throughput strategy to produce array of metal NPs on surfaces for several applications such as in plasmonic and nanoelectronics [123–130]. Recently, this strategy was applied to thin metal films (such as Au, Ag) for the production of arrays of metal NPs on graphene which were, then, used, for example, in SERS applications [68–71]. We discuss the results of such a strategy pointing out the microscopic parameters involved in the dewetting process of metal films on graphene.

Section 4, shortly discuss some aspects related to the metal-graphene contacts to draw the general requirements for a metal contacts to be suitable to be efficiently used as an electrode to grapheme in nanoelectronics devices.

Finally, the last paragraph (Section 5) summarizes conclusions, open points and perspectives in the graphene-metal NPs composites field of study.

2. Adsorption and Diffusion of Metal Atoms on/in Graphene and Nanoparticles Nucleation and Growth

2.1. Adsorption and Diffusion of Metals Atoms on/in Graphene: Theoretical Results

2.1.1. General Considerations

Liu et al. [57,100] systematically studied metal adatoms adsorption on graphene by ab initio calculations, ranging from alkali metals, to sp-simple, transition, and noble metals. In these works, the main aim was the correlation between the adatom adsorption properties and the growth morphology of the metals on the graphene. The authors main finding lies in the fact that the metal growth morphology is determined by the E_a/E_c parameter (with E_a the adsorption energy of the metal on graphene and E_c the bulk metal cohesive energy) and by the ΔE parameter (i.e., the activation energy for the metal adatom diffusion on graphene). First of all, experimental data (as we will see in Section 2.2) show that different metals on graphene exhibit very different growth morphologies even if deposited in similar conditions and at similar coverage. For example, considering a single-layer graphene obtained by thermal annealing of SiC, a 0.8 ML deposition of Pb with the sample at 40 K, results in the formation of large crystalline Pb islands [100]; deposition of Fe, in the same conditions, results, instead, in continuous nucleation of large, medium, and small size islands [100]; further experimental data [65] concern deposition of metals on single-layer graphene grown on Ru(0001): Pt and Rh result in finely dispersed small clusters, Pd and Co in larger clusters at similar coverages. To complicate further the situation, for example, Gd atoms deposited on graphene/SiC at room temperature nucleate in two-dimensional islands of fractal morphology [100]. In general, therefore, even if the various metals follow a Volmer-Weber growth mode (three-dimensional growth without a wetting layer) on the graphene, a wide-range of morphology for metals nanostructures deposited on graphene are observed. Within this mess of data, Liu et al. [57,100] performed a systematic theoretical study to understand, quantitatively, the key parameters governing the metal clusters growth morphology during deposition. They start from the idea that the interaction of the metal atoms with free-standing graphene determines the specific adatoms diffusion mechanisms establishing, then, how the adatoms nucleate and growth. So, the authors, performed first-principles calculations based on the density functional theory to evaluate the interaction of the metals with graphene. Several results were inferred by these simulations which can be summarized as follows:

(a) The adsorption site of metal atom on graphene is the more energetically stable and it depends on the chemical nature of the atom. So, for Mg, Al, In, Mn, Fe, Co, Ni, Gd the adsorption site is the hexagonal center site in the graphene lattice, named the hollow site (H). The adsorption site for Cu, Pb, Au atoms is at the top of a carbon atom, named T site. The adsorption site for Ag, Cr, Pd, Pt atoms is at the middle of a carbon–carbon bond, named B site. The second column in Table 1 summarizes the results of Liu et al. [57,100] about the energetic stable sites in graphene for all the investigated atoms.

(b) The results for the adsorption energy E_a of the atoms adsorbed on graphene are plotted in Figure 1a and listed in the third column of Table 1. The value of the adsorption energy ranges from less than 1.0 kcal/mol to 45.0 kcal/mol depending on the chemical nature of the atom. This value is an indication of the strength of the interaction between the adsorbed atom and graphene. For example, the interaction of Mg and Ag atoms with graphene is very weak since the corresponding adsorption energies are in the 0.5–0.6 kcal/mol range. On the contrary, the binding of Pd and Pt atoms on graphene is much stronger since the corresponding adsorption energies are 26.5 and 39.3 kcal/mol, respectively. On the other hand, in general, the adsorption energy of I–IV metals on graphene is intermediate, in the 6–27 kcal/mol.

(c) The calculated values for diffusion barrier energies for several atoms on free-standing graphene are plotted in Figure 1b and listed in the fourth column of Table 1. In general, the following correlation between the adsorption energy and diffusion barrier energy exists: the diffusion barrier increases as a consequence of the increase of the adsorption energy (even if some exceptions are present as in the case of Ni and Pt).

(d) Liu et al. [100] calculated several other parameters related to the atoms-graphene interactions as summarized in the other columns of Table 1: E_a/E_c (metal adsorption energy on graphene to bulk metal cohesive energy and E_c-E_a.

In particular, the growth morphology of metals on graphene is connected to the E_a/E_c and ΔE parameters characterizing the metal atoms-graphene system. Figure 2 reports E_a/E_c for the analyzed atoms on graphene. So, the combination of ΔE (Figure 1b) and E_a/E_c (Figure 2) is claimed by Liu et al. [100] as the main reason establishing the growth morphology of the specific metal species on the free-standing graphene. From a general point of view, the occurring of the three-dimensional Volmer-Weber growth mode (i.e., growth of three-dimensional clusters typically almost spherical or semispherical directly on the substrate surface) is determined by the energetic condition $E_c > E_a$ (i.e., $(E_a/E_c) < 1$): in fact, in this condition the bonding between the deposited atoms is higher than the bonding to the graphene.

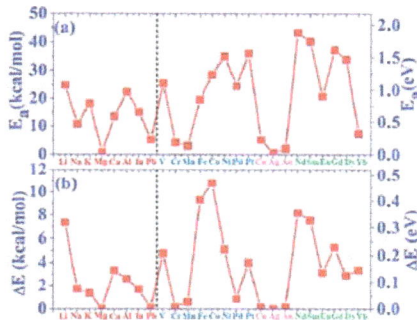

Figure 1. (**a**) Adsorption energy (E_a) and (**b**) diffusion barrier (ΔE) for several adatoms on graphene as calculated by Liu et al. using density functional theory. Reproduced from Reference [100] with permission from the Royal Society of Chemistry.

Table 1. E_a (adsorption energy of the metal atom on graphene, kcal/mol), ΔE (diffusion barrier of the metal adatom on graphene, kcal/mol), E_a/E_c (with E_c the bulk metal cohesive energy), and E_c-E_a (kcal/mol). Reproduced from Reference [100] with permission from the Royal Society of Chemistry.

Adatoms	Sites	E_a	ΔE	E_a/E_c	E_c-E_a
Li	H	24.77	7.33	0.659	12.82
Na	H	10.70	1.71	0.417	14.97
K	H	18.10	1.36	0.818	4.036
Mg	H	0.65	0.02	0.019	34.18
Ca	H	13.44	3.34	0.317	28.99
Al	H	22.41	2.58	0.287	55.70
In	H	15.15	1.68	0.261	42.90
Pb	T	5.28	0.09	0.113	41.47
V	H	25.44	4.77	0.208	96.86
Cr	B	4.34	0.14	0.046	90.22
Mn	H	3.04	0.60	0.045	64.30
Fe	H	19.65	9.32	0.199	79.08
Co	H	28.53	10.79	0.282	72.65
Ni	H	35.01	5.12	0.342	67.37
Pd	B	24.47	0.85	0.273	65.24
Pt	B	36.09	3.99	0.268	98.59
Cu	T	5.17	0.12	0.090	52.26
Ag	B	0.51	0.00	0.007	67.53
Au	T	2.08	0.14	0.024	85.79
Nd	H	43.31	8.16	0.552	35.15
Sm	H	40.15	7.52	0.814	9.20
Eu	H	20.85	3.14	0.486	22.05
Gd	H	37.17	5.28	0.389	58.39
Dy	H	33.94	2.88	0.484	36.19
Yb	H	7.40	3.37	0.201	29.43

However, ΔE establishes the adatoms hopping probability. So, it dictates the rate of the adatoms joining to the closest preformed metal cluster with respect to the rate of adatoms joining to other adatoms to form a new cluster. Therefore, ΔE establishes the surface density of the metal clusters on the graphene. For example, the small value of E_a/E_c for Fe on graphene establishes a standard three-dimensional Volmer-Weber growth mode for Fe clusters on graphene consistent with the experimental observations [100].

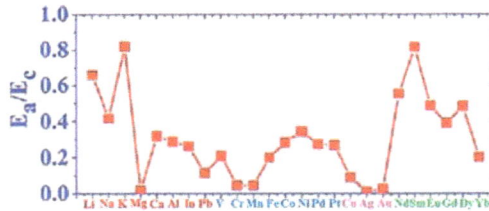

Figure 2. Ratio of adsorption energy to bulk cohesive energy for various materials calculated by Liu et al. Reproduced from Reference [100] with permission from the Royal Society of Chemistry.

In addition, the diffusion barrier ΔE for Fe on graphene is high so that a high Fe clusters density is produced with respect, for example, the clusters density for Pb deposited on graphene. In fact, ΔE for Pb is lower than for Fe. As a consequence, the Pb adatoms diffuse faster than the Fe adatoms resulting in larger clusters but with a lower surface density. On the other hand, Gd has (E_a/E_c) but higher than that of Fe or Pb and a diffusion barrier intermediate between that of Fe and Pb. This results in fractal-like morphology of the Gd islands on graphene. A further observation concerns, for example, Dy and Eu: despite similar values of E_a/E_c and ΔE, their growth morphologies are different. Dy forms small three-dimensional clusters while Eu forms flat top crystalline islands with well-defined facets. Therefore, Liu et al. [100] suggest that other factors affect the growth morphology in addition to E_a/E_c ratio and ΔE and identify the main factor in specific characteristics of the adatom-adatom interaction. For example, the repulsive interaction between Dy adatoms, arising from a large electric dipole moment, is larger than Eu adatoms resulting in a higher effective barrier for diffusion.

We observe that these results can be regarded as a general rough guide in understand the growth morphology of metals on graphene. However, these results neglect some effects which are, instead, observed by experimental analyses such as the difference in the growth morphology of deposited metals by changing the substrate supporting the graphene layer (highlighting, so, an effect of the adatoms interaction with the substrate supporting the graphene) or by changing the number of graphene layer. The theoretical results by Liu et al. [57,100] are, in fact, obtained for free-standing single layer graphene sheets. As we will see in the next sections, some theoretical and experimental works studied the effect of supporting substrate on the adatoms-graphene interaction and its impact on the metals growth morphology.

Besides these general considerations, in the following subsections we focus our attention on the model Au-graphene system since it is, surely, the main studied system from a technological point of view due to its exceptional performances in technological devices ranging from sensors and biosensors to transistors and solar cells.

2.1.2. Mobility and Clustering of Au on Graphene

Srivastava et al. [92] performed density functional calculations to investigate the bonding properties of Au_n ($n = 1$–5) clusters on perfect free-standing single-layer graphene. In synthesis, their results show that the Au_n clusters are bonded to graphene through an anchor atom and that the geometries of the clusters on graphene are similar to their free-standing counterparts. Figure 3 shows, in particular, the results for the stable geometry configurations of the Au_1, Au_2, Au_3, Au_4,

Au_5 on the graphene. According to these results: (a) the energetically stable site for the Au atom on graphene is atop to C atom (at an equilibrium distance of 2.82 Å), in agreement with the finding of Liu et al. [100]; (b) for $n > 1$, each of the Au_n cluster is bonded to the graphene by one Au atom which is closer to the graphene and the overall geometry of the cluster remembers its freestanding configuration. Concerning the Au_5 cluster two different stable configurations are found, i.e., the last two rows in Figure 3. These two configurations differ for taking into in account or not van der Waals interaction: the last configuration for the Au_5 cluster (named $Au_5(P)$) is obtained taking into in account the van der Waals interaction. The overall results of the calculations performed by Srivastava et al. are summarized in Table 2. This table reports, for each Au_n cluster: h_a which is the distance of the Au anchor atom of the cluster from the graphene plane; d_{ac} which is the distance of the Au anchor atom from the nearest-neighbor C atom of the graphene layer; the binding energies BE^1, BE^2, BE^3 of the Au_n clusters with the graphene, being these energies defined by $BE^1 = (E_{G+Au\_n} - E_G - nE_{Au})/n$, $BE^2 = E_{G+Aun} - E_{G+Au\_n-1} - E_{Au\_1}$, $BE^3 = E_{G+Au\_n} - E'_G - E_{Au\_n}$ with E_G the energy of the free-standing graphene, E'_G the energy of the graphene after adsorbing the Au_n cluster, E_{Au_n} the energy of the isolated Au_n cluster, E_{G+Aun} the energy of the system formed by the free-standing graphene and the isolated Au_n cluster, n the number of Au atoms in the cluster. With these definitions, BE^1 represents the cohesive energy of the cluster affected by the interaction with the graphene, BE^2 is the energy gained by the system in consequence of the addition of one more atom to the already existing cluster, BE^3 is the energy gained by the system resulting from the interaction of graphene and cluster.

In particular, analyzing the binding energies, the following conclusions can be drawn: the Au_n cluster is bonded to the graphene by the Au anchor atom and the bonding energy is dependent both on h_a and d_{ac}. Furthermore, the small values of BE^3 are the signature of the weak bond between the Au_n clusters and the graphene. This should favor high mobility of Au adatoms and Au_n cluster on perfect free-standing graphene. However, this mobility is, also, determined by the diffusion barrier. To analyze this point, we discuss the theoretical findings of Amft et al. [93]. They used density functional calculations to study the Au_n ($n = 1–4$) mobility on free-standing single layer graphene and their clustering properties. In particular, they studied the mobility of the Au atoms (Au_1) and the mobility of the Au_2, Au_3, and Au_4 clusters finding that the diffusion barrier of all studied clusters ranges from 4 to 36 meV. On the other hand, they found that the Au_n adsorption energy ranges from -0.1 to -0.59 eV. The diffusion barrier, therefore, results much lower than the adsorption energies. These results confirm the high mobility of the Au_{1-4} clusters on graphene along the C–C bonds. The Au_4 cluster shows a peculiarity with respect to the other clusters: it can present two distinct structure, i.e., the diamond-shaped Au_4^D cluster and the Y-shaped Au_4^Y cluster. From the vapor phase, these clusters are formed on the graphene surface by two distinct clustering processes: $Au_1 + Au_3 \rightarrow Au_4^D$, $2Au_2 \rightarrow Au_4^Y$. On the graphene surface they are characterized by different adsorption energies and diffusion barriers. In particular, the authors conclude that Au_4^Y has the highest adsorption energy on graphene, -0.59 eV, while the adsorption energy of the Au_4^D is -0.41 eV. Au_1 has the lowest adsorption energy, -0.1 eV. The adsorption energy of Au_2 is about -0.45 eV and of Au_3 is about -0.50 eV. To complete, their calculations about the activation energy for the Au_n clusters diffusion on graphene (i.e., the diffusion barrier) along the C–C bonds indicate the values of 15 meV for Au_1, 4 meV for Au_2, 36 meV for Au_3, 4 meV for Au_4^D, and 24 meV for Au_4^Y. Comparing the calculated values for the adsorption energy and for the diffusion barrier of the Au_{1-4} clusters, Amft et al. [93] conclude that the low diffusion barriers for the Au_n clusters (with respect to the adsorption energy) suggest a high mobility of the clusters on the graphene also at low temperatures. So, the adsorbed Au_n clusters can easily diffuse on the graphene and, upon merging, they form larger clusters to minimize the total energy of the system (since the Au–Au bonding energy is higher than the Au–C one).

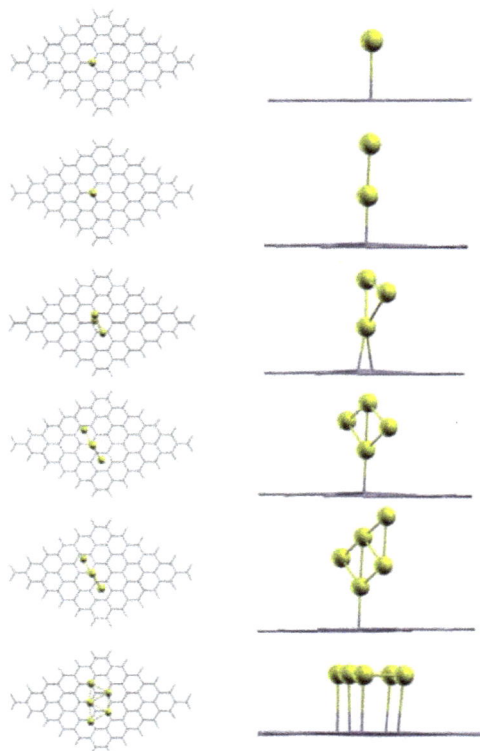

Figure 3. Stable geometries of Au clusters adsorbed on perfect graphene. Au_1–Au_5 and $Au_5(P)$ clusters are shown from top to bottom rows. Left and right columns show top and side views, respectively. Reproduced from Reference [92] with permission from the American Physical Society.

Table 2. Anchor atom's distance above graphene plane (h_a), distance from nearest-neighbor C atom (d_{ac}), binding energies (BE^1–BE^3) of Au_n clusters adsorbed on perfect graphene. Reproduced from Reference [92] with permission from the American Physical Society.

System	h_a (Å)	d_{ac} (Å)	BE^1 (eV)	BE^2 (eV)	BE^3 (eV)
Au1	2.89	2.82	−0.107	−0.107	−0.122
Au2	2.45	2.32	−1.373	−2.639	−0.526
Au3	2.43	2.33	−1.345	−1.288	−0.654
Au4	2.49	2.34	−1.608	−2.397	−0.515
Au5	2.57	2.45	−1.681	−1.975	−0.218

2.1.3. Adsorption and Diffusion of Au on Graphene/Ru(0001)

The theoretical results illustrated in the previous Sections 2.1.1 and 2.1.2 are derived for atoms and cluster on free-standing graphene. However, as we will see in Section 2.2, some experimental results pointed out some differences in the growth morphology of metals deposited on graphene by changing the substrate supporting the graphene. So, in the present section we review a theoretical analysis (as model system analyses) about diffusion and mobility of Au atoms on graphene taking into in account the effect of the substrate supporting the graphene sheet. These theoretical data, so, can be directly compared to the theoretical data for adsorption and diffusion of Au atoms on free-standing graphene as reported in the previous sections.

Semidey-Flecha et al. [99] used density functional theory calculations to investigate the adsorption and diffusion of Au adatom on the graphene moiré superstructure on Ru(0001). Their results can be synthesized as follows: (a) the FCC region on the graphene moiré is the most stable adsorption site for Au_1; (b) the diffusion barrier for Au_1 is determined to be 0.71 eV (much higher than the value of 15 meV evaluated by Amft et al. [93] for Au_1 on free-standing graphene).

The epitaxial growth of graphene on Ru(0001) is usually used to produce supported high-quality large area graphene sheets [65,99,131]. In this case, the graphene layer presents the moiré super-structure due to mismatch between the graphene and Ru(0001). In addition, from an experimental point of view, the study of metal atoms (Pd [65], Co [65], Au [65,132], Fe [133], Pt [134]) deposited on graphene/Ru(0001) is a very active field of study in view of catalytic applications. So, the theoretical study of metal atoms bonding and mobility on graphene/Ru(0001) is crucial in reach a control on the metal growth process. In particular, the theoretical analysis by Semidey-Flecha et al. [99] are focused on the diffusion properties of Au atoms on graphene/Ru(0001).

First of all, Figure 4 reports the graphene structures taken into considerations by the authors to run the simulations: (a) free-standing grapheme; (b) graphene on fcc Ru(0001); (c) graphene on hcp Ru(0001); (d) graphene on ridge Ru(0001). Each image reports, also, the indication of the notable sites.

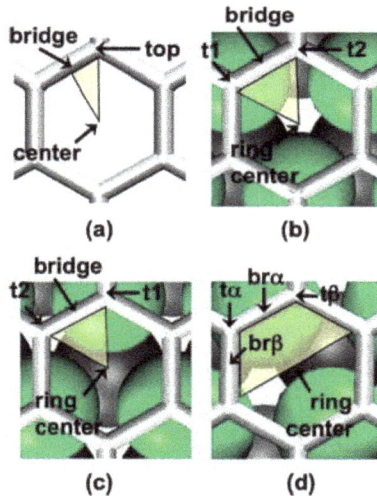

Figure 4. Structures of the (3 × 3) surfaces used for the simulations: (**a**) freestanding graphene; (**b**) graphene on fcc Ru(0001); (**c**) graphene on hcp Ru(0001); and (**d**) graphene on ridge Ru(0001). Graphene is shown as bonds only. Top and second layer Ru atoms are shown as green and grey spheres, respectively. Reproduced from Reference [99] with permission from the American Institute of Physics.

Figure 5 reports, according to the calculations of Semidey-Flecha et al. [99], the potential surface energy for Au_1 calculated on the same set of (3 × 3) surfaces. These potential surfaces energy furnish the preferential diffusion path for Au_1 as well as the global diffusion barrier.

Figure 5a refers to Au_1 adsorbed on free-standing graphene: it shows that, in this configuration, the most stable sites are the top site on the free-standing graphene (see Figure 4a), for which the adsorption energy is $\Delta E = -0.11$ eV. In addition, on the free-standing graphene, Au_1 diffusion preferentially occurs between adjacent top sites with a barrier of $E_a = 0.002$ eV. Figure 5b refers to Au_1 adsorbed on graphene supported on the fcc version of Ru(0001): it shows that the most stable sites are the t2 ones (see Figure 4b), for which the adsorption energy is $\Delta E = -1.42$ eV. In this case, Au_1 preferentially diffuses between adjacent t2 sites via the t1 site, with barrier of $E_a = 0.76$ eV. Figure 5c refers to Au_1 adsorbed on graphene supported on the hcp version of Ru(0001): it shows that the most stable sites are the t2 ones (see Figure 4c), for which the adsorption energy is $\Delta E = -1.13$ eV. In this case, Au_1 preferentially diffuses between adjacent t2 sites via the t1 site, with barrier of $E_a = 0.66$ eV. Finally, Figure 5d refers to Au_1 adsorbed on graphene supported on the ridge version of Ru(0001): it shows that the most stable sites are the tβ ones (see Figure 4d), for which the adsorption energy is $\Delta E = -0.92$ eV. In this case, Au_1 preferentially diffuses between adjacent tβ sites with barrier of $E_a = 0.32$ eV.

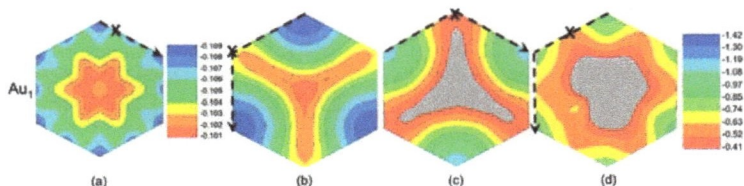

Figure 5. Potential energy surfaces for Au_1 on the (3×3) surfaces: (**a**) freestanding graphene; (**b**) graphene on fcc Ru(0001); (**c**) graphene on hcp Ru(0001); and (**d**) graphene on ridge Ru(0001). The hexagon identifies the standard graphene hexagon. In each image, the dashed line signs the adatom minimum-energy diffusion path. "X" marks the transition state from a local minimum energy site to another. The energy scale is in eV. Reproduced from Reference [99] with permission from the American Institute of Physics.

To conclude, Semidey-Flecha et al. [99] report, also, the resulting coarse-grained potential energy surface for Au_1 on graphene/Ru(0001), see Figure 6: it allows the determination of the minimum-energy diffusion path (the dashed line) for Au_1 from the global minimum-energy adsorption site in the fcc region of one moiré cell to that in an adjacent moiré. For this diffusion path, the authors were able to calculate the Au_1 diffusion barrier as $E_a = 0.71$ eV. So, using this value in the Arrhenius law of the hopping rate $r = A\exp(-E_a/kT)$ and the value $A = 10^{12}$ s^{-1} for the pre-exponential factor, the room-temperature hopping rate is estimated in about 0.1 s^{-1}.

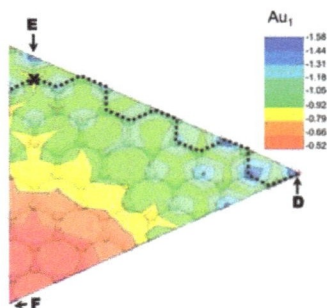

Figure 6. Potential energy surfaces for Au_1 sampled at the top and ring center sites in the symmetry-irreducible zone of the full graphene/Ru(0001) surface. The dashed line signs the adatom minimum-energy diffusion path. The minimum energy diffusion path for the adatom is marked by a dashed line. "D", "E", and "F" mark the preferential adsorption sites, and "X" marks the highest-energy site. Reproduced from Reference [99] with permission from the American Institute of Physics.

2.1.4. In-Plane Adsorption and Diffusion of Au in Graphene

Another interesting aspect studied by means of theoretical analyses concerns the in-plane diffusion of Au atoms in graphene. Malola et al. [98], in particular, studied this phenomenon using density functional calculations motivated by the experimental data of Gan et al. [58] which experimentally observed in-plane adsorption of Au atoms in vacancies of graphene sheets and measured the rate for the in-plane Au diffusion (as we will discuss in Section 2.2).

The analysis of Malola et al. [98] starts considering that the vacancies formation in the graphene sheets is the essential condition for the Au in-plane adsorption and diffusion since the Au in-plane diffusion is mediated by these vacancies. So, first of all, the authors calculated the carbon vacancy formation energy in free-standing graphene as a function of the number of vacancies corresponding to some selected geometries, see in Figure 7 the empty points. In addition, they calculated the formation energy for Au adsorbed in these graphene vacancies, see in Figure 7 the full points.

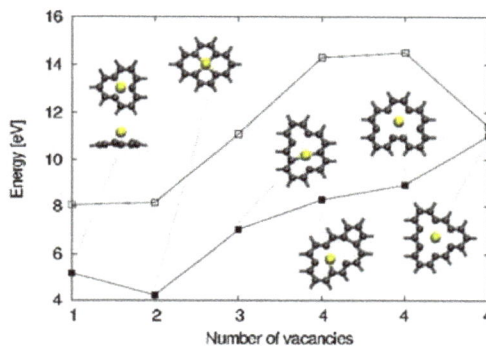

Figure 7. Carbon vacancies formation energy in graphene (empty squares), and formation energies for Au adsorbed in graphene vacancies (full points). For each vacancy, the insets show the selected geometry for the vacancy. Reproduced from Reference [98] with permission from the American Institute of Physics.

For example, the single and double vacancies formation energy is about 8 eV, and then it increases by a rate of about 2 eV/C increasing the number of C atoms to remove. The difference between the two curves in Figure 7 is the Au adsorption energy and it is in the 3–6 eV range on the basis of the number of vacancies being formed. Considering these data, the authors observe that the in- and out-plane bonding energy for Au is higher when adsorbed in double vacancies concluding, so, that the Au-double vacancy should be the most stable configuration. Therefore, Malola et al. [98] used molecular dynamics simulations to simulate the four different diffusion paths presented in Figure 8 for the Au atom in the double vacancy and for each of them calculated the value of the diffusion barrier.

The diffusion barrier of 4.0 eV (diffusion path I) corresponds to the out-of-plane motion of Au. Diffusion path II with 5.8 eV barrier involves out-of-plane motion of C, instead. A diffusion barrier of 7.0 eV corresponds to the in-plane diffusion path III while the path IV has a 7.5 eV barrier. These values are not able to explain the 2.5 eV value experimentally measured by Gan et al. [58] for the in-plane diffusion of Au atoms in graphene by using in-situ transmission electron microscopy (operating at 300 kV) analyses. Then, Malola et al. [98] conclude that the 2.5 eV corresponds to the in-plane radiation enhanced diffusion of the Au atoms in the sense that the in-plane Au atoms diffusion is enhanced by electrons irradiation arising from the electron beam of transmission electron microscopy. The electrons radiation should cause displacement of C atoms generating vacancies which favor Au to overcome the large 4 eV (or higher) energy barrier, resulting in the effective 2.5 eV measured by Gan et al. [58].

I 4.0 eV / II 5.8 eV

III 7.0 eV

IV 7.5 eV

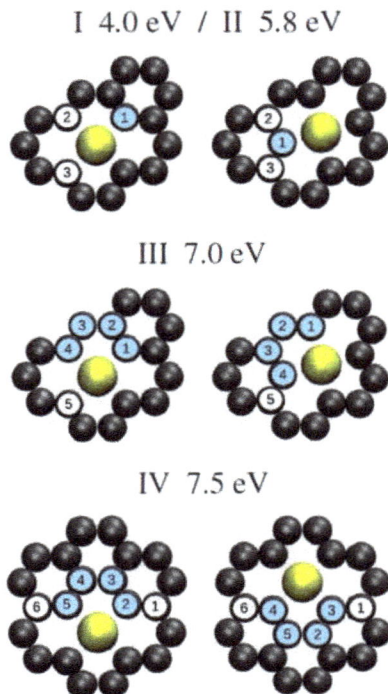

Figure 8. Au in double vacancies in graphene: simulations of different diffusion paths (path I, II, III, IV) of the Au atom (yellow sphere), whereas the blue dots indicate the C atoms which change position as result of the Au atom jump. In addition, each path is accompanied by the estimated diffusion barrier for the Au jump. Reproduced from Reference [98] with permission from the American Institute of Physics.

2.2. Adsorption, Diffusion, Nucleation and Growth of Metal Atoms on/in Graphene: Experimental Results

2.2.1. General Considerations

A set of experimental data on the growth of a range of metal NPs by vapor-phase depositions of metal atoms on graphene was reported by Zhou et al. [65]. In this work, the authors deposited, by thermal evaporation, Pt, Rh, Pd, Co, and Au on a graphene moiré pattern on Ru(0001). Then they performed systematic scanning tunneling microscopy studies to analyze the growth mode of the resulting NPs as a function of the amount (in unity of monolayers, ML) of deposited material and as a function of the annealing temperature of a subsequent annealing process. The authors, in particular, tried to highlight the differences observed for the various metals: in fact, their experimental data show that Pt and Rh form small particles sited at fcc sites on graphene. Instead, in similar coverage conditions, Pd and Co form larger particles. Analyzing these results, the authors conclude that the metal-carbon bond strength and metal cohesive energy are the main parameters in determining the metal clusters formation process and the morphology of the clusters in the initial stages of growth. On the other hand, experimental data on the growth of Au show a further different behavior (Au forms a single-layer film on graphene) suggesting, in this case, that other factors affect the growth of the Au cluster. Figures 9–11 summarize some scanning tunneling microscopy analyses of various metals deposited on the graphene/Ru(0001) substrate, as reported by Zhou et al. [65].

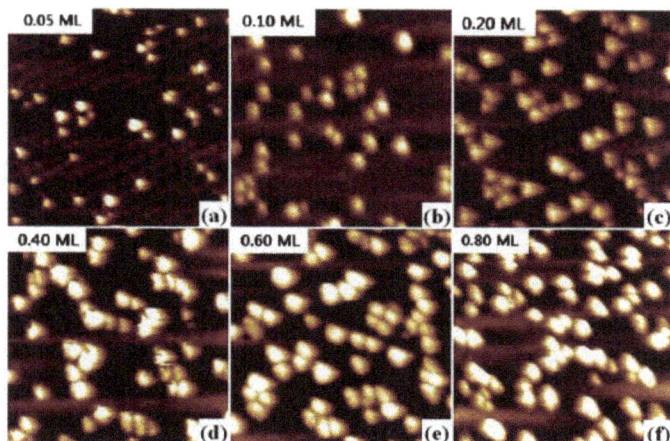

Figure 9. Scanning Tunneling Microscopy images (50 nm × 50 nm) of (**a**) 0.05 ML; (**b**) 0.1 ML; (**c**) 0.2 ML; (**d**) 0.4 ML; (**e**) 0.6 ML and (**f**) 0.8 ML Rh deposited on graphene/Ru(0001) at room temperature. Reproduced from Reference [65] with permission from the Elsevier.

In particular, Figure 9 reports Scanning Tunneling Microscopy images for Rh deposited at room-temperature on the graphene/Ru(0001) substrates and increasing the amount of deposited Rh (from 0.05 to 0.80 ML). From a quantitative point of view, using these analyses, the authors inferred that until 0.6 ML the average Rh clusters size increases by increasing the amount of deposited Rh: the Rh cluster size and height significantly increase when the amount of deposited material increase but, correspondingly, a much lower increases of the particles density is observed. Similar is the behavior of Pt: for a coverage of 0.1 ML, 2 nm-diameter highly dispersed Pt particles are formed at fcc sites; for a coverage of 1 ML, instead, 5 nm-diameter Pt particles are formed and characterized by a narrow size distribution. Figure 10 shows other Scanning Tunneling Microscopy images: (a) and (b) report images of 0.1 and 0.4 ML Pd deposited on graphene/Ru(0001), respectively. In this case, at a coverage of 0.1 ML, 8–14 nm-diameter three-dimensional Pd particles are formed at fcc sites and with a lower surface density compared to Rh and Pt. (c) and (d) report images of 0.2 and 0.4 ML of Co on graphene/Ru(0001). At a coverage of 0.2 ML, 10 nm-diameter three-dimensional Co particles are formed, while, at a coverage of 0.4 ML, 12 nm-diameter clusters are observed. (e) and (f) report images of 0.2 and 0.6 ML Au on graphene/Ru(0001). At 0.2 ML, small two-dimensional Au particles are formed at fcc sites. However, differently from the previous metals, increasing the coverage (0.6 ML, for example), Au forms a film of NPs covering the graphene moiré pattern. Finally, Figure 11 serves as an example to analyze the thermal stability of the nucleated NPs: it presents images of the Rh NPs on the graphene/Ru(0001) substrate after annealing process from 600 to 1100 K for 600 s. These images show that no significant change can be recognized in the Rh NPs below 900 K. Instead, a NPs coalescence process starts at ∼900 K as indicated by the decreased cluster density and larger dimensions. The NPs coalescence process is more evident after the annealing of the sample at 1100 K.

On the basis of their experimental results, Zhou et al. [65] draw the following conclusions about the growth processes for the investigated metal NPs on the graphene/Ru(0001) substrate:

(a) Pt, Rh, Pd and Co: these metals should grow on the graphene as three-dimensional clusters due to the high difference in the surface energy of graphene (46.7 mJ/cm^2) and of these metals (in the 1–2 J/cm^2 range). However, the interaction between the metals adatoms and the graphene strongly influences this situation by determining the adatoms mobility. Only a small interaction energy of the adatoms with the graphene (with respect to the adatom-adatom interaction energy) will assure a high adatoms mobility and, so, the occurrence of the three-dimensional growth of the clusters.

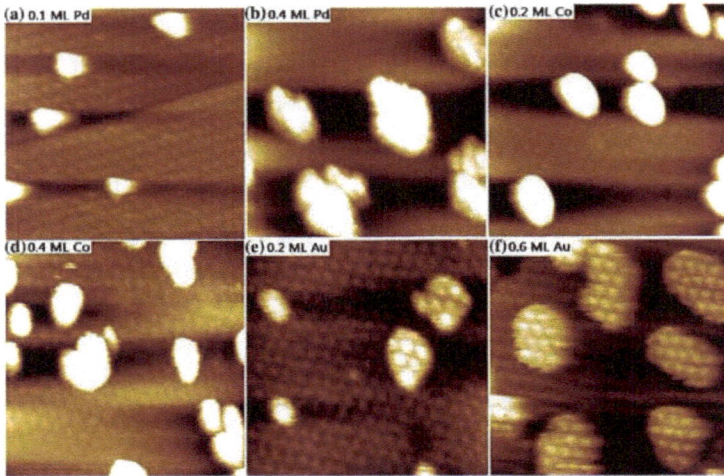

Figure 10. Scanning Tunneling Microscopy images (50 nm × 50 nm) of (**a**) 0.1 ML Pd; (**b**) 0.4 ML Pd; (**c**) 0.2 ML Co; (**d**) 0.4 ML Co; (**e**) 0.2 ML Au and (**f**) 0.6 ML Au deposited on graphene/Ru(0001) at room temperature. Reproduced from Reference [65] with permission from the Elsevier.

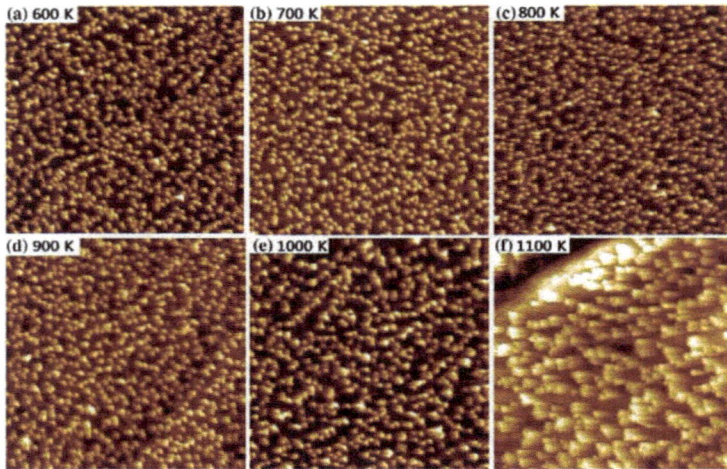

Figure 11. Scanning Tunneling Microscopy images (50 nm × 50 nm) of 0.8 ML Rh on graphene/Ru(0001) acquired after annealing the samples to (**a**) 600 K; (**b**) 700 K; (**c**) 800 K; (**d**) 900 K; (**e**) 1000 K and (**f**) 1100 K for 10 min. Reproduced from Reference [65] with permission from the Elsevier.

On the basis of this consideration, the authors attribute the observed differences in the Pt, Rh, Pd and Co NPs growth morphologies to the different strengths of the metal-carbon bond. The increase of the strength of the metal-carbon bond will result in the decreasing of the diffusion coefficient for the metal on graphene at a given flux. As a consequence, the decrease of the diffusion coefficient will result in the increase of the metal clusters nucleation rate allowing to obtain, thus, uniformly dispersed the two-dimensional clusters at the initial growth stage. So, the authors note that the relevant metal-carbon dissociation energies are: 610 kJ/mol for Pt-C, 580 kJ/mol for Rh-C, 436 kJ/mol for

Pd-C, and 347 kJ/mol for Co-C, so that the metals with higher bond dissociation energies (Pt and Rh) form highly dispersed clusters while those with lower bond dissociation energies (Pd and Co) form large three-dimensional clusters with low surface densities. On the other hand, however, with the continued atoms deposition, the pre-formed cluster on the graphene surface start to growth in size by incorporating the new incoming atoms and this process is competitive to the nucleation of new clusters on the surface. The joining of two or more metal atoms is characterized by the metals cohesive energy which establishes the strength of the metallic bonds. So, now, the metal-carbon dissociation energy and the metal cohesive energy become competitive parameters in establishing the final cluster growth mode and morphology. So, the authors' picture is improved as follows [65]: the C atoms of the graphene strongly interact with Pt and Rh atoms, largely influencing the initial growth stage leading to the formation of uniformly distributed small particles. On the other hand, the bond strength of Pd and Co atoms to the C atoms is much weaker, so that the metals cohesive energy drive the NPs formation and growth, resulting in the formation of large three-dimensional clusters at initial growth stage.

(b) Effect of the substrate supporting the graphene: in their analysis, Zhou et al. [65] compared their results with other literature results. For example, they compared their results on the growth of Pt on graphene/Ru(0001) with the results of N'Daye et al. [61,64] on the growth of Pt on graphene/Ir(111) in similar conditions of depositions. They highlight some crucial differences in the growth morphology of the Pt clusters and impute these differences to the specific interaction of the metal atoms with the substrate supporting the graphene layer. In summary, Zhou et al. [65] report that the equilibrium spacing between graphene and the Ir(111) surface has been calculated to be 0.34 nm. Instead, the equilibrium spacing between the graphene and the Ru(0001) surface has been calculated to be 0.145 nm. This difference arises from the higher interaction of the graphene with the Ru(0001) than with Ir(111). Thus, in general, increasing the interaction energy between the C atoms of the graphene layer with the substrate on which it is supported, will lead to a decrease in the interaction energy between the C atoms and the deposited metal adatoms. This will result in an increased metal adatoms diffusivity. The consequence is that the metal clusters grown on graphene/Ir(111) are spatially more ordered than on graphene/Ru(0001) and that the transition from two-dimensional to three-dimensional morphology of clusters on graphene/Ru(0001) occurs at much lower amount of deposited material.

(c) Au: due to the weak interaction between Au and C, Au is expected, so, to grow on graphene as three-dimensional isolated Au clusters. Instead, Zhou et al. [65] observed that Au on graphene/Ru(0001) forms a continuous nano-granular film. They attribute this behavior, mainly, to the low Au cohesive energy (i.e., Au tends to wet a metal surface with a larger cohesive energy. Note that the Au cohesive energy is 3.81 eV whereas, for example, the Pt cohesive energy is 5.84 eV). In addition, the nearest-neighbor distance for Au is 0.288 nm which is larger than the graphene lattice parameter (0.245 nm). N'Diaye et al. [61,64] inferred that metal with a nearest-neighbor distance of 0.27 nm can perfectly fit the graphene lattice. So, Au atoms do not fit the graphene lattice, contributing to the lowering of the Au-C interaction energy. Therefore, the Au low cohesive energy and the low Au-C interaction energy contribute in determining the atypical Au growth.

In addition to Zhou et al. [65], N'Diaye et al. [61,64] reported another set of experimental analyses on the growth morphologies of Ir, Pt, W, and Re on graphene/Ir(111) and then Feibelman [75,76] reported additional theoretical analyses on the experimental results of N'Diaye et al.

The main results of N'Diaye et al. [64] rely in the establishment of the condition for which a metal form a superlattice on the graphene/Ir(111) substrate: (1) A large metal cohesive energy; (2) a high interaction energy of the deposited metal atoms with graphene established by the large extension of a localized valence orbital of the deposited metal; and (3) the fitting between the graphene lattice parameter and the nearest-neighbor distance of the deposited metal. In the course of their studies, N'Diaye et al. [64] were able, in addition, to infer several characteristics on the metals growth morphology. From an experimental point of view, first of all, the authors choose to deposit materials with very different cohesive energy so to study the impact of this parameter on their growth morphology. In fact, the cohesive energy for W, Re, Ir and Pt is, respectively, 8.90, 8.03, 6.94, 5.84 eV.

Figure 12 shows, for example, Scanning Tunneling Microscopies of graphene flakes grown on Ir(111) after deposition, at room-temperature, of 0.2–0.8 ML of various metals. In the areas without graphene, metals form some isolated islands of monolayer height. All deposited materials are pinned to graphene flakes forming NPs. Ir and Pt form similar very ordered superlattices of clusters on the graphene flakes (compare Figure 12a,b). At 0.2 ML both materials exhibit two distinct height levels of the clusters. Also W forms an ordered cluster superlattice (see Figure 12c), however with higher height than that obtained for Ir. These W clusters present distinct height levels. A lower spatial order is obtained, instead, for Re clusters as visible by Figure 12d. For Fe (Figure 12e) and Au (Figure 12f) clusters the spatial order is completely absent so that no superlattice is obtained. The authors attribute the absence of the regular cluster superlattice for these metals to their small cohesive energy and/or small binding energy to graphene: metals with small cohesive energy present a more pronounced wetting behavior on graphene with respect to metal with higher cohesive energy (i.e., metals with small cohesive energy have lower surface energy than the metals with higher cohesive energy). Metals with low bonding strength to graphene present high mobility (with respect to metals with higher bonding strength) so that graphene is not able to trap efficiently these adatoms and small clusters). The authors verified [64] these conclusions by depositing Re, Au and Fe on the graphene/Ir(111) substrate at lower temperatures (200 K), so to decrease the adatoms diffusivity. In this case the formation of the superlattices structures for the Re, Au, and Fe clusters was observed.

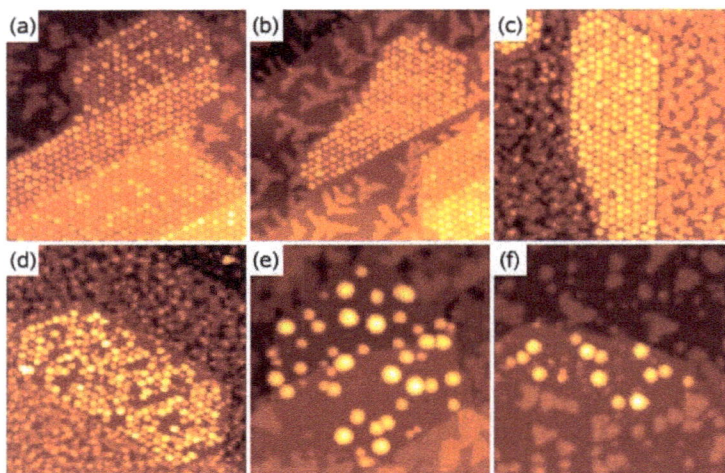

Figure 12. Scanning Tunneling Microscopy images (70 nm × 70 nm) of graphene flakes on Ir(111) after deposition, maintaining the substrate at 300 K, of: (**a**) 0.20 ML Ir; (**b**) 0.25 ML Pt; (**c**) 0.44 ML W; (**d**) 0.53 ML Re; (**e**) 0.77 ML Fe; (**f**) 0.25 ML Au. Reproduced from Reference [64] with permission from IOPscience.

Then, the authors investigated the effect of a subsequent annealing process on the morphology and order of the deposited metal clusters. Some results are reported in Figure 13: it reports the Scanning Tunneling Microscopies of Pt deposited on the graphene/Ir(111) substrate and annealed for 300 s from 350 K to 650 K. Figure 13g quantifies the annealing effect by plotting the temperature dependence of the moiré unit cell occupation probability n as a function of the annealing temperature T for all the investigated metals.

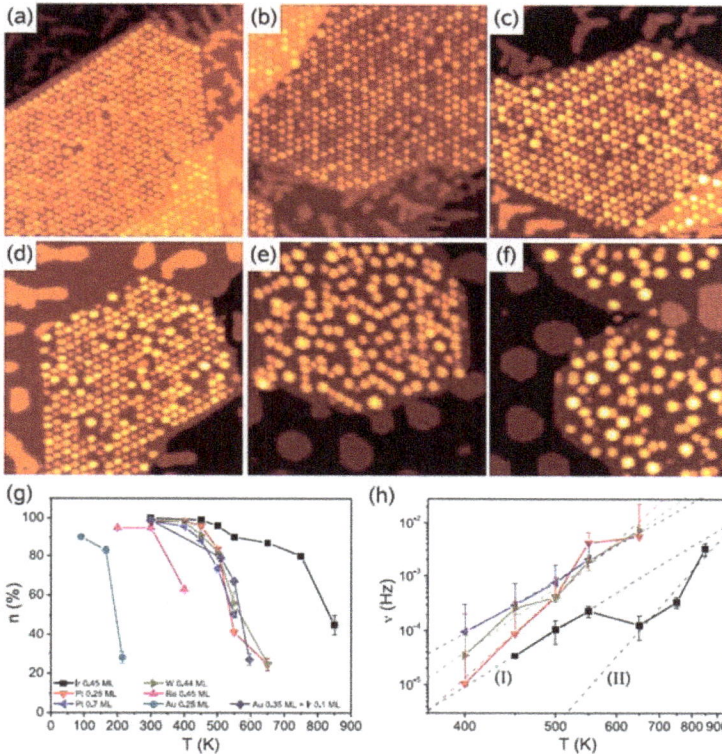

Figure 13. Scanning Tunneling Microscopy images (70 nm × 70 nm) of (**a**) 0.25 ML Pt deposited on graphene/Ir(111) maintaining the substrate at 300 K. This sample was then annealed for 300 s at (**b**) 400 K; (**c**) 450 K (**d**) 500 K; (**e**) 550 K and (**f**) 650 K; (**g**) Plot of n (occupation probability of the moiré cell by a particle) versus the annealing temperature T; (**h**) Arrhenius plot of particle jumping rate ν(T). Lines represent fits for the hopping rate with diffusion parameters as shown in Table 3. Reproduced from Reference [64] with permission from IOPscience.

The evolution of the cluster superlattice (i.e., decay) is due to the thermally activated diffusion of clusters. The clusters perform a random motion around their equilibrium positions and two or more cluster can coalesce if the temperature is high enough to enough increase the diffusion length. The cluster diffusion, and so the probability for two or more cluster to join, is dictated by the activation barrier E_a which the cluster has to overpass to leave its moiré unit cell. This effect is illustrated by Figure 14 showing a sequence of images taken at 390 K (a–e) or at 450 K (f–j). White circles in the images sequences indicate locations of thermally activated changes, i.e., clusters that having overpassed the activation barrier for diffusion and perform a coalescence process.

In addition, N'Diaye et al. [64] were able to infer quantitative evaluations on the parameters involved in this process: supposing the clusters attempt frequency to overpass the diffusion barrier (i.e., the clusters joining frequency) expressed by an Arrhenius law, i.e., $\nu = \nu_0 \exp(-E_a/kT)$, and supposing the probability that one cluster encounters another one is proportional to n, the data in Figure 13h can be fitted to extract the clusters activation energy for diffusion (E_a) with the corresponding deviation (ΔE_a), and the pre-exponential factor ν_0. All these evaluated parameters are summarized in Table 3.

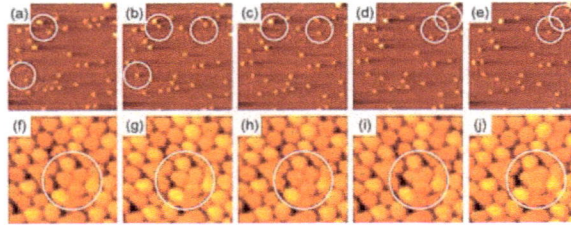

Figure 14. (a–e) Scanning Tunneling Microscopy images (25 nm × 25 nm) of 0.01 ML Ir deposited at 350 K on graphene/Ir(111) (a) and annealed at 390 K for 120 s (b); 240 s (c); 360 s (d); 480 s (e). Circles indicate where modification occur in successive images; (f–j) Scanning Tunneling Microscopy images (15 nm × 15 nm) of 1.5 ML Ir deposited on graphene/Ir(111) at 350 K (f) and annealed at 450 K for 120 s (g); 240 s (h); 360 s (i); 480 s (j). Reproduced from Reference [64] with permission from IOPscience.

Table 3. Activation energy for diffusion (E_a) with the corresponding deviation (ΔE_a), and the pre-exponential factor v_0 with the corresponding errors (fifth and sixth columns) for the cases of Ir, Pt and W deposited on graphene/Ir(111). Reproduced from Reference [64] with permission from IOPscience.

Clusters	E_a (eV)	ΔE_a (eV)	v_0 (Hz)
Ir, 0.45 ML (I)	0.41	0.02	1.4
Ir, 0.45 ML (II)	0.75	0.2	67
Ir, 0.45 ML	0.28	0.08	0.06
Pt, 0.25 ML	0.60	0.08	500
Pt, 0.70 ML	0.38	0.02	6.2
W, 0.44 ML	0.47	0.04	33

2.2.2. Au Nanoparticles on Graphene

Zan et al. [66] used Transmission Electron Microscopy to study the morphological and structural evolution of Au NPs on free-standing single-layer graphene sheet changing the effective deposited Au film thickness from less than 0.1 nm to 2.12 nm.

Figure 15 shows the results of the Au depositions: the preferential sites for the Au clusters nucleation are in correspondence of the Au hydrocarbon contamination, as revealed by the wormlike contrast in the high-resolution Transmission Electron Microscopy images. This is a signature of the very high diffusivity of Au atoms on graphene. Furthermore, the images show that the Au cluster number per unit area increases with increasing evaporated amount of Au, and at a nominal Au thickness larger than 1 nm clusters start to joining by coalescence.

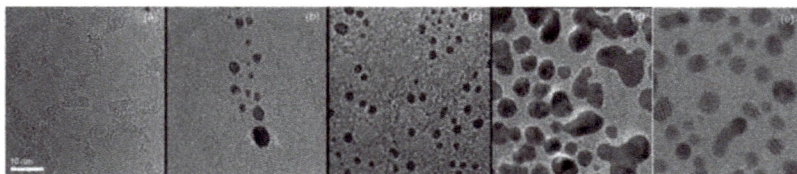

Figure 15. (a–d) Transmission Electron Microscopy images of Au deposited on free-standing graphene increasing the amount of deposited metal; (a) Sparse coverage; (b) sparse groups of clusters at Au thickness lower than 0.1 nm; (c) Higher cluster densities at 0.12 nm of Au thickness; (d) Coalescence of clusters occurring for 2.12 nm-thick deposited Au; (e) Scanning Transmission Electron Microscopy bright-field image of 0.5 nm-thick evaporated Au. Scale bar: 10 nm in all images. Reproduced from Reference [66] with permission from Wiley.

Figure 16 shows the observed in-situ coalescence process of some Au clusters. The lighter areas within the clusters correspond to clean graphene patches overlaid by the clusters. As examples two of these overlaid regions are marked by the white lines in Figure 16a: the left one occurs at the coalescence front of two coalescing clusters, the right-hand one in the middle of a cluster.

In addition, Zan et al. [66] motivated by the fact that a standard method to modify and functionalize graphene is by hydrogenation, studied the Au growth morphology on intentionally-hydrogenated free-standing graphene.

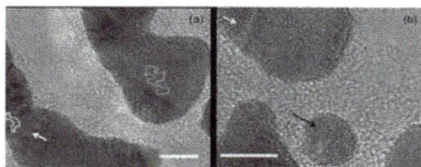

Figure 16. Coalesced Au clusters corresponding to the deposition of 2.12 nm Au on graphene. (**a**) shows variations in thickness and relative crystallographic orientations and (**b**) planar faults such as stacking faults (white arrows) and twin boundaries (black arrow). Scale bars: 5 nm. Reproduced from Reference [66] with permission from Wiley.

Hydrogenation breaks graphene sp^2 bonds and leads to sp^3 bond formation. Au depositions, 0.2 nm in nominal thickness, were, so, carried out on graphene surfaces that had been hydrogenated and the results compared to those obtained for 0.2 nm Au deposited on pure graphene. As can be seen in Figure 17a, the hydrogenated sample presents a higher Au clusters density and cluster sizes are less dispersed than in the pure graphene sample, as shown in the image in Figure 17b. However, similar to pristine graphene, Au clusters nucleate in the defects represented by the contaminations sites where the hydrogenation occurred. So, the increased hydrogenation of the graphene leads to a more effective adhesion of Au, enhancing the nucleation probability of Au clusters in the contaminations. This picture is confirmed by the observation of the occurring of coalescence of Au clusters under the electron beam of the Transmission Electron Microscopy (a process which is not observed for the Au on the pristine graphene). An example of this process in the hydrogenated sample is shown in Figure 17c,d: these Transmission Electron Microscopies present the evolution of the Au clusters under the electron beam at temporal distance of about 10 s. The agglomeration of the Au clusters (marked by the solid circles and dashed rectangles in Figure 17c,d) occurs rapidly, in the 10 s time range. In contrast, the Au clusters formed on the pristine graphene perform a coalescence process on the graphene during the Au deposition and not in few seconds under exposure to the electron beam. So, evidently, the hydrogenation process of the graphene lowers the diffusion barrier for the pre-formed Au clusters, the electron beam furnishes enough energy to the clusters to overcome this diffusion barrier, and the Au clusters coalescence starts and rapidly occurs (~seconds).

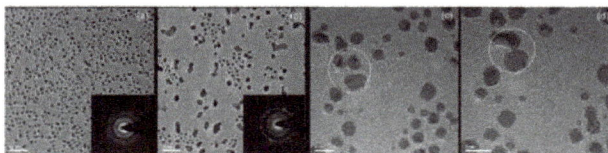

Figure 17. (**a**,**b**): Transmission Electron Microscopy images of 0.2 nm Au evaporated onto hydrogenated and pristine graphene (scale bar: 20 nm). The corresponding diffraction patterns are shown as insets; (**c**,**d**) Images of Au evaporated onto hydrogenated graphene, taken in a sequence of scans, and showing the Au clusters merging by coalescence as indicated by the solid circles and dashed rectangles (scale bar: 5 nm). Reproduced from Reference [66] with permission from Wiley.

2.2.3. Au and Pt Nanoparticles in Graphene

Another aspect related to the kinetic processes of metal atoms interacting with graphene was analyzed by Gan et al. [58]: they studied, experimentally, the in-plane diffusion characteristics of Au and Pt atoms in graphene and the corresponding nucleation process towards the formation of NPs by using in-situ Transmission Electron Microscopy analyses at high temperature. The analysis by the authors starts by the consideration that carbon vacancies in the graphene layers favor the atoms in-plane diffusion with respect to the on-plane diffusion.

So, to perform the experiments, the authors mixed powders of Au or Pt with graphite powder. Then they obtained a mixed fine deposit by an electric arc discharge system. After dispersing and sonicating the resulting deposit, it was placed on standard grids for in-situ Transmission Electron Microscopy analysis. During the Transmission Electron Microscopy studies, the samples were annealed in the 600–700 °C range to induce the metal atoms diffusion. The used fabrication method produces layers consisting of one or few graphene layers characterized by crystal vacancies allowing the in-plane metal atoms diffusion. As an example, Figure 18a,b show Pt atoms in a four-layers graphene structure held at 600 °C. The image in Figure 18b was acquired 60 s after Figure 18a. Two Pt atoms (indicated by the arrows) merge and form a nucleus. Such nuclei of two or several Au or Pt atoms were often observed by the authors. Then they acquired several images with the viewing direction along the graphene layers. In this condition, the observed metal atom apparently remains immobile during the annealing and overlaps with the contrast of the outermost graphene layers: this fact excludes that the metal atom is located on top of the layer. So, after several observations, the authors conclude that the metal atoms are located in-plane with the graphene sheet occupying vacancies on the carbon sites.

To analyze the atoms diffusion, Figure 19 shows the temporal evolution by reporting plan-view Transmission Electron Microscopy images acquired in the same region of the sample which is held at 600 °C and increasing the time. These images follow, in particular, the evolution of Pt atoms. The arrows in the first images identify some Pt atoms and by the images sequence how these atoms change their position by diffusion can be recognized. Atoms diffusing within the layer are marked by "L". It can be concluded that metal atoms prefer edge locations rather than in-plane sites. It is also visible how the atoms at the edge (marked by "E") move along the edge. Using these real-time analyses, the authors, in particular, were able to measure the diffusion length for several of Au and Pt atoms (quantified along the layer) versus time at different temperatures, obtaining data which follow the square-root law connecting the diffusion length to the diffusion time.

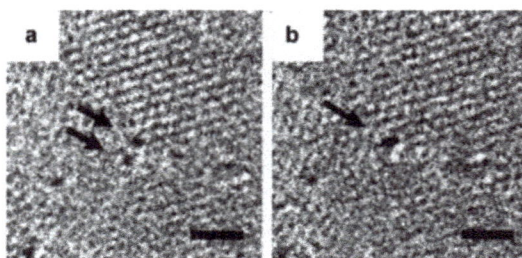

Figure 18. (**a,b**) Plan-view Transmission Electron Microscopy images of Pt atoms in a four-layer graphitic sheet held at 600 °C. The image (**b**) was acquired 60 s after (**a**). Two Pt atoms (arrowed) merge and form a cluster. The scale bar is 1 nm. Reproduced from Reference [58] with permission from Wiley.

So, the mean diffusion length x is connected to the diffusion time t by $D = x^2/4t$, with D the atomic diffusion coefficient. Using the experimental data, the authors derived values for the Pt and Au atoms in-plane diffusion coefficient: $D = 6 \times 10^{-22}$–2×10^{-21} m^2/s for Au at 600 °C, $D = 4 \times 10^{-22}$–1×10^{-21} m^2/s for Pt at 600 °C, $D = 1 \times 10^{-21}$–7×10^{-21} m^2/s for Pt at 700 °C.

Using these values, Gan. et al. [58] evaluated the activation energy for the graphene in-plane diffusion of the Pt and Au atoms: in fact, considering that $D = ga^2\nu_0 exp[-E_a/kT]$, with $g \approx 1$ a geometrical factor, a the graphene lattice constant, ν_0 the attempt frequency which can be assumed to be the Debye frequency, then E_a is estimated, both for Pt and Au, in about 2.5 eV.

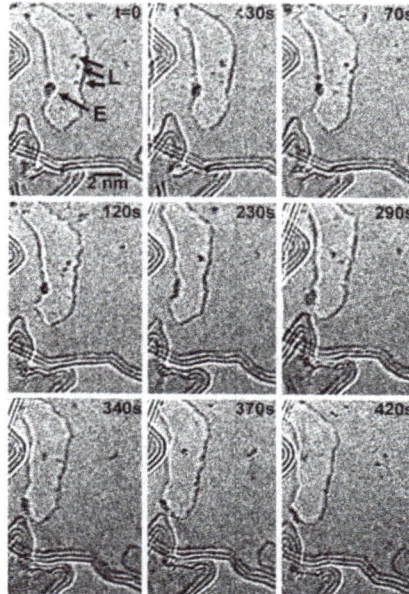

Figure 19. Series of Transmission Electron Microscopies showing the diffusion of Pt atoms in graphene at 600 °C as a function of time. "L" marks the region in a two-three layer graphene where Pt atoms are diffusing in two dimensions. "E" marks a Pt cluster located at the edge of a graphene layer and where Pt atoms are observed one-dimensionally diffuse along the edge. Reproduced from Reference [58] with permission from Wiley.

This value arises by the combined effect from the covalent bonding between Pt or Au and C atoms and from the activation energy for site exchange of carbon atoms that is given by the vacancy migration energy in graphene (1.2 eV). However, a question arises about these results: the role of the electron beam used for the in-situ Transmission Electron Microscopy analyses on the observed metal atoms diffusion process. In fact, it could determine an enhanced radiation diffusion. This point was, in particular, addressed, from a theoretical point of view, by Malola et al. [84] as discussed in Section 2.1.4. Their theoretical simulations indicate that the lowest-energy path with 4.0 eV barrier involves out-of-plane motion of Au (see Figure 8). Other diffusion paths are characterized by higher energy barriers. So, the 2.5 eV barrier value measured by Gan et al. [58] for the in-plane diffusion of Au atoms in graphene should arise as an electron (300 keV) radiation enhanced diffusion: in fact, assuming Au in double vacancy, at least one of the 14 neighboring C atoms should be removed every 10 s as result of the electron beam interaction. This generation of vacancies favor Au to overcome the large 4 eV (or higher) energy barrier, resulting in the effective 2.5 eV. The radiation enhanced diffusion interpretation is in agreement with the experimental result that the 2.5 eV barrier is found both for Au and Pt which is not expected a-priori considering that C–Pt interaction is stronger than the C–Au one. In fact, on the basis of this fact, the activation energy for the Pt diffusion should be higher. Instead, the C-metal energy interaction is substantially negligible in the diffusion process if it is dominated by radiation enhancement.

2.2.4. Au Nanoparticles on Graphene Supported on Different Substrates

Liu et al. [60] investigated, from an experimental point of view, the nucleation phenomenon of Au NPs on graphene. In particular, they focused the attention on the effect of the substrate supporting the graphene and of the graphene layer number on the NPs nucleation kinetics. The experimental data were discussed within the mean field theory of diffusion-limited aggregation, allowing to evaluate the Au adatom effective diffusion constants and activation energies.

Liu et al. [60], so, prepared graphene samples by mechanical exfoliation of graphite onto SiO_2/Si substrates or hexagonal boron nitride substrates. Raman spectroscopy was used to analyze the number of graphene layers. Au was deposited on the graphene layers by electron beam evaporation, having care, in addition, to produce reference samples were by depositing Au on graphite substrates. To induce morphological evolution of the Au on the substrates, subsequent annealing processes were performed. At each step of evolution, the authors performed Atomic Force Microscopy analyses to study the samples surface morphology, i.e., the Au NPs morphology, size, surface density and surface roughness.

First of all, the authors deposited 0.5 nm of Au on single-layer (1 L) graphene and bilayer (2 L) graphene supported onto SiO_2/Si, and onto graphite surfaces maintaining the substrates at room temperature. Then, the Atomic Force Microscopy analyses allowed infer the following conclusions: on the graphite surface, Au NPs coalesce to form ramified islands. The large Au-Au binding energy (~3.8 eV), drives the Au adatoms diffusion towards the joining and formation of small compact NPs. Once formed, these very small NPs diffuse slowly on the graphite and then they coalescence to form islands. Under the same deposition conditions on the 1 L graphene, Au NPs with a narrower-size distribution and higher surface density are obtained. Instead, concerning the Au NPs obtained on the 2 L graphene, some of these evidence an ongoing evolution from elongated islands structures to ramified structures. This difference with the Au NPs obtained on graphite is the signature of the lower diffusion coefficient of the Au adatoms on 1 L and 2 L graphene than on graphite. Further results are summarized by Figure 20: by depositing 0.1 nm of Au, the observed density of Au NPs is about 1200 μm^{-2} (Figure 20a) on 1 L graphene. A 350 °C-2 h thermal process leads to the decrease of the surface density of the Au NPs to about 130 μm^{-2} (Figure 20c). Instead, on the graphite substrate, the thermal process causes a decrease of the Au NPs density from about 180 μm^{-2} (Figure 20b) to about 3 μm^{-2} (Figure 20d). These data confirm the thermal-activated nature of the NPs growth mechanism.

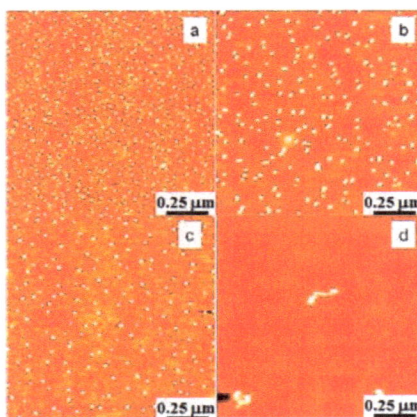

Figure 20. Atomic Force Microscopy images (1 $\mu m \times$ 1 μm) of 0.1 nm Au deposited on a single-layer graphene (**a**) and on graphite (**b**). (**c,d**) show the same samples (Au on single-layer graphene in (**c**) and Au on graphite in (**d**)) after 2 h of thermal annealing at 350 °C. Reproduced from Reference [60] with permission from the American Chemical Society.

As a comparison, the authors performed similar studies for Au deposited on graphene supported onto hexagonal boron nitride (h-BN): this choice is dictated by the fact that graphene is known to be flatter on single-crystal h-BN than on SiO_2. So, the upper part of the image 21a shows the surface of h-BN presenting a roughness of 47 pm, while the bottom shows the h-BN surface supporting 1 L graphene, with a roughness of 54 pm. The other Atomic Force Microscopy images in Figure 21 show the resulting Au NPs obtained by the deposition of 0.1 nm of Au on the surface of bare h-BN, on 1 L graphene supported on h-BN and on 1 L graphene supported on SiO_2. The comparison of these images allow us to conclude that the NPs growth is faster on h-BN and 1 L graphene supported on h-BN than on 1 L graphene supported on SiO_2.

At this point, once recorded these experimental data, Liu et al. [60] exploited the mean-field nucleation theory to analyze these data so to extract quantitative information on the parameters involved in Au NPs morphological evolution processes. As the amount of deposited materials increases, three kinetic regimes for the Au clusters growth can be recognized: clusters nucleation, clusters growth, and steady-state. At the early stages of deposition, moving adatoms on the substrate explore a certain area in a certain time so that they can encounter each other and and, so, they have some finite probability to join (nucleation process) and form stable nucleii. The number of nuclei increases with time. However, in the same time, new atoms arrive from the vapor-phase and they can be captured by the preexisting nuclei. At enough high deposition time, so, the nuclei growth in cluster of increasing size and new nucleii are not formed: a steady state is reached. In this condition, the mean Au adatoms diffusion length is equal to the mean Au NP spacing and a saturation density for the nuclei is obtained. The authors, then, considered that, according to the mean-field nucleation theory, the nuclei saturation density n is predicted as $n(Z) \sim N_0 \eta(Z)(F/N_0 v)^{i/(i+2.5)} \exp[(E_i + iE_d)/(i + 2.5)kT]$ being Z a parameter depending on the total deposition time, N_0 the substrate atomic density (cm^{-2}), $\eta(Z)$ a dimensionless parameter, F is the rate of arriving atoms from the vapor phase ($cm^{-2} s^{-1}$), v an effective surface vibration frequency ($\sim 10^{11}$–$10^{13} s^{-1}$), i the number of Au atoms in the critical cluster, E_i the Au atom binding energy in the critical cluster, and E_d the activation energy for the Au atom diffusion.

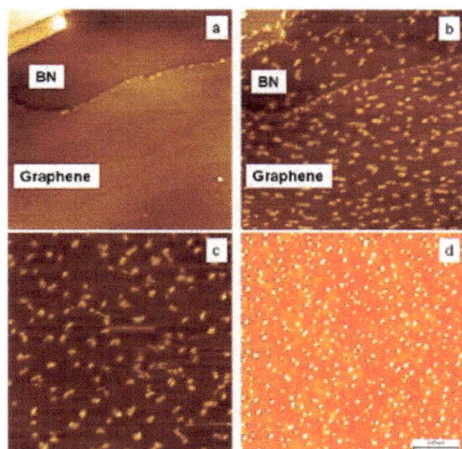

Figure 21. (**a**) Atomic Force Microscopy images (1 µm × 1 µm) of 1 layer graphene on hexagonal boron nitride (h-BN); and (**b**) Atomic Force Microscopy images (1 µm × 1 µm) of 0.1 nm Au deposited on 1 layer graphene supported on h-BN; (**c**) Atomic Force Microscopy images (1 µm × 1 µm) of 0.1 nm Au deposited directly on h-BN; (**d**) Atomic Force Microscopy images (1 µm × 1 µm) of 0.1 nm Au deposited on bilayer graphene supported on SiO_2. Reproduced from Reference [60] with permission from the American Chemical Society.

Clusters of size smaller than *i* shrinks while clusters larger than size *i* grow and form the stable NPs. The authors consider that in the examined experiments, *i* should be small and, so, they analyze their experimental data on n(z) for *i* = 1 and *i* = 2 obtaining the values reported in Figure 22: the Au adatom diffusion energy E_d and the corresponding diffusion coefficient D calculated as $D = (a^2 v_d/4) \exp[-E_d/kT]$ with a the graphene lattice parameter (0.14 nm) and v_d the adatom attempt frequency ($\sim 10^{12}$ s^{-1}).

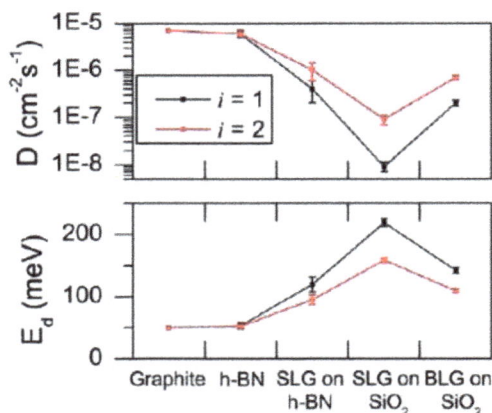

Figure 22. The calculated diffusion energy, E_d, and diffusion constant, D, of Au adatoms on various surfaces: graphite, hexagonal boron nitride (h-BN), single-layer graphene (SLG) on h-BN, SLG on SiO$_2$, and bilayer graphene (BLG) on SiO$_2$. Calculations using different critical sizes *i* are given in black (*i* = 1) and red (*i* = 2) curves for comparison. For graphite, E_d and D are assumed as 50 meV and 7×10^{-6} cm^2 s^{-1}, independent of critical size *i*. Reproduced from Reference [60] with permission from the American Chemical Society.

On the basis of Figure 22 it is clear that the activation energy for the Au adatom diffusion process is higher on 1 L graphene on h-BN, 1 L and 2 L graphene on SiO$_2$ than on graphite and bare h-BN. In addition, it is higher on 1 L graphene on SiO$_2$ than on 2 L graphene on SiO$_2$ which is, in turn, higher than 1 L graphene on h-BN.

Now, the question concerning why these differences are observed arises. In this sense, the authors, first of all, note that adatom diffusion is affected by the surface strains which is, in turn, related to the surface roughness. A compressive strain of the surface reduces the energy barrier for the adatom diffusion while a tensile strain tends to increase it. In its free-standing configuration, 1 L graphene displays ripples with about 1 nm height variation. In contrast, when supported and annealed on SiO$_2$/Si substrates, graphene follows the local SiO$_2$ roughness: the graphene−SiO$_2$ van interaction energy is balanced by the elastic deformation energy of graphene with the consequent increase of the graphene roughness with respect to its free-standing configuration. The Atomic Force Microscopy measurements by Liu et al. [60] show that 1 L graphene on SiO$_2$ presents a roughness 8 times higher than bulk graphite. On this rough graphene surface, there will be regions of both concave and convex curvature. The Au adatoms have, locally, different mobility on these different-curvature regions: within regions where they have a lower mobility then their nucleation in small clusters is favored with respect to defect-free graphite. A further aspect is that the energy barrier for the adatoms diffusion increases as the bonding strength of Au with the C atom increases: 2 L graphene is more stable than 1 L graphene due to the π bonding between the layers. In addition, the 2 L graphene has a lower roughness compared to 1 L graphene. Both factors should create weaker Au bonding and, thus, faster diffusion of the Au adatoms on 2 L graphene. About the diffusion of Au atoms on 1 L graphene on h-BN:

according to the experimental data the mobility of the Au adatoms on 1 L graphene supported on h-BN should be higher than on 1 L graphene supported on a SiO$_2$ substrate. However, the calculations in Figure 22 lead to the opposite conclusion which the authors impute to increased van der Waals forces between 1 L graphene and SiO$_2$ with respect to 1 L graphene and h-BN. This condition should lead to the increased mobility of the Au adatoms on 1 L graphene supported on h-BN.

3. Thin Metal Films Deposition on Graphene and Nanoparticles Formation by Dewetting Processes

3.1. The Dewetting Process

Thin metal films deposited on a non-metal susbstrate are, generally, thermodinamically unstable. Then, if enough energy is furnished to the film so that atomic diffusion occurs, the system tends to minimize the total surface and interface energy: the result is the break-up of the film and the formation of spherical metal particles minimizing the total exposed surface [123–130]. The dewetting process starts in structural defects of the films: these are the locations in which holes in the film, reaching the underlaying substrate, nucleate. The holes grow with time and two or more holes join (i.e. coalesce) with the result of leaving the film in filaments structures. These filaments, then, being unstable, decay in spherical particles by a Raileigh-like instability process. The overall result is, so, the formation of an array of metal NPs. In a certain range, the mean size and mean spacing of the formed NPs can be controlled by the thickness of the deposited film or by the characteristic parameters of the process inducing the dewetting phenomenon such as the temperature or time of an annealing process [123–130]. The energetic budget needed to start the dewetting process of the film (i.e., to activate the atomic diffusion) can be furnished to the film by standard thermal annealing, or, alternatively, by laser, ion, electron beam irradiations. In addition, for metal films, the dewetting process can occur both in the solid or molten state.

Nowadays, in the nanotechnology working framework, the dewetting of ultrathin metal films on surface is routinely exploited to produce arrays of metal NPs on surfaces in view of technological applications [123–130] such as those based on plasmonic effects (Surface Enhanced Raman Scattering), magnetic recording, nanoelectronics, catalysis, etc.

Due to these peculiarities, the dewetting process was, also, exploited to produce, in a controlled way, metal NPs on graphene surface from deposited thin metal films. This approach is effective in the production of shape- and size-selected metal NPs on the graphene surface for some interesting applications involving, for example, the Surface Enhanced Raman Scattering of the NPs as modified by the interaction with the graphene layer.

3.2. Dewetting of Au Films on Graphene

Zhou et al. [68] investigated the possibility to produce and to control size, density and shape of Au NPs on graphene by the dewetting process of deposited thin films. So, after depositing the Au films, they performed annealing processes to induce the evolution of the films in NPs and, interestingly, they found that the shape, size and density of the obtained NPs can be controlled by the number of graphene layers and by the annealing temperature.

First of all, the authors [68] transferred n-layer graphene on a SiO$_2$ substrate after having obtained the n-layer graphene by standard mechanical exfoliation. The number of the graphene layers on the SiO$_2$ substrate was determined by crossing optical microscope and micro-Raman spectroscopy. After depositing (by thermal evaporation) thin Au films onto the n-layers graphene and onto the SiO$_2$ surface as reference, annealing processes were performed in the 600 °C–900 °C temperature range for 2 h. Then the authors used Scanning Electron Microscopy analysis to study shape, size and density of the observed NPs as a function of the annealing temperature, thickness of the starting deposited Au film, number of graphene layers supporting the Au film. The following general considerations are drawn by the authors on the basis of the results of these analysis: firstly, if the annealing temperature is in the 600–700 °C range, then the Au film on n-layer graphenes can be tuned into hexagon-shaped Au NPs.

Secondly, annealing at 800 °C produces, instead, coexistence of hexagonal and triangular Au NPs on graphenes. Thirdly, annealing at 900 °C produces irregular-shaped Au NPs on graphenes. Moreover, the density and size of the formed Au NPs on n-layer graphenes are strictly dependent on the number n of graphene layers. In particular, increasing n the NPs mean size increases and the NPs surface density decreases. As an example, Figure 23 reports Scanning Electron Microscopy images showing the dewetting of 2 nm-thick Au film into hexagonal Au NPs on graphene after thermal annealing at 700 °C for 2 h being the Au film supported directly on SiO_2 (Figure 23a left) and on monolayer graphene (Figure 23a right), supported directly on SiO_2 (Figure 23b left) and on bilayer graphene (Figure 23b right), supported directly on SiO_2 (Figure 23c right) and on trilayer graphene (Figure 23c left), supported on bilayer graphene (Figure 23d left) and on four-layer graphene (Figure 23d right). It can be recognized that the size and density of hexagonal Au NPs is established by the number n of graphene layers. In fact, it is observed that the increase of n produces an increase of the the size of the Au NPs increases and a decrease of their surface density.

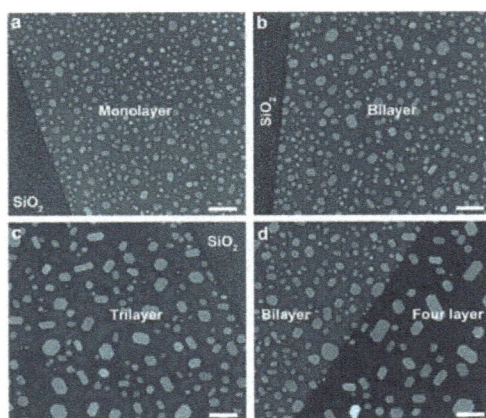

Figure 23. Scanning Electron Microscopy images showing the dewetting of Au films into hexagonal Au NPs on graphene after thermal annealing at 700 °C for 2 h. Film thickness: 2.0 nm. Scale bar: 200 nm. (**a**) Hexagonal Au NPs on SiO_2 (**left**) and monolayer graphene (**right**); (**b**) Hexagonal Au NPs on SiO_2 (**left**) and bilayer graphene (**right**); (**c**) Hexagonal Au NPs on trilayer graphene (**left**) and SiO_2 (**right**); (**d**) Hexagonal Au NPs on bilayer (**left**) and four layer graphene (**right**). Reproduced from Reference [68] with permission from Elsevier.

Another aspect is that the Au film dewetting process on the n-layer graphenes is thickness-dependent. The influence of the Au film thickness on the shape of the obtained Au NPs is described by the images in Figure 24: it presents Scanning Electron Microscopy images of 1 nm, 1.5 nm and 2 nm thick Au films on n-layers graphene and annealed at 600 °C for 2 h. With the increase of the Au film thickness, the effect of the graphene layers number on the shape of Au NPs becomes more and more weak. When film thickness is below 2.0 nm, after thermal annealing at 600 or 700 °C, almost all the Au NPs show hexagonal shape. Whereas for 5.0 nm Au film or more, although hexagon-shaped Au NPs still exist after annealing at 600 °C, the Au NPs are not well faceted.

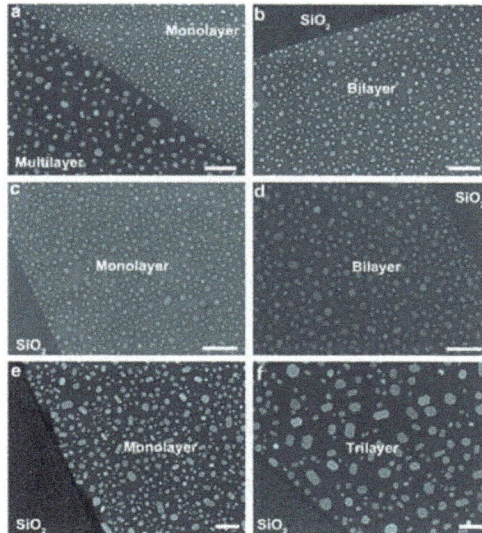

Figure 24. Scanning Electron Microscopy images showing the effects of the starting thickness of the deposited Au film on the shape of the resulting NPs after the annealing process at 600 °C for 2 h. Scale bar: 200 nm. (**a,b**) 1.0 nm thick Au; (**c,d**) 1.5 nm thick Au; (**e,f**) 2.0 nm thick Au. It is obvious to find that with the increase of Au film thickness, the modulation becomes less effective. Reproduced from Reference [68] with permission from Elsevier.

All these experimental data highlight the key role of the graphene layers number in determining size, density and shape of the Au NPs clearly indicates that n establishes the interaction strength between the graphene and the Au atoms affecting, as a consequence, the Au diffusivity and the final Au NPs morphology. To infer information on the parameters governing the Au NPs shape, size and density evolution, the authors [68] take into considerations the following main factors: the Au adatoms are weakly bonded with C atoms on graphene surface (interpreted as a physical adsorption rather than a chemical bonding) and the strength of this bonding is largely influenced by the number of graphene layers [101–104]. So, with the increase of layer number the inter-layer interaction strength decreases and, consequently, the interaction between Au adatoms and n-layer graphene becomes much weaker, resulting in the thickness-dependent particle size and density of Au NPs on graphenes by the different Au mobility on the graphenes. The surface diffusion of metal adatoms on graphenes can be described by these two equations: $D \propto \exp(-E_{a,n}/kT)$ and $N \propto (1/D)^{1/3}$, being D the adatoms surface diffusion, N the NPs surface density, $E_{a,n}$ the activation energy for the adatoms surface diffusion on n-layers graphene. Combining these two equations, the relation $N \propto \exp(E_{a,n}/3kT)$ is obtained. So, Zhou et al. [68] conclude that the decrease of surface diffusion barrier with increasing the number of graphene layers n explains the observed experimental data: the diffusion coefficient establishing the diffusion length, determines the joining probability for the adatoms. Therefore, concerning the thermal annealing post-growth processes, it establishes the size and density of the formed NPs by the competition between nucleation and growth phenomena [69]. Therefore, different surface diffusion coefficients (by different activation energies $E_{a,n}$) of the Au adatoms on the n-layers graphene can result in n-dependent morphologies, sizes, and density of the Au NPs on n-layer graphenes. To support quantitatively these considerations, in a further study, Zhou et al. [69], proceeded to the quantification of the size and density of the Au NPs after the thermal treatment. In particular, the authors proceeded to the following experiment: after depositing a Au film on the SiO₂ substrate, on 1-layer, 2-layers,

3-layers, and 4-layers graphene supported on the SiO$_2$ substrate, the authors performed a 1260 °C-30 s annealing to obtain round-shaped Au NPs as shown in Figure 25a but with a different size and surface density N of the NPs on the basis of the number n of the graphene layers.

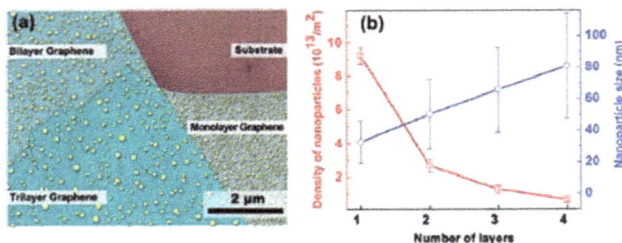

Figure 25. Morphologies, size, and density of Au nanoparticles on n-layer graphenes after annealing at 1260 °C in vacuum for 30 s (false-color image). Note that no Au NPs are found in the substrate. (**a**) Au NPs on monolayer, bilayer, and trilayer graphene, respectively; (**b**) Statistics of the size and density of gold nanoparticles on n-layer graphenes. Reproduced from Reference [69] with permission from the American Chemical Society.

As reported in Figure 25b the authors quantified the size and the surface density of the Au NPs as a function of n. In particular, N *versus* n was analyzed by the N∝exp (E$_{a,n}$/3kT) relation. Although it is difficult to obtain the absolute value of barriers due to the lack of the pre-exponential factor, the authors were able in evaluate the barrier difference between n-layer graphene by the density ratios: E$_{a,1}$ − E$_{a,2}$ = 3kTln(N$_1$/N$_2$) = 504 ± 44 meV, and similarly, E$_{a,2}$ − E$_{a,3}$ = 291 ± 31 meV, E$_{a,3}$ − E$_{a,4}$ = 242 ± 22 meV.

3.3. Dewetting of Ag Films on Graphene

Zhou et al. [71] extended their work to the dewetting of Ag films on n-layers graphene. In this case, in addition, a detailed study of the Surface Enhanced Raman Scattering of the Ag NPs was also conducted. The authors deposited Ag films onto n-layer graphenes (supported on SiO$_2$). In this case experiments were conducted maintaining the substrate temperature at 298, 333, and 373 K during the Ag depositions and Scanning Electron Microscopy images were used to study the morphology, size, surface density of the produced Ag NPs on the graphene layers as a function of the substrate temperature. In addition, also in this case, a strict dependence of the Ag NPs morphology, size and surface density on the number of graphene layers n supporting the Ag film was found. Similarly to Au, this was attributed by the authors to the changes in the surface diffusion coefficient of Ag on n-layer graphenes at different temperatures (the substrate temperature during Ag depositions, in this case). In addition, the authors observed that Raman scattering of n-layer graphenes is greatly enhanced by the presence of the Ag NPs. In particular, they found that the enhancement factors depend on the number n of graphene layers. Monolayer graphene has the largest enhancement factors, and the enhancement factors decrease with layer number increasing. Obviously, this is due to the specific structural characteristics of the Ag NPs as determined by n.

In particular, the authors [71] thermally evaporated 2 or 5 nm Ag films onto n-layer graphenes supported on the 300 nm-thick SiO$_2$ layer grown on Si. During the Ag depositions, the substrate is kept at 298 K, or 333 K, or 373 K. On the basis of the substrate temperature and number of graphene layers, different shapes, sizes, and surface density are obtained for the resulting Ag NPs. As an example, Figure 26 reports Scanning Electron Microscopy images of 5 nm-thick Ag film deposited on SiO$_2$, on 1-layer, and 2-layers graphene with the substrate kept at 298 K (a–b), 333 K (c–d), 373 K (e–f). The differences in the formed Ag NPs are just evident at 333 K: on one layer graphene, the density of Ag NPs larger than that on bilayer graphene, the NPs spacing is lower, but the NPs diameter are similar in the two samples. At 373 K these differences are enhanced: the Ad NPs present very different sizes, spacing, and surface density as a function of the number n of the graphene layers. For example, the NPs on monolayer graphene are much smaller than those on bilayer graphene.

Then, using Raman spectroscopy, the authors found different Surface Enhanced Raman Spectroscopy (SERS) effects of Ag on n-layer graphenes [71], as summarized in Figures 27 and 28. In Figure 27, the authors compare the enhancement effects of 2 and 5 nm Ag deposited at 298 K on the graphene samples. Raman spectra of n-layer graphenes with 5 nm are enhanced with respect to 2 nm Ag (and pristine graphene): the G and 2D bands are more intense. This can be attribute to the fact that the deposition of the 5 nm Ag leads to the formation of NPs with higher surface density and lower spacing with the result to increase the SERS hot spots number per unit area. As a consequence, the increased density of hot spots causes a higher electric filed localization and, so, higher enhancement factors. To further analyze the Raman scattering properties of the graphene supporting the Ag NPs, the authors measured the Raman spectra of n-layer graphenes supporting NPs obtained by the deposition of 5 nm-thick Ag maintaining the substrate at 298, 333, and 373 K, see Figure 28. A higher SERS enhancement factor is obtained from graphene covered by Ag NPs obtained by depositing 5 nm Ag maintaining the substrate at 333 K than at 298 K: in fact, at 333 K larger Ag NPs are obtained with the same spacing of those obtained at 298 K. However, when the 5 nm Ag film is deposited maintaining the substrate at 373 K, the particles are larger but, also, the NPs spacing increases, resulting in a decrease of the enhancement factor.

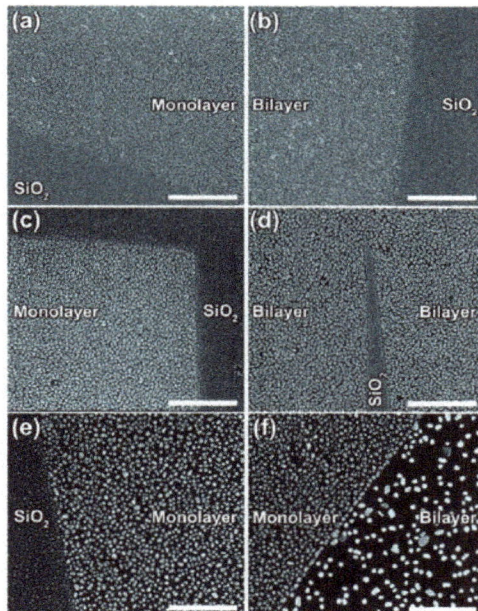

Figure 26. Scanning Electron Microscopy images (1 μm scale bar) of monolayer and bilayer graphene on SiO$_2$ after deposition of 5 nm Ag maintaining the substrate at different temperature during the deposition: (**a,b**) 298 K; (**c,d**) 333 K; (**e,f**) 373 K. Reproduced from Reference [71] with permission from the American Chemical Society.

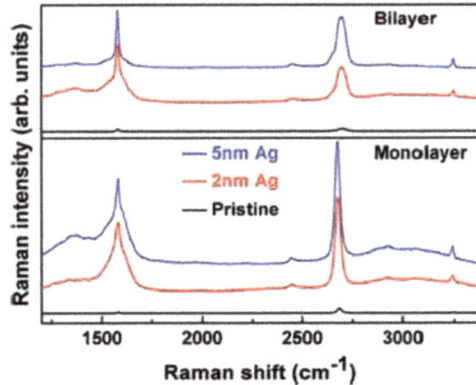

Figure 27. Raman spectra from monolayer and bilayer graphene on SiO$_2$ having deposited on the graphene 2 or 5 nm Ag film. Reproduced from Reference [71] with permission from the American Chemical Society.

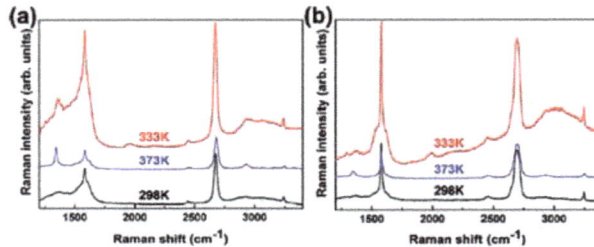

Figure 28. Raman spectra of monolayer (**a**) and bilayer (**b**) graphene covered by 5 nm Ag deposited maintaining the substrate at 298 K (black line), 333 K (red line), and 373 K (blue line). Reproduced from Reference [71] with permission from the American Chemical Society.

4. Some Considerations on the Electrical Behavior of Metal-Graphene Contacts

As discussed in the introductory section, metal NPs/graphene hybrid systems present properties which are exploited for applications in areas such as Surface Enhanced Raman Scattering (SERS), nanoelectronics, photovoltaics, catalysis, electrochemical sensing, hydrogen storage, etc. [17–19,31–71]. All these applications, however, are connected to the specific interaction occurring at the metal NP/graphene interface because characteristics like metal adhesion and electrical contact properties are strongly influenced by the interface structure. In this sense, the theoretical and experimental study of the metal-graphene interface structure and how the metal contact influences the graphene electronic properties is an active field of study [75–113]. From a more general point of view, any application of graphene in building electronic devices requires the graphene contacting by metal layers and it is widely recognized that the metal-graphene interaction strongly influences the graphene electrical conduction properties being, often, a limiting factor in produce high-efficiency electronic devices. For example, several theoretical and experimental analysis suggest that the difference in the work functions of the metal and graphene leads to the charge transfer and doping of the graphene layer [101–104]. In general, to realize high-performance devices, it is very important to produce metallic contacts on graphene which show a very low contact resistance. In principle, an Ohmic contact is obtained without any difficulty by the contact of a metal with graphene layer due to the graphene lack of a band gap but it is concerned that a very small density of states (DOS) for graphene might suppress the current injection from the metal to graphene [78,107]. In general [107], a metal/metal contact has no potential barrier and the carrier is transferred directly through the metal/metal interface to cancel the difference in work functions. Since graphene has not band-gap, the case of the metal/graphene contact should be similar to the metal/metal contact. However, differently from the metal/metal contact, in the metal/graphene contact the effects of the very small DOS for graphene have to be considered: in particular, the amount of charge transfer gradually decreases from the metal/graphene interface. This charge transfer forms the dipole layer at the interface and the very small DOS around the Fermi level of graphene increases produces a high screening length. As a result of the long charge transfer region, a p–n junction arises near the metal/graphene contact. On the other hand, the graphene Fermi level position with respect to the conical point is strongly influenced by the adsorption of metal atoms on the graphene surface [101–104] causing the graphene doping. As a consequence, metal/graphene contacts show different electrical behaviors depending on the specific graphene doping induced by the peculiar contacting metal. As summarized in Table 4, theoretical calculations by Giovannetti et al. [101,102], for example, show that different metals, by their specific electronic interaction with graphene, causes different shifts of the graphene Fermi level with respect to the Dirac point: those metals (Au, Pt) which interacting with graphene causes the shift the graphene Fermi level below the Dirac point, are p-type doping the graphene. These are the metals which causes an increase of the free-standing graphene work-function (4.48 eV), see Table 4. Those metals (Ni, Co, Pd, Al, Ag, Cu) which interacting with graphene causes the shift the graphene Fermi level above the Dirac point, are n-type doping the graphene. These are the metals which causes a decrease of the free-standing graphene work-function, see Table 4.

Table 4. Results of the calculations of Giovannetti et al. for the electronic characteristics of metals/graphene contacts: d_{eq} equilibrium distance for the metal atom-graphene system, ΔE metal atom-graphene binding energy, W_M metal work-function, W graphene work-function in the free-standing configuration (4.48 eV) and when in contact with the metal. Reproduced from Reference [102] with permission from the American Physical Society.

	Gr	Ni	Co	Pd	Al	Ag	Cu	Au	Pt
d_{eq} (Å)		2.05	2.05	2.30	3.41	3.33	3.26	3.31	3.30
ΔE (eV)		0.125	0.160	0.084	0.027	0.043	0.033	0.030	0.038
W_M (eV)		5.47	5.44	5.67	4.22	4.92	5.22	5.54	6.13
W (eV)	4.48	3.66	3.78	4.03	4.04	4.24	4.40	4.74	4.87

So, it is clear, also from an experimental point of view, the importance to study the electrical characteristics of several metal-graphene systems. In this regard, an interesting analysis was reported by Watanabe et al. [105]: in this work the authors studied, experimentally, the contact resistance R_C of several metals (Ti, Ag, Co, Cr, Fe, Ni, Pd) to graphene with the results summarized in Figure 29: it reports the contact resistance (the square marks the mean value for the specific metal) for several metal films deposited on graphene. It is interesting to note that is not strongly related to the metal work function. Instead, analyzing the microstructure of the deposited metal films, the authors conclude that the contact resistance is significantly affected by this microstructure (as determined by the deposition conditions) according to the pictorial scheme reported in Figure 30.

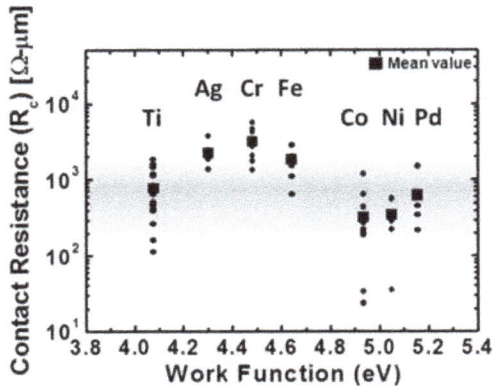

Figure 29. Metal-graphene contact resistance versus metal work-function. The square indicates the mean value. Reproduced from Reference [105] with permission from Elsevier.

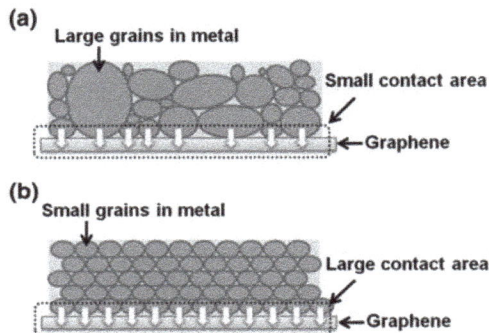

Figure 30. Schematic picture of metal contact to graphene. (**a,b**) indicate a schematic model of the metal/graphene junction for the large and small contact resistance values, respectively. The model shows that the contact resistance becomes smaller with increasing contact area between the metal grain and the graphene. Reproduced from Reference [105] with permission from Elsevier.

Connecting the analysis on the contact resistance to the microstructure of the metal films, the authors draw the following conclusions: for the large contact resistance metals (Ag, Fe, and Cr) the films result to be formed by large grains and to present rough surfaces, while for the small contact resistance metals (Pd, Ni, Co) the films are formed by small grains and present uniform surfaces. The effects of these different situations are pictured in Figure 30: large grains and rough surface of a metal films lead to a small contact are between the metal and the graphene, resulting in high contact resistance; small grains and uniform surface of a metal film lead to a large contact are between the metal and the graphene, resulting in a low contact resistance. These results clearly indicate, as stressed throughout the entire paper, the importance of the control of the kinetics and thermodynamics nucleation and growth processes for metals deposited on graphene so to reach the optimum nano- and micro-scale structure/morphology of the growing films/NPs for specific functional applications.

5. Conclusions, Open Points, and Perspectives

The next developments for metal NPs/Graphene nanocomposites are conditioned to the atomic scale control of the fabrication of the metal NPs and optimization of the techniques for reaching the wide-range control of the nano-architecture. Nowadays, several properties and applications of metal NPs/Graphene nanocomposites have been explored. As a non-exhaustive synthesis, Table 5 reports some examples of the properties and technological applications for several metal NPs/graphene systems, ranging from sensing and biosensing to nanoelectronics, catalysis and solar devices [135–139].

Surely, new insights and perspectives are related to the nanoscale control of the spatial organization and shape of the NPs. In this sense, the use of techniques to self-assembly the metal NPs on the graphene in spatially ordered arrays will be the key approach. So, in general, the key step towards real engineering of the metal NPs/graphene nanocomposites is the development of methodologies to produce complex nanoscale architectures. Towards this end, the vapor-deposition based techniques can open new perspectives.

Fine control of the morphology of the metal NPs on graphene is also a very interesting challenge. By the possibility to grow a range of geometric shapes at the nanoscale, the production of complex-morphology metal NPs on graphene is an interesting area of research, especially with regard to the resulting plasmonic properties.

Another interesting point concerns the use of new metal NPs (with specific functionalities) in the mixing with graphene. Probably, alloys of metals and core-shell type NPs (Ag/Au, Au/Pd, Pd/Pt, Pt/Rh, Pt/Ru) could be very useful tool, particularly in information storage and biomedicine applications.

A recent field of investigation for metal NPs/graphene nanocomposites is that related to photocatalysis [140]. Towards this application, however, the key requirement is the development of procedures allowing the preparation of composites which are biocompatible, biodegradable, and non-toxic and assuring, also, the control of the NPs size and shape. Notably, the long-term efficiencies of the metal NPs/graphene in real photocatalytic applications composites represents an important practical issue to be resolved.

Table 5. Table summarizing some specific metal NPs-graphene composite systems with the corresponding exploited properties and/or applications.

System	Property	Application	Reference
Au NPs/Graphene	Sensitivity Enhancement	Clinical Immunoassays	[31]
Pd NPs/Graphene	Electrochemical Activity	Glucose Biosensor	[32]
Ag NPs/Graphene	Raman Scattering Electrochemical activity	Surface Enhanced Raman scattering H_2O_2 Sensing Glucose Sensing	[35]
Pd NPs/Graphene	Electrical Conduction	Hydrogen Sensing	[36]
Ag NPs/Graphene	Thermal Conductivity	Thermal Interface Materials	[37]
Au NPs/Graphene	Localized Surface Plasmon Resonance	Flexible and Transparent Optoelectronics	[40]
Au, Pd, Pt NPs/Graphene	Electrochemical Activity	H_2S Sensing	[41]
Pd NPs/Graphene	-	Heterogeneous Catalysis	[48]
Au NPs/Graphene	Plasmon Absorption	-	[51]
Au, Ag NPs/Graphene	Plasmonic Properties	Surface Enhanced Raman Spectroscopy	[52]
Au, Ag, Pd, Pt NPs/Graphene	Plasmonic Properties	Raman Spectroscopy	[53]
Ag NPs/Graphene	Plasmonic Properties	Surface Enhanced Raman Spectroscopy	[54]
Au, Co, Pd, Pt, Rh NPs/Graphene	-	Catalysis	[65]
Au NPs/Graphene	-	Surface Enhanced Raman Spectroscopy	[68]
Ag NPs/Graphene	-	Surface Enhanced Raman Spectroscopy	[71]
AuAg NPs/Graphene	Plasmonic properties	Solar Cell	[135]
Ag NPs/Graphene	Plasmonic properties	Photodetection	[136]
Al NPs/Graphene	Plasmonic properties	Solar Cell	[137]
Au NPs/Graphene	-	Catalysis	[138]
Ni NPs/Graphene	-	Photocatalysis	[139]
Pd NPs/Graphene	-	Hydrogen Storage	[140]
Pb NPs/Graphene	-	Thermoelectric Devices	[141]

In the renewable energy production field, metal NPs-graphene composites are attracting great interest. The graphene can be used in a solar cell as a transparent conductive electrode and the metal NPs as plasmonic scattering elements [40,135,137]. Significant results have been already achieved. However, in addition solar cell devices, thermoelectric devices are attracting much attention [141].

Towards these perspectives and developments, the physical vapor deposition processes-based techniques to produce the metal NPs-graphene composites will acquire, surely, more and more importance due to their simplicity, versatility, and high throughput. For these reasons such techniques are, in perspective, the main candidates to be implemented in the industry market for the large-area production and commercialization of functional devices based on the metal NPs-graphene composites. Toward this end, the present paper highlighted the key importance of the understanding and controlling the microscopic thermodynamics and kinetics mechanisms involved in the nucleation and growth processes of atoms on/in graphene. So, crossed theoretical and experimental studies characterizing these mechanisms and quantifying the involved parameters such as adsorption energies, activation energies, diffusion constants, etc. will acquire more and more importance. In fact, the fine control of these parameters will allow the superior control on the morphological/structural characteristics of the composites and so, as a consequence, the tuning of all the physico-chemical properties of the composites for high-efficiency functional applications.

Acknowledgments: This work has been supported, in part, by the project GraNitE "Graphene heterostructures with Nitrides for high frequency Electronics" (Grant No. 0001411), in the framework of the EU program "FET Flagship ERA-NET" (FLAG-ERA)

Conflicts of Interest: The authors declare no conflict of interest.

References

1. Novoselov, K.S.; Geim, A.K.; Morozov, S.V.; Jiang, D.; Zhang, Y.; Dubonos, S.V.; Grigorieva, I.V.; Firsov, A.A. Electric Field Effect in Atomically Thin Carbon Films. *Science* **2004**, *306*, 666–669. [CrossRef] [PubMed]
2. Geim, A.K.; Novoselov, K.S. The rise of graphene. *Nat. Mater.* **2007**, *6*, 183–191. [CrossRef] [PubMed]
3. Du, X.; Skachko, I.; Barker, A.; Andrei, E. Approaching ballistic transport in suspended graphene. *Nat. Nanotechnol.* **2008**, *3*, 491–495. [CrossRef] [PubMed]
4. Balandin, A.A.; Ghosh, S.; Bao, W.; Calizo, I.; Teweldebrhan, D.; Miao, F.; Lau, C.N. Superior Thermal Conductivity of Single-Layer Graphene. *Nano Lett.* **2008**, *8*, 902–907. [CrossRef] [PubMed]
5. Balandin, A.A. Thermal properties of graphene and nanostructured carbon materials. *Nat. Mater.* **2011**, *10*, 569–581. [CrossRef] [PubMed]
6. Nika, D.L.; Balandin, A.A. Two-dimensional phonon transport in graphene. *J. Phys. Condens. Matter.* **2012**, *24*, 233203. [CrossRef] [PubMed]
7. Nair, R.R.; Blake, P.; Grigorenko, A.N.; Novoselov, K.S.; Booth, T.J.; Stauber, T.; Peres, N.M.R.; Geim, A.K. Fine structure constant defines visual transparency of graphene. *Science* **2008**, *320*, 1308. [CrossRef] [PubMed]
8. Stoller, M.D.; Park, S.; Zhu, Y.; An, J.; Ruoff, R.S. Graphene-Based Ultracapacitors. *Nano Lett.* **2008**, *8*, 3498–3502. [CrossRef] [PubMed]
9. Obradovic, B.; Kotlyar, R.; Heinz, F.; Matagne, P.; Rakshit, T.; Giles, M.D.; Stettler, M.A.; Nikonov, D.E. Analysis of graphene nanoribbons as a channel material for field-effect transistors. *Appl. Phys. Lett.* **2006**, *88*, 142102. [CrossRef]
10. Hass, J.; Feng, R.; Li, T.; Li, X.; Zong, Z.; de Heer, W.A.; First, P.N.; Conrad, E.H.; Jeffrey, C.A.; Berger, C. Highly ordered graphene for two dimensional electronics. *Appl. Phys. Lett.* **2006**, *89*, 143106. [CrossRef]
11. Katsnelson, M.I. Graphene: Carbon in two dimensions. *Mater. Today* **2007**, *10*, 20–27. [CrossRef]
12. Gilje, S.; Han, S.; Wang, M.S.; Wang, K.L.; Kaner, R.B. A Chemical Route to Graphene for Device Applications. *Nano Lett.* **2007**, *7*, 3394–3398. [CrossRef] [PubMed]
13. Cao, X.H.; Shi, Y.M.; Shi, W.H.; Lu, G.; Huang, X.; Yan, Q.Y.; Zhang, Q.; Zhang, H. Preparation of Novel 3D Graphene Networks for Supercapacitor Applications. *Small* **2011**, *7*, 3163–3168. [CrossRef] [PubMed]
14. Qi, X.Y.; Li, H.; Lam, J.W.Y.; Yuan, X.T.; Wei, J.; Tang, B.Z.; Zhang, H. Graphene Oxide as a Novel Nanoplatform for Enhancement of Aggregation-Induced Emission of Silole Fluorophores. *Adv. Mater.* **2012**, *24*, 4191–4195. [CrossRef] [PubMed]
15. Choi, W.; Lee, J. *Graphene: Synthesis and Applications*; CRC Press: New York, NY, USA, 2012.
16. Warner, J.H.; Schäffel, F.; Bachmatiuk, A.; Rümmel, M.H. *Graphene: Fundamentals and Emergent Applications*; Elsevier: Oxford, UK, 2013.

17. Tiwari, A.; Syväjärvi, M. *Graphene Materials-Fundamental and Emerging Applications*; Scrivener Publishing-Wiley: Hoboken, NJ, USA, 2015.

18. Huang, X.; Qi, X.Y.; Boey, F.; Zhang, H. Graphene-based composites. *Chem. Soc. Rev.* **2012**, *41*, 666–686. [CrossRef] [PubMed]

19. Huang, X.; Yin, Z.Y.; Wu, S.X.; Qi, X.Y.; He, Q.Y.; Zhang, Q.C.; Yan, Q.; Boey, F.; Zhang, H. Graphene-Based Materials: Synthesis, Characterization, Properties, and Applications. *Small* **2011**, *7*, 1876–1902. [CrossRef] [PubMed]

20. Mittal, V. *Polymer-Graphene Nanocomposites*; RSC Publishing: Cambridge, UK, 2012.

21. Potts, G.R.; Dreyer, D.R.; Bielawski, C.W.; Ruoff, R.S. Graphene-based polymer nanocomposites. *Polymer* **2011**, *52*, 5–25. [CrossRef]

22. Rafiee, M.A.; Lu, W.; Thomas, A.V.; Zandiatashbar, A.; Rafiee, J.; Tour, J.M.; Koratkar, N.A. Graphene nanoribbon composites. *ACS Nano* **2010**, *4*, 7415–7420. [CrossRef] [PubMed]

23. Cheng, J.; Du, J. Facile synthesis of germanium–graphene nanocomposites and their application as anode materials for lithium ion batteries. *Cryst. Eng. Commun.* **2012**, *14*, 397–400. [CrossRef]

24. Yin, Z.Y.; Wu, S.X.; Zhou, X.Z.; Huang, X.; Zhang, Q.C.; Boey, F.; Zhang, H. Electrochemical Deposition of ZnO Nanorods on Transparent Reduced Graphene Oxide Electrodes for Hybrid Solar Cells. *Small* **2010**, *6*, 307–312. [CrossRef] [PubMed]

25. Hsu, Y.-W.; Hsu, T.-K.; Sun, C.-L.; Nien, Y.-T.; Pu, N.-W.; Ger, M.-D. Synthesis of CuO/graphene nanocomposites for nonenzymatic electrochemical glucose biosensor applications. *Electrochim. Acta* **2012**, *82*, 152–157. [CrossRef]

26. Jiang, Z.; Wang, J.; Meng, L.; Huang, Y.; Liu, L. A highly efficient chemical sensor material for ethanol: Al_2O_3/Graphene nanocomposites fabricated from graphene oxide. *Chem. Commun.* **2011**, *47*, 6350–6352. [CrossRef] [PubMed]

27. An, X.; Yu, J.C.; Wang, Y.; Hu, Y.; Yu, X.; Zhang, G. WO_3 nanorods/graphene nanocomposites for high-efficiency visible-light-driven photocatalysis and NO_2 gas sensing. *J. Mater. Chem.* **2012**, *22*, 8525–8531. [CrossRef]

28. Su, J.; Cao, M.; Ren, L.; Hu, C. Fe_3O_4–Graphene Nanocomposites with Improved Lithium Storage and Magnetism Properties. *J. Phys. Chem. C* **2011**, *115*, 14469–14477. [CrossRef]

29. Williams, G.; Seger, B.; Kamat, P.V. TiO_2–Graphene Nanocomposites. UV-Assisted Photocatalytic Reduction of Graphene Oxide. *ACS Nano* **2008**, *2*, 1487–1491. [CrossRef] [PubMed]

30. Zhang, Y.; Tang, Z.-R.; Fu, X.; Xu, Y.-J. TiO_2–Graphene Nanocomposites for Gas-Phase Photocatalytic Degradation of Volatile Aromatic Pollutant: Is TiO_2–Graphene Truly Different from Other TiO_2–Carbon Composite Materials? *ACS Nano* **2010**, *4*, 7303–7314. [CrossRef] [PubMed]

31. Zhong, Z.; Wu, W.; Wang, D.; Wang, D.; Shan, J.; Qing, Y.; Zhang, Z. Nanogold-enwrapped graphene nanocomposites as trace labels for sensitivity enhancement of electrochemical immunosensors in clinical immunoassays: Carcinoembryonic antigen as a model. *Biosens. Bioelectron.* **2010**, *25*, 2379–2383. [CrossRef] [PubMed]

32. Zeng, Q.; Cheng, J.-S.; Liu, X.-F.; Bai, H.-T.; Jiang, J.-H. Palladium nanoparticle/chitosan-grafted graphene nanocomposites for construction of a glucose biosensor. *Biosens. Bioelectron.* **2011**, *26*, 3456–3463. [CrossRef] [PubMed]

33. Wang, C.; Li, J.; Amatore, C.; Chen, Y.; Jiang, H.; Wang, X.-M. Gold Nanoclusters and Graphene Nanocomposites for Drug Delivery and Imaging of Cancer Cells. *Angew. Chem. Int. Ed.* **2011**, *50*, 11644–11648. [CrossRef] [PubMed]

34. Gao, H.; Xiao, F.; Ching, C.B.; Duan, H. One-Step Electrochemical Synthesis of PtNi Nanoparticle-Graphene Nanocomposites for Nonenzymatic Amperometric Glucose Detection. *ACS Appl. Mater. Interfaces* **2011**, *3*, 3049–3057. [CrossRef] [PubMed]

35. Zhang, Y.; Liu, S.; Wang, L.; Qin, X.; Tian, J.; Lu, W.; Chang, G.; Sun, X. One-pot green synthesis of Ag nanoparticles-graphene nanocomposites and their applications in SERS, H_2O_2, and glucose sensing. *RSC Adv.* **2012**, *2*, 538–545. [CrossRef]

36. Johnson, J.L.; Behnam, A.; Pearton, S.J.; Ural, A. Hydrogen Sensing Using Pd-Functionalized Multi-Layer Graphene Nanoribbon Networks. *Adv. Mater.* **2010**, *22*, 4877–4880. [CrossRef] [PubMed]

37. Goyal, V.; Balandin, A.A. Thermal properties of the hybrid graphene-metal nano-micro-composites: Applications in thermal interface materials. *Appl. Phys. Lett.* **2012**, *100*, 073113. [CrossRef]

38. Wang, Z.J.; Zhang, J.; Yin, Z.Y.; Wu, S.X.; Mandler, D.; Zhang, H. Fabrication of nanoelectrode ensembles by electrodepositon of Au nanoparticles on single-layer graphene oxide sheets. *Nanoscale* **2012**, *4*, 2728–2733. [CrossRef] [PubMed]

39. Lee, J.; Shim, S.; Kim, B.; Shin, H.S. Surface-Enhanced Raman Scattering of Single- and Few-Layer Graphene by the Deposition of Gold Nanoparticles. *Chem. Eur. J.* **2011**, *17*, 2381–2387. [CrossRef] [PubMed]

40. Lee, S.; Lee, M.H.; Shin, H.-J.; Choi, D. Control of density and LSPR of Au nanoparticles on graphene. *Nanotechnology* **2013**, *24*, 275702. [CrossRef] [PubMed]

41. Gutés, A.; Hsia, B.; Sussman, A.; Mickelson, W.; Zettl, A.; Carraro, C.; Maboudian, R. Graphene decoration with metal nanoparticles: Towards easy integration for sensing applications. *Nanoscale* **2012**, *4*, 438–440. [CrossRef] [PubMed]

42. Kamat, P.V. Graphene-Based Nanoarchitectures. Anchoring Semiconductor and Metal Nanoparticles on a Two-Dimensional Carbon Support. *J. Phys. Chem. Lett.* **2010**, *1*, 520–527. [CrossRef]

43. Vedala, H.; Sorescu, D.C.; Kotchey, G.P.; Star, A. Chemical Sensitivity of Graphene Edges Decorated with Metal Nanoparticles. *Nano Lett.* **2011**, *11*, 2342–2347. [CrossRef] [PubMed]

44. Huang, X.; Li, H.; Li, S.Z.; Wu, S.X.; Boey, F.; Ma, J.; Zhang, H. Synthesis of gold square-like plates from ultrathin gold square sheets: the evolution of structure phase and shape. *Angew. Chem. Int. Ed.* **2011**, *50*, 12245–12248. [CrossRef] [PubMed]

45. Zhou, H.; Yang, H.; Qiu, C.; Liu, Z.; Yu, F.; Chen, M.; Hu, L.; Xia, X.; Yang, H.; Gu, C.; et al. Experimental evidence of local magnetic moments at edges of n-layer graphenes and graphite. *J. Phys. Chem. C* **2011**, *115*, 15785–15792. [CrossRef]

46. Huang, X.; Li, S.Z.; Wu, S.X.; Huang, Y.Z.; Boey, F.; Gan, C.L.; Zhang, H. Graphene Oxide-Templated Synthesis of Ultrathin or Tadpole-Shaped Au Nanowires with Alternating *hcp* and *fcc* Domains. *Adv. Mater.* **2012**, *24*, 979–983. [CrossRef] [PubMed]

47. Zaniewski, A.M.; Schriver, M.; Lee, J.G.; Crommie, M.F.; Zettl, A. Electronic and optical properties of metal-nanoparticle filled graphene sandwiches. *Appl. Phys. Lett.* **2013**, *102*, 023108. [CrossRef]

48. Jin, Z.; Nackashi, D.; Lu, W.; Kittrell, C.; Tour, J.M. Decoration, Migration, and Aggregation of Palladium Nanoparticles on Graphene Sheets. *Chem. Mater.* **2010**, *22*, 5695–5699. [CrossRef]

49. Huang, X.; Zhou, X.Z.; Wu, S.X.; Wei, Y.Y.; Qi, X.Y.; Zhang, J.; Boey, F.; Zhang, H. Reduced Graphene Oxide-Templated Photochemical Synthesis and in situ Assembly of Au Nanodots to Orderly Patterned Au Nanodot Chains. *Small* **2010**, *6*, 513–516. [CrossRef] [PubMed]

50. Huang, X.; Li, S.Z.; Huang, Y.Z.; Wu, S.X.; Zhou, X.Z.; Li, S.Z.; Gan, C.L.; Boey, F.; Mirkin, C.A.; Zhang, H. Synthesis of hexagonal close-packed gold nanostructures. *Nat. Commun.* **2011**, *2*, 292. [CrossRef] [PubMed]

51. Muszynski, R.; Seger, B.; Kamat, P.V. Decorating Graphene Sheets with Gold Nanoparticles. *J. Phys. Chem. C* **2008**, *112*, 5263–5266. [CrossRef]

52. Sidorov, A.N.; Sławiński, G.W.; Jayatissa, A.H.; Zamborini, F.P.; Sumanasekera, G.U. A surface-enhanced Raman spectroscopy study of thin graphene sheets functionalized with gold and silver nanostructures by seed-mediated growth. *Carbon* **2012**, *50*, 699–705. [CrossRef]

53. Subrahmanyam, K.S.; Manna, A.K.; Pati, S.K.; Rao, C.N.R. A study of graphene decorated with metal nanoparticles. *Chem. Phys. Lett.* **2010**, *497*, 70–75. [CrossRef]

54. Lee, J.; Novoselov, K.S.; Shin, H.S. Interaction between Metal and Graphene: Dependence on the Layer Number of Graphene. *ACS Nano* **2011**, *5*, 608–612. [CrossRef] [PubMed]

55. Zhou, H.; Yu, F.; Yang, H.; Qiu, C.; Chen, M.; Hu, L.; Guo, Y.; Yang, H.; Gu, C.; Sun, L. Layer-dependent morphologies and charge transfer of Pd on *n*-layer graphenes. *Chem. Commun.* **2011**, *47*, 9408–9410. [CrossRef] [PubMed]

56. Zan, R.; Bangert, U.; Ramasse, Q.; Novoselov, K.S. Metal−Graphene Interaction Studied via Atomic Resolution Scanning Transmission Electron Microscopy. *Nano Lett.* **2011**, *11*, 1087–1092. [CrossRef] [PubMed]

57. Liu, X.; Wang, C.-Z.; Hupalo, M.; Lin, H.-Q.; Ho, K.-M.; Tringides, M.C. Metals on Graphene: Interactions, Growth Morphology, and Thermal Stability. *Crystals* **2013**, *3*, 79–111. [CrossRef]

58. Gan, Y.; Sun, L.; Banhart, F. One- and Two-Dimensional Diffusion of Metal Atoms in Graphene. *Small* **2008**, *4*, 587–591. [CrossRef] [PubMed]

59. Pandey, P.A.; Bell, G.R.; Rourke, J.P.; Sanchez, A.M.; Elkin, M.D.; Hickey, B.J.; Wilson, N.R. Physical vapor deposition of metal nanoparticles on chemically modified graphene: observations on metal-graphene interactions. *Small* **2011**, *7*, 3202–3210. [CrossRef] [PubMed]

60. Liu, L.; Chen, Z.; Wang, L.; Polyakova, E.; Taniguchi, T.; Watanabe, K.; Hone, J.; Flynn, G.W.; Brus, L.E. Slow Gold Adatom Diffusion on Graphene: Effect of Silicon Dioxide and Hexagonal Boron Nitride Substrates. *J. Phys. Chem. B* **2013**, *117*, 4305–4312. [CrossRef] [PubMed]

61. N'Diaye, A.T.; Bleikamp, S.; Feibelman, P.J.; Michely, T. Two-dimensional Ir cluster lattice on a graphene moiré on Ir(111). *Phys. Rev. Lett.* **2006**, *97*, 215501. [CrossRef] [PubMed]

62. Pan, Y.; Gao, M.; Huang, L.; Liu, F.; Gao, H.J. Directed self-assembly of monodispersed platinum nanoclusters on graphene Moiré template. *Appl. Phys. Lett.* **2009**, *95*, 093106. [CrossRef]

63. Zhang, H.; Fu, Q.; Cui, Y.; Tan, D.L.; Bao, X.H. Fabrication of metal nanoclusters on graphene grown on Ru(0001). *Chin. Sci. Bull.* **2009**, *54*, 2446–2450. [CrossRef]

64. N'Diaye, A.T.; Gerber, T.; Busse, C.; Myslivecek, J.; Coraux, J.; Michely, T. A versatile fabrication method for cluster superlattices. *New J. Phys.* **2009**, *11*, 103045. [CrossRef]

65. Zhou, Z.; Gao, F.; Goodman, D.W. Deposition of metal clusters on single-layer graphene/Ru(0001): Factors that govern cluster growth. *Surf. Sci.* **2010**, *604*, L31–L38. [CrossRef]

66. Zan, R.; Bangert, U.; Ramasse, Q.; Novoselov, K.S. Evolution of Gold Nanostructures on Graphene. *Small* **2011**, *7*, 2868–2872. [CrossRef] [PubMed]

67. Wang, B.; Yoon, B.; König, M.; Fukamori, Y.; Esch, F.; Heiz, U.; Landman, U. Size-Selected Monodisperse Nanoclusters on Supported Graphene: Bonding, Isomerism, and Mobility. *Nano Lett.* **2012**, *12*, 5907–5912. [CrossRef] [PubMed]

68. Zhou, H.; Yu, F.; Chen, M.; Qiu, C.; Yang, H.; Wang, G.; Yu, T.; Sun, L. The transformation of a gold film on few-layer graphene to produce either hexagonal or triangular nanoparticles during annealing. *Carbon* **2013**, *52*, 379–387. [CrossRef]

69. Zhou, H.; Qiu, C.; Liu, Z.; Yang, H.; Hu, L.; Liu, J.; Yang, H.; Gu, C.; Sun, L. Thickness-Dependent Morphologies of Gold on *N*-Layer Graphenes. *J. Am. Chem. Soc.* **2010**, *132*, 944–946. [CrossRef] [PubMed]

70. Luo, Z.; Somers, L.A.; Dan, Y.; Ly, T.; Kybert, N.J.; Mele, E.J.; Johnson, A.T.C. Size-Selective Nanoparticle Growth on Few-Layer Graphene Films. *Nano Lett.* **2010**, *10*, 777–781. [CrossRef] [PubMed]

71. Zhou, H.; Qiu, C.; Yu, F.; Yang, H.; Chen, M.; Hu, L.; Sun, L. Thickness-Dependent Morphologies and Surface-Enhanced Raman Scattering of Ag Deposited on *n*-Layer Graphenes. *J. Phys. Chem. C* **2011**, *115*, 11348–11354. [CrossRef]

72. Feldheim, D.L.; Foss, C.A., Jr. *Metal Nanoparticles-Synthesis, Characterizations, and Applications*; Marcel Dekker Inc.: New York, NY, USA, 2002.

73. Johnston, R.L.; Wilcoxon, J. *Metal Nanoparticles and Nanoalloys*; Elsevier: Oxford, UK, 2012.

74. Sau, T.K.; Rogach, A.L. *Complex-Shaped Metal Nanoparticles-Bottom-up Syntheses and Applications*; Wiley-VCH: Weinheim, Germany, 2012.

75. Knoch, J.; Chen, Z.; Appenzeller, J. Properties of Metal–Graphene Contacts. *IEEE Trans. Nanotechnol.* **2012**, *11*, 513–519. [CrossRef]

76. Russo, S.; Craciun, M.F.; Yamamoto, M.; Morpurgo, A.F.; Tarucha, S. Contact resistance in graphene-based devices. *Phys. E* **2010**, *42*, 677–679. [CrossRef]

77. Venugopal, A.; Colombo, L.; Vogel, E.M. Contact resistance in few and multilayer graphene devices. *Appl. Phys. Lett.* **2010**, *96*, 013512. [CrossRef]

78. Nagashio, K.T.; Nishimura, T.; Kita, K.; Toriumi, A. Contact resistivity and current flow path at metal/graphene contact. *Appl. Phys. Lett.* **2010**, *97*, 143514. [CrossRef]

79. Schwierz, F. Graphene transistors. *Nature Nanotechnol.* **2010**, *5*, 487–496. [CrossRef] [PubMed]

80. Fiori, G.; Bonaccorso, F.; Iannaccone, G.; Palacios, T.; Neumaier, D.; Seabaugh, A.; Banerjee, S.K.; Colombo, L. Electronics based on two-dimensional materials. *Nat. Nanotechnol.* **2014**, *9*, 768–779. [CrossRef] [PubMed]

81. Fisichella, G.; Schilirò, E.; Di Franco, S.; Fiorenza, P.; Lo Nigro, R.; Roccaforte, F.; Ravesi, S.; Giannazzo, F. Interface Electrical Properties of Al_2O_3 Thin Films on Graphene Obtained by Atomic Layer Deposition with an in Situ Seedlike Layer. *ACS Appl. Mater. Interfaces* **2017**, *9*, 7761–7771. [CrossRef] [PubMed]

82. Vaziri, S.; Smith, A.D.; Östling, M.; Lupina, G.; Dabrowski, J.; Lippert, G.; Mehr, W.; Driussi, F.; Venica, S.; Di Lecce, V.; et al. Going ballistic: Graphene hot electron transistors. *Solid State Commun.* **2015**, *224*, 64–75. [CrossRef]

83. Giannazzo, F.; Fisichella, G.; Greco, G.; La Magna, A.; Roccaforte, F.; Pecz, B.; Yakimova, R.; Dagher, R.; Michon, A.; Cordier, Y. Graphene integration with nitride semiconductors for high power and high frequency electronics. *Phys. Status Solidi A* **2017**, *214*, 1600460. [CrossRef]

84. Wei, W.; Pallecchi, E.; Haque, S.; Borini, S.; Avramovic, V.; Centeno, A.; Amaia, Z.; Happy, H. Mechanically robust 39 GHz cut-off frequency graphene field effect transistors on flexible substrates. *Nanoscale* **2016**, *8*, 14097–14103. [CrossRef] [PubMed]

85. Fisichella, G.; Lo Verso, S.; Di Marco, S.; Vinciguerra, V.; Schilirò, E.; Di Franco, S.; Lo Nigro, R.; Roccaforte, F.; Zurutuza, A.; Centeno, A.; et al. Advances in the fabrication of graphene transistors on flexible substrates. *Beilstein J. Nanotechnol.* **2017**, *8*, 467–474. [CrossRef] [PubMed]

86. Moon, J.S.; Antcliffe, M.; Seo, H.C.; Curtis, D.; Lin, S.; Schmitz, A.; Milosavljevic, I.; Kiselev, A.A.; Ross, R.S.; Gaskill, D.K.; et al. Ultra-low resistance ohmic contacts in graphene field effect transistors. *Appl. Phys. Lett.* **2012**, *100*, 203512. [CrossRef]

87. Politou, M.; Asselberghs, I.; Radu, I.; Conard, T.; Richard, O.; Lee, C.S.; Martens, K.; Sayan, S.; Huyghebaert, C.; Tokei, Z.; et al. Transition metal contacts to graphene. *Appl. Phys. Lett.* **2015**, *107*, 153104. [CrossRef]

88. Smith, J.T.; Franklin, A.D.; Farmer, D.B.; Dimitrakopoulos, C.D. Reducing Contact Resistance in Graphene Devices through Contact Area Patterning. *ACS Nano* **2013**, *7*, 3661–3667. [CrossRef] [PubMed]

89. Feibelman, P.J. Pinning of graphene to Ir(111) by flat Ir dots. *Phys. Rev. B* **2008**, *77*, 165419. [CrossRef]

90. Feibelman, P.J. Onset of three-dimensional Ir islands on a graphene/Ir(111) template. *Phys. Rev. B* **2009**, *80*, 085412. [CrossRef]

91. Varns, R.; Strange, P. Stability of gold atoms and dimers adsorbed on graphene. *J. Phys. Condens. Matter* **2008**, *20*, 225005. [CrossRef]

92. Srivastava, M.K.; Wang, Y.; Kemper, A.F.; Cheng, H.-P. Density functional study of gold and iron clusters on perfect and defected graphene. *Phys. Rev. B* **2012**, *85*, 165444. [CrossRef]

93. Amft, M.; Sanyal, B.; Eriksson, O.; Skorodumova, N.V. Small gold clusters on graphene, their mobility and clustering: a DFT study. *J. Phys. Condens. Matter* **2011**, *23*, 205301. [CrossRef] [PubMed]

94. Mao, Y.; Yuan, J.; Zhong, J. Density functional calculation of transition metal adatom adsorption on graphene. *J. Phys. Condens. Matter* **2008**, *20*, 115209. [CrossRef] [PubMed]

95. Sevinçli, H.; Topsakal, M.; Durgun, E.; Ciraci, S. Electronic and magnetic properties of 3d transition-metal atom adsorbed graphene and graphene nanoribbons. *Phys. Rev. B* **2008**, *77*, 195434. [CrossRef]

96. Chan, K.T.; Neaton, J.B.; Cohen, M.L. First-principles study of metal adatom adsorption on graphene. *Phys. Rev. B* **2008**, *77*, 235430. [CrossRef]

97. Zhang, W.; Sun, L.; Xu, Z.; Krasheninnikov, A.V.; Huai, P.; Zhu, Z.; Banhart, F. Migration of gold atoms in graphene ribbons: Role of the edges. *Phys. Rev. B* **2010**, *81*, 125425. [CrossRef]

98. Malola, S.; Häkkinen, H.; Koskinen, P. Gold in graphene: In-plane adsorption and diffusion. *Appl. Phys. Lett.* **2009**, *94*, 043106. [CrossRef]

99. Semidey-Flecha, L.; Teng, D.; Habenicht, B.F.; Sholl, D.S.; Xu, Y. Adsorption and Diffusion of the Rh and Au Adatom on Graphene Moiré/Ru(0001). *J. Chem. Phys.* **2013**, *138*, 184710. [CrossRef] [PubMed]

100. Liu, X.; Wang, C.Z.; Hupalo, M.; Lu, W.C.; Tringides, M.C.; Yao, Y.X.; Ho, K.M. Metals on graphene: Correlation between adatom adsorption behavior and growth morphology. *Phys. Chem. Chem. Phys.* **2012**, *14*, 9157–9166. [CrossRef] [PubMed]

101. Khomyakov, P.A.; Giovannetti, G.; Rusu, P.C.; Brocks, G.; van den Brink, J.; Kelly, P.J. First-principles study of the interaction and charge transfer between graphene and metals. *Phys. Rev. B* **2009**, *79*, 195425. [CrossRef]

102. Giovannetti, G.; Khomyakov, P.A.; Brocks, G.; Karpan, V.M.; van den Brink, J.; Kelly, P.J. Doping graphene with metal contacts. *Phys. Rev. Lett.* **2008**, *101*, 026803. [CrossRef] [PubMed]

103. Barraza-Lopez, S.; Vanević, M.; Kindermann, M.; Chou, M.Y. Effects of Metallic Contacts on Electron Transport through Graphene. *Phys. Rev. Lett.* **2010**, *104*, 076807. [CrossRef] [PubMed]

104. Sławińska, J.; Wlasny, I.; Dabrowski, P.; Klusek, Z.; Zasada, I. Doping domains in graphene on gold substrates: First-principles and scanning tunneling spectroscopy studies. *Phys. Rev. B* **2012**, *85*, 235430. [CrossRef]

105. Watanabe, E.; Conwill, A.; Tsuya, D.; Koide, Y. Low contact resistance metals for graphene based devices. *Diam. Relat. Mater.* **2012**, *24*, 171–174. [CrossRef]

106. Leong, W.S.; Gong, H.; Thong, J.T.L. Low-contact-resistance graphene devices with nickel-etched-graphene contacts. *ACS Nano* **2014**, *8*, 994–1001. [CrossRef] [PubMed]

107. Nagashio, K.; Toriumi, A. Density-of-States Limited Contact Resistance in Graphene Field-Effect Transistors. *Jpn. J. Appl. Phys.* **2011**, *50*, 070108. [CrossRef]

108. Liu, W.; Wei, J.; Sun, X.; Yu, H. A Study on Graphene—Metal Contact. *Crystals* **2013**, *3*, 257–274. [CrossRef]

109. Wang, X.; Xie, W.; Du, J.; Wang, C.; Zhao, N.; Xu, J.-B. Graphene/Metal Contacts: Bistable States and Novel Memory Devices. *Adv. Mater.* **2012**, *24*, 2614–2619. [CrossRef] [PubMed]

110. Kathami, Y.; Li, H.; Xu, C.; Banerjee, K. Metal-to-multilayer-graphene contact-Part I: Contact resistance modeling. *IEEE Trans. Nanotechnol.* **2012**, *59*, 2444–2452. [CrossRef]

111. Robinson, J.A.; LaBella, M.; Zhu, M.; Hollander, M.; Kasarda, R.; Hughes, Z.; Trumbull, K.; Cavalero, R.; Snyder, D. Contacting graphene. *Appl. Phys. Lett.* **2011**, *98*, 053103. [CrossRef]

112. Xia, F.; Perebeinos, V.; Lin, Y.-M.; Wu, Y.; Avouris, P. The origins and limits of metal–graphene junction resistance. *Nat. Nanotechnol.* **2011**, *6*, 179–184. [CrossRef] [PubMed]

113. Sundaram, R.S.; Steiner, M.; Chiu, H.-Y.; Engel, M.; Bol, A.A.; Krupke, R.; Burghard, M.; Kern, K.; Avouris, P. The Graphene–Gold Interface and Its Implications for Nanoelectronics. *Nano Lett.* **2011**, *11*, 3833–3837. [CrossRef] [PubMed]

114. Venables, J.A. *Introduction to Surface and Thin Film Processes*; Cambridge University Press: Cambridge, UK, 2000.

115. Lin, Y.; Chen, X. *Advanced Nano Deposition Methods*; Wiley-VCH: Weinheim, Germany, 2016.

116. Campbell, C.T. Ultrathin metal films and particles on oxide surfaces: Structural, electronic and chemisorptive properties. *Surf. Sci. Rep.* **1997**, *27*, 1–111. [CrossRef]

117. Ruffino, F.; Torrisi, V.; Marletta, G.; Grimaldi, M.G. Effects of the embedding kinetics on the surface nano-morphology of nano-grained Au and Ag films on PS and PMMA layers annealed above the glass transition temperature. *Appl. Phys. A* **2012**, *107*, 669–683. [CrossRef]

118. Ruffino, F.; De Bastiani, R.; Grimaldi, M.G.; Bongiorno, C.; Giannazzo, F.; Roccaforte, F.; Spinella, C.; Raineri, V. Self-organization of Au nanoclusters on the SiO_2 surface induced by 200 keV-Ar$^+$ irradiation. *Nucl. Instrum. Meth. Phys. Res. B* **2007**, *257*, 810–814. [CrossRef]

119. Ruffino, F.; Torrisi, V.; Marletta, G.; Grimaldi, M.G. Atomic force microscopy investigation of the kinetic growth mechanisms of sputtered nanostructured Au film on mica: Towards a nanoscale morphology control. *Nanoscale Res. Lett.* **2011**, *6*, 112. [CrossRef] [PubMed]

120. Ruffino, F.; Crupi, I.; Irrera, A.; Grimaldi, M.G. Pd/Au/SiC Nanostructured Diodes for Nanoelectronics: Room Temperature Electrical Properties. *IEEE Trans. Nanotechnol.* **2010**, *9*, 414–421. [CrossRef]

121. Ruffino, F.; Grimaldi, M.G. Island-to-percolation transition during the room-temperature growth of sputtered nanoscale Pd films on hexagonal SiC. *J. Appl. Phys.* **2010**, *107*, 074301. [CrossRef]

122. Ruffino, F.; Torrisi, V.; Marletta, G.; Grimaldi, M.G. Kinetic growth mechanisms of sputter-deposited Au films on mica: From nanoclusters to nanostructured microclusters. *Appl. Phys. A* **2010**, *100*, 7–13. [CrossRef]

123. Herminghaus, S.; Brinkmann, M.; Seemann, R. Wetting and Dewetting of Complex Surface Geometries. *Annu. Rev. Mater. Res.* **2008**, *38*, 101–121. [CrossRef]

124. Thompson, C.V. Solid-State Dewetting of Thin Films. *Annu. Rev. Mater. Res.* **2012**, *42*, 399–434. [CrossRef]

125. Giermann, A.L.; Thompson, C.V. Solid-state dewetting for ordered arrays of crystallographically oriented metal particles. *Appl. Phys. Lett.* **2005**, *86*, 121903. [CrossRef]

126. Henley, S.J.; Carrey, J.D.; Silva, S.R.P. Pulsed-laser-induced nanoscale island formation in thin metal-on-oxide films. *Phys. Rev. B* **2005**, *72*, 195408. [CrossRef]

127. Ye, J.; Thompson, C.V. Templated Solid-State Dewetting to Controllably Produce Complex Patterns. *Adv. Mater.* **2011**, *23*, 1567–1571. [CrossRef] [PubMed]

128. Ruffino, F.; Grimaldi, M.G. Self-organized patterned arrays of Au and Ag nanoparticles by thickness-dependent dewetting of template-confined films. *J. Mater. Sci.* **2014**, *49*, 5714–5729. [CrossRef]

129. Ruffino, F.; Grimaldi, M.G. Template-confined dewetting of Au and Ag nanoscale films on mica substrate. *Appl. Surf. Sci.* **2013**, *270*, 697–706. [CrossRef]

130. Ruffino, F.; Pugliara, A.; Carria, E.; Bongiorno, C.; Spinella, C.; Grimaldi, M.G. Formation of nanoparticles from laser irradiated Au thin film on SiO_2/Si: Elucidating the Rayleigh-instability role. *Mater. Lett.* **2012**, *84*, 27–30. [CrossRef]

131. Pan, Y.; Zhang, H.G.; Shi, D.X.; Sun, J.T.; Du, S.X.; Liu, F.; Gao, H.J. Highly Ordered, Millimeter-Scale, Continuous, Single-Crystalline Graphene Monolayer Formed on Ru (0001). *Adv. Mater.* **2009**, *21*, 2777–2780. [CrossRef]

132. Xu, Y.; Semidey-Flecha, L.; Liu, L.; Zhou, Z.; Goodman, D.W. Exploring the structure and chemical activity of 2-D gold islands on graphene moiré/Ru(0001). *Faraday Discuss.* **2011**, *152*, 267–276. [CrossRef] [PubMed]

133. Gyamfi, M.; Eelbo, T.; Wasniowska, M.; Wiesendanger, R. Fe adatoms on graphene/Ru(0001): Adsorption site and local electronic properties. *Phys. Rev. B* **2011**, *84*, 113403. [CrossRef]

134. Donner, K.; Jakob, P. Structural properties and site specific interactions of Pt with the graphene/Ru(0001) moiré overlayer. *J. Chem. Phys.* **2009**, *131*, 164701. [CrossRef] [PubMed]
135. Li, X.; Jia, C.; Ma, B.; Wang, W.; Fang, Z.; Zhang, G.; Guo, X. Substrate-induced interfacial plasmonics for photovoltaic conversion. *Sci. Rep.* **2015**, *5*, 14497. [CrossRef] [PubMed]
136. Maiti, R.; Sinha, T.K.; Mukherjee, S.; Adhikari, B.; Ray, S.M. Enhanced and Selective Photodetection Using Graphene-Stabilized Hybrid Plasmonic Silver Nanoparticles. *Plasmonics* **2016**, *11*, 1297–1304. [CrossRef]
137. Chen, X.; Jia, B.; Zhang, Y.; Gu, M. Exceeding the limit of plasmonic light trapping in textured screen-printed solar cells using Al nanoparticles and wrinkle-like graphene sheets. *Light Sci. Appl.* **2013**, *2*, e92. [CrossRef]
138. Li, Y.; Fan, X.; Qi, J.; Ji, J.; Wang, S.; Zhang, G.; Zhang, F. Gold nanoparticles–graphene hybrids as active catalysts for Suzuki reaction. *Mater. Res. Bull.* **2010**, *45*, 1413–1418. [CrossRef]
139. Zhang, W.; Li, Y.; Zeng, X.; Peng, S. Synergetic effect of metal nickel and graphene as a cocatalyst for enhanced photocatalytic hydrogen evolution via dye sensitization. *Sci. Rep.* **2015**, *5*, 10589. [CrossRef] [PubMed]
140. Zhou, C.; Szpunar, J.A. Hydrogen Storage Performance in Pd/Graphene Nanocomposites. *ACS Appl. Mater. Interface* **2016**, *8*, 25933–25940. [CrossRef] [PubMed]
141. Liang, Y.; Lu, C.; Ding, D.; Zhao, M.; Wang, D.; Hu, C.; Qiu, J.; Xie, G.; Tang, Z. Capping nanoparticles with graphene quantum dots for enhanced thermoelectric performance. *Chem. Sci.* **2015**, *6*, 4103–4108. [CrossRef]

crystals

MDPI

Review

Advanced Scanning Probe Microscopy of Graphene and Other 2D Materials

Chiara Musumeci [ID]

Department of Materials Science and Engineering and NU*ANCE* Center, Northwestern University, Evanston, IL 60208, USA; chiara.musumeci@northwestern.edu

Academic Editor: Filippo Giannazzo
Received: 17 May 2017; Accepted: 7 July 2017; Published: 11 July 2017

Abstract: Two-dimensional (2D) materials, such as graphene and metal dichalcogenides, are an emerging class of materials, which hold the promise to enable next-generation electronics. Features such as average flake size, shape, concentration, and density of defects are among the most significant properties affecting these materials' functions. Because of the nanoscopic nature of these features, a tool performing morphological and functional characterization on this scale is required. Scanning Probe Microscopy (SPM) techniques offer the possibility to correlate morphology and structure with other significant properties, such as opto-electronic and mechanical properties, in a multilevel characterization at atomic- and nanoscale. This review gives an overview of the different SPM techniques used for the characterization of 2D materials. A basic introduction of the working principles of these methods is provided along with some of the most significant examples reported in the literature. Particular attention is given to those techniques where the scanning probe is not used as a simple imaging tool, but rather as a force sensor with very high sensitivity and resolution.

Keywords: scanning probe microscopy; 2D materials; opto-electronic properties; mechanical properties; nanoscale characterization

1. Introduction

The growing interest in atomically thin two-dimensional (2D) materials, driven by the continuous discovery of new properties and low-dimensional physics, provides fertile ground for revolutionary post-silicon electronics [1–3]. Graphene is the most studied among 2D materials [4] because of its high ambipolar mobility, unique band structure, and a wealth of interesting properties such as the presence of massless Dirac fermions, the room temperature quantum hall effect, quasiparticle symmetry, chirality, and pseudospin [5–9]. Other 2D materials, such as transition metal dichalcogenides (TMDs) [10], have practical applications and fundamental properties complementary to those of graphene [11]. They exhibit atomically sharp interfaces, ultrathin dimensions, flexibility, and large optical effects [12]. Molybdenum disulfide (MoS_2), for example, has been tested in proof-of-concept ultrafast field-effect transistors (FETs), optical devices, and flexible electronics [13–15]. The presence of a band gap in a 6.5 Å thin monolayer MoS_2 makes it suitable for applications in nanoelectronics, allowing for the fabrication of transistors with low power dissipation, high current on/off ratios, and high charge mobility. Various memory devices have been fabricated with 2D materials, showing low power and energy consumption [16] as well as the possibility to be integrated in flexible devices [17]. Memristors based on grain boundaries in single layer MoS_2 devices have shown switching ratios up to ~10^3 [18].

Because of its high resolution and its ability to correlate several properties with the sample morphology at nanoscale [19,20], scanning probe microscopy has given a vast and valuable contribution to the understanding of the fundamental properties of graphene and other 2D materials [21]. Despite being a single-atom thick sheet, graphene is not perfectly flat. Corrugations up to 1 nm normal to the plane of the sheet, ripples, are commonly observed and thought to impart stability to the 2D lattice [22,23]. Only single-layer graphene is a zero-gap semiconductor, with one type of electron and one type of hole, while for three or more layers, several charge carriers appear and the conduction and valence bands start overlapping [7]. The aspect ratio of the graphene flakes influences the minimum conductivity [24]; and ripples also play a role in its electronic properties by inducing charge inhomogeneity as a consequence of the rehybridization of the π-σ bonding [23]. Large area films of 2D materials are polycrystalline. Consequently, grain boundaries, i.e., the interfaces between single-crystalline domains, inevitably affect their electronic transport, optical, mechanical, and thermal properties [25]. Optimizing large-scale growth processes for increasing the size of single-crystalline graphene is one of the main vectors of research. However, purposefully introducing and manipulating topological disorder is expected to become another important research objective to tailor 2D materials. Exceptional magnetic properties, for example, arise from the interplay of dislocation-induced localized states, doping, and locally unbalanced stoichiometry in grain boundaries in TMDs [26]. The number of layers, the size of the flakes, deformations, and the presence of defects or adsorbed molecules thus hugely affect these materials' properties, and because of the nanoscopic nature of these features, a tool addressing morphological and functional characterization on this scale is fundamental [27].

The aim of this review is to present an overview of different Scanning Probe Microscopy techniques used for the characterization of 2D materials (Figure 1), going from Scanning Tunneling Microscopy (STM) to Atomic Force Microscopy (AFM), and from Electrostatic Force Microscopy (EFM) and Kelvin Probe Force Microscopy (KPFM) to Conductive Atomic Force Microscopy (C-AFM) and Photoconductive Atomic Force Microscopy (PC-AFM). A basic introduction of the principles of operation and several among the most significant examples in the literature are shown. Particular attention is then given to those modes enabling an accurate control of the mechanical forces involved when an AFM tip is interacting with a 2D crystal sheet, where the AFM is not used as a simple imaging tool, but rather as a force sensor with very high sensitivity. In this respect, Force Spectroscopy modes, Friction Force Microscopy (FFM), and Piezoresponse Force Microscopy (PFM) are identified as very valuable tools to get quantitative information on single and multilayer 2D materials, ultimately enabling the tuning of their properties through strain engineering [28].

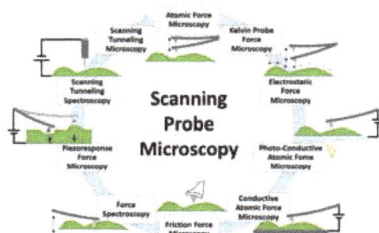

Figure 1. Scheme showing the different Scanning Probe Microscopy techniques described in this review for the characterization of two-dimensional (2D) materials.

2. Scanning Tunneling Microscopy

In a Scanning Tunneling Microscopy (STM) [29] experiment, a sharp metallic tip is separated by a few angstroms from a conductive sample. When a voltage is applied between the tip and the sample, electrons tunnel between them, producing an electric current, which decays exponentially with increasing tip–sample separation. In a standard operation, the current is kept constant during scanning by a feedback circuit, so that the vertical displacement of the scanner reflects the surface topography and gives true atomic resolution. Tip shape and sharpness are the two most important parameters in imaging surfaces, particularly those with significant topography. STM images invariably include contributions from specimen structure and tip geometry. Thus, the study of the tip's geometry is indispensable in distinguishing between the apparent and the true structure, or to establish the relationship among the tip's geometry, the true surface structure, and the STM image [30–32].

The crystal lattice of single-layer graphene has been observed by STM measurements on a wide variety of substrates, such as SiO_2 [33–35], SiC [36,37], Ir [38], Pd [39], Cu [40], Ru(0001) [41], and h-BN [42], showing different degrees of corrugation (Figure 2a,b). Moreover, STM can also provide information about the density of states of the 2D samples by Scanning Tunneling Spectroscopy (STS) and differential conductance (dI/dV) measurements [43,44]. STS has allowed for the measurement of the Dirac point of graphene, and has been valuable in demonstrating the correlation between atomic structure, defects, and grain boundaries in the electronic properties of these single-layer crystals [45–47]. Figure 2c,d, for example, shows the energy position of the Dirac point, ED, as a function of applied gate voltage, which could be extracted from the conductance minimum in dI/dV measurements of graphene deposited on SiO_2 [34]. An STM tip has been recently used to strain a graphene sample locally, in the form of a small Gaussian bump, and at the same time to map the imbalance of the local density of states (LDOS) at the sublattice level, demonstrating the pseudospin polarization by a pseudomagnetic field [48].

The applications of 2D materials for thermal management and thermoelectric energy conversion is also an emerging field of investigation. Appropriate nanostructuring and bandgap engineering of graphene can strongly reduce the lattice thermal conductance and enhance the Seebeck coefficient without dramatically degrading the electronic conductance [49]. Atomic-scale mapping of the thermopower of epitaxial graphene has been performed using STM, revealing that the spatial distributions of thermovoltage have a direct correspondence to the electronic density of states, and local thermopower distortions result from the modification of the electronic structure induced by individual defects, such as wrinkles, at the monolayer-bilayer interfaces [50].

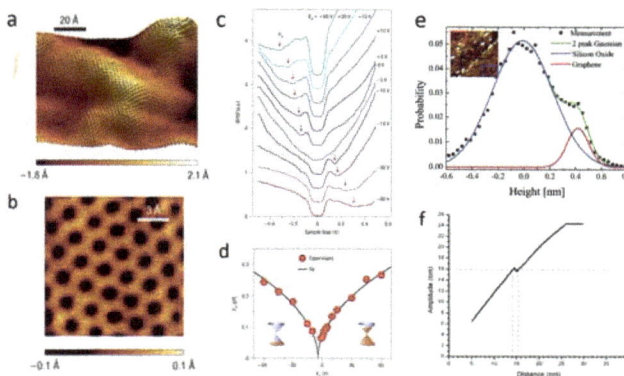

Figure 2. (**a–c**) Scanning Tunneling Microscopy (STM) of a graphene flake on a SiO_2 substrate. (**a**) Large scale constant current topography; (**b**) Close-up showing the honeycomb lattice; (**c**) dI/dV spectrum of graphene for different gate voltages, Vg, with red arrows indicating the gate-dependent positions of the conductance minimum outside the gap feature; (**d**) Energy position of the Dirac point, ED, as a function of applied gate voltage (extracted from the conductance minimum in (c) (adapted from [34]); (**e**) Height histogram acquired across the graphene-substrate boundary (see inset) of an Atomic Force Microscopy (AFM) image acquired in Non-Contact mode in ultrahigh vacuum (UHV). The data are fit by two Gaussian distributions with means separated by 4.2 Å (adapted from [35]); (**f**) The damping of the cantilever oscillation as a function of piezo displacement, recorded by approaching the tip towards the surface of a single graphene flake on silicon oxide substrate (adapted from [51]).

3. Atomic Force Microscopy

Atomic Force Microscopy images surfaces using the force exerted between the AFM probe and the sample as the feedback parameter [52]. To obtain an AFM topographic image, the sample is scanned by a tip mounted on a cantilever spring. While scanning, a feedback loop maintains the force between the tip and the sample constant by adjusting pixel by pixel the scanner's height, so that the image is obtained by plotting the height position versus its position on the sample. There are different modes of operation, which differ for the nature of tip motion and tip–sample interaction. Interactions can be attractive or repulsive, ultimately setting the distance between the tip and the sample. In a static mode, i.e., *Contact* mode, the tip is raster scanned over the sample's surface by maintaining its deflection constant. In dynamic modes, such as *Tapping*, *Non-Contact*, and *PeakForce Tapping*, the tip oscillates and the feedback is given by the amplitude, frequency, or maximum force at the contact point.

The high resolution obtained with Non-Contact AFM has allowed the visualization of ultra-flat graphene monolayers deposited on mica. The apparent roughness in these graphene layers was less than 25 picometres over micrometer lateral length scales, indicating that intrinsic ripples can be strongly suppressed by interfacial van der Waals interactions when this material is supported on an appropriate atomically-flat substrate [35]. Despite the high resolution, which makes AFM able to visualize nanostructures and defects, such as induced nanoripples [53] and adsorbates [54] on graphene, some limitations are still present for the determination of the monolayer thickness.

Contact mode AFM has been used to determine the number of layers of graphene films, but differences in height have been observed between forward and reverse scans. These differences have been attributed to the high lateral forces, such as friction, which play a non-negligible role in influencing the cantilever bending, ultimately resulting in an inaccurate estimation of the thickness. Such forces are negligible in dynamic modes, which are therefore preferred over *Contact* mode [51]. However, a notable discrepancy in the values for a single-layer thickness measured by dynamic modes AFM is present in the literature, with values ranging from 0.3 to 1.7 nm for single-layer graphene [51,55–57] and from 0.6 to 1 nm for MoS_2 [13,58,59] being attributed to tip–surface interactions and the experimental environment (e.g., physisorbed water and impurities). Using *Non-contact* atomic force microscopy, Ishigami et al. measured the thickness of a graphene film in ultrahigh vacuum (UHV) and in ambient conditions, showing that the large height measured in ambient conditions is due to a significant presence of atmospheric species under and/or on the graphene film (Figure 2e) [35]. The most common mode used to image and measure the thickness of these layered materials is certainly *Tapping* mode (TM) AFM. Similarly to what generally occurs for nanostructure imaging [60], the optimization of the free amplitude of oscillation in tapping mode was shown to be critical to a correct single-layer thickness assessment. In this mode, long range attractive forces are responsible for the oscillation damping. When the tip starts to approach the sample, the amplitude decreases linearly. At a certain tip–sample separation, a jump occurs in the amplitude, marking the onset of a region where, upon decreasing the tip–sample distance further, both long range attractive forces and short range repulsive forces act on the tip (Figure 2f). By looking at the amplitude-displacement curves of single layer graphene, a jump could be indeed observed where two different piezo displacement values corresponded to the same amplitude, the difference being about 1 nm [51]. This implies that if the measurement setpoint is selected in such a way as to coincide with the jump in amplitude, the feedback electronics may produce random switching from one displacement value to the other. After the jump, the damping of the oscillation increases further, and net repulsive forces characterize the tip–sample interaction [61]. Nemes-Incze et al. showed that the amplitude of the tapping cantilever greatly influences the measured height of the very same graphene platelet, so that differences of as much as 1 nm could be observed. They also demonstrated that to gain reliable thickness data, one needs to use a setpoint where the tip scans in the net repulsive regime, where the damping of the cantilever is largely due to the topography of the sample [51]. Likewise, a reversible decrease of the measured height from 1.69 to 0.43 nm was observed when imaging in *PeakForce Tapping* with loading forces from 1 to 10 nN, with the true value being obtained at higher forces [62]. Significantly, the measured thickness of multilayer graphene flakes was found to be independent of the applied force, with a constant step of 0.3 nm. It was speculated that the water layer between the flakes and the substrate is squeezed when a higher force, that is higher pressure, is exerted on the single layer during imaging. This minimizes the artifact, and allows for a more accurate measurement of the thickness. Both approaches, i.e., *Tapping* mode in a net repulsive regime and *PeakForce Tapping* at high contact forces, relying on the use of high forces during imaging, give a method to overcome the limitation of measuring single layers in ambient atmosphere, and therefore achieve a more accurate thickness measurement also in routine lab measurements.

Another fundamental issue is to image grain boundaries in 2D materials. Observing and engineering grain boundaries have been key in controlling the grain sizes, their electronic properties, and the related device performances. A valuable and easy method to observe grain boundaries rely on the combination of selective oxidation and AFM imaging. Selective oxidation is obtained by exposing directly the layers to ultraviolet light irradiation under moisture-rich conditions. The generated oxygen and hydroxyl radicals selectively functionalize defective grain boundaries, causing clear morphological changes at the boundaries, which can be clearly visualized by AFM imaging [63,64].

4. Electrical Modes

When AFM is operated in one of its electrical modes, it is possible to measure local electrical properties together with the sample's topography. These modes of operation make use of metal or metal-coated probes, and enable the application of an additional voltage between the tip and the sample. Electrostatic Force Microscopy (EFM) [65] and Kelvin Probe Force Microscopy (KPFM) [66] measures the contact potential difference or surface potential (SP) of a sample by recording long range electrostatic forces resulting from tip–sample interactions. These techniques provide a contactless electrical mapping of 2D flakes, allowing the extraction of crucial information about thickness, the distribution of the electrical potential and charge, as well as work function at nanoscale [20]. While the EFM method allows mainly for the qualitative mapping of surface potential, the KPFM technique provides quantitative values of the work function difference:

$$\Phi_s = \Phi_{tip} - eV_{CPD}, \tag{1}$$

where Φ_s and Φ_{tip} are the work functions of the sample and probe, respectively, and V_{CPD} is the contact potential difference directly measured by KPFM. In Conductive Atomic Force Microscopy (C-AFM) [67,68], the conductive tip acts as a movable electrode. The voltage is applied between the tip and a counter-electrode in contact with the sample, and a current is measured with high sensitivity, giving information on the local conductivity of the sample. In the simplest configuration, the sample is deposited on top of a conductive substrate and the conductive tip is scanned over such a surface by measuring point by point the current flowing vertically. Conversely, in a horizontal configuration, the sample is deposited on an insulating support and the electrical connection is obtained by laterally patterning a metal electrode. Current can in this way flow through the material, from the biased lateral contact to the movable metal-coated scanning probe tip [19]. In a similar configuration, named Photoconductive Atomic Force Microscopy (PC-AFM) [69], a light source is additionally used to excite the sample, so that the resulting photocurrent is measured by the AFM probe.

For electrical measurements, preventing probe-induced artifacts is very important. To obtain reliable data, the probe should be uniform, i.e., it should not have significant work function variations, and tip changes through tip–sample contact should be avoided. Optimal resolution in KPFM maps is obtained by long and slender but slightly blunt tips on cantilevers of minimal width and surface area [70]. However, tips modified with gold nanoparticles have also shown good resolution and sensitivity for graphene imaging [71]. Conductive tips fabricated by coating commercially available metal-varnished tips with graphene showed very high resistance to both high currents and frictions, leading to much longer lifetimes and preventing false imaging due to tip–sample interaction [72,73].

KPFM and EFM have been successfully used to identify the number of layers in epitaxial graphene [74]. Whereas an accurate topographical characterization is hindered by the presence of adsorbates in ambient conditions (see previous section), EFM can provide straightforward identification of the number of layers on the substrate [75]. Quantitative KPFM has revealed that graphene's work function is comparable to that of graphite, that is ~4.6 eV, and depends sensitively on the number of layers [76].

Theoretical studies have shown that the differences in surface potential between monolayers and bilayers can be ascribed to different substrate-induced doping levels [77]. Substrate characteristics, such as terrace width in SiC, can also be a dominating factor in determining the unintentional doping of monolayers [78]. Unique work function variations of graphene line defects, grain boundaries, standing-collapsed wrinkles, and folded wrinkles could be clearly identified by high-resolution KPFM (Figure 3a–c). Classical and quantum molecular dynamics simulations reveal that the work function distribution of each type of line defect is again originated from the doping effect induced by the substrate [79]. The abrupt change of the cantilever phase (fraction of phase shift >0.9, see Figure 3f) in the EFM images across a bisecting grain boundary (GB) in MoS_2 memory devices indicated that the electrostatic potential drops primarily at the grain boundary, i.e., the GB is resistive, consistently with the overall higher resistance of a bisecting-GB memristor compared to a bridge-GB memristor. Because the local surface potential and thus resistivity varies as a square root of the variation in the EFM phase signal, it was evaluated that more than 94% of the total device resistance would come from the grain boundary [18].

The doping caused by adsorbed molecules was also investigated by Pearce et al. [80], who showed a different sensitivity of monolayers and bilayers to chemical gating by exposing graphene samples to electron donating and withdrawing gases, and monitoring the change in work function via KPFM (Figure 3d,e). The larger shift in surface potential upon exposure to electron withdrawing and donating gases observed in monolayers rather than double layers was ascribed to the narrower energy dispersion around the Dirac point in graphene single sheets. The stepwise chemical reduction of individual graphene oxide flakes could be observed by monitoring the change of surface charge distribution, which revealed that the oxidized nanoscale domains are reduced by the leaching of sharp oxidized asperities from the surface followed by gradual thinning and the formation of uniformly mixed oxidized and graphitic domains across the entire flake [81]. Finally, electric field-induced changes in the work function of single layers were observed in gate modulated measurements, due to the Fermi level tuning induced by the gate voltage [82].

C-AFM has been successfully used to obtain spatial mapping of the conductivity of graphene on different substrates [83,84]. A high imaging contrast was used to distinguish domains of epitaxial graphene from the adjacent SiC surface thanks to strong differences in the tip–sample contact resistance [85]. The local conductance degradation in epitaxial graphene over the SiC substrate steps or at the junction between monolayer and bilayer regions could also be visualized, the degradation at the substrate steps being due to a lower substrate-induced electrostatic doping of graphene over the step sidewall, while that at the junction between the mono- and bilayer regions to the weak wave-function coupling between the monolayer and bilayer bands [86]. Also, by operating in current spectroscopy mode, i.e., by performing local I-V measurements, the Schottky barrier height (SBH) of epitaxial graphene grown on H-SiC was estimated to be 0.36 ± 0.1 eV, which is 0.49 eV lower than the barrier of graphene exfoliated from HOPG and deposited on the same substrate (0.85 ± 0.06 eV). The result was explained as a Fermi-level pinning effect above the Dirac point in epitaxial graphene due to the presence of positively charged states [87]. Similarly, C-AFM allowed the mapping of the spatial inhomogeneities of the SBH and the ideality factor of contacts on MoS_2, due to spatial variations in the density and energy of MoS_2 surface states, and to correlate local resistivity with local SBH [88]. Spatially resolved SBH maps revealed a substantial conductivity difference between MoS_2 with and without subsurface metal-like defects depending on the tip's work function, with high work function tips showing large spatial variations up to ~40% [89]. The nanoscale Schottky barrier distribution at the surface of multilayer MoS_2 could be tailored by varying the incorporated oxygen concentration by O_2 plasma functionalization. Whereas a narrow SBH distribution (0.2–0.3 eV) was measured for pristine MoS_2, a broader distribution (from 0.2 to 0.8 eV) in the modified one allowed both electrons and holes injection (Figure 4a–c) [90]. An attractive application of electrical mode SPMs is the use of conductive probes to induce electrochemical reactions and to pattern materials in electric field-induced nanolithography processes [91]. Local AFM-tip-induced electrochemical reduction processes were used to pattern conductive pathways on insulating graphene oxide to fabricate micropatterned graphene field-effect transistors featuring high charge-carrier mobilities

(Figure 4d,e) [92,93]. By changing the polarity of the applied voltage between graphene and a conductive AFM tip, hydrogenation and oxidation could be controlled at the nanoscale, and used to fabricate nanostructures such as graphene nanoribbons [94]. Changes due to the electro-reduction process could be monitored directly on a device by KPFM even at single sheet level [95].

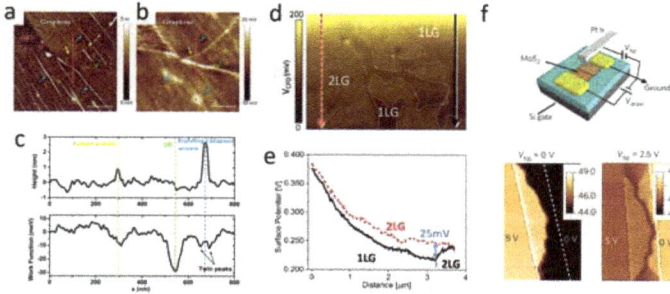

Figure 3. AFM topography (**a**) and Kelvin Probe Force Microscopy (KPFM) mapping (**b**) of standing collapsed wrinkles with various sizes; (**c**) Cross section profiles along the red line marked in (a) and (b). The dashed lines indicate the locations of folded wrinkle, grain boundaries, and standing collapsed wrinkle, respectively, with color coding corresponding to that in (a) and (b). Scale bars in (a) and (b) are 500 nm (adapted from [79]); (**d**) Surface potential map showing decreasing surface potential of single and double layer graphene with NO_2 exposure. The arrows represent the direction of the scan; (**e**) Section profiles along the arrows marked in (d) for single (black) and double (red) layer (adapted from [80]); (**f**) Device scheme and Electrostatic Force Microscopy (EFM) phase images of a bisecting grain boundary (GB) MoS_2 memristor at tip biases $V_{tip} = 0$ V and 2.5 V. Color scale bars show the EFM phase in degrees. Device bias conditions: $V_{drain} = 5$ V and $V_{source} = V_g = 0$ V. The dotted lines highlight the metal–MoS2 junctions with less contrast (adapted from [18]).

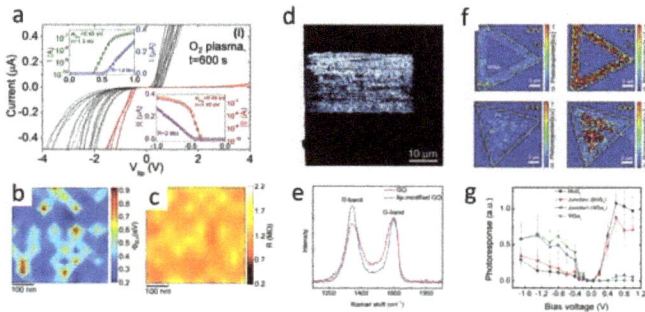

Figure 4. (**a**) Arrays of local current-voltage (I-V_{tip}) characteristics measured by Conductive Atomic Force Microscopy (C-AFM) on MoS2 subjected to O_2 plasma. Insets in (a) show fittings of representatives curves on linear- and semilog-scale to extract the local Schottky barrier height ($\Phi_{B,n}$) and the series resistance (R); (**b,c**) Two-dimensional (2D) maps of the local $\Phi_{B,n}$ and R (adapted from [90]); (**d**) C-AFM current map of a tip-modified rectangular region of a few-layer GO film with a top-contact gold electrode on the left side (current range 50 nA); (**e**) Raman spectra obtained on modified and unmodified GO (adapted from [92]); (**f**) Spatially resolved photoresponse maps and derived plot (**g**) of a $WSe_2 - MoS_2$ heterostructure crystal in both forward (0.2 and 0.6 V) and reverse voltage (−0.3 V and −0.8 V) regimes under illumination of $\lambda = 550$ nm. The photoresponse maps are generated by subtracting a photocurrent map under illumination (Photoconductive Atomic Force Microscopy (PC-AFM) map) from a dark current map (C-AFM map), and normalized by the incident laser power at a selected wavelength. (adapted from [96]).

The combination of C-AFM and PC-AFM was demonstrated to be convenient and versatile to efficiently examine layer-dependent electronic and optoelectronic characteristics in 2D crystals containing regions of different thicknesses [97,98]. Current transport mechanisms and photoresponse of mono- and multilayers, as well as heterostructures, could be investigated at the nanoscale junctions. For example, in a fascinating experiment, Son et al. investigated WSe_2-MoS_2 heterostructures by C-AFM and PC-AFM. By modulating the polarity and magnitude of the applied voltage, the photoresponse could be selectively switched on and off in a portion of the heterostructure crystal, demonstrating the possibility of fabricating high-resolution pixel arrays of switchable photodiodes (Figure 4f,g) [96].

5. Friction Force Microscopy

Friction Force microscopy (FFM) [99], also known as Lateral Force Microscopy (LFM), can detect lateral force variations on the atomic scale when sliding a sharp tip over a flat surface [100]. The essential feature of the method is that AFM is operated and controlled in the conventional contact mode, but that torsional deformations of the cantilever are monitored. Calculations of quantities such as lateral contact stiffness, friction force, and shear strength are possible after proper calibration procedures [101,102]. LFM was used to identify graphene on rough substrates, and to map the crystallographic orientation of the domains nondestructively, reproducibly, and at high resolution [103]. The atomic-scale friction of an MoS_2 surface was studied by Fujisawa et al. [104], confirming the existence of two-dimensionally discrete friction, due to spatially discrete adhesion and jumps corresponding to the lattice periodicity. The nanoscale frictional characteristics of atomically-thin sheets of different 2D materials exfoliated onto a weakly adherent substrate, such as silicon oxide, were compared to those of their bulk counterparts by Lee et al, showing a monotonically increased friction as the number of layers decreased [105]. Interestingly, the use of a strongly adherent substrate, such as mica, suppressed the trend. Different domains differing by their friction characteristics, and having a periodicity of 180°, were observed on exfoliated monolayer graphene by using angle-dependent scanning, with the friction anisotropy decreasing with an increased applied load (Figure 5a,b) [106]. It was proposed that the domains arise from ripple distortions as a result of anisotropic puckering deformation, due to the tip pushing the ripple crests forward along the scanning direction [105].

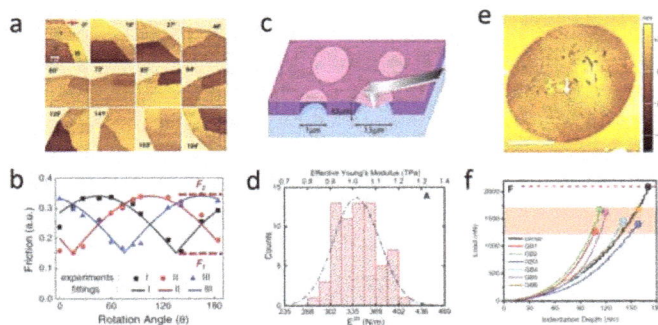

Figure 5. (**a**) Friction force images showing the changing friction contrast of different graphene domains as the sample is rotated from 0° to 184° relative to the horizontal scan direction (red arrow); The plot in (**b**) represents the normalized friction force vs. the rotation angle for three different domains, showing a 180° periodicity (reproduced from [106]); (**c**) Schematic of nanoindentation on a suspended graphene membrane and (**d**) histogram of the elastic modulus measured from force-distance curves on the graphene membrane (adapted from [107]); (**e**) AFM topography of a suspended graphene membrane showing grain boundaries (marked by dashed lines); (**f**) AFM indentation curves showing that fracture occurs at a slightly lower load when AFM tip indents on the grain boundary (reproduced from [108]).

6. Force Spectroscopy

In most AFM applications, the image contrast is obtained from the short range repulsion occurring when the electron orbitals of the tip and the sample overlap, in the so-called Born repulsion regime. However, further interactions can occur and can be used to investigate different properties of the materials when AFM is used in Force Spectroscopy mode. The probe is moved towards the sample in the normal direction and the vertical position and the deflection of the cantilever are recorded and converted to force vs. distance curves, from which several kinds of information on the mechanical properties of the samples, such as elastic modulus and breaking strength, can be obtained [109,110].

The elastic properties of three-dimensional (3D) materials are commonly described by the elastic modulus E. E is also called Young's modulus when the applied strain is uniaxial, with E = σ/ε, where σ is stress and ε is strain. The maximum tension that a material can withstand represents its tensile strength. Elastic modulus and tensile strength have units of J/m^3 or Pa. In 2D, however, these parameters are normalized by the planar elastic energy, leading to units of J/m^2 or N/m. Although 2D modulus and strength are more suitable to describe 2D materials, for the purpose of comparison between 2D and 3D materials, these 2D parameters are normally converted to 3D ones by dividing the 2D values with the thickness of the sample [111].

The elastic properties and intrinsic breaking strengths of monolayer graphene membranes were measured for the first time by AFM by Lee et al. in 2008 (see Figure 5c,d) [107]. For this purpose, graphene flakes were deposited onto a substrate patterned with an array of circular wells. Free-standing monolayer membranes could be obtained to be probed by nanoindentation with the AFM probe. This method has been applied extensively and the principle is described here below.

When the tip radius is $r_{tip} < r_{hole}$, the load force (F) applied by the AFM tip is related to the deformation geometry of the membrane:

$$F = \left(\sigma_0^{2D}\pi\right)\delta + \left(E^{2D}\frac{q^3}{r^2}\right)\delta^3, \tag{2}$$

being δ the indentation depth at the center of the membrane, r the hole radius, q a dimensionless constant determined by the Poisson's ratio ν of the membrane, and E^{2D} and σ_0^{2D} the 2D modulus and the 2D pretension respectively. The indentation depth is determined by the displacement of the scanning piezo-tube of the AFM (Δz_p) and the deflection of the AFM tip (Δz_t), and the applied load is obtained by multiplying the deflection of the AFM cantilever with its spring constant. E^{2D} and σ_0^{2D} can be derived by a least-square fitting of the experimental force-displacement curve (F(d)). With the same setup, also the maximum stress for a tightly clamped, linear elastic, and circular membrane under a spherical indenter can be calculated as:

$$\sigma_m^{2D} = \left(\frac{FE^{2D}}{4\pi r_{tip}}\right)^{1/2}, \tag{3}$$

where F is the breaking force, and r_{tip} is the tip radius. Accurate quantitative measurements require the calibration of geometrical and mechanical properties of the tip, as well as the choice of a suitable model for describing the cantilever-tip-sample system [112].

The force-displacement behavior for a monolayer graphene was interpreted within a framework of nonlinear elastic stress-strain response and showed a breaking strength of a defect-free layer of 42 N m^{-1} and a Young's modulus of E = 1.0 TPa (Figure 5d), representing graphene as the strongest material ever measured, and showing that atomically perfect nanoscale materials can be mechanically tested to deformations well beyond the linear regime [107]. The elastic properties of multilayer flakes with thicknesses (h) varying from 2.4 to 33 nm (8 to 100 layers) could also be extracted from force-distance curves [113]. The extracted bending rigidity was found to increase strongly (proportional to h^3) for thicknesses below 10 nm. Thicker flakes were found to have a smaller bending rigidity, possibly because of the presence of stacking defects in the flakes. The stable sp^2 bonds forming the graphene lattice compete against changes in length and angle, yielding a very high tensile energy when strained; at the same time, bending a graphene layer does not lead to significant deformation of the sp^2 bonds, thus resulting in its impressive Young's modulus [114]. However, these properties can be modified by the presence of defects or chemistry. The dependence of the elastic properties of graphene on the presence of defects was addressed by Zandiatashbar et al., who employed a modified oxygen plasma technique to induce defects in pristine graphene in a controlled manner [115]. By looking at the evolution of Raman spectra as a function of sheet defectiveness, they were able to categorize the defects as being predominantly sp^3-type (partial oxidation), or predominantly vacancy-type. The 2D elastic modulus (E^{2D}) was found to remain constant over the sp^3-type defect regions, indicating that these defects do not appreciably affect the stiffness. In the vacancy-type defect regions, instead, E^{2D} decreased with increasing defect density, reaching about 30% of the stiffness of the pristine sheet at the maximum exposure time. The breaking strength was found to decrease only about 14% with respect to pristine graphene, meaning that in the sp^3-type defect regime, the elastic stiffness of defective graphene is not significantly diminished in comparison with its pristine counterpart. Under longer plasma exposure, a significant number of carbon atoms were expected to be physically removed from the graphene lattice, as the density of defects increases and adjacent defects coalesce to form bigger voids or extended cavities, so that a dramatic drop in elastic stiffness and strength was expected. Interestingly, the elastic modulus of highly defective, plasma treated graphene (0.3 TPa) was found to be comparable to that measured by AFM indentation for graphene oxide (GO) (0.256 ± 0.028 TPa) [116] and by topographical AFM imaging of wrinkled flakes (0.23 ± 0.07 TPa) [117]. The latter approach revealed significant local heterogeneity in the in-plane elastic modulus of such materials, which is also evidenced by a certain variability of the value reported by different groups [118,119]. The discrepancy in the elastic modulus of GO was attributed to defect concentration or clustering of different functional groups. The ring opening of the epoxide functions and the subsequent formation of ether groups in the basal plane of GO was the origin of plasticity and ductility in n-butylamine-modified GO, which showed an elastic modulus 13% lower than that of the original GO [116]. The conductivity of reduced graphene oxide sheets was shown to scale inversely with the elastic modulus. Theoretical predictions confirmed that the sheets with higher elastic modulus and lower conductivity could be assigned to those of higher oxygen content [120].

The strength of graphene was found to be only slightly reduced by the presence of grain boundaries. Nanoindentation tests showed that fracture loads at the grain boundaries are 20 to 40% smaller than in pristine graphene, representing at most a 15% reduction of the intrinsic strength (Figure 5e,f) [108]. Lee et al. used hierarchical patterning to obtain conformal wrinkling with a soft skin layer [121]. The wrinkle wavelength could be finely controlled by tuning the skin layer's thickness. Force curves measured locally on the peaks of the wrinkles suggested an increase of stiffness with the wavelength, from 14.67 nN/nm for λ ~160 nm to 96.97 nN/nm for λ ~450 nm. Adhesion forces between the surface and tip were nearly invariant (~10 nN on valleys and ~18 nN on peaks) for all of the graphene wrinkles, while a much smaller adhesive force (~3 nN) was expectedly measured on crumples since these features are delaminated from the surface.

Similar studies have also been performed on other 2D membranes. MoS_2 monolayers measured by AFM nanoindentation showed a Young's modulus of 300 GPa [122], comparable to that of steel, only one third lower than exfoliated graphene, and higher than other 2D crystals such as reduced graphene oxide (0.2 TPa) [120], hexagonal boron nitride (0.25 TPa) [123], or carbon nanosheets (10–50 GPa) [124]. The average breaking strength of 23 GPa was found to correspond to the theoretical intrinsic strength of the Mo-S bond, indicating that the material can be highly crystalline and almost free of defects and dislocations [125]. The elastic properties of freely suspended MoS_2 sheets with thicknesses ranging from 5 to 25 layers were also investigated [122]. The thinnest sheets (up to 8 layers) presented strongly nonlinear force-displacement traces, while in sheets thicker than 10 layers the force-displacement traces were linear, indicating a trade-off between a bending-dominated and a stretching-dominated behavior.

It is worth mentioning that SPM nanoindentations do not leave the atomically thin membrane mechanically undisturbed during measurements [126,127]. Local membrane deformations at the location of the scanning tip are produced. When localized strains are induced in suspended few-layer graphene, the strain distribution under and around the AFM tip could be indeed mapped in situ using hyperspectral Raman imaging via the strain-dependent frequency shifts of the few-layer graphene's G and 2D Raman bands [128]. The contact of the nm-sized scanning probe tip resulted in a two-dimensional strain field with μm dimensions in the suspended membrane. Such deformations and the resulting localized strain distribution in the two-dimensional material can complicate SPM measurement interpretation, and also lead to degradation in the two-dimensional material upon measurement. From a different perspective, in most two-dimensional materials, the application of strain could also lead to changes in opto-electronic properties, which allows for the strain engineering of the material's properties. This opens a way to probe such strain-dependent opto-electronic properties based on the application of localized strain through SPM-based techniques.

7. Strain Engineering and Piezoresponse Force Microscopy

The possibility of finely tuning material properties is highly desirable for a wide range of applications, and strain engineering has been introduced as an interesting way to achieve it [28]. In bulk materials, however, the tuning range is limited by plastic behavior and low fracture limit due to the presence of defects and dislocations. Atomically-thin membranes instead exhibit high elasticity and breaking strength, which makes them sound candidates for engineering their properties via strain.

When graphene is deformed, the unit cell area of graphene would be changed by stretching or compressing. The strain-induced shift of carbon atoms would affect the band structure and therefore the electronic properties. The carrier density would be decreased or increased, and the Fermi level and the work function would change accordingly. The work function of graphene, measured by KPFM, was found to increase with increasing uniaxial strain, reaching a variation as large as 161 meV under a 7% strain, which can be explained by the strain-induced change of the density of states [129]. The dependence of the work function on strain is closely related to graphene's topography, and could be weakened by the presence of ripples (Figure 6a,b) [129]. The conductivity of graphene, however, could be preserved even under structural deformation. C-AFM measurements on wrinkled graphene showed that the current at a fixed bias was similar across the boundaries between crumples and wrinkle domains with different wavelengths, despite the presence of such defective structures having shown a clear effect on stiffness and adhesion [121]. The effect of mechanical strain on the electrical conductivity of suspended MoS_2 membranes was probed by positioning the AFM tip in the center of a flake connected to two microfabricated electrodes. The nanosheet was deformed by the AFM tip while measuring the current flowing between the electrodes (Figure 6c,d). The current increased as soon as the membrane began to deform, reaching a value around four times higher at maximum deformation. The effect was reversible, as the current followed the opposite trend and returned to its pre-deformation value as the tip was fully retracted [130]. This observed piezoresistive behavior can be understood in terms of band gap reduction under tensile strain. At different reduction/strain rates, the piezoresistive effect could be also tuned for monolayer, bilayer, and trilayer MoS_2, because of the different orbital contributions of the band-edge states and their different hybridization [130]. MoS_2 is expected to exhibit piezoelectric effects because of the non-centrosymmetric arrangement of the Mo and S atoms, which develop asymmetrical electrical dipoles when the material is subjected to external stress.

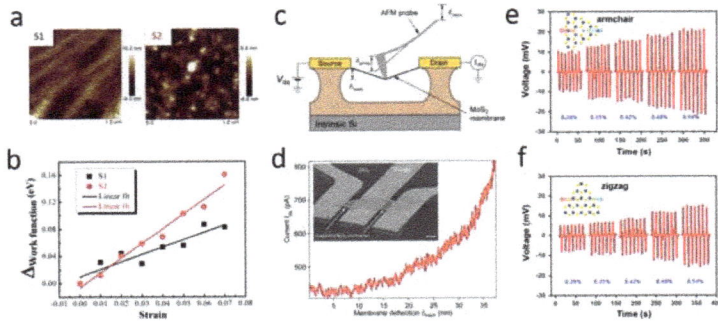

Figure 6. (**a**) AFM images of graphene with (S1) and without (S2) ripples; (**b**) Work function variation of graphene with (S1) and without (S2) ripples as a function of uniaxial strain measured by KPFM. (adapted from [129]); (**c**) Schematic drawing of the setup for direct current electrical characterization of suspended channel MoS$_2$ devices under strain. The suspended atomically thin membrane (see inset in d) is deformed at the center using an AFM probe; (**d**) Plot of the current vs. membrane deformation measured when a voltage is applied between source and drain electrodes (adapted from [130]). Output voltage obtained from a monolayer MoS$_2$ as a function of applied strain applied along the armchair direction (**e**) and the zigzag direction (**f**) (reproduced from [131]).

For piezoelectric materials, an applied voltage causes an expansion or a contraction of polarized domains, which in turn results in a measurable deflection of the cantilever. This phenomenon is used in Piezoresponse Force Microscopy (PFM) [132]. In PFM, the tip is brought in contact with the surface and the electromechanical response of the surface is detected as the first-harmonic component of the bias induced tip deflection:

$$d = d_0 + A\cos(\omega t + \varphi),\qquad(4)$$

where φ is the phase, which yields information on the polarization direction below the tip. For c-domains, i.e., when the polarization vector is pointing downward, the application of a positive tip bias results in the expansion of the sample, and the surface oscillations are in phase with the tip voltage, $\varphi = 0$. For c+ domains, $\varphi = 180°$. The amplitude A defines the local electromechanical response and depends on the geometry of the tip-surface system and the material's properties [133,134].

Vertical piezoresponse from single-layer graphene was observed by PFM. The calculated vertical piezocoefficient was found to be about 1.4 nm/V, which is much higher than that of conventional piezoelectric materials [135]. Atomically-thin graphene nitride also exhibits anomalous piezoelectricity, due to the fact that a stable phase of a sheet features regularly spaced triangular holes, as indicated by ab initio calculations [136]. Directional dependent piezoelectric effects in chemical vapor deposited MoS$_2$ monolayers were measured through lateral PFM when an electric field was applied laterally across the flakes [131]. The piezoelectric coefficient was found to be 3.78 pm/V in the armchair direction, and 1.38 pm/V in the zigzag direction, clearly revealing its distinct anisotropic piezoresponse in single-crystalline monolayers (Figure 6d,e).

8. Conclusions

Two-dimensional (2D) nanosheets are an emerging new class of materials, and new ways to define and quantify their structure-function relationships are required. Features such as average flake size, shape, concentration, and density of defects present at the chemical level are among the most significant properties affecting these materials' function. The wealth of scientific reports described in this review and beyond confirms the usefulness of Scanning Probe Microscopy as an efficient and valuable multilevel tool for atomic- and nanoscale 2D materials characterization, from structural and morphological, to (opto-)electronic and mechanical perspectives.

Because mechanical properties play vital roles in the design of flexible, stretchable, and epidermal electronics that may potentially dominate the future electronics industry [111], particular attention has been dedicated to this aspect here. The mechanics of 2D atomically-thin materials, their behavior under stress and electrical fields, and the interactions between adjacent sheets, between sheets and a substrate, or between sheets and their environment should be better understood to optimize the level of performance that shall be achieved. Thus, new concepts are required for modeling these materials. The coupling between piezoelectric polarization and semiconductor properties, such as electronic transport and photoresponse, is expected to give rise to unprecedented device characteristics [137,138]. The emerging fields of piezotronics and piezo-phototronics, which propose new means of manipulating charge-carrier transport in the operation of flexible devices through the application of external mechanical stimuli [139], demonstrate the need for the correlative characterization of mechanical and electronic properties at the single layer level, therefore opening new opportunities for the unceasing development of SPMs.

Conflicts of Interest: The author declares no conflict of interest.

References

1. Schwierz, F. Graphene transistors. *Nat. Nanotechnol.* **2010**, *5*, 487–496. [CrossRef] [PubMed]
2. Jariwala, D.; Sangwan, V.K.; Lauhon, L.J.; Marks, T.J.; Hersam, M.C. Emerging device applications for semiconducting two-dimensional transition metal dichalcogenides. *ACS Nano* **2014**, *8*, 1102–1120. [CrossRef] [PubMed]
3. Cain, J.D.; Hanson, E.D.; Shi, F.; Dravid, V.P. Emerging opportunities in the two-dimensional chalcogenide systems and architecture. *Curr. Opin. Solid State Mater. Sci.* **2016**, *20*, 374–387. [CrossRef]
4. Allen, M.J.; Tung, V.C.; Kaner, R.B. Honeycomb carbon: A review of graphene. *Chem. Rev.* **2010**, *110*, 132–145. [CrossRef] [PubMed]
5. Novoselov, K.S.; Geim, A.K.; Morozov, S.V.; Jiang, D.; Katsnelson, M.I.; Grigorieva, I.V.; Dubonos, S.V.; Firsov, A.A. Two-dimensional gas of massless dirac fermions in graphene. *Nature* **2005**, *438*, 197–200. [CrossRef] [PubMed]
6. Zhang, Y.; Tan, Y.-W.; Stormer, H.L.; Kim, P. Experimental observation of the quantum hall effect and berry's phase in graphene. *Nature* **2005**, *438*, 201–204. [CrossRef] [PubMed]
7. Geim, A.K.; Novoselov, K.S. The rise of graphene. *Nat. Mater.* **2007**, *6*, 183–191. [CrossRef] [PubMed]
8. Castro Neto, A.H.; Guinea, F.; Peres, N.M.R.; Novoselov, K.S.; Geim, A.K. The electronic properties of graphene. *Rev. Mod. Phys.* **2009**, *81*, 109–162. [CrossRef]
9. Geim, A.K. Graphene: Status and prospects. *Science* **2009**, *324*, 1530–1534. [CrossRef] [PubMed]
10. Butler, S.Z.; Hollen, S.M.; Cao, L.; Cui, Y.; Gupta, J.A.; Gutiérrez, H.R.; Heinz, T.F.; Hong, S.S.; Huang, J.; Ismach, A.F.; et al. Progress, challenges, and opportunities in two-dimensional materials beyond graphene. *ACS Nano* **2013**, *7*, 2898–2926. [CrossRef] [PubMed]
11. Geim, A.K.; Grigorieva, I.V. Van der Waals heterostructures. *Nature* **2013**, *499*, 419–425. [CrossRef] [PubMed]
12. Wang, F.; Wang, Z.; Xu, K.; Wang, F.; Wang, Q.; Huang, Y.; Yin, L.; He, J. Tunable gate-MoS$_2$ Van der Waals p–n junctions with novel optoelectronic performance. *Nano Lett.* **2015**, *15*, 7558–7566. [CrossRef] [PubMed]
13. Radisavljevic, B.; Radenovic, A.; Brivio, J.; Giacometti, V.; Kis, A. Single-layer MoS$_2$ transistors. *Nat. Nanotechnol.* **2011**, *6*, 147–150. [CrossRef] [PubMed]
14. Wang, H.; Yu, L.; Lee, Y.-H.; Shi, Y.; Hsu, A.; Chin, M.L.; Li, L.-J.; Dubey, M.; Kong, J.; Palacios, T. Integrated circuits based on bilayer MoS$_2$ transistors. *Nano Lett.* **2012**, *12*, 4674–4680. [CrossRef] [PubMed]
15. Krasnozhon, D.; Lembke, D.; Nyffeler, C.; Leblebici, Y.; Kis, A. MoS$_2$ transistors operating at gigahertz frequencies. *Nano Lett.* **2014**, *14*, 5905–5911. [CrossRef] [PubMed]
16. Lee, S.; Sohn, J.; Jiang, Z.; Chen, H.-Y.; Philip Wong, H.S. Metal oxide-resistive memory using graphene-edge electrodes. *Nat. Commun.* **2015**, *6*, 8407. [CrossRef] [PubMed]
17. Han, S.-T.; Zhou, Y.; Chen, B.; Wang, C.; Zhou, L.; Yan, Y.; Zhuang, J.; Sun, Q.; Zhang, H.; Roy, V.A.L. Hybrid flexible resistive random access memory-gated transistor for novel nonvolatile data storage. *Small* **2016**, *12*, 390–396. [CrossRef] [PubMed]

18. Sangwan, V.K.; Jariwala, D.; Kim, I.S.; Chen, K.-S.; Marks, T.J.; Lauhon, L.J.; Hersam, M.C. Gate-tunable memristive phenomena mediated by grain boundaries in single-layer MoS_2. *Nat. Nano* **2015**, *10*, 403–406. [CrossRef] [PubMed]

19. Musumeci, C.; Liscio, A.; Palermo, V.; Samorì, P. Electronic characterization of supramolecular materials at the nanoscale by conductive atomic force and kelvin probe force microscopies. *Mater. Today* **2014**, *17*, 504. [CrossRef]

20. Liscio, A.; Palermo, V.; Samorì, P. Nanoscale quantitative measurement of the potential of charged nanostructures by electrostatic and kelvin probe force microscopy: Unraveling electronic processes in complex materials. *Acc. Chem. Res.* **2010**, *43*, 541–550. [CrossRef] [PubMed]

21. Deshpande, A.; LeRoy, B.J. Scanning probe microscopy of graphene. *Phys. E* **2012**, *44*, 743–759. [CrossRef]

22. Meyer, J.C.; Geim, A.K.; Katsnelson, M.I.; Novoselov, K.S.; Booth, T.J.; Roth, S. The structure of suspended graphene sheets. *Nature* **2007**, *446*, 60–63. [CrossRef] [PubMed]

23. Fasolino, A.; Los, J.H.; Katsnelson, M.I. Intrinsic ripples in graphene. *Nat. Mater.* **2007**, *6*, 858–861. [CrossRef] [PubMed]

24. Miao, F.; Wijeratne, S.; Zhang, Y.; Coskun, U.C.; Bao, W.; Lau, C.N. Phase-coherent transport in graphene quantum billiards. *Science* **2007**, *317*, 1530–1533. [CrossRef] [PubMed]

25. Yazyev, O.V.; Chen, Y.P. Polycrystalline graphene and other two-dimensional materials. *Nat. Nano* **2014**, *9*, 755–767. [CrossRef] [PubMed]

26. Zhang, Z.; Zou, X.; Crespi, V.H.; Yakobson, B.I. Intrinsic magnetism of grain boundaries in two-dimensional metal dichalcogenides. *ACS Nano* **2013**, *7*, 10475–10481. [CrossRef] [PubMed]

27. Zou, X.; Yakobson, B.I. An open canvas—2D materials with defects, disorder, and functionality. *Acc. Chem. Res.* **2015**, *48*, 73–80. [CrossRef] [PubMed]

28. Rafael, R.; Andrés, C.-G.; Emmanuele, C.; Francisco, G. Strain engineering in semiconducting two-dimensional crystals. *J. Phys. Condens. Matter* **2015**, *27*, 313201.

29. Binnig, G.; Rohrer, H.; Gerber, C.; Weibel, E. Surface studies by scanning tunneling microscopy. *Phys. Rev. Lett.* **1982**, *49*, 57–61. [CrossRef]

30. Zhang, R.; Ivey, D.G. Preparation of sharp polycrystalline tungsten tips for scanning tunneling microscopy imaging. *J. Vac. Sci. Technol. B* **1996**, *14*, 1–10. [CrossRef]

31. Khan, Y.; Al-Falih, H.; Zhang, Y.; Ng, T.K.; Ooi, B.S. Two-step controllable electrochemical etching of tungsten scanning probe microscopy tips. *Rev. Sci. Instrum.* **2012**, *83*, 063708. [CrossRef] [PubMed]

32. Yamada, T.K.; Abe, T.; Nazriq, N.M.K.; Irisawa, T. Electron-bombarded <110>-oriented tungsten tips for stable tunneling electron emission. *Rev. Sci. Instrum.* **2016**, *87*, 033703. [CrossRef] [PubMed]

33. Stolyarova, E.; Rim, K.T.; Ryu, S.; Maultzsch, J.; Kim, P.; Brus, L.E.; Heinz, T.F.; Hybertsen, M.S.; Flynn, G.W. High-resolution scanning tunneling microscopy imaging of mesoscopic graphene sheets on an insulating surface. *Proc. Natl. Acad. Sci. USA* **2007**, *104*, 9209–9212. [CrossRef] [PubMed]

34. Zhang, Y.; Brar, V.W.; Wang, F.; Girit, C.; Yayon, Y.; Panlasigui, M.; Zettl, A.; Crommie, M.F. Giant phonon-induced conductance in scanning tunnelling spectroscopy of gate-tunable graphene. *Nat. Phys.* **2008**, *4*, 627–630. [CrossRef]

35. Ishigami, M.; Chen, J.H.; Cullen, W.G.; Fuhrer, M.S.; Williams, E.D. Atomic structure of graphene on SiO_2. *Nano Lett.* **2007**, *7*, 1643–1648. [CrossRef] [PubMed]

36. Berger, C.; Song, Z.; Li, X.; Wu, X.; Brown, N.; Naud, C.; Mayou, D.; Li, T.; Hass, J.; Marchenkov, A.N.; et al. Electronic confinement and coherence in patterned epitaxial graphene. *Science* **2006**, *312*, 1191–1196. [CrossRef] [PubMed]

37. Moran-Meza, J.A.; Cousty, J.; Lubin, C.; Thoyer, F. Understanding the STM images of epitaxial graphene on a reconstructed 6H-SiC(0001) surface: The role of tip-induced mechanical distortion of graphene. *Phys. Chem. Chem. Phys.* **2016**, *18*, 14264–14272. [CrossRef] [PubMed]

38. Coraux, J.; N'Diaye, A.T.; Busse, C.; Michely, T. Structural coherency of graphene on Ir(111). *Nano Lett.* **2008**, *8*, 565–570. [CrossRef] [PubMed]

39. Kwon, S.-Y.; Ciobanu, C.V.; Petrova, V.; Shenoy, V.B.; Bareño, J.; Gambin, V.; Petrov, I.; Kodambaka, S. Growth of semiconducting graphene on palladium. *Nano Lett.* **2009**, *9*, 3985–3990. [CrossRef] [PubMed]

40. Gao, L.; Guest, J.R.; Guisinger, N.P. Epitaxial graphene on Cu(111). *Nano Lett.* **2010**, *10*, 3512–3516. [CrossRef] [PubMed]

41. Dubout, Q.; Calleja, F.; Sclauzero, G.; Etzkorn, M.; Lehnert, A.; Claude, L.; Papagno, M.; Natterer, F.D.; Patthey, F.; Rusponi, S.; et al. Giant apparent lattice distortions in STM images of corrugated sp²-hybridised monolayers. *New J. Phys.* **2016**, *18*, 103027. [CrossRef]

42. Xue, J.; Sanchez-Yamagishi, J.; Bulmash, D.; Jacquod, P.; Deshpande, A.; Watanabe, K.; Taniguchi, T.; Jarillo-Herrero, P.; LeRoy, B.J. Scanning tunnelling microscopy and spectroscopy of ultra-flat graphene on hexagonal boron nitride. *Nat. Mater.* **2011**, *10*, 282–285. [CrossRef] [PubMed]

43. Harners, R.J. Atomic-resolution surface spectroscopy with the scanning tunneling microscope. *Annu. Rev. Phys. Chem.* **1989**, *40*, 531–559.

44. Tromp, R.M. Spectroscopy with the scanning tunnelling microscope: A critical review. *J. Phys. Condens. Matter* **1989**, *1*, 10211. [CrossRef]

45. Liu, H.; Zheng, H.; Yang, F.; Jiao, L.; Chen, J.; Ho, W.; Gao, C.; Jia, J.; Xie, M. Line and point defects in MoSe₂ bilayer studied by scanning tunneling microscopy and spectroscopy. *ACS Nano* **2015**, *9*, 6619–6625. [CrossRef] [PubMed]

46. Park, J.H.; Vishwanath, S.; Liu, X.; Zhou, H.; Eichfeld, S.M.; Fullerton-Shirey, S.K.; Robinson, J.A.; Feenstra, R.M.; Furdyna, J.; Jena, D.; et al. Scanning tunneling microscopy and spectroscopy of air exposure effects on molecular beam epitaxy grown WSe₂ monolayers and bilayers. *ACS Nano* **2016**, *10*, 4258–4267. [CrossRef] [PubMed]

47. Huang, Y.L.; Chen, Y.; Zhang, W.; Quek, S.Y.; Chen, C.-H.; Li, L.-J.; Hsu, W.-T.; Chang, W.-H.; Zheng, Y.J.; Chen, W.; et al. Bandgap tunability at single-layer molybdenum disulphide grain boundaries. *Nat. Commun.* **2015**, *6*, 6298. [CrossRef] [PubMed]

48. Georgi, A.; Nemes-Incze, P.; Carrillo-Bastos, R.; Faria, D.; Viola Kusminskiy, S.; Zhai, D.; Schneider, M.; Subramaniam, D.; Mashoff, T.; Freitag, N.M.; et al. Tuning the pseudospin polarization of graphene by a pseudomagnetic field. *Nano Lett.* **2017**, *17*, 2240–2245. [CrossRef] [PubMed]

49. Philippe, D.; Viet Hung, N.; Jérôme, S.-M. Thermoelectric effects in graphene nanostructures. *J. Phys. Condens. Matter* **2015**, *27*, 133204.

50. Park, J.; He, G.; Feenstra, R.M.; Li, A.-P. Atomic-scale mapping of thermoelectric power on graphene: Role of defects and boundaries. *Nano Lett.* **2013**, *13*, 3269–3273. [CrossRef] [PubMed]

51. Nemes-Incze, P.; Osváth, Z.; Kamarás, K.; Biró, L.P. Anomalies in thickness measurements of graphene and few layer graphite crystals by tapping mode atomic force microscopy. *Carbon* **2008**, *46*, 1435–1442. [CrossRef]

52. Binnig, G.; Quate, C.F.; Gerber, C. Atomic force microscope. *Phys. Rev. Lett.* **1986**, *56*, 930–933. [CrossRef] [PubMed]

53. Wang, Y.; Yang, R.; Shi, Z.; Zhang, L.; Shi, D.; Wang, E.; Zhang, G. Super-elastic graphene ripples for flexible strain sensors. *ACS Nano* **2011**, *5*, 3645–3650. [CrossRef] [PubMed]

54. Burnett, T.L.; Yakimova, R.; Kazakova, O. Identification of epitaxial graphene domains and adsorbed species in ambient conditions using quantified topography measurements. *J. Appl. Phys.* **2012**, *112*, 054308. [CrossRef]

55. Kim, J.-S.; Choi, J.S.; Lee, M.J.; Park, B.H.; Bukhvalov, D.; Son, Y.-W.; Yoon, D.; Cheong, H.; Yun, J.-N.; Jung, Y.; et al. Between Scylla and Charybdis: Hydrophobic graphene-guided water diffusion on hydrophilic substrates. *Sci. Rep.* **2013**, *3*, 2309. [CrossRef] [PubMed]

56. Novoselov, K.S.; Jiang, D.; Schedin, F.; Booth, T.J.; Khotkevich, V.V.; Morozov, S.V.; Geim, A.K. Two-dimensional atomic crystals. *Proc. Natl. Acad. Sci. USA* **2005**, *102*, 10451–10453. [CrossRef] [PubMed]

57. Russo, P.; Compagnini, G.; Musumeci, C.; Pignataro, B. Raman monitoring of strain induced effects in mechanically deposited single layer graphene. *J. Nanosci. Nanotechnol.* **2012**, *12*, 8755–8758. [CrossRef] [PubMed]

58. Robinson, B.J.; Giusca, C.E.; Gonzalez, Y.T.; Kay, N.D.; Kazakova, O.; Kolosov, O.V. Structural, optical and electrostatic properties of single and few-layers MoS₂: Effect of substrate. *2D Mater.* **2015**, *2*, 015005. [CrossRef]

59. Lee, Y.-H.; Zhang, X.-Q.; Zhang, W.; Chang, M.-T.; Lin, C.-T.; Chang, K.-D.; Yu, Y.-C.; Wang, J.T.-W.; Chang, C.-S.; Li, L.-J.; et al. Synthesis of large-area MoS₂ atomic layers with chemical vapor deposition. *Adv. Mater.* **2012**, *24*, 2320–2325. [CrossRef] [PubMed]

60. Mechler, Á.; Kopniczky, J.; Kokavecz, J.; Hoel, A.; Granqvist, C.-G.; Heszler, P. Anomalies in nanostructure size measurements by AFM. *Phys. Rev. B* **2005**, *72*, 125407. [CrossRef]

61. García, R.; San Paulo, A. Attractive and repulsive tip-sample interaction regimes in tapping-mode atomic force microscopy. *Phys. Rev. B* **1999**, *60*, 4961–4967. [CrossRef]
62. Cameron, J.S.; Ashley, D.S.; Andrew, J.S.; Joseph, G.S.; Christopher, T.G. Accurate thickness measurement of graphene. *Nanotechnology* **2016**, *27*, 125704.
63. Ly, T.H.; Chiu, M.-H.; Li, M.-Y.; Zhao, J.; Perello, D.J.; Cichocka, M.O.; Oh, H.M.; Chae, S.H.; Jeong, H.Y.; Yao, F.; et al. Observing grain boundaries in CVD-grown monolayer transition metal dichalcogenides. *ACS Nano* **2014**, *8*, 11401–11408. [CrossRef] [PubMed]
64. Duong, D.L.; Han, G.H.; Lee, S.M.; Gunes, F.; Kim, E.S.; Kim, S.T.; Kim, H.; Ta, Q.H.; So, K.P.; Yoon, S.J.; et al. Probing graphene grain boundaries with optical microscopy. *Nature* **2012**, *490*, 235–239. [CrossRef] [PubMed]
65. Leng, Y.; Williams, C.C. *Molecular Charge Mapping with Electrostatic Force Microscope*; SPIE: Bellingham, WA, USA, 1993; pp. 35–39.
66. Nonnenmacher, M.; O'Boyle, M.P.; Wickramasinghe, H.K. Kelvin probe force microscopy. *Appl. Phys. Lett.* **1991**, *58*, 2921–2923. [CrossRef]
67. O'Shea, S.J.; Atta, R.M.; Murrell, M.P.; Welland, M.E. Conducting atomic force microscopy study of silicon dioxide breakdown. *J. Vac. Sci. Technol. B* **1995**, *13*, 1945–1952. [CrossRef]
68. Avila, A.; Bhushan, B. Electrical measurement techniques in atomic force microscopy. *Crit. Rev. Solid State Mater. Sci.* **2010**, *35*, 38–51. [CrossRef]
69. Coffey, D.C.; Reid, O.G.; Rodovsky, D.B.; Bartholomew, G.P.; Ginger, D.S. Mapping local photocurrents in polymer/fullerene solar cells with photoconductive atomic force microscopy. *Nano Lett.* **2007**, *7*, 738–744. [CrossRef] [PubMed]
70. Jacobs, H.O.; Leuchtmann, P.; Homan, O.J.; Stemmer, A. Resolution and contrast in kelvin probe force microscopy. *J. Appl. Phys.* **1998**, *84*, 1168–1173. [CrossRef]
71. Hormeno, S.; Penedo, M.; Manzano, C.V.; Luna, M. Gold nanoparticle coated silicon tips for kelvin probe force microscopy in air. *Nanotechnology* **2013**, *24*, 395701. [CrossRef] [PubMed]
72. Lanza, M.; Bayerl, A.; Gao, T.; Porti, M.; Nafria, M.; Jing, G.Y.; Zhang, Y.F.; Liu, Z.F.; Duan, H.L. Graphene-coated atomic force microscope tips for reliable nanoscale electrical characterization. *Adv. Mater.* **2013**, *25*, 1440–1444. [CrossRef] [PubMed]
73. Hui, F.; Vajha, P.; Shi, Y.; Ji, Y.; Duan, H.; Padovani, A.; Larcher, L.; Li, X.R.; Xu, J.J.; Lanza, M. Moving graphene devices from lab to market: Advanced graphene-coated nanoprobes. *Nanoscale* **2016**, *8*, 8466–8473. [CrossRef] [PubMed]
74. Kazakova, O.; Panchal, V.; Burnett, T. Epitaxial graphene and graphene-based devices studied by electrical scanning probe microscopy. *Crystals* **2013**, *3*, 191. [CrossRef]
75. Burnett, T.; Yakimova, R.; Kazakova, O. Mapping of local electrical properties in epitaxial graphene using electrostatic force microscopy. *Nano Lett.* **2011**, *11*, 2324–2328. [CrossRef] [PubMed]
76. Panchal, V.; Pearce, R.; Yakimova, R.; Tzalenchuk, A.; Kazakova, O. Standardization of surface potential measurements of graphene domains. *Sci. Rep.* **2013**, *3*, 2597. [CrossRef] [PubMed]
77. Ziegler, D.; Gava, P.; Güttinger, J.; Molitor, F.; Wirtz, L.; Lazzeri, M.; Saitta, A.M.; Stemmer, A.; Mauri, F.; Stampfer, C. Variations in the work function of doped single- and few-layer graphene assessed by kelvin probe force microscopy and density functional theory. *Phys. Rev. B* **2011**, *83*, 235434. [CrossRef]
78. Eriksson, J.; Pearce, R.; Iakimov, T.; Virojanadara, C.; Gogova, D.; Andersson, M.; Syväjärvi, M.; Lloyd Spetz, A.; Yakimova, R. The influence of substrate morphology on thickness uniformity and unintentional doping of epitaxial graphene on SiC. *Appl. Phys. Lett.* **2012**, *100*, 241607. [CrossRef]
79. Long, F.; Yasaei, P.; Sanoj, R.; Yao, W.; Král, P.; Salehi-Khojin, A.; Shahbazian-Yassar, R. Characteristic work function variations of graphene line defects. *ACS Appl. Mater. Interfaces* **2016**, *8*, 18360–18366. [CrossRef] [PubMed]
80. Pearce, R.; Eriksson, J.; Iakimov, T.; Hultman, L.; Lloyd Spetz, A.; Yakimova, R. On the differing sensitivity to chemical gating of single and double layer epitaxial graphene explored using scanning kelvin probe microscopy. *ACS Nano* **2013**, *7*, 4647–4656. [CrossRef] [PubMed]
81. Kulkarni, D.D.; Kim, S.; Chyasnavichyus, M.; Hu, K.; Fedorov, A.G.; Tsukruk, V.V. Chemical reduction of individual graphene oxide sheets as revealed by electrostatic force microscopy. *J. Am. Chem. Soc.* **2014**, *136*, 6546–6549. [CrossRef] [PubMed]
82. Yu, Y.-J.; Zhao, Y.; Ryu, S.; Brus, L.E.; Kim, K.S.; Kim, P. Tuning the graphene work function by electric field effect. *Nano Lett.* **2009**, *9*, 3430–3434. [CrossRef] [PubMed]

83. Fisichella, G.; Greco, G.; Roccaforte, F.; Giannazzo, F. Current transport in graphene/AlGaN/GaN vertical heterostructures probed at nanoscale. *Nanoscale* **2014**, *6*, 8671–8680. [CrossRef] [PubMed]

84. Fisichella, G.; Di Franco, S.; Fiorenza, P.; Lo Nigro, R.; Roccaforte, F.; Tudisco, C.; Condorelli, G.G.; Piluso, N.; Spartà, N.; Lo Verso, S.; et al. Micro- and nanoscale electrical characterization of large-area graphene transferred to functional substrates. *Beilstein J. Nanotechnol.* **2013**, *4*, 234–242. [CrossRef] [PubMed]

85. Kellar, J.A.; Alaboson, J.M.P.; Wang, Q.H.; Hersam, M.C. Identifying and characterizing epitaxial graphene domains on partially graphitized SiC(0001) surfaces using scanning probe microscopy. *Appl. Phys. Lett.* **2010**, *96*, 143103. [CrossRef]

86. Giannazzo, F.; Deretzis, I.; La Magna, A.; Roccaforte, F.; Yakimova, R. Electronic transport at monolayer-bilayer junctions in epitaxial graphene on SiC. *Phys. Rev. B* **2012**, *86*, 235422. [CrossRef]

87. Sonde, S.; Giannazzo, F.; Raineri, V.; Yakimova, R.; Huntzinger, J.R.; Tiberj, A.; Camassel, J. Electrical properties of the graphene/4H-SiC(0001) interface probed by scanning current spectroscopy. *Phys. Rev. B* **2009**, *80*. [CrossRef]

88. Giannazzo, F.; Fisichella, G.; Piazza, A.; Agnello, S.; Roccaforte, F. Nanoscale inhomogeneity of the Schottky barrier and resistivity in MoS_2 multilayers. *Phys. Rev. B* **2015**, *92*. [CrossRef]

89. Bampoulis, P.; van Bremen, R.; Yao, Q.; Poelsema, B.; Zandvliet, H.J.W.; Sotthewes, K. Defect dominated charge transport and fermi level pinning in MoS_2/metal contacts. *ACS Appl. Mater. Interfaces* **2017**, *9*, 19278–19286. [CrossRef] [PubMed]

90. Giannazzo, F.; Fisichella, G.; Greco, G.; Di Franco, S.; Deretzis, I.; La Magna, A.; Bongiorno, C.; Nicotra, G.; Spinella, C.; Scopelliti, M.; et al. Ambipolar MoS_2 transistors by nanoscale tailoring of Schottky barrier using oxygen plasma functionalization. *ACS Appl. Mater. Interfaces* **2017**. [CrossRef] [PubMed]

91. Liu, H.; Hoeppener, S.; Schubert, U.S. Nanoscale materials patterning by local electrochemical lithography. *Adv. Eng. Mater.* **2016**, *18*, 890–902.

92. Mativetsky, J.M.; Treossi, E.; Orgiu, E.; Melucci, M.; Veronese, G.P.; Samorì, P.; Palermo, V. Local current mapping and patterning of reduced graphene oxide. *J. Am. Chem. Soc.* **2010**, *132*, 14130–14136. [CrossRef] [PubMed]

93. Faucett, A.C.; Mativetsky, J.M. Nanoscale reduction of graphene oxide under ambient conditions. *Carbon* **2015**, *95*, 1069–1075. [CrossRef]

94. Byun, I.-S.; Yoon, D.; Choi, J.S.; Hwang, I.; Lee, D.H.; Lee, M.J.; Kawai, T.; Son, Y.-W.; Jia, Q.; Cheong, H.; et al. Nanoscale lithography on monolayer graphene using hydrogenation and oxidation. *ACS Nano* **2011**, *5*, 6417–6424. [CrossRef] [PubMed]

95. Mativetsky, J.M.; Liscio, A.; Treossi, E.; Orgiu, E.; Zanelli, A.; Samorì, P.; Palermo, V. Graphene transistors via in situ voltage-induced reduction of graphene-oxide under ambient conditions. *J. Am. Chem. Soc.* **2011**, *133*, 14320–14326. [CrossRef] [PubMed]

96. Son, Y.; Li, M.-Y.; Cheng, C.-C.; Wei, K.-H.; Liu, P.; Wang, Q.H.; Li, L.-J.; Strano, M.S. Observation of switchable photoresponse of a monolayer WSe_2–MoS_2 lateral heterostructure via photocurrent spectral atomic force microscopic imaging. *Nano Lett.* **2016**, *16*, 3571–3577. [CrossRef] [PubMed]

97. Ruzmetov, D.; Zhang, K.; Stan, G.; Kalanyan, B.; Bhimanapati, G.R.; Eichfeld, S.M.; Burke, R.A.; Shah, P.B.; O'Regan, T.P.; Crowne, F.J.; et al. Vertical 2D/3D semiconductor heterostructures based on epitaxial molybdenum disulfide and gallium nitride. *ACS Nano* **2016**, *10*, 3580–3588. [CrossRef] [PubMed]

98. Son, Y.; Wang, Q.H.; Paulson, J.A.; Shih, C.-J.; Rajan, A.G.; Tvrdy, K.; Kim, S.; Alfeeli, B.; Braatz, R.D.; Strano, M.S. Layer number dependence of MoS_2 photoconductivity using photocurrent spectral atomic force microscopic imaging. *ACS Nano* **2015**, *9*, 2843–2855. [CrossRef] [PubMed]

99. Mate, C.M.; McClelland, G.M.; Erlandsson, R.; Chiang, S. Atomic-scale friction of a tungsten tip on a graphite surface. *Phys. Rev. Lett.* **1987**, *59*, 1942–1945. [CrossRef] [PubMed]

100. Bennewitz, R. Friction force microscopy. *Mater. Today* **2005**, *8*, 42–48. [CrossRef]

101. Cain, R.G.; Biggs, S.; Page, N.W. Force calibration in lateral force microscopy. *J. Colloid Interface Sci.* **2000**, *227*, 55–65. [CrossRef] [PubMed]

102. Gibson, C.T.; Watson, G.S.; Myhra, S. Lateral force microscopy—A quantitative approach. *Wear* **1997**, *213*, 72–79. [CrossRef]

103. Marsden, A.J.; Phillips, M.; Wilson, N.R. Friction force microscopy: A simple technique for identifying graphene on rough substrates and mapping the orientation of graphene grains on copper. *Nanotechnology* **2013**, *24*, 255704. [CrossRef] [PubMed]

104. Fujisawa, S.; Kishi, E.; Sugawara, Y.; Morita, S. Atomic-scale friction observed with a two-dimensional frictional-force microscope. *Phys. Rev. B* **1995**, *51*, 7849–7857. [CrossRef]

105. Lee, C.; Li, Q.; Kalb, W.; Liu, X.-Z.; Berger, H.; Carpick, R.W.; Hone, J. Frictional characteristics of atomically thin sheets. *Science* **2010**, *328*, 76–80. [CrossRef] [PubMed]

106. Choi, J.S.; Kim, J.-S.; Byun, I.-S.; Lee, D.H.; Lee, M.J.; Park, B.H.; Lee, C.; Yoon, D.; Cheong, H.; Lee, K.H.; et al. Friction anisotropy–driven domain imaging on exfoliated monolayer graphene. *Science* **2011**, *333*, 607–610. [CrossRef] [PubMed]

107. Lee, C.; Wei, X.; Kysar, J.W.; Hone, J. Measurement of the elastic properties and intrinsic strength of monolayer graphene. *Science* **2008**, *321*, 385–388. [CrossRef] [PubMed]

108. Lee, G.-H.; Cooper, R.C.; An, S.J.; Lee, S.; van der Zande, A.; Petrone, N.; Hammerberg, A.G.; Lee, C.; Crawford, B.; Oliver, W.; et al. High-strength chemical-vapor–deposited graphene and grain boundaries. *Science* **2013**, *340*, 1073–1076. [CrossRef] [PubMed]

109. Butt, H.-J.; Cappella, B.; Kappl, M. Force measurements with the atomic force microscope: Technique, interpretation and applications. *Surf. Sci. Rep.* **2005**, *59*, 1–152. [CrossRef]

110. Cappella, B.; Dietler, G. Force-distance curves by atomic force microscopy. *Surf. Sci. Rep.* **1999**, *34*, 1–104. [CrossRef]

111. Liu, K.; Wu, J. Mechanical properties of two-dimensional materials and heterostructures. *J. Mater. Res.* **2016**, *31*, 832–844. [CrossRef]

112. Hutter, J.L.; Bechhoefer, J. Calibration of atomic-force microscope tips. *Rev. Sci. Instrum.* **1993**, *64*, 1868–1873. [CrossRef]

113. Poot, M.; van der Zant, H.S.J. Nanomechanical properties of few-layer graphene membranes. *Appl. Phys. Lett.* **2008**, *92*, 063111. [CrossRef]

114. Palermo, V.; Kinloch, I.A.; Ligi, S.; Pugno, N.M. Nanoscale mechanics of graphene and graphene oxide in composites: A scientific and technological perspective. *Adv. Mater.* **2016**, *28*, 6232–6238. [CrossRef] [PubMed]

115. Zandiatashbar, A.; Lee, G.-H.; An, S.J.; Lee, S.; Mathew, N.; Terrones, M.; Hayashi, T.; Picu, C.R.; Hone, J.; Koratkar, N. Effect of defects on the intrinsic strength and stiffness of graphene. *Nat. Commun.* **2014**, *5*, 3186. [CrossRef] [PubMed]

116. Wei, X.; Mao, L.; Soler-Crespo, R.A.; Paci, J.T.; Huang, J.; Nguyen, S.T.; Espinosa, H.D. Plasticity and ductility in graphene oxide through a mechanochemically induced damage tolerance mechanism. *Nat. Commun.* **2015**, *6*, 8029. [CrossRef] [PubMed]

117. Kunz, D.A.; Feicht, P.; Gödrich, S.; Thurn, H.; Papastavrou, G.; Fery, A.; Breu, J. Space-resolved in-plane moduli of graphene oxide and chemically derived graphene applying a simple wrinkling procedure. *Adv. Mater.* **2013**, *25*, 1337–1341. [CrossRef] [PubMed]

118. Cao, C.; Daly, M.; Singh, C.V.; Sun, Y.; Filleter, T. High strength measurement of monolayer graphene oxide. *Carbon* **2015**, *81*, 497–504. [CrossRef]

119. Suk, J.W.; Piner, R.D.; An, J.; Ruoff, R.S. Mechanical properties of monolayer graphene oxide. *ACS Nano* **2010**, *4*, 6557–6564. [CrossRef] [PubMed]

120. Gómez-Navarro, C.; Burghard, M.; Kern, K. Elastic properties of chemically derived single graphene sheets. *Nano Lett.* **2008**, *8*, 2045–2049. [CrossRef] [PubMed]

121. Lee, W.-K.; Kang, J.; Chen, K.-S.; Engel, C.J.; Jung, W.-B.; Rhee, D.; Hersam, M.C.; Odom, T.W. Multiscale, hierarchical patterning of graphene by conformal wrinkling. *Nano Lett.* **2016**, *16*, 7121–7127. [CrossRef] [PubMed]

122. Castellanos-Gomez, A.; Poot, M.; Steele, G.A.; van der Zant, H.S.J.; Agraït, N.; Rubio-Bollinger, G. Elastic properties of freely suspended MoS$_2$ nanosheets. *Adv. Mater.* **2012**, *24*, 772–775. [CrossRef] [PubMed]

123. Song, L.; Ci, L.; Lu, H.; Sorokin, P.B.; Jin, C.; Ni, J.; Kvashnin, A.G.; Kvashnin, D.G.; Lou, J.; Yakobson, B.I.; et al. Large scale growth and characterization of atomic hexagonal boron nitride layers. *Nano Lett.* **2010**, *10*, 3209–3215. [CrossRef] [PubMed]

124. Turchanin, A.; Beyer, A.; Nottbohm, C.T.; Zhang, X.; Stosch, R.; Sologubenko, A.; Mayer, J.; Hinze, P.; Weimann, T.; Gölzhäuser, A. One nanometer thin carbon nanosheets with tunable conductivity and stiffness. *Adv. Mater.* **2009**, *21*, 1233–1237. [CrossRef]

125. Bertolazzi, S.; Brivio, J.; Kis, A. Stretching and breaking of ultrathin MoS$_2$. *ACS Nano* **2011**, *5*, 9703–9709. [CrossRef] [PubMed]

126. Klimov, N.N.; Jung, S.; Zhu, S.; Li, T.; Wright, C.A.; Solares, S.D.; Newell, D.B.; Zhitenev, N.B.; Stroscio, J.A. Electromechanical properties of graphene drumheads. *Science* **2012**, *336*, 1557–1561. [CrossRef] [PubMed]

127. Xu, P.; Neek-Amal, M.; Barber, S.D.; Schoelz, J.K.; Ackerman, M.L.; Thibado, P.M.; Sadeghi, A.; Peeters, F.M. Unusual ultra-low-frequency fluctuations in freestanding graphene. *Nat. Commun.* **2014**, *5*, 3720. [CrossRef] [PubMed]

128. Elibol, K.; Bayer, B.C.; Hummel, S.; Kotakoski, J.; Argentero, G.; Meyer, J.C. Visualising the strain distribution in suspended two-dimensional materials under local deformation. *Sci. Rep.* **2016**, *6*, 28485. [CrossRef] [PubMed]

129. He, X.; Tang, N.; Sun, X.; Gan, L.; Ke, F.; Wang, T.; Xu, F.; Wang, X.; Yang, X.; Ge, W.; et al. Tuning the graphene work function by uniaxial strain. *Appl. Phys. Lett.* **2015**, *106*, 043106. [CrossRef]

130. Manzeli, S.; Allain, A.; Ghadimi, A.; Kis, A. Piezoresistivity and strain-induced band gap tuning in atomically thin MoS_2. *Nano Lett.* **2015**, *15*, 5330–5335. [CrossRef] [PubMed]

131. Kim, S.K.; Bhatia, R.; Kim, T.-H.; Seol, D.; Kim, J.H.; Kim, H.; Seung, W.; Kim, Y.; Lee, Y.H.; Kim, S.-W. Directional dependent piezoelectric effect in cvd grown monolayer MoS_2 for flexible piezoelectric nanogenerators. *Nano Energy* **2016**, *22*, 483–489. [CrossRef]

132. Christman, J.A., Jr.; Woolcott, R.R., Jr.; Kingon, A.I.; Nemanich, R.J. Piezoelectric measurements with atomic force microscopy. *Appl. Phys. Lett.* **1998**, *73*, 3851–3853. [CrossRef]

133. Güthner, P.; Dransfeld, K. Local poling of ferroelectric polymers by scanning force microscopy. *Appl. Phys. Lett.* **1992**, *61*, 1137–1139. [CrossRef]

134. Kalinin, S.V.; Bonnell, D.A. Imaging mechanism of piezoresponse force microscopy of ferroelectric surfaces. *Phys. Rev. B* **2002**, *65*, 125408. [CrossRef]

135. Da Cunha Rodrigues, G.; Zelenovskiy, P.; Romanyuk, K.; Luchkin, S.; Kopelevich, Y.; Kholkin, A. Strong piezoelectricity in single-layer graphene deposited on SiO_2 grating substrates. *Nat. Commun.* **2015**, *6*, 7572. [CrossRef] [PubMed]

136. Zelisko, M.; Hanlumyuang, Y.; Yang, S.; Liu, Y.; Lei, C.; Li, J.; Ajayan, P.M.; Sharma, P. Anomalous piezoelectricity in two-dimensional graphene nitride nanosheets. *Nat. Commun.* **2014**, *5*, 4284. [CrossRef] [PubMed]

137. Bunch, J.S.; van der Zande, A.M.; Verbridge, S.S.; Frank, I.W.; Tanenbaum, D.M.; Parpia, J.M.; Craighead, H.G.; McEuen, P.L. Electromechanical resonators from graphene sheets. *Science* **2007**, *315*, 490–493. [CrossRef] [PubMed]

138. Chen, C.; Lee, S.; Deshpande, V.V.; Lee, G.-H.; Lekas, M.; Shepard, K.; Hone, J. Graphene mechanical oscillators with tunable frequency. *Nat. Nanotechnol.* **2013**, *8*, 923–927. [CrossRef] [PubMed]

139. Wu, W.; Wang, Z.L. Piezotronics and piezo-phototronics for adaptive electronics and optoelectronics. *Nat. Rev. Mater.* **2016**, *1*, 16031. [CrossRef]

crystals

MDPI

Review

Synthesis Methods of Two-Dimensional MoS₂: A Brief Review

Jie Sun [1,2,*], Xuejian Li [1], Weiling Guo [1], Miao Zhao [3], Xing Fan [1], Yibo Dong [1], Chen Xu [1], Jun Deng [1] and Yifeng Fu [4,*]

1　Key Laboratory of Optoelectronics Technology, College of Microelectronics, Beijing University of Technology, Beijing 100124, China; xuejianli@emails.bjut.edu.cn (X.L.); guoweiling@bjut.edu.cn (W.G); fanxing111@emails.bjut.edu.cn (X.F.); donyibo@emails.bjut.edu.cn (Y.D.); xuchen58@bjut.edu.cn (C.X.); dengsu@bjut.edu.cn (J.D.)
2　Quantum Device Physics Laboratory, Department of Microtechnology and Nanoscience, Chalmers University of Technology, Göteborg 41296, Sweden
3　High-Frequency High-Voltage Device and Integrated Circuits Center, Institute of Microelectronics, Chinese Academy of Sciences, Beijing 10029, China; zhaomiao@ime.ac.cn
4　Electronics Material and Systems Laboratory, Department of Microtechnology and Nanoscience, Chalmers University of Technology, Göteborg 41296, Sweden
*　Correspondence: jie.sun@bjut.edu.cn (J.S.); yifeng.fu@chalmers.se (Y.F.)

Academic Editor: Filippo Giannazzo
Received: 19 May 2017; Accepted: 28 June 2017; Published: 1 July 2017

Abstract: Molybdenum disulfide (MoS₂) is one of the most important two-dimensional materials after graphene. Monolayer MoS₂ has a direct bandgap (1.9 eV) and is potentially suitable for post-silicon electronics. Among all atomically thin semiconductors, MoS₂'s synthesis techniques are more developed. Here, we review the recent developments in the synthesis of hexagonal MoS₂, where they are categorized into top-down and bottom-up approaches. Micromechanical exfoliation is convenient for beginners and basic research. Liquid phase exfoliation and solutions for chemical processes are cheap and suitable for large-scale production; yielding materials mostly in powders with different shapes, sizes and layer numbers. MoS₂ films on a substrate targeting high-end nanoelectronic applications can be produced by chemical vapor deposition, compatible with the semiconductor industry. Usually, metal catalysts are unnecessary. Unlike graphene, the transfer of atomic layers is omitted. We especially emphasize the recent advances in metalorganic chemical vapor deposition and atomic layer deposition, where gaseous precursors are used. These processes grow MoS₂ with the smallest building-blocks, naturally promising higher quality and controllability. Most likely, this will be an important direction in the field. Nevertheless, today none of those methods reproducibly produces MoS₂ with competitive quality. There is a long way to go for MoS₂ in real-life electronic device applications.

Keywords: Molybdenum disulfide; transition metal dichalcogenide; two-dimensional materials; micromechanical exfoliation; chemical vapor deposition

---

1. Introduction

Recently, graphene has received considerable attention in scientific communities. It is recognized as a new wonder material (e.g., record carrier mobility, thermal conductivity, mechanical strength and flexibility), and applications in nanocomposites, paints, transparent electrodes, heat spreaders, etc. are being explored [1]. However, one should not forget that the original motivation for studying graphene was in its field effect [2], with a hope that it might play a key role in nanoelectronics, rather than serving as a passive element such as electrodes [3]. Nevertheless, at this stage, due to the difficulties in opening an energy bandgap in graphene, it is not clear whether graphene will become a key material in electronics at all. Therefore, the search for other two-dimensional (2D) materials (materials with one or few atomic layers) with an inherent energy bandgap that is suitable to make transistors is essential. Black phosphorus [4], transition metal dichalcogenide (TMD) [5], gallium nitride [6], etc. are extensively explored. For any type of application, the material production is always the first experimental step. Among these 2D semiconductors, molybdenum disulfide (MoS_2) is relatively more mature in terms of its synthesis technologies. Hence, in this paper, we provide a short literature survey of recent progresses in the synthesis of MoS_2. The paper focuses only on the production techniques rather than serving as a complete review on MoS_2-related physics, materials and electronics. Also, we put special emphasis on the recent achievements in metalorganic chemical vapor deposition (MOCVD) and atomic layer deposition (ALD), where gaseous precursors are used to grow MoS_2. It has a high potential for producing MoS_2 with better quality and controllability for nanoelectronic device applications.

2. Synthesis Methods of MoS_2

MoS_2 has several polymorphs, where the 2H MoS_2 has a layered hexagonal structure with lattice parameters of $a = 0.316$ nm and $c = 1.229$ nm [7], yielding an inter-layer spacing of 0.615 nm [8], as schematically shown in Figure 1. Bulk 2H MoS_2 is an indirect bandgap semiconductor (~1.29 eV). With decreasing thickness, the bandgap increases. When it comes to monolayer MoS_2, it becomes a direct bandgap material, and the energy gap can be as large as ~1.90 eV [9], suitable for fabricating most electronic devices. The synthesis methods of MoS_2 can be divided into top-down and bottom-up approaches. We hereby describe them in turn.

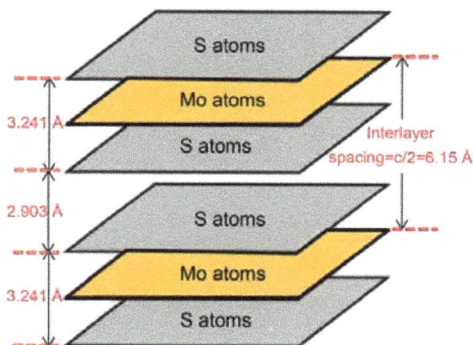

Figure 1. Distances between different layers in 2H MoS_2.

2.1. Top-down Approaches

2.1.1. Micromechanical Exfoliation

Similar to the micromechanical exfoliation of graphene (a vivid lab demonstration can be found in [10]), MoS_2 flakes can be produced on a substrate using sticky tapes. The starting material is bulk MoS_2, where some parts are peeled off with the tape and pressed into the substrate. Upon the release of the tape, some parts stay with the substrate rather than the tape due to the van der Waals force to the substrate. Repeating the process may produce flakes of MoS_2 with random shapes, sizes and numbers of layers. This method offers 2D materials of the highest quality, enabling to study the pristine properties and ultimate device performances. For example, Kis et al. have used micro-exfoliation to produce MoS_2 monolayers that are suitable for ultrasensitive photodetectors [11], analogue [12] and digital circuits [13], and observation of interesting physics such as mobility engineering and a metal–insulator transition [14]. However, it was found that the van der Waals adhesion of many TMDs to SiO$_2$, a typical substrate, is much weaker than that of graphene. Therefore, the produced flakes are very small (typically <10 µm) in their lateral size [15]. These results notwithstanding, large area MoS$_2$ flakes can be exfoliated by slightly modifying the technique. Magda et al. obtained mono MoS$_2$ layers with lateral size of several hundreds of microns by improving the MoS$_2$-substrate adhesion, as shown in Figure 2 [15]. The mechanism is that sulfur atoms can bind more strongly to gold than to SiO$_2$. The drawback is that such as-exfoliated MoS$_2$ flakes have to be transferred to another new substrate, since in most applications the material is used on insulating substrates. In addition, like their graphene counterpart, the tape-assisted micromechanical exfoliation method offers very low yield, and it's not possible to scale up for large volume production, therefore MoS$_2$ synthesized by this method is limited to fundamental study at lab scale.

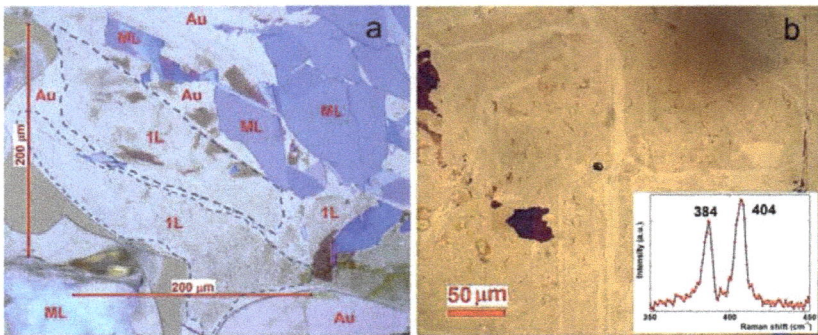

Figure 2. Large area MoS$_2$ flakes exfoliated with chemically enhanced adhesion to the substrate [15].

2.1.2. Liquid Phase Exfoliation

Liquid phase exfoliation also starts from bulk MoS$_2$, producing flakes with random shapes, sizes and number of layers. The quantities, however, are much larger, and the quality lower. Roughly, there are 2 routes to exfoliate MoS$_2$ in solution. The first is exfoliation by mechanical means such as sonication, shearing, stirring, grinding and bubbling [16–21]. The essence is purely physical, although some chemistry may still be involved [22–24]. For example, surfactants such as sodium deoxycholate (SDC) bile salt [22] and chitosan [23] may be added to the solution to prevent the exfoliated flakes from recombination. Electrolysis is employed to generate bubbles in some cases [24]. It is known that fine bubbles can squeeze into material interfaces and exfoliate them [25]. The yield of this method is dramatically improved from tape-assisted exfoliation, but the efficiency is still low for industrial applications. Therefore, a second route by atomic intercalation via solution chemistry [26] or electrochemistry [27] has been proposed. Lithium is typically used to intercalate between the MoS$_2$ layers and enlarge the interlayer spacing, easing the following exfoliation by mechanical treatment (e.g., sonication). Fan et al. developed a process to exfoliate MoS$_2$ nanosheets very efficiently using sonication assisted Li intercalation. They found that the complete Li intercalation with butylithium can be effected within 1.5 hours [28]. An example of the MoS$_2$ flakes synthesized by this method is shown in Figure 3. However, this method results in the loss of semiconducting properties due to the structural changes during Li intercalation. It is reported that above an annealing temperature of 300 °C, these MoS$_2$ flakes can be somewhat recovered [26].

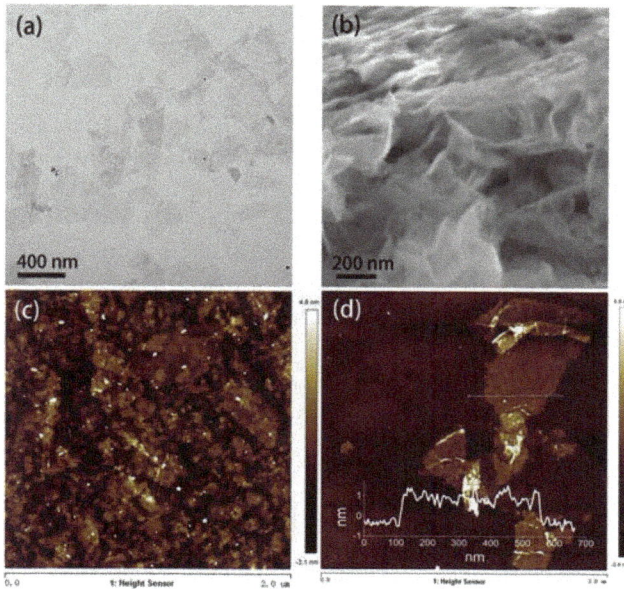

Figure 3. TEM, SEM and AFM images of MoS$_2$ flakes exfoliated by Li intercalation and sonication [28].

Liquid phase exfoliation is a simple technology with low cost, producing large quantity 2D nanosheets with relatively high quality. Therefore, it is regarded as the most suitable method for the large-scale industrial production of one- to few-layer 2D material flakes at low expense [29]. It is worth noting that liquid phase exfoliation can produce not only wet suspensions of flakes or dry powders, but macroscopically continuous thin films as well [30–32]. On substrates, through drop casting of the suspension and drying, MoS$_2$ thin films are prepared by Walia et al. [30–32], where optical and electrical properties can be studied.

2.2. Bottom-up Approaches

2.2.1. Physical Vapor Deposition

Although molecular beam epitaxy (MBE) is an advanced technology for growing single crystal semiconductor thin films, its application in producing 2D materials needs further development, and the grain size of the 2D material is not as high as expected [33]. In fact, even ordinary physical vapor deposition is rare. There is a report on growing MoS$_2$-Ti composite by vacuum sputtering, using Ti and MoS$_2$ as the targets [34]. However, the MoS$_2$ is an amorphous structure in this case.

2.2.2. Solution Chemical Process

There are various methods for producing MoS$_2$ by solution chemistry. Hydrothermal synthesis [35] and solvothermal synthesis [36] typically use molybdate to react with sulfide or just sulfur in a stainless steel autoclave, where a series of physicochemical reactions take place under relatively high temperature (e.g., 200 °C) and high pressure for several hours or longer. The resultant is MoS$_2$ powders of different shapes. The size of individual particles can be adjusted to some extent. Very frequently, the powders are high-temperature post-annealed, to improve their crystalline quality and purity. The only difference between hydrothermal and solvothermal synthesis is that the precursor solution in the latter case is usually not aqueous. Other solution chemical processes start at almost room temperature and atmospheric pressure, where post-annealing is often used anyway [37–41].

The products can be either a powder or thin film, depending on the preparation details. The most commonly-used precursor is $(NH_4)_2MoS_4$ (ammonium tetrathiomolybdate), or something similar. $(NH_4)_2MoS_4$ decomposes to form MoO_3 at 120–360 °C, which can be further converted into MoS_2. In some cases, electrochemical deposition [42–44] and photo-assisted deposition are used [45]. For instance, Li et al. fabricated MoS_2 nanowires and nanoribbons using MoO_x as the precursor [43]. The MoO_x precursor is then exposed to H_2S at high temperature (500–800 °C) to synthesize MoS_2. Depending on the temperature for processing, both 2H and 3R phase of MoS_2 can be synthesized. The MoS_2 nanoribbons can be 50 μm in length and are aligned in parallel arrays which are preferable for device fabrication, as shown in Figure 4.

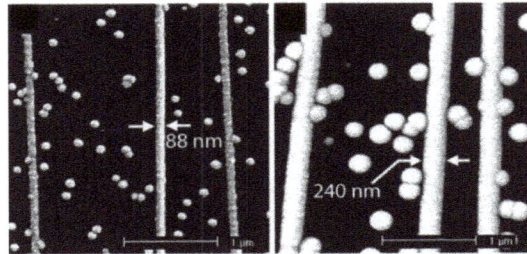

Figure 4. MoS_2 nanowires/nanoribbons synthesized from MoO_x precursor [43].

2.2.3. Chemical Vapor Deposition (CVD)

Among all the synthesis methods, CVD is the most compatible to existing semiconductor technology. Typically, it forms the film on a substrate, where chemical reactions are involved in the growth mechanism. The large-area growth of highly uniform MoS_2 will enable the batch fabrication of atomically thin devices and circuitry. Currently, most works start with solid state precursors of molybdenum, such as MoO_3 [46,47] or Mo [48]. If the Mo precursors are not vaporized in the process, often the technique is also called sulfurization [46–48]. Typical S precursors are H_2S gas [49] or vaporized S [46–48,50]. It is argued that using $MoCl_5$ precursors are easier to achieve monolayer MoS_2 over large area [50]. Earlier, the MoS_2 could be noncontinuous [47], which is not a big problem nowadays. The usual growth temperatures are 700–1000 °C, and the inclusion of a metal catalyst such as Au is favorable for the film quality [49]. Typical reactions taking place in the CVD are one of these:

$$MoO_3 + 2H_2S + H_2 = MoS_2 + 3H_2O \tag{1}$$

$$8MoO_3 + 24H_2S = 8MoS_2 + 24H_2O + S_8 \tag{2}$$

$$16MoO_3 + 7S_8 = 16MoS_2 + 24SO_2 \tag{3}$$

With plasma-enhanced chemical vapor deposition (PECVD), the growth temperature of MoS2 can be lowered to 150–300 °C, making it possible to directly deposit MoS_2 even on plastic substrates [51].

Metal-organic CVD is a special case of CVD where organometallic precursors are used. It is used in the modern semiconductor industry to produce single crystal expitaxial films. Recently, a few reports have appeared on the application of this method in the growth of MoS_2 thin films [49,52–54]. Kang et al. [52] have grown MoS_2 and WS_2 on 4-inch oxidized silicon wafers by MOCVD. The MoS_2 is grown with gaseous precursors of $Mo(CO)_6$ (boiling point 156 °C) and $(C_2H_5)_2S$. At 550 °C, monolayer MoS_2 with full coverage of the substrate (typically SiO_2/Si and fused silica) is deposited for $t_0 \sim 26$ h. Figure 5a shows scanning electron microscopy (SEM) images of the samples, demonstrating the effect of adding H_2 in the gas mixture. It is argued that H_2 enhances the decomposition of $(C_2H_5)_2S$ (increasing nucleation due to hydrogenolysis) and etches MoS_2 (preventing intergrain connection) [52]. It seems that a low H_2 rate is beneficial for growing continuous monolayer MoS_2. Nevertheless, H_2 is necessary for removing the carbonaceous species generated during the MOCVD. Therefore, the H_2 flow should be optimized. To have successful growth, dehydrating the growth environment is also necessary. Figure 5b indicates that the growth mode is layer-by-layer under this specific condition. When $t < t_0$ there is almost no nucleation of the second MoS_2 layer, which takes place mainly at the grain boundaries of the first layer when $t > t_0$. Figure 5c shows a top gated (V_{TG}) MoS_2 transistor with a field-effect mobility of 29 cm^2 V^{-1} s^{-1} and transconductance 2 µSµm, determined from the transfer properties. The current saturation occurs at relatively low V_{SD} (see the inset). These features compete well with the-state-of-the-art MoS_2 transistors. Figure 5d–f demonstrates the optical absorption, Raman and photoluminescence spectra measured from the MOCVD grown MoS_2 (and WS_2) films, respectively, where the features agree well with the characteristics of exfoliated monolayer counterparts (denoted by diamonds).

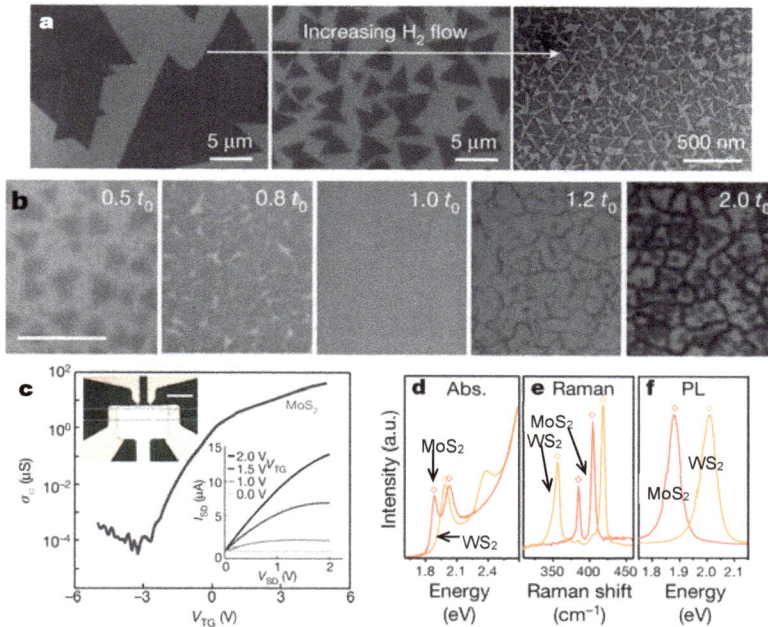

Figure 5. (**a**) Grain size evolution of monolayer MoS_2 as a function of H_2 flow. From left to right, 5, 20 and 200 sccm H_2. (**b**) Optical images of MoS_2 films as a function of growth time. t_0 is the optimal time for full monolayer coverage (~26 h). (**c**) Transfer and output properties of the field-effect transistor fabricated in the MOCVD grown MoS_2 with top gate configuration. $\sigma_\square$ denotes sheet conductance. Scale bar: 10 µm. (**d–f**) Optical absorption, Raman and normalized photoluminescence spectra of as-grown monolayer MoS_2 (red) and WS_2 (orange) films. The peak positions in (**d–f**) are consistent with exfoliated flakes (denoted by diamonds) [52].

At this stage, the material quality in [53] is still too low to be competitive. The large-area deposition of monolayer MoS_2 as well as other TMDs with spatial homogeneity and high performance is still challenging. MoS_2 films today are polycrystalline with the grain size typically ranging from 1 to 10 micrometers. This notwithstanding, the MOCVD synthesis of 2D semiconductors indeed opens a pathway towards future exciting nanoelectronic materials. The most important reason for this is that, in MOCVD, ultrapure gaseous precursors are injected into the reactor and finely dosed to deposit a very thin layer of atoms onto the substrate. Even if a solid or liquid precursor is used, it will be vaporized in situ in the chamber [53]. Unlike solid and liquid precursors, here the building-blocks for growing the film are much finer, naturally leading to better quality and controllability. The heated metalorganic precursor molecules decompose via pyrolysis, leaving the atoms on the substrate. The atoms bond to the surface, and an ultrathin MoS_2 layer is grown.

Based on this argument, one can easily think of another promising synthesis method: ALD [55], which can be seen as a subclass of CVD. In ALD, the film is grown by exposing the substrate surface to alternating gas precursors. In contrast to ordinary CVD, the precursors are never simultaneously present in the chamber. Instead, they are introduced as sequential and non-overlapping pulses, separated by purges in between. In each pulse, molecules react with the surface in a self-limiting way. Hence, it is possible to grow materials with high-thickness precision (layer by layer). Indeed, very recently, people have used ALD to grow MoS_2, such as in [56]. Using $MoCl_5$ (vaporized) and H_2S precursors, monolayer to few-layer MoS_2 on large SiO_2/Si or quartz substrates is obtained [56]. In the photos of two 150 mm quartz wafers before and after the ALD, the substrate changes to dark yellow uniformly (~2 layers of MoS_2) [56]. It is found that photoresist residues on the surface lead to more nucleation sites for MoS_2, but this mechanism remains to be confirmed. Figure 6a plots Raman signals for 50 cycle ALD MoS_2, where the E^1_{2g} and A_{1g} characteristic peaks are sharper and higher for 475 °C deposition temperature, indicating its higher material quality. After the ALD, some samples are annealed in a sulfur environment (H_2S or S atmosphere). Figure 6b shows that after sulfur annealing at increasingly high temperature, the characteristic peaks get narrower, stronger and more symmetric. After annealing at 920 °C, the full width at half maximum of the E^1_{2g} peak is close to that of exfoliated MoS_2 monolayer (4.2 cm^{-1} vs. 3.7 cm^{-1}). By analyzing the separation Δ between E^1_{2g} and A_{1g} peaks, one can tentatively determine the number of layers in MoS_2 films [56]. In Figure 6c, the sample grown at 475 °C and annealed at 920 °C seems to be 1–2 monolayers. Future electronic measurements are needed to further examine the film quality.

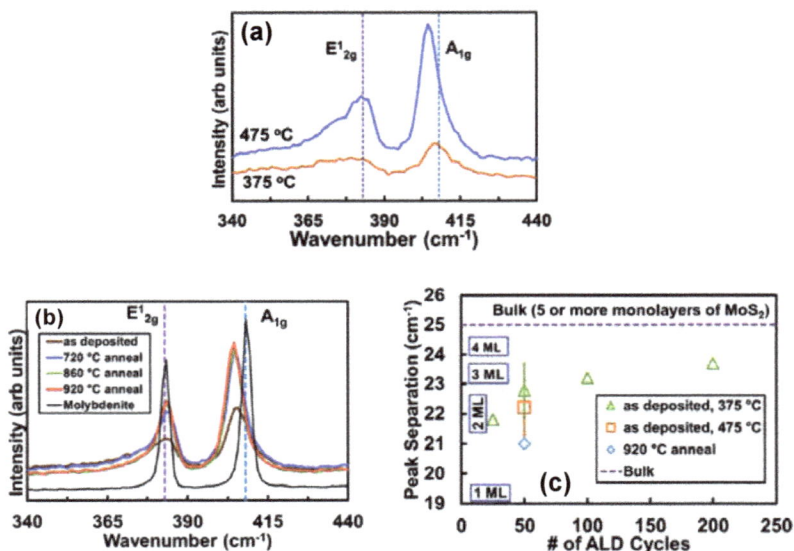

Figure 6. (a) Raman spectroscopy measurements of ALD MoS_2 at 375 or 475 °C, where the latter shows higher quality. (b) Raman spectra of 475 °C deposited MoS_2 films (some are post annealed in sulfur environment), where 920 °C annealing leads to the best quality. The curve for bulk MoS_2 is for reference. (c) E^1_{2g} to A_{1g} peak separation Δ plotted against the cycle number in ALD. Δ~19 cm^{-1} tentatively indicates MoS_2 monolayer, whereas bulk (>5 layers) material has Δ~25 cm^{-1} [56].

3. Conclusions

In this paper, we have reviewed the synthesis methods of MoS_2 in terms of top-down and bottom-up approaches. Micromechanical exfoliation offers the highest material quality but is limited by the yield so it is mainly for basic research, whereas liquid phase exfoliation and solution chemical process are of low cost, with decent material quality and suitable for large-scale industrial production. MoS_2 thin films aiming for high-end electronic applications are produced by CVD. In this review, we have put a special emphasis on the MOCVD and ALD methods, because they are more sophisticated technology with gas phase precursors and will most likely be the dominant techniques in producing MoS_2 films oriented towards real electronic device applications in the future. In most cases the MoS_2 films are transfer-free, which is vital for industrial applications that require a high process controllability and reproducibility [57]. At the moment, production of MoS_2 as a high quality electronic material is still challenging. This notwithstanding, most of the reviewed processes can be extended to other 2D materials (with some modification), and with future technological advances, the successful production of advanced 2D nanoelectronic materials can be envisioned.

Acknowledgments: We would like to thank National Natural Science Foundation of China (11674016), National key R & D program (2017YFB0403100), Beijing Natural Science Foundation (4152003), Beijing Municipal Commission of Education (PXM2016_014204_500018), and Swedish Foundation for International Cooperation in Research and Higher Education (CH2015-6202).

Author Contributions: J. Sun, X. Li and Y. Fu wrote the paper, led the discussion and summarized the literature; W. Guo, M. Zhao, C. Xu and J. Deng participated in the discussion; X. Fan and Y. Dong helped to summarize the literature.

Conflicts of Interest: The authors declare no conflict of interest.

References

1. Ferrari, A.C.; Bonaccorso, F.; Fal'Ko, V.; Novoselov, K.S.; Roche, S.; Bøggild, P.; Borini, S.; Koppens, F.H.; Palermo, V.; Pugno, N.; et al. Science and technology roadmap for graphene, related two-dimensional crystals, and hybrid systems. *Nanoscale* **2015**, *7*, 4598–4810. [CrossRef] [PubMed]
2. Novoselov, K.S.; Geim, A.K.; Morozov, S.V.; Jiang, D.; Zhang, Y.; Dubonos, S.V.; Grigorieva, I.V.; Firsov, A.A. Electric field effect in atomically thin carbon films. *Science* **2004**, *306*, 666–669. [CrossRef] [PubMed]
3. Xu, K.; Xu, C.; Deng, J.; Zhu, Y.; Guo, W.; Mao, M.; Zheng, L.; Sun, J. Graphene transparent electrodes grown by rapid chemical vapor deposition with ultrathin indium tin oxide contact layers for GaN light emitting diodes. *Appl. Phys. Lett.* **2013**, *102*, 162102.
4. Li, L.; Yu, Y.; Ye, G.J.; Ge, Q.; Ou, X.; Wu, H.; Feng, D.; Chen, X.H.; Zhang, Y. Black phosphorus field-effect transistors. *Nature Nanotechnol.* **2014**, *9*, 372–377. [CrossRef] [PubMed]
5. Chhowalla, M.; Shin, H.S.; Eda, G.; Li, L.J.; Loh, K.P.; Zhang, H. The chemistry of two-dimensional layered transition metal dichalcogenide nanosheets. *Nature Chem.* **2013**, *5*, 263–275. [CrossRef] [PubMed]
6. Al Balushi, Z.Y.; Wang, K.; Ghosh, R.K.; Vilá, R.A.; Eichfeld, S.M.; Caldwell, J.D.; Qin, X.; Lin, Y.C.; DeSario, P.A.; Stone, G.; et al. Two-dimensional gallium nitride realized via graphene encapsulation. *Nature Mater.* **2016**, *15*, 1166–1171. [CrossRef] [PubMed]
7. Stewart, J.A.; Spearot, D.E. Atomistic simulations of nanoindentation on the basal plane of crystalline molybdenum disulfide (MoS_2). *Model. Simul. Mater. Sci. Eng.* **2013**, *21*, 045003. [CrossRef]
8. Ai, K.; Ruan, C.; Shen, M.; Lu, L. MoS_2 nanosheets with widened interlayer spacing for high-efficiency removal of mercury in aquatic systems. *Adv. Funct. Mater.* **2016**, *26*, 5542–5549. [CrossRef]
9. Splendiani, A.; Sun, L.; Zhang, Y.; Li, T.; Kim, J.; Chim, C.-Y.; Galli, G.; Wang, F. Emerging photoluminescence in monolayer MoS_2. *Nano Lett.* **2010**, *10*, 1271–1275. [CrossRef] [PubMed]
10. Vivid Lab Demonstration. Available online: https://www.youtube.com/watch?v=9l5d0YLZgec&feature= youtu.be (accessed on 30 June 2017).
11. Lopez-Sanchez, O.; Lembke, D.; Kayci, M.; Radenovic, A.; Kis, A. Ultrasensitive photodetectors based on monolayer MoS_2. *Nature Nanotechnol.* **2013**, *8*, 497–501. [CrossRef] [PubMed]

12. Radisavljevic, B.; Whitwick, M.B.; Kis, A. Small-signal amplifier based on single-layer MoS$_2$. *Appl. Phys. Lett.* **2012**, *101*, 043103. [CrossRef]

13. Radisavljevic, B.; Whitwick, M.B.; Kis, A. Integrated circuits and logic operations based on single-layer MoS$_2$. *ACS Nano* **2011**, *5*, 9934–9938. [CrossRef] [PubMed]

14. Radisavljevic, B.; Kis, A. Mobility engineering and a metal–insulator transition in monolayer MoS$_2$. *Nat. Mater.* **2013**, *12*, 815–820. [CrossRef] [PubMed]

15. Magda, G.; Petö, J.; Dobrik, G.; Hwang, C.; Biró, L.; Tapasztó, L. Exfoliation of large-area transition metal chalcogenide single layers. *Sci. Rep.* **2015**, *5*, 14714. [CrossRef] [PubMed]

16. Forsberg, V.; Zhang, R.; Bäckström, J.; Dahlström, C.; Andres, B.; Norgren, M.; Andersson, M.; Hummelgård, M.; Olin, H. Exfoliated MoS$_2$ in Water without Additives. *PLoS ONE* **2016**, *11*, 0154522. [CrossRef] [PubMed]

17. Gupta, A.; Arunachalam, V.; Vasudevan, S. Liquid-phase exfoliation of MoS$_2$ nanosheets: The critical role of trace water. *J. Phys. Chem. Lett.* **2016**, *7*, 4884–4890. [CrossRef] [PubMed]

18. Yao, Y.; Lin, Z.; Li, Z.; Song, X.; Moon, K.-S.; Wong, C.-p. Large-scale production of two-dimensional nanosheets. *J. Mater. Chem.* **2012**, *22*, 13494–13499. [CrossRef]

19. Yu, Y.; Jiang, S.; Zhou, W.; Miao, X.; Zeng, Y.; Zhang, G.; Liu, S. Room temperature rubbing for few-layer two-dimensional thin flakes directly on flexible polymer substrates. *Sci. Rep.* **2013**, *3*, 2697. [CrossRef] [PubMed]

20. Varrla, E.; Backes, C.; Paton, K.; Harvey, A.; Gholamvand, Z.; McCauley, J.; Coleman, J. Large-scale production of size-controlled MoS$_2$ nanosheets by shear exfoliation. *Chem. Mater.* **2015**, *27*, 1129–1139. [CrossRef]

21. Paton, K.; Varrla, E.; Backes, C.; Smith, R.J.; Khan, U.; O'Neill, A.; Boland, C.; Lotya, M.; Istrate, O.M.; King, P.; et al. Scalable production of large quantities of defect-free few-layer graphene by shear exfoliation in liquids. *Nature Mater.* **2014**, *13*, 624–630. [CrossRef] [PubMed]

22. Zhang, M.; Howe, R.; Woodward, R.; Kelleher, E.; Torrisi, F.; Hu, G.; Popov, S.; Taylor, J.; Hasan, T. Solution processed MoS2-PVA composite for sub-bandgap mode-locking of a wideband tunable ultrafast Er:fiber laser. *Nano Res.* **2015**, *8*, 1522–1534. [CrossRef]

23. Zhang, W.; Wang, Y.; Zhang, D.; Yu, S.; Zhu, W.; Wang, J.; Zheng, F.; Wang, S.; Wang, J. A one-step approach to the large-scale synthesis of functionalized MoS$_2$ nanosheets by ionic liquid assisted grinding. *Nanoscale* **2015**, *7*, 10210–10217. [CrossRef] [PubMed]

24. Liu, N.; Kim, P.; Kim, J.; Ye, J.; Kim, S.; Lee, C. Large-area atomically thin MoS$_2$ nanosheets prepared using electrochemical exfoliation. *ACS Nano* **2014**, *8*, 6902–6910. [CrossRef] [PubMed]

25. Liu, L.; Liu, X.; Zhan, Z.; Guo, W.; Xu, C.; Deng, J.; Chakarov, D.; Hyldgaard, P.; Schröder, E.; Yurgens, A.; Sun, J. A mechanism for highly efficient electrochemical bubbling delamination of CVD-grown graphene from metal substrates. *Adv. Mater. Interf.* **2016**, *3*, 1500492. [CrossRef]

26. Eda, G.; Yamaguchi, H.; Voiry, D.; Fujita, T.; Chen, M.; Chhowalla, M. Photoluminescence from chemically exfoliated MoS$_2$. *Nano Lett.* **2011**, *11*, 5111–5116. [CrossRef] [PubMed]

27. Zeng, Z.; Yin, Z.; Huang, X.; Li, H.; He, Q.; Lu, G.; Boey, F.; Zhang, H. Single layer semiconducting nanosheets: High-yield preparation and device fabrication. *Angew. Chem. Int. Ed.* **2011**, *50*, 11093–11097. [CrossRef] [PubMed]

28. Fan, X.; Xu, P.; Zhou, D.; Sun, Y.; Li, Y.; Nguyen, M.; Terrones, M.; Mallouk, T. Fast and efficient preparation of exfoliated 2H MoS$_2$ nanosheets by sonication-assisted lithium intercalation and infrared laser-induced 1T to 2H phase reversion. *Nano Lett.* **2015**, *15*, 5956–5960. [CrossRef] [PubMed]

29. Coleman, J.; Lotya, M.; O'Neill, A.; Bergin, S.D.; King, P.J.; Khan, U.; Young, K.; Gaucher, A.; De, S.; Smith, R.J.; et al. Two-dimensional nanosheets produced by liquid exfoliation of layered materials. *Science* **2011**, *331*, 568–571. [CrossRef] [PubMed]

30. Rezk, A.R.; Walia, S.; Ramanathan, R.; Nili, H.; Ou, J.Z.; Bansal, V.; Friend, J.R.; Bhaskaran, M.; Yeo, L.Y.; Sriram, S. Acoustic–excitonic coupling for dynamic photoluminescence manipulation of quasi-2D MoS$_2$ nanoflakes. *Adv. Opt. Mater.* **2015**, *3*, 888–894. [CrossRef]

31. Balendhran, S.; Walia, S.; Nili, H.; Ou, J.Z.; Zhuiykov, S.; Kaner, R.B.; Sriram, S.; Bhaskaran, M.; Kalantar-zadeh, K. Two-dimensional molybdenum trioxide and dichalcogenides. *Adv. Funct. Mater.* **2013**, *23*, 3952–3970. [CrossRef]

32. Walia, S.; Balendhran, S.; Wang, Y.; Kadir, R.A.; Zoolfakar, A.S.; Atkin, P.; Ou, J.Z.; Sriram, S.; Kalantar-zadeh, K.; Bhaskaran, M. Characterization of metal contacts for two-dimensional MoS_2 nanoflakes. *Appl. Phys. Lett.* **2013**, *103*, 232105. [CrossRef]

33. Vishwanath, S.; Liu, X.; Rouvimov, S.; Mende, P.; Azcatl, A.; McDonnell, S.; Wallace, R.; Feenstra, R.; Furdyna, J.; Jena, D.; Xing, H. Comprehensive structural and optical characterization of MBE grown $MoSe_2$ on graphite, CaF_2 and graphene. *2D Mater.* **2015**, *2*, 024007. [CrossRef]

34. Qin, X.; Ke, P.; Wang, A.; Kim, K. Microstructure, mechanical and tribological behaviors of MoS_2-Ti composite coatings deposited by a hybrid HIPIMS method. *Surf. Coat. Technol.* **2013**, *228*, 275–281. [CrossRef]

35. Zhou, X.; Xu, B.; Lin, Z.; Shu, D.; Ma, L. Hydrothermal synthesis of flower-like MoS_2 nanospheres for electrochemical supercapacitors. *J. Nanosci. Nanotechnol.* **2014**, *14*, 7250–7254. [CrossRef] [PubMed]

36. Feng, X.; Tang, Q.; Zhou, J.; Fang, J.; Ding, P.; Sun, L.; Shi, L. Novel mixed–solvothermal synthesis of MoS_2 nanosheets with controllable morphologies. *Cryst. Res. Technol.* **2013**, *48*, 363–368. [CrossRef]

37. Liao, H.; Wang, Y.; Zhang, S.; Qian, Y. A solution low-temperature route to MoS_2 fiber. *Chem. Mater.* **2011**, *13*, 6–8. [CrossRef]

38. Li, X.; Zhang, W.; Wu, Y.; Min, C.; Fang, J. Solution-processed MoS_x as an efficient anode buffer layer in organic solar cells. *ACS Appl. Mater. Interf.* **2013**, *5*, 8823–8827. [CrossRef] [PubMed]

39. Liu, K.K.; Zhang, W.; Lee, Y.-H.; Lin, Y.-C.; Chang, M.-T.; Su, C.-Y.; Chang, C.-S.; Li, H.; Shi, Y.; Zhang, H.; Lai, C.-S.; Li, L.-J. Growth of large-area and highly crystalline MoS_2 thin layers on insulating substrates. *Nano Lett.* **2012**, *12*, 1538–1544. [CrossRef] [PubMed]

40. Bezverkhy, I.; Afanasiev, P.; Lacroix, M. Aqueous preparation of highly dispersed molybdenum sulfide. *Inorg. Chem.* **2000**, *39*, 5416–5417. [CrossRef] [PubMed]

41. Afanasiev, P.; Geantet, C.; Thomozeau, C.; Jouget, B. Molybdenum polysulfide hololow microtubules grown at room temperature from solution. *Chem. Commun.* **2000**, *12*, 1001–1002. [CrossRef]

42. Maijenburg, A.; Regis, M.; Hattori, A.; Tanaka, H.; Choi, K.-S.; ten Elshof, J. MoS_2 nanocube structures as catalysts for electrochemical H_2 evolution from acidic aqueous solutions. *ACS Appl. Mater. Interf.* **2014**, *6*, 2003–2010. [CrossRef] [PubMed]

43. Li, Q.; Walter, E.; van der Veer, W.; Murray, B.; Newberg, J.; Bohannan, E.; Switzer, J.; Hemminger, J.; Penner, R. Molybdenum disulfide nanowires and nanoribbons by electrochemical/chemical synthesis. *J. Phys. Chem. B* **2005**, *109*, 3169–3182. [CrossRef] [PubMed]

44. Kibsgaard, J.; Chen, Z.; Reinecke, B.; Jaramillo, T. Engineering the surface structure of MoS_2 to preferentially expose active edge sites for electrocatalysis. *Nature Mater.* **2012**, *11*, 963–969. [CrossRef] [PubMed]

45. Nguyen, M.; Tran, P.; Pramana, S.; Lee, R.; Batabyal, S.; Mathews, N.; Wong, L.; Graetzel, M. In situ photo-assisted deposition of MoS_2 electrocatalyst onto zinc cadmium sulphide nanoparticle surfaces to construct an efficient photocatalyst for hydrogen generation. *Nanoscale* **2013**, *5*, 1479–1482. [CrossRef] [PubMed]

46. Lin, Y.-C.; Zhang, W.; Huang, J.-K.; Liu, K.-K.; Lee, Y.-H.; Liang, C.-T.; Chu, C.-W.; Li, L.-J. Wafer-scale MoS_2 thin layers prepared by MoO_3 sulfurization. *Nanoscale* **2012**, *4*, 6637–6641. [CrossRef] [PubMed]

47. Cai, J.; Jian, J.; Chen, X.; Lei, M.; Wang, W. Regular hexagonal MoS_2 microflakes grown from MoO_3 precursor. *Appl. Phys. A* **2007**, *89*, 783–788. [CrossRef]

48. Zhan, Y.; Liu, Z.; Najmaei, S.; Ajayan, P.; Lou, J. Large-area vapor-phase growth and characterization of MoS_2 atomic layers on a SiO_2 substrate. *Small* **2014**, *8*, 966–971. [CrossRef] [PubMed]

49. Song, I.; Park, C.; Hong, M.; Baik, J.; Shin, H.-J.; Choi, H. Patternable large-scale molybdenium disulfide atomic layers grown by gold-assisted chemical vapor deposition. *Angew. Chem. Int. Ed.* **2014**, *53*, 1266–1269. [CrossRef] [PubMed]

50. Yu, Y.; Li, C.; Liu, Y.; Su, L.; Zhang, Y.; Cao, L. Controlled scalable synthesis of uniform, high-quality monolayer and few-layer MoS_2 Films. *Sci. Rep.* **2013**, *3*, 1866. [CrossRef] [PubMed]

51. Ahn, C.; Lee, J.; Kim, H.-U.; Bark, H.; Jeon, M.; Ryu, G.; Lee, Z.; Yeom, G.; Kim, K.; Jung, J.; Kim, Y.; Lee, C.; Kim, T. Low-temperature synthesis of large-scale molybdenum disulfide thin films directly on a plastic substrate using plasma-enhanced chemical vapor deposition. *Adv. Mater.* **2015**, *27*, 5223–5229. [CrossRef] [PubMed]

52. Kang, K.; Xie, S.; Huang, L.; Han, Y.; Huang, P.; Mak, K.; Kim, C.-J.; Muller, D.; Park, J. High-mobility three-atom-thick semiconducting films with wafer-scale homogeneity. *Nature* **2015**, *520*, 656–660. [CrossRef] [PubMed]

53. Olofinjana, B.; Egharevba, G.; Taleatu, B.; Akinwunmi, O.; Ajayi, E. MOCVD of molybdenum sulphide thin film via single solid source precursor bis-(morpholinodithioato-s,s')-Mo. *J. Mod. Phys.* **2011**, *2*, 341–349. [CrossRef]

54. Kumar, V.; Dhar, S.; Choudhury, T.; Shivashankar, S.; Raghavan, S. A predictive approach to CVD of crystalline layers of TMDs: The case of MoS_2. *Nanoscale* **2015**, *7*, 7802–7810. [CrossRef] [PubMed]

55. Sun, J.; Larsson, M.; Maximov, I.; Hardtdegen, H.; Xu, H. Gate-defined quantum-dot devices realized in InGaAs/InP by incorporating a HfO_2 layer as gate dielectric. *Appl. Phys. Lett.* **2009**, *94*, 042114. [CrossRef]

56. Valdivia, A.; Tweet, D.; Conley, J., Jr. Atomic layer deposition of two dimensional MoS_2 on 150 mm substrates. *J. Vac. Sci. Technol. A* **2016**, *34*, 021515. [CrossRef]

57. Sun, J.; Cole, M.; Lindvall, N.; Teo, K.; Yurgens, A. Noncatalytic chemical vapor deposition of graphene on high-temperature substrates for transparent electrodes. *Appl. Phys. Lett.* **2012**, *100*, 022102.

Review

Role of the Potential Barrier in the Electrical Performance of the Graphene/SiC Interface

Ivan Shtepliuk [1,*], Tihomir Iakimov [1], Volodymyr Khranovskyy [1], Jens Eriksson [1], Filippo Giannazzo [2] and Rositsa Yakimova [1]

[1] Department of Physics, Chemistry and Biology, Linköping University, Linköping SE-58183, Sweden; tihomir.iakimov@liu.se (T.I.); volkh@ifm.liu.se (V.K.); jenser@ifm.liu.se (J.E.); roy@ifm.liu.se (R.Y.)
[2] CNR-IMM, Strada VIII, 5 Zona Industriale, Catania 95121, Italy; filippo.giannazzo@imm.cnr.it
[*] Correspondence: ivan.shtepliuk@liu.se; Tel.: +46-707-6652-4089

Academic Editor: Helmut Cölfen
Received: 1 May 2017; Accepted: 31 May 2017; Published: 2 June 2017

Abstract: In spite of the great expectations for epitaxial graphene (EG) on silicon carbide (SiC) to be used as a next-generation high-performance component in high-power nano- and micro-electronics, there are still many technological challenges and fundamental problems that hinder the full potential of EG/SiC structures and that must be overcome. Among the existing problems, the quality of the graphene/SiC interface is one of the most critical factors that determines the electroactive behavior of this heterostructure. This paper reviews the relevant studies on the carrier transport through the graphene/SiC, discusses qualitatively the possibility of controllable tuning the potential barrier height at the heterointerface and analyses how the buffer layer formation affects the electronic properties of the combined EG/SiC system. The correlation between the sp^2/sp^3 hybridization ratio at the interface and the barrier height is discussed. We expect that the barrier height modulation will allow realizing a monolithic electronic platform comprising different graphene interfaces including ohmic contact, Schottky contact, gate dielectric, the electrically-active counterpart in p-n junctions and quantum wells.

Keywords: graphene; SiC; interface; buffer layer; barrier height; carrier transport

1. Introduction

Due to the never-ending miniaturization of electronic devices and integrated circuits, the spatial sizes of the materials become sufficiently small for new size-dependent physical limitations to effective carrier and heat transport to occur. Such constraints are governed to a large extent by the formation of an unstable transition layer at the heterointerface between two different materials. This is particularly the case of epitaxial graphene grown on silicon carbide (SiC), where complete decoupling of the graphene from the SiC surface is still a great challenge, and the interface significantly impacts many properties of graphene. Thus, a reliable control of heteroboundary quality and deep understanding of the physical nature of interface formation are imperative in order to produce device-quality graphene having the realistic chance to reach the market.

There exists evidence that up to 30% of the carbon atoms in the semiconductor-like transition carbon layer (called also the buffer layer, zero graphene layer or interfacial layer) are covalently bonded to the Si atoms (belonging to SiC) by sp^3 hybridized bonds [1–3]. It is believed that the first graphene layer fits into a $(6\sqrt{3} \times 6\sqrt{3})\,R30°$ surface reconstruction on SiC, and the $(6\sqrt{3} \times 6\sqrt{3})\,R30°$ unit cell ideally coincides with a graphene unit. However, recent research findings with more sensitive and precise techniques raise additional concerns about buffer layer formation and have revealed that the buffer layer is not commensurate with SiC surface reconstruction [4]. From the thermodynamic and chemical points of view, such features (namely, the formation of additional C-Si bonds at the graphene/SiC interface) are originating from the natural necessity to saturate the remaining Si dangling bonds (after high temperature sublimation). In this context, many attempts at growing epitaxial graphene were dedicated to breaking up the covalent bonds between SiC and zero layer graphene and to saturating as-formed dangling bonds by guest species, so-called intercalants [5–14]. Experimental studies clearly reveal a strong effect of the buffer layer on the electronic properties of graphene; specifically, it was documented that due to the charge transfer through interfacial dangling bonds, the buffer layer is found to pin the Fermi level to ≈0.49 eV in the conduction band, making the material *n*-type [15]. As a consequence of the charge transfer from the interface states to graphene and spontaneous appearance of the interface dipole moment, the Fermi level and work function of graphene can be modulated, thereby determining its electroactive behavior [16]. Furthermore, the existence of a giant inelastic tunneling (50% of total tunneling current) caused by localized states at the interface layer of graphene/SiC was confirmed by atomically-resolved scanning tunneling microscopy and spectroscopy [17]. Another surprising fact related to the buffer layer effect was the observation of a small gap (~0.26–0.5 eV) in epitaxial graphene on SiC induced by breaking the sublattice symmetry, but the fundamental nature of this band gap opening is controversial [18,19]. New insights into the origin of the band-gap opening induced by the structural periodicity in the epitaxial graphene buffer layer have been recently reported by Nair et al [20]. Taking the aforementioned aspects into account, it is reasonable to assume that a control of the sp^2/sp^3 hybridization ratio in epitaxial graphene is a good strategy towards the atomistic-level engineering of the graphene/SiC heterointerface to tailor the electronic properties of graphene. Indeed, it was recently reported that epitaxial graphene, depending on the material quality, can play different roles when being interfaced with SiC, such as the ohmic contact [21], the Schottky contact [22], the gate electrode [21], the heterojunction counterpart [23] and/or even the quantum well component [24]. Although a great deal of attention has been paid to buffer layer effects, the physical reasons why the same material exhibits such diverse electronic features are not fully understood. There is still a point to be discussed: the correlation between buffer layer "physics" and mechanisms/possible scenarios underlying the electroactive behavior of epitaxial graphene.

Undoubtedly, for a deep understanding and correct interpretation of experimental data on the electronic properties of epitaxial graphene, one should pay proper attention to the role of the buffer layer. On the other hand, the presence of interfacial states at the heterointerface may be responsible for other physical processes and phenomena underlying the heat transport and ferromagnetism. In particular, perturbation of the ballistic heat transport caused by strong phonon scattering at the graphene/SiC interface was discussed in [25]. Thermal transport through graphene/SiC depending on the kind of SiC polytype, face polarity and atomic bond has been intensively investigated in [26–28]. These results show that the heat transfer is highly sensitive to the kind of interface between the graphene and SiC. Thus, solving the so-called thermal management problem can be achieved via controlling the geometry and the chemical nature of the interface region, i.e., the buffer layer.

As another example of the crucial role of the interfacial layer on the physical properties of graphene, Giesbers et al. [29] have reported on strong room temperature ferromagnetism (with magnetic moment of 0.9 µB per carbon hexagon projected area) and suggested that such ferromagnetic behavior may be attributed to an exchange interaction between the Coulomb-induced localized silicon dangling bonds (belonging to the buffer layer) and the localized mid-gap state. Zhou et al. [30] have also proposed some ideas towards using graphene/SiC interface-induced magnetism for spintronic applications.

The aim of this paper is to review the current status of the main experimental and theoretical studies of graphene on SiC towards understanding the physical nature of the interfacial layer formation and how this layer manifests itself in the carrier transport. In the next section, key features of the buffer layer structure will be described. We will discuss possible scenarios regarding the buffer layer-assisted interaction between graphene and SiC: we will show that the electroactive behavior of epitaxial graphene is strongly dependent on the interface chemistry and quality of the buffer layer. Finally, we make some concluding remarks regarding the relation between the quality of the graphene-SiC heteroboundary and the expected behavior of epitaxial graphene on SiC.

2. Electrical Properties of the Graphene/SiC Interface

Being combined in a single system, graphene and silicon carbide exhibit unique behavior under the influence of an external electric field, which differs from the behavior of the contact between the metal and the semiconductor under classical considerations. In the first place, the difference in the properties is caused by a possibility to control the work function of graphene and the polarizability of its π orbitals [31,32]. Therefore, the energy properties of the heterojunction may be purposefully altered by changing the interfacial chemistry between the materials.

Since the presence of the buffer layer significantly influences the electrical properties of the graphene/SiC structure, first of all, it is important to understand how current flows through a buffer-free graphene/SiC structure. From the theoretical point of view, there are, at least, three cases when we can avoid the formation of the buffer layer:

(1) The vertical structure for electrical measurements can be prepared by simple mechanical contact between exfoliated graphene and the desired SiC substrate. The starting point of the sample preparation in this case is a mechanical exfoliation of highly-oriented pyrolytic graphite (HOPG) by sticky tape, followed by the application of the exfoliated graphene films onto the SiC surface. From the literature analysis, we know that there are many techniques for the exfoliation of graphite based on common mechanical mechanisms [33].

(2) Graphene formed on the carbon-face SiC by high-temperature thermal decomposition of the SiC substrate can be also chosen as a sample for electrical characterization. Indeed, it has been repeatedly shown that the growth of graphene on the carbon-face SiC substrate does not promote the formation of the buffer layer [34].

(3) Another way to avoid the undesirable buffer layer is intercalation of graphene grown on the Si-face SiC substrate by high-temperature Si sublimation. The main scenario for this is to break the covalent bonds between the buffer layer and Si atoms on the SiC surface and to saturate the silicon dangling bonds. Then, the buffer layer can be converted into a new graphene layer with graphene symmetry and typical electronic structure. As was confirmed by experimental studies, H [9,35], Na [36], O [37,38], Li [39], Si [40], Au [12], F [11] and Ge [41] intercalation can transform the buffer layer into a graphene layer with enhanced electrical performance in comparison with untreated monolayer graphene, which exists on the buffer layer. Intercalant species can penetrate into the interface between the buffer layer and the Si-face SiC substrate and create the chemical bonds with topmost Si atoms, thereby causing the transformation of the buffer layer to quasi-free-standing graphene. It is interesting to note that non-metallic (for example, fluorine, oxygen and hydrogen) intercalations are expected to be more effective since they can strongly covalently interact with Si species of the SiC.

Let us consider a physical model of the contact between exfoliated graphene and uniformly-doped n-type silicon carbide, assuming the absence of any intermediate phases or surface states between them. In such an ideal case, the Fermi level is located at the Dirac point of graphene. Since the electron affinity of 4H-SiC (3.7 eV) is less than the work function of graphene (4.2 eV), the flow of electrons from the semiconductor exceeds the flow of electrons from the graphene (Figure 1). As a result, graphene acquires a negative charge, whereas the silicon carbide acquires a positive charge. Consequently, the built-in electric field between the contacting materials will prevent the further charge transfer from the silicon carbide to graphene. The exchange of charges between the semiconductor and graphene will proceed until the Fermi energies of the two materials and thermionic emission currents reach thermal equilibrium. As a result, near the surface of the semiconductor, energy bands are bent upward, and the contact resistance increases significantly. Chen et al. [42] reported a direct experimental observation of the band bending at the interface between epitaxial graphene and 6H-SiC by using in situ synchrotron-based photoemission spectroscopy. It was revealed that the band bending depends strongly on the polarity of the surface of the underlying substrate (Figure 2) and increases from 0.4 eV (for graphene on Si-terminated 6H-SiC) to 1.3 eV (in the case of graphene on C-terminated

6H-SiC). Obviously, this difference is caused by the unique nature of the growth kinetics and structural features of epitaxial graphene on silicon carbide substrates with different polarity faces. It is generally accepted that in the case of graphenization on a C-terminated 6H-SiC, the buffer layer is absent, and the interaction between the substrate and the graphene is largely unabated. It suggests that this graphene possesses a single linearly-dispersing π-band with the Dirac point located close to the Fermi level. However, according to some experimental data, the energy of the Dirac point may vary from +33 meV (*p*-doped) to −14 meV (*n*-doped) and even more [43]. Additional small doping may be caused by the presence of dangling bonds at the interface. Obviously, it could also affect the barrier height and electron transport through the interface. As has been shown in the work of Jayasekera and others [44], in the case of an unpassivated interface, the main source of electron doping is dangling bonds on the remaining carbon atoms (Figure 3a), while the energy states associated with dangling bonds on the silicon atoms are located in the valence band 1 eV above the Fermi level, and their effect can be neglected. A partial passivation of the dangling bonds on the carbon atoms (dangling bonds on silicon atoms still remain) leads to the strong interaction of silicon species with graphene (Figure 3b), and thus, graphene becomes like a buffer layer, similarly to the case of the graphenization of the Si-face SiC substrate. At the same time, the total surface passivation leads to the fact that interaction between the graphene surface and silicon carbide is weakened so that the completely detached carbon layer acquires the properties of a neutral graphene (Figure 3c).

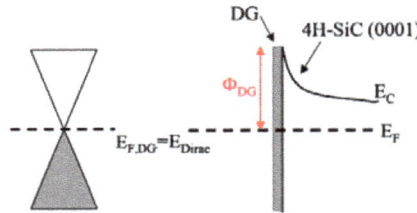

Figure 1. Energy band diagram of the graphene Schottky contact to silicon carbide (4H-SiC) [9]. To exclude the buffer layer effect on Schottky barrier formation, the authors used the graphene exfoliated from highly oriented pyrolytic graphite and deposited on 4H-SiC. DG denotes the deposited graphene on the SiC substrate, without the buffer layer [15]. E_C represents the energy of the conduction band edge; E_F is the Fermi energy for the bulk 4H-SiC; $E_{F,DG}$ is the Fermi energy of exfoliated graphene, which was directly deposited on the SiC surface. Φ_{DG} is the Schottky barrier height. E_{Dirac} corresponds to the Dirac point energy. Due to the absence of the buffer layer (only weak van der Waals-like interaction between DG and the Si substrate occurs), $E_{F,gr}$ corresponds to the Dirac point energy. Reprinted from Sonde et al. [15]. Copyright (2009) with permission from The American Physical Society.

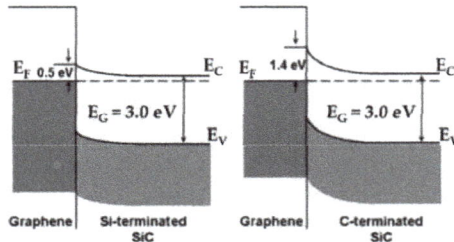

Figure 2. Band line-up at the interfaces between epitaxial graphene and (left panel) the Si-terminated 6H-SiC (0001) and (right panel) the C-terminated 6H-SiC [42]. E_C and E_V represent the energies of the conduction and valence band edge, respectively. E_F is the Fermi energy. The Fermi levels of the two materials are aligned. E_G is the band gap energy of the 6H-SiC. Reprinted from Chen et al. [42]. Copyright (2010) with permission from The Japan Society of Applied Physics.

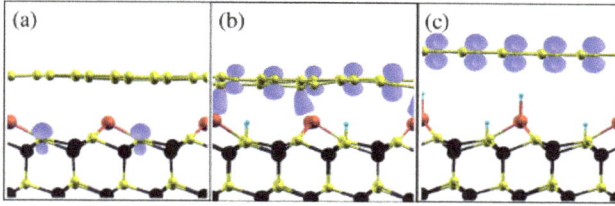

Figure 3. Possible configurations of the graphene/SiC(0001) interface and corresponding localized density of states (at the Dirac point): (**a**) unpassivated, (**b**) half-passivated and (**c**) fully-passivated systems [44]. Red atoms correspond to silicon adatoms; black atoms represent silicon atoms belonging to SiC bulk; yellow balls are C atoms; and cyan atoms are H species. Electron density isosurfaces (blue color) correspond to 0.15×10^{-3} electrons. Reprinted from Jayasekera et al. [44]. Copyright (2011) with permission from The American Physical Society.

In the third case of the absence of a buffer layer, after exposure to intercalants, we consider how hydrogen intercalation effects the electrical properties of the vertical graphene/SiC structure. As has been reported by Dharmaraj et al. [45], the vertical structure composed of as-grown epitaxial graphene and the Si-face SiC substrate exhibits a rectifying behavior with large leakage current under reverse bias (the Schottky barrier height in this case is approximately equal to 0.55 ± 0.05 eV). The authors ascribed this quite low value of the Schottky barrier height to the enhanced density of unsaturated silicon bonds in the vicinity of the interface. For correctness, the covalently bound C atoms should be also taken into account. Due to these reasons, Fermi level pinning induced by charge transfer occurs. The band diagram presented in Figure 4a clearly demonstrates that both the presence of silicon dangling bonds and the buffer layer leads to strong *n*-type doping of graphene and, as a consequence, an increase of the work function and reduction of the Schottky barrier height. During hydrogen intercalation, hydrogen species simultaneously saturate the silicon dangling bonds and break partially the covalent bonds between the buffer layer and the Si-terminated surface of SiC, thereby decoupling the buffer layer. As a result of the removal of the buffer layer and the partial saturation of unsaturated Si bonds, the hydrogen-intercalated epitaxial graphene/SiC structure demonstrates improved rectifying behavior with low leakage current in the reverse bias regime (the Schottky barrier height after intercalation procedure was estimated to be 0.75 ± 0.05 eV). As can be seen from Figure 4b, an increase in the value of the Schottky barrier height can be explained by Fermi level depinning induced by the reduction of the density of the unsaturated Si bonds and *n*-type doping of graphene. A complete passivation of the Si-terminated surface may lead to a change in the conductivity type of graphene. The change from *n*-type to *p*-type results in the modification of the band bending from upward bending to downward bending, and thus, the electrical properties of the graphene/SiC structure can be also modified.

In the case of epitaxial graphene films on the Si-face of SiC substrates, a carbon-rich buffer layer with partial sp^3 hybridization is always formed [46]. As mentioned above, the buffer layer substantially affects the electronic properties of graphene and causes pinning of the Fermi level and the subsequent reduction of the Schottky barrier height (Figure 5). In particular, it was shown that in the absence of the buffer layer, the barrier height is 0.85 ± 0.06 eV (buffer-free deposited graphene/4H-SiC) [15] and 0.75 ± 0.05 eV (buffer-free hydrogen-intercalated graphene/4H-SiC) [45], whereas the structure comprising a buffer layer exhibits a significantly reduced barrier height of 0.36 ± 0.1 eV (epitaxial graphene/4H-SiC) [15] and 0.55 ± 0.05 eV (epitaxial graphene/4H-SiC) [45].

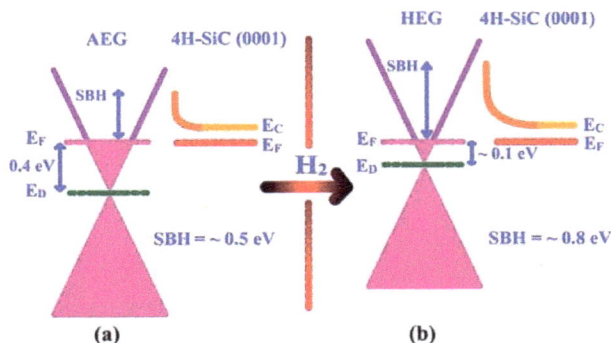

Figure 4. Energy band diagram of vertical Schottky barrier diodes: (**a**) as-deposited epitaxial graphene/4H-SiC(0001) and (**b**) hydrogen-intercalated graphene/4H-SiC(0001). AEG and HEG denote as-grown epitaxial graphene and hydrogen intercalated epitaxial graphene, respectively. During hydrogen intercalation, the Dirac point of graphene (E_D) is shifted towards E_F, thereby leading to the increase in the work function of graphene and the Schottky barrier height (SBH) [45]. Reprinted from Dharmaraj et al. [45]. Copyright (2016) with permission from The American Institute of Physics (AIP).

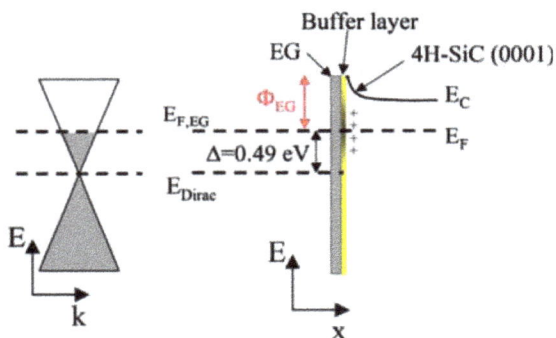

Figure 5. Band diagram of Schottky contact between graphene and SiC in the presence of a buffer layer at the interface [15]. EG denotes the epitaxial graphene on SiC substrate, with the buffer layer. E_C represents the energy of the conduction band edge; E_F is the Fermi energy for the bulk 4H-SiC; $E_{F,EG}$ is the Fermi energy of epitaxial graphene grown on the SiC surface by the thermal decomposition technique. Φ_{EG} is the Schottky barrier height. E_{Dirac} corresponds to the Dirac point energy. Due to the presence of the buffer layer, the Fermi level pinning above the Dirac point occurs. Reprinted from Sonde et al. [15]. Copyright (2009) with permission from The American Physical Society.

3. Experimental Control of the Barrier Height at the Graphene/SiC Interface

In some cases, the height and the width of the potential barrier become low/narrow enough for the electrons to easily tunnel through or overcome the barrier. In this case, the epitaxial graphene demonstrates an ohmic behavior with respect to the silicon carbide, and a linear current-voltage characteristic has been observed [21,45,47]. Apparently, it is caused by increasing the density of unsaturated dangling bonds contributing to the pinning of the Fermi level and leading to a significant doping of epitaxial graphene. In fact, increased sp^3 hybridization at the interface will result in the appearance of additional conduction channels, whereby there is a significant decrease of the contact resistance. Hertel et al. [47] have studied the electrical properties of the graphene/SiC heterointerface by a linear transfer length method. It was shown that the formation of an ohmic contact to the weakly-doped 6H-SiC is associated with the low energy barrier $\Phi_B = 0.3$ eV between the epitaxial graphene and 6H-SiC due to a small mismatch between the work functions of both materials. As a consequence, the electrons can overcome this barrier at ambient temperature, causing a current flow. The contact resistance can be improved by increasing the donor concentration through ion implantation under the contact. This reduces the barrier width, and tunneling through the barrier creates an additional conduction channel. In the same study, a comparative analysis of the electrical properties of two partners (graphene-4H-SiC and graphene-6H-SiC) was carried out. Since 4H-SiC has a lower value of the work function (by 0.3 eV), compared with 6H-SiC, a higher Schottky barrier of $\Phi_B = 0.6$ eV is formed at the interface graphene/4H-SiC, thereby contributing to an increase of the contact resistance. In turn, the work function of graphene varies significantly with the number of layers (Figure 6) as was reported in a recent paper by Mammadov et al. [48]. In particular, it was found that the buffer layer has a reduced work function of 3.89 ± 0.05 eV, and every subsequent layer leads to increasing the work function, reaching a value of 4.43 ± 0.05 eV for the case of trilayer graphene. On the other hand, annealing of the zero graphene layer in an ultra-pure hydrogen environment leads to the growth of quasi-free-standing monolayer (QFMLG) graphene with a lack of buffer layer. QFMLG has a much greater work function of about 4.79 ± 0.05 eV. At that, the work function decreases with increasing numbers of layers to a value of 4.63 ± 0.05 eV for the quasi-free-standing trilayer graphene (QFTLG). From a practical point of view, this is a very important result, because the optimum performance of devices based on Schottky diodes requires a constant and uniform barrier height across the interface, not varying over the interfacial surface. Samples with a non-uniform thickness will demonstrate a wide variation of values of the work function and Schottky barrier height for the entire sample area. Indeed, as can be seen from the histogram in Figure 7, different authors reported on the large spread of this parameter for Schottky diodes based on nominally the same material [22].

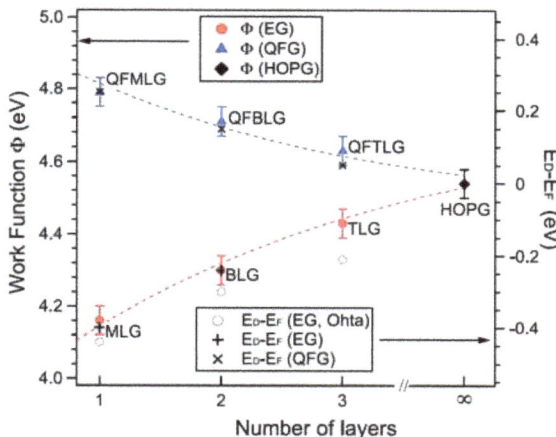

Figure 6. Dependence of the work function of epitaxial graphene (EG) with the buffer layer (red symbols) and buffer-free quasi-free-standing graphene (blue symbols) on the number of layers [48]. Abbreviation of QFG means quasi-free-standing graphene. The black symbols correspond to the position of the Dirac point with respect to the Fermi level for epitaxial graphene and quasi-free standing graphene. Reprinted from Mammadov et al. [48]. Copyright (2017) with permission from Institute of Physics Publishing Ltd. HOPG, highly-oriented pyrolytic graphite.

As can be seen from this histogram, the Schottky barrier height at the graphene/SiC heterointerface is strongly sensitive to the growth method and the graphene thickness. Furthermore, as was reported earlier, the unintentional presence of unavoidable natural ripples and ridges in epitaxial graphene on SiC may also cause the fluctuations in the Schottky barrier height [49–54]. The key factors influencing the uniformity of the Schottky barrier height for graphene/SiC structures are the homogeneity of the graphene thickness, the quality of the grown interface (defects, pits, dislocations, surface roughness), the kind of grown interface (SiC polytypism, face polarity) and the growth conditions. We found that the Schottky junctions formed by the high-temperature Si sublimation approach [55] exhibit the smallest standard deviation of the mean value of the Schottky barrier height. This can be explained by the fact that the graphenization process via thermal decomposition of SiC promotes the formation of large-scale homogeneous epitaxial graphene layers [56–59].

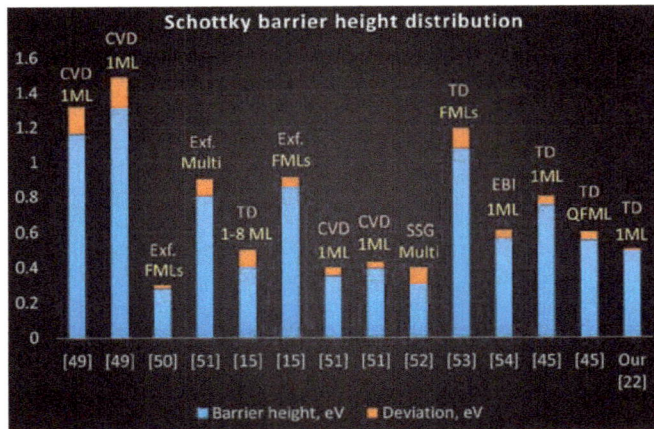

Figure 7. Histogram of the Schottky barrier height distribution for graphene on SiC made by different techniques and different thicknesses.

Indeed, our *I-V* measurements of the graphene (99% of the total coverage is monolayer)/4H-SiC(0001) vertical device revealed the very stable rectifying behavior of the graphene/SiC diode (Figure 8). In line with the statistical distribution, the determined values of the Schottky barrier height range from 0.46 to 0.503 eV for the graphene/SiC junction, while the ideality factor ranges from 1.011 to 1.026. The standard deviations yield 0.013 eV and 0.0049 for the two parameters, respectively. We determined that the mean values of Schottky barrier height and ideality factor of the Schottky diode are 0.4879 eV and 1.018, respectively. The extracted value of the Schottky barrier height coincides well with the theoretical value of 0.5 eV. In comparison to previously-reported results, our sample (by virtue of the high thickness uniformity) demonstrates the smallest reported value of the standard deviation for the Schottky barrier height.

Multifunctional properties of the graphene-silicon carbide interface have been used to create a monolithic transistor based on a single platform (Figure 9) [21]. It should be noted that in this case, the graphene plays a dual role. On the one hand, the graphene forms ohmic contacts to silicon carbide, thus acting as the source and drain. On the other hand, the graphene forms a Schottky barrier and plays the role of the gate contact. Interestingly, the presence or absence of the buffer layer is the key factor that determines the role of graphene. Epitaxial graphene monolayer (electron density and carrier mobility are equal to $n = 10^{13}$ cm^{-2}, $\mu_e = 900$ cm$^2\cdot$V$\cdot$s^{-1}, respectively) with a carbon-rich buffer layer underneath (Figure 9b) exhibits an ideal ohmic behavior, despite the weak doping level of silicon carbide. Upon hydrogen intercalation (in H$_2$ atmosphere at 850 °C), the buffer layer is decoupled from the substrate, and quasi-free-standing bilayer graphene is formed (Figure 9c) with the following parameters: hole density $p = 10^{13}$ cm^{-2}, $\mu_h = 2.000$ cm$^2\cdot$V$\cdot$s^{-1} at room temperature. Indeed, in this case, breaking of the covalent bonds between the graphene and silicon carbide occurs, and the remaining dangling bonds are saturated with hydrogen atoms. As a result, the buffer layer is converted to an additional layer of graphene. Since quasi-free-standing bilayer graphene (QFBLG) on SiC exhibits rectifying behavior inherent to the Schottky diode, it can be used as the gate contact. Analysis of the capacitance-voltage and current-voltage characteristics allowed estimating both the lower limit of barrier height ($\phi_{B,IV}$ = 0.9 eV) and the upper limit ($\phi_{B,CV}$ = 1.1 eV ... 1.6 eV). The high value of the barrier height and negligible leakage current meet the main requirements of the Schottky diodes. Despite the fact that QFBLG forms a Schottky contact to *n*-type SiC, large fluctuations in the Schottky barrier height over the entire area of contact between the two materials are still a significant problem, limiting the fabrication of high-performance transistors. Importantly, a complete deactivation of the buffer layer due to the hydrogen-induced intercalation could lead to the conductivity type switching in graphene from *n*-type to *p*-type. This phenomenon is due to the intrinsic spontaneous polarization of the hexagonal silicon carbide substrate and results in both a substantial increase in the Schottky barrier height of more than >1 eV [45] and even the activation of another kind of current in a *p-n* junction, which is governed by recombination or generation processes within the *p-n* diode structure.

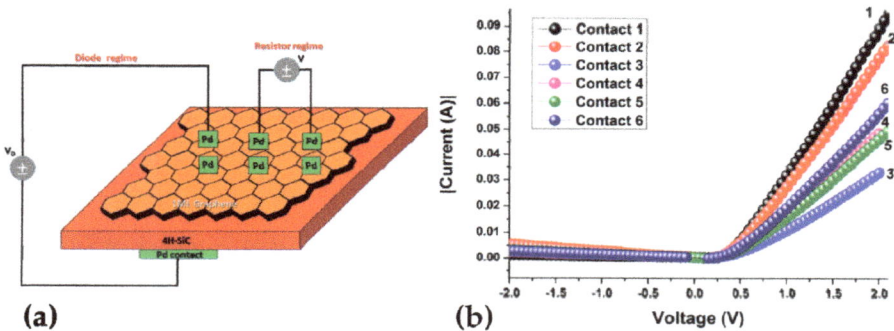

Figure 8. (**a**) Sketch of the graphene/4H-SiC vertical device and (**b**) current-voltage characteristics of the vertical graphene-4H-SiC device. The *y*-axis indicates the absolute values of current. Palladium contacts (1–6) are positioned at different places on the graphene surface [22]. Reprinted from Shtepliuk et al. [22]. Copyright (2016) Shtepliuk et al.; licensee Beilstein-Institut.

Butt et al. [60] demonstrated the important role of the buffer layer beneath graphene in the process of the photogeneration of carriers in a transistor. It is known that the strong interaction between the substrate and the buffer layer contributes to lowering the energy barrier, while the complete deactivation of the buffer layer will lead to a significant increase in the Schottky barrier. The aforementioned work confirmed that the absence of an energy barrier at the interface causes the carrier injection rate to become equal to the rate of the thermal generation of carriers and becomes dependent on the thickness and barrier height at the interface. Figure 10 shows the qualitative phenomenon of photo-induced electrostatic doping of graphene. The energy diagrams of the SiC/graphene interface along the direction perpendicular to the channel are shown for different injection rate limits in dark and light conditions with the illustrated charge density in SiC. As a result of the photogeneration of carriers in SiC in the light regime, holes drift in the direction of the source and drain contacts due to the voltage applied to the gate (V_{bg} = 20 V). In the absence of a barrier at the graphene/SiC interface, the injection rate is equal to the thermal generation rate, and therefore, holes do not accumulate at the interface. In this situation, spatial charge separation does not occur at the interface, and thus, the substrate-induced electrostatic effect in graphene is negligible, as shown in Figure 10b. In the presence of the energy barrier at the interface (the injection rate is less than the thermal generation rate), the drift of the photogenerated carriers from the substrate results in charge accumulation in the vicinity of the surface of the SiC (Figure 10c,d).

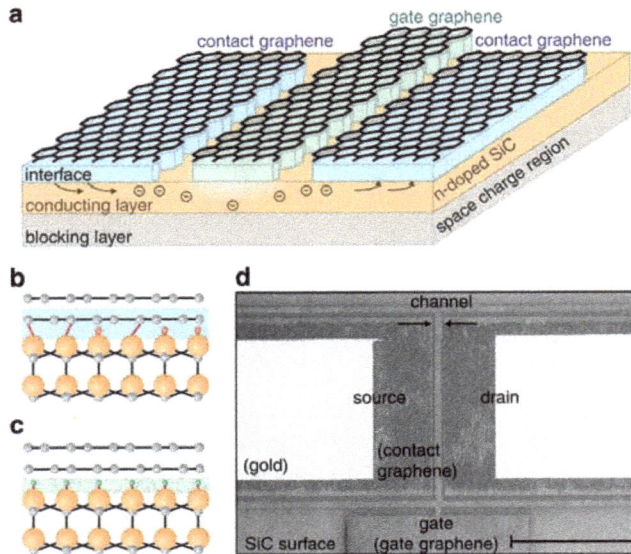

Figure 9. (**a**) Schematic of the device with two kinds of graphene: graphene as the ohmic contact for the source/drain and (**b**) and graphene as a Schottky-like gate (**c**). (**d**) An electron micrograph illustrating the realistic device configuration [21]. In (**b,c**), the orange atoms correspond to silicon adatoms; grey balls are C atoms; and red/green atoms are H species. Reprinted from Hertel et al. [21]. Copyright (2012) with permission from Macmillan Publishers Limited.

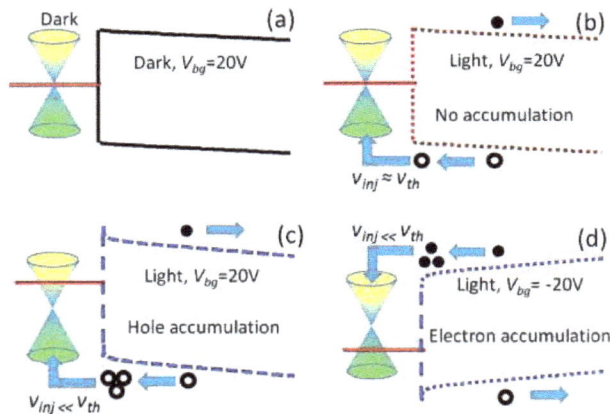

Figure 10. Band diagram of graphene/SiC interface (**a**) in dark and (**b–d**) under illumination. No electrostatic doping in graphene is observed in dark conditions. V_g is the voltage applied to the gate; V_{inj} is velocity of carrier injection from the SiC into the graphene; and V_{th} is the carrier's thermal velocity, respectively. Graphene becomes *n*-type after applying forward bias in the "light" regime (**c**) and *p*-type after applying a reverse bias in the light "regime" (**d**) [60]. Reprinted from Butt et al. [60]. Copyright (2015) with permission from IEEE.

At this point, we would like to draw the reader's attention to the difference between the aforementioned current transport mechanisms through the graphene/SiC heterointerface:

(1) If the potential barrier is wide and high (as in the case of stable Schottky diode, formed on the buffer-layer free graphene and SiC), the current is driven by thermal excitation of the electrons and their transfer from the silicon carbide into graphene (thermionic emission).

(2) If the potential barrier at the interface is rather narrow (due to the strong interaction between the buffer layer and the topmost layers of silicon carbide), then current flows due to tunneling through energy barriers regardless of their width and energy height (ohmic contact formation).

(3) In some intermediate cases, epitaxial graphene may consist of sp^2-bonded carbon atoms with a small fraction of sp^3 hybridized carbon species bound to SiC (less than 30%). It is clear that under such conditions, the currents through the interface are regulated by two competing mechanisms: thermal excitation of carriers and their tunneling through the top of the barrier. In this scenario, a Schottky barrier is high enough to provide rectifying behavior. On the other hand, the leakage current will be increased substantially, thus degrading the performance of the diode.

4. Observations of Uncommon Phenomena at the Graphene/SiC Interface

It is important to note that some heterostructures consisting of epitaxial graphene on silicon carbide can exhibit behavior that significantly differs from both the ohmic contact and the Schottky contact. Contrary to the Schottky diodes, the potential barrier height (2.53 eV [23], 2.70 eV [61] and 2.90 eV [24]) in such structures can be as large as the band gap energy of silicon carbide (E_g = 3.23 eV). There are possible explanations for these phenomena, which are connected with specific features of the graphenization of SiC, unusual polarization-induced p-type doping of the epitaxial graphene or even more likely the formation of the p-n junction between graphene and SiC.

Indeed, Andersson et al. [23] found that the *I-V* characteristics across the n-type epitaxial graphene/p-SiC interface are better described by a p-n diode model than by thermionic emission. According to this theory, the carrier transport mechanism through the anisotype graphene/silicon carbide p-n junction is governed by diffusion and recombination of carriers in the quasi-neutral region. In order to confirm the applicability of the theory, electroluminescence studies were performed, allowing one to obtain a direct proof of the radiative recombination in the silicon carbide layers. If a forward or reverse-biasing voltage is applied across the p-n junction, a visible emission at the edges of the graphene mesa is observed, as shown in the in Figure 11b. The observed peak at 410 nm (3.02 eV) is associated with an optical radiative transition between the acceptor level and the conduction band in SiC. This proves the injection of minority carriers in silicon carbide from the graphene side. In principle, the injection of minority carriers is also possible in the case of a Schottky diode, but the injection ratio is negligible because of the low carrier density in the SiC [62]. Therefore, these experimental observations are more likely associated with the formation of a heterojunction than a Schottky barrier. It is assumed that in this case, an anisotype (p-n) transition type 1 is formed, with a dominant current due to recombination and diffusion in the forward bias regime and reverse bias leakage caused by thermal generation. The band diagram of the system being studied is shown in the Figure 11a.

In another work, Anderson et al. [61] have performed a comparative analysis of the electrical properties of the anisotype p-n junction (n-type epitaxial graphene/p-type SiC) and isotype p-p junction (p-type epitaxial graphene/p-type SiC) in the dark and light regimes. Band diagrams for these devices are shown in Figure 12.

Expectedly, the p-n junction exhibits well-pronounced rectifying behavior with small leakage current (Figure 13a), suggesting the dominating character of the drift-diffusion mechanism of current transport. Within this model, the leakage current at the reverse bias is negligible, because it consists of only diffusion and thermal generation current components. On the other hand, the p-type graphene/p-type SiC structure behaves like the typical isotype p-p junction or the Schottky diode (Figure 13b). The charge transport through the device with the dominant role of the majority carriers in the forward bias regime is governed by thermionic emission, whereas the leakage current in the reverse bias mode increases linearly with increasing reverse voltage.

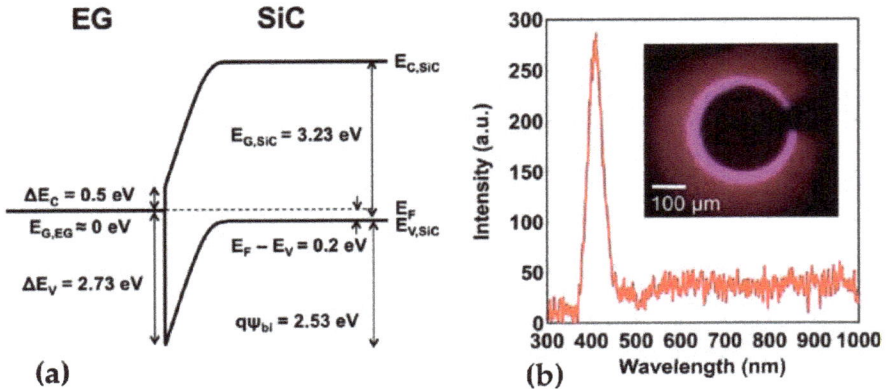

Figure 11. (a) Energy diagram of the EG/p-SiC structure; (b) Electroluminescence (EL) spectrum and EL image collected from EG/p-SiC at high forward bias [23]. $E_{G,SiC}$ is the band gap energy of the bulk SiC; $E_{G,EG}$ is the band gap energy of the epitaxial graphene. E_F is the Fermi energy; E_C (E_V) is the energy of the conduction (valence) band edge; and qψ_{bi} is the barrier height, respectively. ΔE_V and ΔE_C are the expected valence and conduction band discontinuities. Reprinted from Anderson et al. [23]. Copyright (2012) with permission from IEEE.

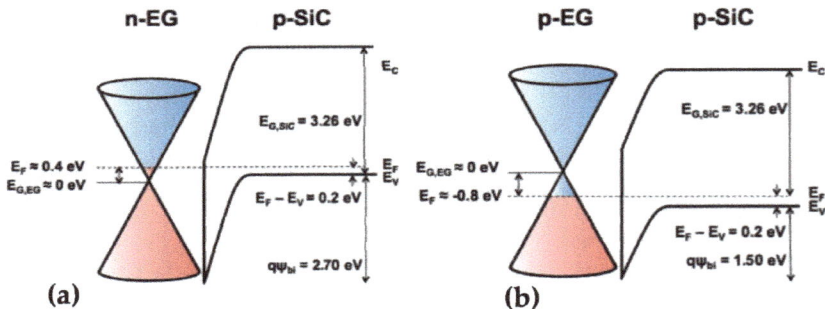

Figure 12. Schemes showing the band diagrams for the n-type epitaxial graphene/p-type SiC (a) and p-type epitaxial graphene/p-type SiC (b) junctions [61]. $E_{G,SiC}$ is the band gap energy of the bulk SiC; $E_{G,EG}$ is the band gap energy of the epitaxial graphene. E_F is the Fermi energy; E_C (E_V) is the energy of the conduction (valence) band edge; and qψ_{bi} is the barrier height, respectively. Reprinted from Anderson et al. [61]. Copyright (2015) with permission from The Japan Society of Applied Physics.

Interestingly, some authors reported that bound quantum states may occur at the graphene/n-type SiC heterointerface [24]. This phenomenon is probably due to the formation of deep (2.9 eV) and narrow (2.15 Å) barriers, the so-called layered nanostructure with the hole quantum well (QW) potential relief (Figure 14a illustrates the band diagram of the investigated structure). The authors emphasized that the carbon-rich buffer layer with surface reconstruction (6$\sqrt{3}$ × 6$\sqrt{3}$) R30° plays the role of the wide band-gap layer, since the chemical interaction between the interfacial layer and SiC diminishes the π-electronic subsystem and opens a gap. According to the literature data, such an energy gap can reach ~2 eV for the (6$\sqrt{3}$ × 6$\sqrt{3}$) R30° layer [63]. Due to these features, the top of the graphene valence band (shaded area in Figure 14a) looks like a hole quantum well, and the quantum confinement of the electrons normal to graphene plane is expected. As can been seen from Figure 14b, three distinguished peaks near the Fermi level at energies of $E_1 = 0.3$ eV, $E_2 = 1.2$ eV and $E_3 = 2.6$ eV are present on the valence band density of states. These singularities were attributed to the quantum well bound states. Taking into account the fact that these peaks are absent on the valence band spectra of graphite and SiC substrate, they have a graphene-like nature. In addition, the authors reasonably assumed that the observed valence band features (E_1, E_2 and E_3) cannot be assigned to the buffer layer because previous studies showed that the ($6\sqrt{3} \times 6\sqrt{3}$) R30° buffer layer is only responsible for the appearance of two additional peaks g_1 and g_2 at binding energies of 0.5 and 1.6 eV [1,64], which do not coincide with the E_1, E_2 and E_3 peaks.

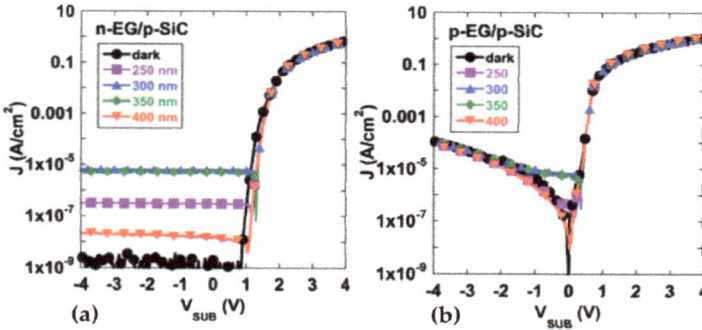

Figure 13. Dependences of the current-voltage characteristics on the illumination wavelength for different devices: *n*-type epitaxial graphene/*p*-type SiC (**a**) and *p*-type epitaxial graphene/*p*-type SiC (**b**) junctions [61]. Reprinted from Anderson et al. [61]. Copyright (2015) with permission from The Japan Society of Applied Physics.

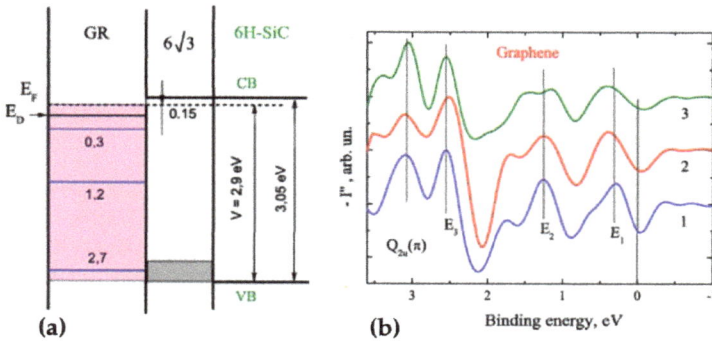

Figure 14. (**a**) Energetic diagram of the hole quantum well formed by graphene on SiC with the carbon-rich buffer layer; (**b**) Inverse second derivatives of the valence band photoemission spectra of epitaxial graphene collected from different sample areas (1–3) [24]. E_D is the Dirac point energy; E_F is the Fermi energy; CB is the bottom of the conduction band; and VB is the top of the valence band, receptively. Reprinted from Mikoushkin et al. [24]. Copyright (2015) with permission from Elsevier.

Table 1 summarizes potential barrier heights for different vertical structures composed of graphene and SiC substrates by different methods.

Table 1. Barrier heights for current transport through vertical graphene/SiC structures depending on the preparation technique. ML: the monolayer; CVD: chemical vapour deposition; HOPG: highly-oriented pyrolytic graphite.

Junction	Growth Method	Thickness	Barrier Height, eV	Reference
Gr/n-4H-SiC	Si sublimation	1–8 ML	0.36 ± 0.1	[15]
Gr/n-4H-SiC	Exfoliation	Few ML	0.85 ± 0.06	[15]
Gr/n-Si-6H-SiC	Thermal decomposition	2 ML	0.9	[21]
Gr/n-Si-4H-SiC	Si sublimation	1 ML	0.487 ± 0.013	[22]
Gr/p-4H-SiC	Si sublimation	1 ML	2.53	[23]
Gr/n-Si-6H-SiC	Si sublimation	1.6 ML	2.90	[24]
p-Gr/p-4H-SiC	Si sublimation	3 ML	1.5	[45]
Gr/n-Si-4H-SiC	Low energy e-beam irradiation	1 ML	0.556 ± 0.05	[45]
Gr/n-Si-4H-SiC	CVD	1 ML	1.16 ± 0.16	[49]
Gr/n-C-4H-SiC	CVD	1 ML	1.306 ± 0.18	[49]
Gr/n-SiC	Exfoliation	Few ML	0.28 ± 0.02	[50]
Gr/n-Si-6H-SiC	CVD	1 ML	0.35 ± 0.05	[51]
Gr/n-C-4H-SiC	CVD	1 ML	0.39 ± 0.04	[51]
Graphite/p-4H-SiC	Solid state graphitization	Multilayer	2.7 ± 0.1	[52]
Graphite/n-4H-SiC	Solid state graphitization	Multilayer	0.3 ± 0.1	[52]
Gr/n-Si-4H-SiC	Thermal decomposition	Few ML	1.066 ± 0.12	[53]
Gr/n-4H-SiC	Exfoliation of HOPG	Multilayer	0.8 ± 0.1	[54]
n-Gr/p-4H-SiC	Si sublimation	3 ML	2.7	[61]
Gr/n-4H-SiC	Si sublimation	Few ML	0.08	[65]
Gr/n-Si-4H-SiC	Electron-beam irradiation	2 ML	0.58	[66]
Gr/n-4H-SiC	CVD	1 ML	0.91	[67]
Gr/n-Si-6H-SiC	Thermal decomposition	2 ML	1.15–1.45	[68]
HOPG/n-SiC	Van der Waals adherence of cleaved HOPG	Multilayer	1.15	[69]

From our previous work [70], we know that the quality of graphene (thickness, uniformity) and its physical properties are highly sensitive to the status of the SiC substrate, including the miscut angle, kind of polytype and face polarity. Therefore, the reliable control of the barrier height requires a clear understanding of the fundamental relation between the properties of the SiC substrate and the quality of the graphene layers.

5. Concluding Remarks

The electrical properties of the graphene/silicon carbide heterointerface have been discussed in order to illustrate the buffer layer role in carrier transport via the heterointerface and the fundamental reasons underlying the barrier height modulation. Although the theoretical value of the barrier height was estimated to be 0.5 eV, it could not be achieved in most experiments, and large deviations from the ideal value, as well as barrier height inhomogeneity have been frequently observed. Such behavior can be understood in terms of structural imperfections (high sp^2/sp^3 hybridization ratio, thickness nonuniformity, domains with different doping levels, ripples, etc.) unintentionally appearing during the graphenization process. In this regard, a reliable and precise control of the barrier height is imperative for further implementation of the epitaxial graphene technology into realistic Schottky diode-based electronic devices. Possible ways to achieve such control might involve the complete decoupling of the carbon-rich buffer layer from the substrate. Indeed, the buffer layer-free samples always offer an enhanced value of the barrier height and a more pronounced Schottky-type rectifying behavior with current transport governed by the thermionic emission mechanism. In the case of the strong interaction between the buffer layer and SiC, deviations from theoretical predictions are observed towards reduced barrier heights. Furthermore, fine tuning of the interfacial chemistry (mainly the sp^2/sp^3 hybridization ratio) makes possible the transition from the Schottky-like contact to the ohmic-like one. An analysis of the literature data led us to deduce that the rarely observed high potential barrier (>2 eV) at the interface seems to be a key factor, which primarily determines the dominant role of the drift-diffusion mechanism in charge transport via the heterostructure. We believe that the barrier height modulation is a good strategy for the development of a monolithic electronic platform comprising the different kinds of graphene behaviors (ohmic contact, Schottky contact, gate dielectric, electrically-active counterpart in *p-n* junction and quantum well) by the most promising and controllable graphitization technique: high-temperature thermal decomposition of the SiC substrate in an argon ambient environment.

Future work should aim at deeper investigations of the relationship "carrier density in graphene-thickness of graphene-barrier height at the interface" to provide more complete physical insights into the current transport via the graphene/SiC interface and facilitate further applications.

Acknowledgments: The project has received funding from the European Union's Horizon 2020 research and innovation program under Grant Agreement No. 696656. The authors would like to thank the financial support via the Vetenskapsrådet (VR) grant 621-2014-5805 and Stiftelsen för strategisk forskning (SSF GMT14-0077, SSF RMA15-0024). Ivan Shtepliuk acknowledges the support from Wallenberg foundation and Ångpanneföreningens Forskningsstiftelse (Grant 16-541). Volodymyr Khranovskyy acknowledges the Swedish Research Council (VR) Marie Skłodowska Curie International Career Grant #2015-00679 "GREEN 2D FOX" and ÅForsk (Grant 14-517).

Author Contributions: Ivan Shtepliuk, Jens Eriksson, Filippo Giannazzo and Rositza Yakimova conceived of the idea and co-wrote the paper. Volodymyr Khranovskyy and Tihomir Iakimov commented on and improved the manuscript. All authors discussed the results.

Conflicts of Interest: The authors declare no competing financial interests.

References

1. Emtsev, K.V.; Speck, F.; Seyller, T.; Ley, L.; Riley, J.D. Interaction, growth, and ordering of epitaxial graphene on SiC(0001) surfaces: A comparative photoelectron spectroscopy study. *Phys. Rev. B* **2008**, *77*, 155303. [CrossRef]
2. Emery, J.D.; Detlefs, B.; Karmel, H.J.; Nyakiti, L.O.; Gaskill, D.K.; Hersam, M.C. Chemically Resolved Interface Structure of Epitaxial Graphene on SiC(0001). *Phys. Rev. Lett.* **2013**, *111*, 215501. [CrossRef] [PubMed]
3. Strupinski, W.; Grodecki, K.; Caban, P.; Ciepielewski, P.; Jozwik-Biala, I.; Baranowski, J.M. Formation mechanism of graphene buffer layer on SiC(0001). *Carbon* **2015**, *81*, 63. [CrossRef]
4. Conrad, M.; Wang, F.; Nevius, M.; Jinkins, K.; Celis, A.; Nair, M.N.; Taleb-Ibrahimi, A.; Tejeda, A.; Garreau, Y.; Vlad, A.; et al. Wide Band Gap Semiconductor from a Hidden 2D Incommensurate Graphene Phase. *Nano Lett.* **2017**, *17*, 341. [CrossRef] [PubMed]
5. Stöhr, A.; Forti, S.; Link, S.; Zakharov, A.A.; Kern, K.; Starke, U.; Benia, H.M. Intercalation of graphene on SiC(0001) via ion implantation. *Phys. Rev. B* **2016**, *94*, 085431. [CrossRef]
6. Visikovskiy, A.; Kimoto, S.; Kajiwara, T.; Yoshimura, M.; Iimori, T.; Komori, F.; Tanaka, S. Graphene/SiC(0001) interface structures induced by Si intercalation and their influence on electronic properties of graphene. *Phys. Rev. B* **2016**, *94*, 245421. [CrossRef]
7. Caffrey, N.M.; Johansson, L.I.; Xia, C.; Armiento, R.; Abrikosov, I.A.; Jacobi, C. Structural and electronic properties of Li-intercalated graphene on SiC(0001). *Phys. Rev. B* **2016**, *93*, 195421. [CrossRef]
8. Caffrey, N.M.; Armiento, R.; Yakimova, R.; Abrikosov, I.A. Charge neutrality in epitaxial graphene on 6H-SiC(0001) via nitrogen intercalation. *Phys. Rev. B* **2015**, *92*, 081409. [CrossRef]
9. Riedl, C.; Coletti, C.; Iwasaki, T.; Zakharov, A.A.; Starke, U. Quasi-Free-Standing Epitaxial Graphene on SiC Obtained by Hydrogen Intercalation. *Phys. Rev. Lett.* **2009**, *103*, 246804. [CrossRef] [PubMed]
10. Sandin, A.; Jayasekera, T.; Rowe, J.E.; Kim, K.W.; Nardelli, M.B.; Dougherty, D.B. Multiple coexisting intercalation structures of sodium in epitaxial graphene-SiC interfaces. *Phys. Rev. B* **2012**, *85*, 125410. [CrossRef]
11. Walter, A.; Jeon, K.-J.; Bostwick, A.; Speck, F.; Ostler, M.; Seyller, T.; Moreschini, L.; Kim, Y.S.; Chang, Y.J.; Horn, K.; et al. Highly *p*-doped epitaxial graphene obtained by fluorine intercalation. *Appl. Phys. Lett.* **2011**, *98*, 184102. [CrossRef]
12. Gierz, I.; Suzuki, T.; Weitz, R.T.; Lee, D.S.; Krauss, B.; Riedl, C.; Starke, U.; Höchst, H.; Smet, J.H.; Ast, C.R.; et al. Electronic decoupling of an epitaxial graphene monolayer by gold intercalation. *Phys. Rev. B* **2010**, *81*, 235408. [CrossRef]

13. Hsu, C.-H.; Ozolins, V.; Chuang, F.-C. First-principles study of Bi and Sb intercalated graphene on SiC(0001) substrate. *Surf. Sci.* **2013**, *616*, 149–154. [CrossRef]

14. Huang, L.; Xu, W.-Y.; Que, Y.-D.; Mao, J.-H.; Meng, L.; Pan, L.-D.; Li, G.; Wang, Y.-L.; Du, S.-X.; Liu, C.; et al. Intercalation of metals and silicon at the interface of epitaxial graphene and its substrates. *Chin. Phys. B* **2013**, *22*, 096803. [CrossRef]

15. Sonde, S.; Giannazzo, F.; Raineri, V.; Yakimova, R.; Huntzinger, J.-R.; Tiberj, A.; Camassel, J. Electrical properties of the graphene/4H-SiC(0001) interface probed by scanning current spectroscopy. *Phys. Rev. B* **2009**, *80*, 241406. [CrossRef]

16. Renault, O.; Pascon, A.M.; Rotella, H.; Kaja, K.; Mathieu, C.; Rault, J.E.; Blaise, P.; Poiroux, T.; Barrett, N.; Fonseca, L.R.C. Charge spill-out and work function of few-layer graphene on SiC(0001). *J. Phys. D Appl. Phys.* **2014**, *47*, 295303. [CrossRef]

17. Červenka, J.; van de Ruit, K.; Flipse, C.F.J. Giant inelastic tunneling in epitaxial graphene mediated by localized states. *Phys. Rev. B* **2010**, *81*, 205403. [CrossRef]

18. Rotenberg, E.; Bostwick, A.; Ohta, T.; McChesney, J.L.; Seyller, T.; Horn, K. Origin of the energy bandgap in epitaxial graphene. *Nature Mater.* **2008**, *7*, 258. [CrossRef] [PubMed]

19. Nevius, M.S.; Conrad, M.; Wang, F.; Celis, A.; Nair, M.N.; Taleb-Ibrahimi, A.; Tejeda, A.; Conrad, E.H. Semiconducting Graphene from Highly Ordered Substrate Interactions. *Phys. Rev. B* **2015**, *115*, 136802. [CrossRef] [PubMed]

20. Nair, M.N.; Palacio, I.; Celis, A.; Zobelli, A.; Gloter, A.; Kubsky, S.; Turmaud, J.-P.; Conrad, M.; Berger, C.; de Heer, W.; et al. Band Gap Opening Induced by the Structural Periodicity in Epitaxial Graphene Buffer Layer. *Nano Lett.* **2017**, *17*, 2681–2689. [CrossRef] [PubMed]

21. Hertel, S.; Waldmann, D.; Jobst, J.; Albert, A.; Albrecht, M.; Reshanov, S.; Schöner, A.; Krieger, M.; Weber, H.B. Tailoring the graphene/silicon carbide interface for monolithic wafer-scale electronics. *Nat. Commun.* **2012**, *3*, 957. [CrossRef] [PubMed]

22. Shtepliuk, I.; Eriksson, J.; Khranovskyy, V.; Iakimov, T.; Lloyd Spetz, A.; Yakimova, R. Monolayer graphene/SiC Schottky barrier diodes with improved barrier height uniformity as a sensing platform for the detection of heavy metals. *Beilstein. J. Nanotechnol.* **2016**, *7*, 1800. [CrossRef] [PubMed]

23. Anderson, T.J.; Hobart, K.D.; Nyakiti, L.O.; Wheeler, V.D.; Myers-Ward, R.L.; Caldwell, J.D.; Bezares, F.J.; Jernigan, G.G.; Tadjer, M.J.; Imhoff, E.A.; et al. Investigation of the Epitaxial Graphene/p-SiC Heterojunction. *IEEE Electron. Device Lett.* **2012**, *33*, 1610. [CrossRef]

24. Mikoushkin, V.M.; Shnitov, V.V.; Lebedev, A.A.; Lebedev, S.P.; Nikonov, S.Y.; Vilkov, O.Y.; Iakimov, T.; Yakimova, R. Size confinement effect in graphene grown on 6H-SiC(0001) substrate. *Carbon* **2015**, *86*, 139. [CrossRef]

25. Xu, Z.; Buehler, M.J. Heat dissipation at a graphene—Substrate interface. *J. Phys. Condens. Matter* **2012**, *24*, 475305. [CrossRef] [PubMed]

26. Hu, M.; Poulikakos, D. Graphene mediated thermal resistance reduction at strongly coupled interfaces. *Int. J. Heat Mass Transfer.* **2013**, *62*, 205. [CrossRef]

27. Li, M.; Zhang, J.; Hu, X.; Yue, Y. Thermal transport across graphene/SiC interface: Effects of atomic bond and crystallinity of substrate. *Appl. Phys. A* **2015**, *119*, 415. [CrossRef]

28. Wang, Z.; Bi, K.; Guan, H.; Wang, J. Thermal Transport between Graphene Sheets and SiC Substrate by Molecular-Dynamical Calculation. *J. Mater.* **2014**, 479808. [CrossRef]

29. Giesbers, A.J.M.; Uhlířová, K.; Konečný, M.; Peters, E.C.; Burghard, M.; Aarts, J.C.; Flipse, F.J. Interface-Induced Room-Temperature Ferromagnetism in Hydrogenated Epitaxial Graphene. *Phys. Rev. Lett.* **2013**, *111*, 166101. [CrossRef] [PubMed]

30. Zhou, P.; He, D. Modulating doping and interface magnetism of epitaxial graphene on SiC(0001). *Chin. Phys. B* **2016**, *25*, 017302. [CrossRef]

31. Yu, Y.J.; Zhao, Y.; Ryu, S.; Brus, L.E.; Kim, K.S.; Kim, P. Tuning the Graphene Work Function by Electric Field Effect. *Nano Lett.* **2009**, *9*, 3430. [CrossRef] [PubMed]

32. Castro Neto, A.H.; Guinea, F.; Peres, N.M.R.; Novoselov, K.S.; Geim, A.K. The electronic properties of graphene. *Rev. Mod. Phys.* **2009**, *81*, 109–162. [CrossRef]

33. Yi, M.; Shen, Z. A review on mechanical exfoliation for the scalable production of graphene. *J. Mater. Chem. A* **2015**, *3*, 11700–11715. [CrossRef]

34. Hass, J.; Varchon, F.; Millán-Otoya, J.E.; Sprinkle, M.; Sharma, N.; de Heer, W.A.; Berger, C.; First, P.N.; Magaud, L.; Conrad, E.H. Why Multilayer Graphene on 4H′SiC(0001) Behaves Like a Single Sheet of Graphene. *Phys. Rev. Lett.* **2008**, *100*, 125504. [CrossRef] [PubMed]

35. Virojanadara, C.; Zakharov, A.A.; Yakimova, R.; Johansson, L.I. Buffer layer free large area bi-layer graphene on SiC(0001). *Surf. Sci.* **2010**, *604*, L4–L7. [CrossRef]

36. Xia, C.; Watcharinyanon, S.; Zakharov, A.A.; Johansson, L.I.; Yakimova, R.; Virojanadara, C. Detailed studies of Na intercalation on furnace-grown graphene on 6H-SiC(0001). *Surf. Sci.* **2013**, *613*, 88–94. [CrossRef]

37. Kowalski, G.; Tokarczyk, M.; Dąbrowski, P.; Ciepielewski, P.; Możdżonek, M.; Strupiński, W.; Baranowski, J.M. New X-ray insight into oxygen intercalation in epitaxial graphene grown on 4H-SiC(0001). *J. Appl. Phys.* **2015**, *117*, 105301. [CrossRef]

38. Oliveira, M.H., Jr.; Schumann, T.; Fromm, F.; Koch, R.; Ostler, M.; Ramsteiner, M.; Seyller, T.; Lopes, J.M.J.; Riechert, H. Formation of high-quality quasi-free-standing bilayer graphene on SiC(0001) by oxygen intercalation upon annealing in air. *Carbon* **2013**, *52*, 83–89. [CrossRef]

39. Virojanadara, C.; Watcharinyanon, S.; Zakharov, A.A.; Johansson, L.J. Epitaxial graphene on 6H-SiC and Li intercalation. *Phys. Rev. B* **2010**, *82*, 205402. [CrossRef]

40. Xia, C.; Watcharinyanon, S.; Zakharov, A.A.; Yakimova, R.; Hultman, L.; Johansson, L.J.; Virojanadara, C. Si intercalation/deintercalation of graphene on 6H-SiC(0001). *Phys. Rev. B* **2012**, *85*, 045418. [CrossRef]

41. Emtsev, K.V.; Zakharov, A.A.; Coletti, C.; Forti, S.; Starke, U. Ambipolar doping in quasifree epitaxial graphene on SiC(0001) controlled by Ge intercalation. *Phys. Rev. B* **2011**, *84*, 125423. [CrossRef]

42. Chen, W.; Chen, S.; Ni, Z.H.; Huang, H.; Qi, D.C.; Gao, X.Y.; Shen, Z.X.; Wee, A.T.S. Band-Bending at the Graphene–SiC Interfaces: Effect of the Substrate. *Jpn. J. Appl. Phys.* **2010**, *49*, 01AH05. [CrossRef]

43. Tejeda, A.; Taleb-Ibrahimi, A.; de Heer, W.; Berger, C.; Conrad, E.H. Electronic structure of epitaxial graphene grown on the C-face of SiC and its relation to the structure. *New J. Phys.* **2012**, *14*, 125007. [CrossRef]

44. Jayasekera, T.; Xu, S.; Kim, K.W.; Nardelli, M.B. Electronic properties of the graphene/6H-SiC(0001) interface: A first-principles study. *Phys. Rev. B* **2011**, *84*, 035442. [CrossRef]

45. Dharmaraj, P.; Justin Jesuraj, P.; Jeganathan, K. Tuning a Schottky barrier of epitaxial graphene/4H-SiC(0001) by hydrogen intercalation. *Appl. Phys. Lett.* **2016**, *108*, 051605. [CrossRef]

46. Hannon, J.B.; Copel, M.; Tromp, R.M. Direct Measurement of the Growth Mode of Graphene on SiC(0001) and SiC(0001). *Phys. Rev. Lett.* **2011**, *107*, 166101. [CrossRef] [PubMed]

47. Hertel, S.; Finkler, A.; Krieger, M.; Weber, H.B. Graphene Ohmic Contacts to n-type Silicon Carbide (0001). *Mater. Sci. Forum* **2015**, *821–823*, 933–936. [CrossRef]

48. Mammadov, S.; Ristein, J.; Krone, J.; Raidel, C.; Wanke, M.; Wiesmann, V.; Speck, F.; Seyller, T. Work function of graphene multilayers on SiC(0001). *2D Mater.* **2017**, *4*, 015043. [CrossRef]

49. Tomer, D.; Rajput, S.; Hudy, L.J.; Li, C.H.; Li, L. Tuning a Schottky barrier of epitaxial graphene/4H-SiC(0001) by hydrogen intercalation. *Appl. Phys. Lett.* **2014**, *105*, 021607. [CrossRef]

50. Zhong, H.; Xu, K.; Liu, Z.; Xu, G.; Shi, L.; Fan, Y.; Wang, J.; Ren, G.; Yang, H. Charge transport mechanisms of graphene/semiconductor Schottky barriers: A theoretical and experimental study. *J. Appl. Phys.* **2014**, *115*, 013701. [CrossRef]

51. Rajput, S.; Chen, M.X.; Liu, Y.; Li, Y.Y.; Weinert, M.; Li, L. Spatial fluctuations in barrier height at the graphene-silicon carbide Schottky junction. *Nat. Commun.* **2013**, *4*, 2752. [CrossRef] [PubMed]

52. Seyller, T.; Emtsev, K.V.; Speck, F.; Gao, K.-Y.; Ley, L. Schottky barrier between 6H-SiC and graphite: Implications for metal/SiC contact formation. *Appl. Phys. Lett.* **2006**, *88*, 242103. [CrossRef]

53. Shivaraman, S.; Herman, L.H.; Rana, F.; Park, J.; Spencer, M.G. Schottky barrier inhomogeneities at the interface of few layer epitaxial graphene and silicon carbide. *Appl. Phys. Lett.* **2012**, *100*, 183112. [CrossRef]

54. Sonde, S.; Giannazzo, F.; Raineri, V.; Rimini, E. Investigation of graphene–SiC interface by nanoscale electrical characterization. *Phys. Status Solidi B* **2010**, *247*, 912–915. [CrossRef]

55. Yakimova, R.; Iakimov, T.; Syväjärvi, M. Process for Growth of Graphene. U.S. Patent US9150417B2, 6 October 2015.

56. Yager, T.; Lartsev, A.; Yakimova, R.; Lara-Avila, S.; Kubatkin, S. Wafer-scale homogeneity of transport properties in epitaxial graphene on SiC. *Carbon* **2015**, *87*, 409–414. [CrossRef]

57. Eriksson, J.; Pearce, R.; Iakimov, T.; Virojanadara, C.; Gogova, D.; Andersson, M.; Syväjärvi, M.; Lloyd Spetz, A.; Yakimova, R. The influence of substrate morphology on thickness uniformity and unintentional doping of epitaxial graphene on SiC. *Appl. Phys. Lett.* **2012**, *100*, 241607. [CrossRef]

58. Yazdi, G.; Vasiliauskas, R.; Iakimov, T.; Zakharov, A.; Syväjärvi, M.; Yakimova, R. Growth of large area monolayer graphene on 3C-SiC and a comparison with other SiC polytypes. *Carbon* **2013**, *57*, 477. [CrossRef]

59. Virojanadara, C.; Syväjarvi, M.; Yakimova, R.; Johansson, L.; Zakharov, A.; Balasubramanian, T. Homogeneous large-area graphene layer growth on 6H-SiC(0001). *Phys. Rev. B* **2008**, *78*, 1. [CrossRef]

60. Butt, N.Z.; Sarker, B.K.; Chen, Y.P.; Alam, M.A. Substrate-Induced Photofield Effect in Graphene Phototransistors. *IEEE Trans. Electron. Devices* **2015**, *62*, 3734. [CrossRef]

61. Anderson, T.J.; Hobart, K.D.; Greenlee, J.D.; Shahin, D.I.; Koehler, A.D.; Tadjer, M.J.; Imhoff, E.A.; Myers-Ward, R.L.; Christou, A.; Kub, F.J. Ultraviolet detector based on graphene/SiC heterojunction. *Appl. Phys. Express* **2015**, *8*, 041301. [CrossRef]

62. Sze, S.M.; Ng, K.K. *Physics of Semiconductor Devices*, 3rd ed.; Wiley: New York, NJ, USA, 2007; pp. 127–172.

63. Cumpson, P.J.; Seah, M.P. Elastic Scattering Corrections in AES and XPS. II. Estimating Attenuation Lengths and Conditions Required for their Valid Use in Overlayer/Substrate Experiments. *Surf. Interface Anal.* **1997**, *25*, 430. [CrossRef]

64. Johansson, L.I.; Owman, F.; Mårtensson, P. High-resolution core-level study of 6H-SiC(0001). *Phys. Rev. B* **1996**, *53*, 13793. [CrossRef]

65. Tadjer, M.J.; Anderson, T.J.; Hobart, K.D.; Nyakiti, L.O.; Wheeler, V.D.; Myers-Ward, R.L.; Gaskill, D.K.; Eddy, C.R., Jr.; Kub, F.J.; Calle, F. Vertical conduction mechanism of the epitaxial graphene/n-type 4H-SiC heterojunction at cryogenic temperatures. *Appl. Phys. Lett.* **2012**, *100*, 193506. [CrossRef]

66. Dharmaraj, P.; Jeganathan, K.; Parthiban, S.; Kwon, J.Y.; Gautam, S.; Chae, K.H.; Asokan, K. Selective area growth of Bernal bilayer epitaxial graphene on 4H-SiC(0001) substrate by electron-beam irradiation. *Appl. Phys. Lett.* **2014**, *105*, 181601. [CrossRef]

67. Tongay, S.; Lemaitre, M.; Miao, X.; Gila, B.; Appleton, B.R.; Hebard, A.F. Rectification at Graphene-Semiconductor Interfaces: Zero-Gap Semiconductor-Based Diodes. *Phys. Rev. X* **2012**, *2*, 011002. [CrossRef]

68. Giannazzo, F.; Hertel, S.; Albert, A.; La Magna, A.; Roccaforte, F.; Krieger, M.; Weber, H.B. Electrical Nanocharacterization of Epitaxial Graphene/Silicon Carbide Schottky Contacts. *Mater. Sci. Forum.* **2014**, *778–780*, 1142–1145. [CrossRef]

69. Tongay, S.; Schumann, T.; Hebard, A.F. Graphite based Schottky diodes formed on Si, GaAs, and 4H-SiC substrates. *Appl. Phys. Lett.* **2009**, *95*, 222103. [CrossRef]

70. Shtepliuk, I.; Khranovskyy, V.; Yakimova, R. Combining graphene with silicon carbide: Synthesis and properties—A review. *Semicond. Sci. Technol.* **2016**, *31*, 113004. [CrossRef]

MDPI
St. Alban-Anlage 66
4052 Basel
Switzerland
Tel. +41 61 683 77 34
Fax +41 61 302 89 18
www.mdpi.com

Crystals Editorial Office
E-mail: crystals@mdpi.com
www.mdpi.com/journal/crystals

www.ingramcontent.com/pod-product-compliance
Lightning Source LLC
Chambersburg PA
CBHW051724210326

41597CB00032B/5597